C000153166

History and Precedent in Environmental Design

History and Precedent in Environmental Design

Amos Rapoport

University of Wisconsin–Milwaukee
Milwaukee, Wisconsin

Plenum Press • New York and London

Library of Congress Cataloging in Publication Data

Rapoport, Amos.
  History and precedent in environmental design / Amos Rapoport.
      p.   cm.
  Includes bibliographical references (p.).
  Includes indexes.
  ISBN 0-306-43429-6. — ISBN 0-306-43445-8 (paperback)
    1. Environmental engineering — History. 2. City planning — History. 3.
  Streets — Design and construction — History. I. Title.
  TA170.R37   1990   90-7271
  628′.09 — dc20   CIP

Drawings and photographs, unless indicated otherwise,
were prepared by the author.

© 1990 Plenum Press, New York
A Division of Plenum Publishing Corporation
233 Spring Street, New York, N.Y. 10013

All rights reserved

No part of this book may be reproduced, stored in a retrieval system, or transmitted
in any form or by any means, electronic, mechanical, photocopying, microfilming,
recording, or otherwise, without written permission from the Publisher

Printed in the United States of America

"For many antiscientists objectivity is not so much impossible as reprehensible."
— **John Passmore,** *Science and Its Critics,*
New Brunswick, NJ: Rutgers University Press, .1978, p. 91.

"With its deconstructionists in literary criticism, its ordinary language and other philosophers, and its novelists, our age may one day come to be known in intellectual history for its role in the advancement of techniques to prove that reality doesn't exist."
— **Joseph Epstein,** "The sunshine girls,"
Commentary, Vol. 77, No. 6, June 1984, p. 67.

Preface

This book is about a new and different way of approaching and studying the history of the built environment and the use of historical precedents in design. However, although what I am proposing is new for what is currently called architectural history, both my approach and even my conclusions are not that new in other fields, as I discovered when I attempted to find supporting evidence.* In fact, of all the disciplines dealing with various aspects of the study of the past, architectural history seems to have changed least in the ways I am advocating.

There is currently a revival of interest in the history of architecture and urban form; a similar interest applies to theory, vernacular design, and culture–environment relations. After years of neglect, the study of history and the use of historical precedent are again becoming important. However, that interest has not led to new approaches to the subject, nor have its bases been examined. This I try to do. In so doing, I discuss a more rigorous and, I would argue, a more valid way of looking at historical data and hence of using such data in a theory of the built environment and as precedent in environmental design. Underlying this is my view of Environment–Behavior Studies (EBS) as an emerging theory rather than as data to help design based on current "theory." Although this will be the subject of another book, a summary statement of this position may be useful.

Since EBS began as a new discipline, I have tried to see it as a basis for developing a new theory of design. In so doing I have differed in several important respects from most of the work in that field. From the beginning I have stressed the need to develop theory even at the level of EBS itself, and I have thus emphasized synthesis and scholarship rather than application and "relevance." In the majority view, EBS is seen as a more or less "scientific" way of helping designers through programming, evaluation, or providing a data base. My own minority view is that EBS can go considerably beyond that: It can become the basis for the development of a new kind of theory of design— one based on a science rather than an art metaphor (Rapoport 1983b, 1986e). That is, it can replace traditional approaches. Note that I do not say *architectural theory* because

---

*"New" is somewhat of a misnomer even in my own case. I proposed a course on architectural history at Berkeley in 1964 which is, in outline, essentially the argument of this book.

this new theory needs to deal with all designed environments, using the term *design* most broadly as any purposeful change to the face of the earth—a matter of great importance in this book. One consequence is that even "natural" landscapes are now effectively designed: Nearly all landscapes are *cultural* landscapes. This is a very different, and much more fundamental, kind of endeavor in which one tries to apply our growing knowledge about humans to an understanding of built environments and ultimately to the design of better environments.

In developing theories, generalization is of great importance. However, I have long argued that valid generalizations about people and environments can only be made if based on the broadest possible evidence. I have thus insisted that this evidence must include the full range of what has been built: preliterate, vernacular, popular, and spontaneous (squatter) environments as well as the more familiar high-style settings; one must also include the relationships among these different environments; all cultures, so that the evidence must be cross-cultural; and the full time span of built environments so that the evidence must, therefore, be historical. Moreover, I have pointed out that the latter means not just the last few centuries in Europe and the United States, or even the 5,000 years of traditional architectural history, but rather a period of close to 2 million years all over the globe. Only in these ways can one be certain of being able to trace patterns and regularities as well as detect significant differences and hence make valid generalizations.

In working at this emerging grand synthesis, I have tended not to engage in empirical work but in the scholarly analysis and synthesis of work already in existence in a wide variety of disciplines as well as the actual environments themselves—or descriptions of them. In that respect, the characteristics of my work seem close to those that are generally accepted as broadly describing the humanities. As I understand it, the subject of the humanities at its broadest is the product of all human culture wherever or whenever created. Each new interpreter builds on the work of those who have preceded him or her, and this enterprise needs to be not only historical but interdisciplinary and multicultural as well. This is the way EBS should proceed, while taking a more explicitly "scientific" approach. As environment–behavior studies have developed, they have tended to diverge from what could be termed *mainstream design theory*. This divergence seemingly has become even more marked during the past few years with the development of so-called *postmodernism* and its successors.* One of the characteristics of these is a revived interest in historicism. Yet, paradoxically, it is precisely this unlikely, and apparently completely contradictory, development that, I believe, may provide an opportunity for some relationship between mainstream design "theory" and EBS-based design theory. This is not only an example of the critical importance in scholarship of establishing commonalities among apparently unrelated areas. It also provides a possibility of providing a valid base for addressing some of the very real criticisms of design raised by postmodernist critics that, however, they have not resolved successfully.

This potential link is the interest in history shared by architectural "theorists" and designers and also central to my approach to EBS to be discussed in this book. I believe

---

*"Fashions" in architecture seem to change as rapidly as in dress; the half-life of current architecture is very short. Moreover, these changes are rather superficial and "cosmetic."

that taking an EBS approach will provide both the rigor and the data that current interest in history lacks.*

The central question being addressed is clearly the relation among past, present, and future—in this case the possible lessons of history for environmental design theory and how these lessons might be learned. Four broad positions are possible on this, as also in looking at vernacular design or other cultures. Such material can be ignored; it can be rejected as irrelevant; it can be copied directly; or one can learn from it by deriving lessons through the application of various models, concepts, and principles to the material in question. My suggestion is that some of the models, concepts, and principles that have been developed in EBS will help provide valid lessons when based on the great body of environmental evidence that we possess.

This is the purpose of this book: to develop more rigorous ways of deriving lessons from historical precedents for the purpose of generalizing more validly about human and humane environments, thus leading to a new theory of design. It had been my original intention to write a single book about both because they are so intimately linked. However, as both projects, particularly the one on theory, grew, this became impossible. Thus the subject of this new theory will be the topic of another book to follow this one.

Consequently, the argument about history will have to stand on its own. I hope that the argument here proposed is sufficiently self-contained to do so. I also hope that the connection to theory will be clear—at least implicitly.

In addition to general questions, I have also been working over the past few years on some more specific studies that try to identify the lessons that the past (in the broad sense described) can have for design theory. These include studies about vernacular design, about appropriate ways of designing for Third World countries (seen as a paradigm for design generally), about the origins of buildings and settlements, and lessons from traditional environments for energy efficiency. There have also been brief studies on the characteristics of urban open spaces and pedestrian streets. The last of these is the topic of the specific case study used to illustrate the general argument of this book.

In order to develop the idea of a new kind of history of the built environment, three major topics needed to be addressed: the subject matter of such history, the purpose of studying it, and how it needs to be studied. These comprise Part I (Chapters 1–3) and were written first. The case study in Part III is intended to be a preliminary example rather than a complete illustration of the approach. I have been concerned with this topic on and off since my undergraduate days and have written much about it. Finally, because an important part of my argument (and my work generally) concerns the need not to start from scratch nor to reinvent the wheel (as architects are want to do), it became necessary to know what comparable work has been done. This comprises the supporting argument in Part II (Chapters 4 and 5) and was written last. In subsequent rewrites, of course, material was partially rearranged.

AMOS RAPOPORT

---

*While doing the final revisions of the manuscript, I came across a literary critic's comment about history as "a source of imaginative quotation in the sketchy, self-centeredly unserious way it is sometimes invoked in postmodern architecture" (Iannone 1987, p. 62).

Acknowledgments

Materials of the case study have been collected for over 20 years. Some research funds from the Royal Institute of British Architects and the University of London, awarded in 1968, helped purchase some city plans and other material. Many people have kindly sent me such materials or suggested sources; they are listed here. Some summer support from the Department of Architecture, University of Wisconsin–Milwaukee, helped provide time to collect much additional material in the library. The rather random collection was efficiently organized before the theoretical work was begun by a work–study student, Cheryl Holzheimer, funded by the Urban Research Center–UWM. The specific development of the theoretical framework began during 1982–1983 while I was a Visiting Fellow at Clare Hall, Cambridge. This was made possible by a fellowship from the Graham Foundation, a Senior Sabbatical Fellowship from the National Endowment for the Arts, and a sabbatical from UWM. The first draft of this theoretical work was also completed at Clare Hall in the summer of 1985 with the help of a grant from the Graduate School of UWM.

Some of the data in the case study had their scales changed and were redrawn by a PhD student, Aniruddha Gupte, who was supported by funds provided both by the Graduate School–UWM and Dean Carl Patton of the School of Architecture and Urban Planning at UWM. While he was redrawing these data, I was writing the second, third, and fourth drafts of this manuscript. Typing was most efficiently and helpfully done by Donna Opper, supported by funds volunteered by the chairman of the Department of Architecture, Robert Greenstreet. Paul Olson, of the departmental photo lab, most ably converted my slides into black and white prints suitable for publication. My wife, Dorothy, undertook the monumental task of preparing the index.

Finally, a number of people and organizations helped with information, slides, plans, and references. In no particular order this is to acknowledge the help of S. K. Chandhoke (New Delhi), Paul English (Austin, Texas), D. E. Goodfriend (New Delhi), Carol Jopling (Alexandria, Virginia), Nancy J. Schmidt (Cambridge, Massachusetts), Hans Hartung (Guadalajara, Mexico), E. E. Calnek (Rochester, New York), P. H. Stahl (Paris), J. Antoniou (London), Charles Correa (Bombay), S. Greenfield, R. Eidt, D. Buck, D. Carozza, H. Van Oudenallen (all at UWM), J. Hansen (Copenhagen), P. T. Han (Taichung, Taiwan), E. Montoulieu, T. Marvel (Puerto Rico), D. Schavelzon (Buenos Aires), P. P. Polo Verano and J. Salcedo Salcedo (Bogotá, Colombia), H. Geertz (Princeton, New Jersey), D. G. Parab (Bombay), K. Noschijs (Lausanne), D. Kornhauser (Honolulu), D. Garbrecht (Zurich), City Planning Department (Zurich), University of Texas, Center for Middle East Studies (Austin), The Melville J. Herskovits Library of African Studies, Northwestern University (Evanston, Illinois), Y. Tzamir (Haifa), K. Jain (Ahmedhabad), L. J. Kamau (Chicago), Y. Ben Aryeh (Jerusalem), M. Turan (Ankara and Pittsburgh), N. Inceoglu (Istanbul), P. L. Shinnie (Calgary), D. Nemeth (Mt. Pleasant, Michigan), S. Raju (Gwalior, India), R. A. Smith (Liverpool), S. I. Hallet (Washington, DC), J. Hardoy (Buenos Aires), R. Ohno (Kobe), and C. Bastidas (Quito).

I am grateful to all the individuals and organizations mentioned as well as those listed in the case study. Without the time and help they made possible I could not have completed the task.

Contents

PART II   THE SUPPORTING ARGUMENTS

PART III   CASE STUDY: PEDESTRIAN STREETS

Introduction

The history of anything is our knowledge of the past of that thing. This may seem like a rather childish definition, but its implications are important in at least two ways. First, it follows that before one can study the history of something, the nature of that something needs to be defined; the subject matter of the domain needs to be identified (Northrop 1947; Shapere 1977, 1984). It also follows that many disciplines that are not generally considered to be historic are so in fact, among these the history of life, evolutionary biology, palaeontology, physical anthropology, geology, cosmology, physics, astrophysics, and so on. We can think of these as *historical sciences*. More generally, there are even recent arguments that all science is inherently historical (Prigogine and Stengers 1984). This at least raises the possibility that the study of history, in the sense of the past of anything, can be related to science and can be scientific.

Because the major distinction between Environment–Behavior Studies (EBS) and the mainstream design professions hinges on the former trying to be scientific, it follows that the increased rigor and validity of the approach being proposed has to do with this very point; in effect, I am arguing for an EBS approach to the history of the built environment.

In fact, the first part of this book deals with that topic. Thus Chapter 1 attempts to define the subject matter of the domain, the body of evidence to be considered by asking: *history of what?* I then turn to the kinds of questions to be asked given that science typically begins with questions or problem definition (Northrop 1947). In fact, the problems regarding the subject matter are part of the definition of the domain (Shapere 1977, 1984). By asking what is the value or the use of such study or knowledge for design, for its theory and, hence, for its practice, a link can be established with the current interest in historical precedents. Chapter 2 therefore asks: *history for what?* These two chapters lead us to the specifics of how the study of the past of the built environment is to be carried out. Chapter 3, therefore, asks: *what history?* In Chapters 4 and 5 some recent developments in a number of other disciplines are examined that are historical in the sense of this discussion. The purpose is to find arguments and conclusions supporting those I have made and reached and also to suggest ways in which the subject might be approached using others' experience as a guide.

Before these themes are developed, it seems useful briefly to state the overall argument. The major purpose of this book is to develop a more rigorous, and hence more valid, way of studying the history of built form and more valid ways of using historical data in any theory of environmental design. This is done in the context of EBS, seen as an emergent theory of the built environment. More generally, the approach is based on a *science metaphor*—with design as a science-based profession—rather than on the commonly used *art metaphor.* That shift is much more fundamental than most of the arguments that occur *within* the art-metaphor-based approach (Rapoport 1983b).

Within a science metaphor, generalization becomes essential—but a rather different type of generalization than commonly found in architecture. There one has found sweeping generalizations not related to any evidence or empirical data and which tend to be suspect because they are not related to any evidence or empirical data. Alternatively, there has been no generalization at all (an "idiographic" approach) with each case seen as unique. The results have been descriptive approaches and a lack of a cumulative body of knowledge.* In EBS also there has been little generalization because it has tended to be nontheoretical. What generalization there has been is inadequate because it has been based on inadequate evidence—mainly empirical work in the present in Western countries.

All science, indeed all scholarship and all research begins with the search for *pattern.* But to see pattern one needs much and varied evidence. Specifying what this evidence should be forms the subject matter of Chapter 1. The next question is: Why study this evidence? Clearly, many answers can be given, depending on the interests of the researcher. However, at a highly general level, two major answers can be given. The first is the inherent interest of the subject matter, that is, for its own sake—because it is there. The second is to derive lessons, to learn from this record of the past. These two are, once again, related but separable. In this book the emphasis will be on the learning aspect, partly because of the current rather misguided interest in "historical precedents" but primarily because of the link that I see between this topic and theory through generalization.

I have already pointed out in the Preface (and will discuss in more detail in Chapter 2) that four broad positions are possible with regard to history—as they are possible regarding any body of evidence. History can be ignored (as it was for some time in architecture). Its existence may be acknowledged, but its lessons or value may be denied or rejected as irrelevant or misleading. Such material can be copied directly in terms of certain formal qualities or elements. Finally, one can learn from evidence by deriving lessons through the application of various models, concepts, or principles—in our case, from EBS. The decision about which of these approaches to adopt has major implications as to how one studies the past, for the method adopted and for the questions asked.

Thus if one is interested in studying environment–behavior interaction, that is, the relation of human behavior (in its broadest sense, including affect, cognition, symbolic behavior, and so on) to the built environment, then very different kinds of questions follow, and different evidence needs to be studied. This has implications for how one goes about studying history; that is the subject of Chapter 3.

At that point one can look at supporting (or disconfirming) arguments in various

---

*These and other related topics will not be developed here because they are more relevant to the discussion of theory, which will be the subject of a second, related, although independent book.

related disciplines; this I do in Chapter 4. One might look at history proper, its theory and philosophy, its methods and presuppositions. As will become apparent, there do seem to be historians who are trying to develop approaches to history not unlike the one here advocated. To give one example, there has been a very small beginning of comparative history, although strangely with no explicit awareness of cross-cultural studies or anthropology. One can look at various social sciences to see to what extent the usefulness of historical data is discussed or acknowledged. One can turn to the so-called "historical sciences" (geology, evolution, cosmology, etc.) that are both historical (in that they study the past) and scientific.* It is also useful to consider two disciplines that are both explicitly historical and concerned with closely related subject matter—cultural landscapes, built environments, and material culture. I refer to historical geography and more particularly to archaeology, which is discussed in Chapter 5. History generally has tended almost completely to ignore material culture and to work with written documents; in fact, history is sometimes distinguished from prehistory in just this way. In historical geography and archaeology, there is awareness of the difficulty of making inferences about behavior when the only data from the past consist of material culture. Historical geography, by and large, like traditional architectural history, has tended to concentrate on periods and places where written sources are available, although this seems to be changing. It will be seen that this is not an adequate approach to the body of evidence we need to consider. Consequently, it is prehistoric archaeology, which has typically dealt with landscapes, built environments, and material culture before the written record, which is the closest analog to the subject matter as I define it. Recent developments and discussions in archaeology as a scientific discipline are thus highly relevant. The result might appear to be an inordinate amount of space devoted to it in Chapter 5—almost a case of the tail wagging the dog.

Archaeology has made two distinct responses to the possibility of inferences about human behavior from material remains from the past. The traditional response was either that it cannot be done or that it is not the province of archaeology to do so. A recent shift becomes of particular interest in the present context: The emphasis is precisely on making inferences about human behavior. In doing so, two approaches that are related but distinct can be discerned. One has been to develop more sophisticated ways of analyzing material culture, clearer and more explicit concepts, models, and theories, so that behavioral inferences can be made, including even inferences about cognitive and symbolic behaviors. The second has been to develop ethnoarchaeology, which is ethnographic work done specifically to be applied to archaeology rather than using the more traditional ethnographic analogies—although even those can be useful.

This brings me back to the earlier references to EBS models. One purpose of studying historical data is to derive concepts about human behavior vis-à-vis the built environment, material culture, and cultural landscapes, which could be tested in present-day contexts. Another purpose is to use such data to test concepts derived from current empirical research whether in EBS, psychology, ethnography, sociology, ethology, and the like. The interest is thus not in history for its own sake but to *test* other

---

*I will not consider another approach which is implicitly historical since it attempts to look at architecture within a *legal metaphor* based on common law; this is, therefore, based on historical precedents (Collins 1971; Hubbard 1980).

things. One can use the past to test concepts and theories and to clarify them rather than to illuminate the past *per se,* although that also tends to illuminate that past. The more such models apply over time, as well as to different environments, different cultures and so on, that is, the more broadly they apply, the more confidence does one have in one's generalizations. At the same time, as we shall see, such contemporary research is essential in order to be able to make inferences about behavior on the basis of past environments. The intention is to use history and historical (and other) evidence not to *illustrate* points but to *test* hypotheses or answer questions (Rapoport 1986c; see Figure I.1).

In this book generally, and the case study specifically, my main interest is in data that properly cover the possible time span. Previously I have tended to emphasize broad cross-cultural coverage and the inclusion of all types of environments. I have also argued for the need to deal with the whole environment and as much of the system of settings as is necessary for any specific study (the extent of which needs to be discovered rather than decided *a priori* [Rapoport 1969a, 1977, 1980c, 1982d, 1986a, 1990a; cf. Vayda 1983]). These four aspects of a broad sample will be discussed in Chapter 1 and will tend to reappear in this book.

Certain assumptions are made when one uses such a broad sample. Some have to do with objectives. For example, one assumes the importance of pattern recognition: the study of regularities and apparent anomalies, the study of regularities behind apparent variety, and of differences behind apparent similarities. Pattern recognition as an essential stage in research is so widely accepted as to be unexceptionable. Its specifics are matters for empirical testing. Some of the other assumptions made may be more controversial. Some will be discussed later but consider just one such fundamental assumption as an example. In studying environment–behavior relations in past environments, it is assumed that inferences about human behavior are feasible. Thus it may be assumed that the same types and forms of social organizations existed before known records as since then (e.g., Dalton 1981). One assumes that there is such a thing as human nature that has not changed in its basics or in principle—for example, in its responses. Cognitive patterns, imposition of meaning, choice processes, and so on are the same. What has changed are the specifics: situation, possibilities, constraints, and the like. Thus one may assume that the primary stimuli that elicit positive emotions (which are sought) or negative ones (which are avoided) are genetic but that associated with these may be emotions (secondary and even tertiary reinforcers) that are culturally variable. Even abstract ideas and symbols can then be argued to have emotional impact dependent on that basic system (Pulliam and Dunford 1980, p. 43; cf. Lopreato 1984).

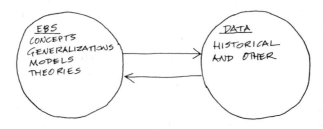

Figure I.1. Relationship between EBS and data (including historical).

Note how similar this is to the nonverbal communication model that I used in my analysis of the meaning of the built environment (Rapoport 1982a).

In essence one assumes *continuity* rather than *discontinuity*. One may even go further by positing evolutionary and sociobiological continuities and baselines. These are certainly rather contentious and lead to rather heated (not to say rude) exchanges. But in all of this it is essentially the old argument about the acceptance of some form of *uniformitarianism* (to be discussed in Chapters 3 and 5). This can be in terms of *process,* that is, patterns of thought, feeling, and behavior rather than design or product, although even in those cases uniformities are possible, either at the level of repertoire (Rapoport 1982a)—in that case, of means available to communicate via the environment—or even at the level of more concrete characteristics—as the case study in this book will attempt to show. More correctly, one probably needs to discover which things are unchanged, which have changed little, which much, and which are brand new.

Without some such uniformitarian assumptions, *no* study of the past can be justified, and it has been rejected (e.g., Popper 1964). Yet such studies can be shown to have worked in the various fields discussed in Chapters 4 and 5 as well as others. The implication is that it may work here also; it is certainly worth a try. Having rejected the position of ignoring the past or the possibility that it has no lessons, I would suggest that it is worth considering the study of history whereby one tries to learn by analysis rather than by copying or imitation. This is the premise of this book—or of its theoretical part that I have been discussing.

That part makes the first of two principal points of this book. It proposes a "new" approach to the study of the built environment, new goals, new ways of tackling the task, and so on. The second principal point is substantive. The theoretical/conceptual approach is illustrated by a case study of certain pedestrian environments; this forms Part III of the book. This case study not only makes more concrete the argument of Parts I and II but shows how current EBS research and a varied body of historical data taken together provide evidence regarding the nature of environments for walking and possible lessons regarding how they should be designed; it provides *precedents for design.*

Together these two principal points are meant to suggest the nature of a more valid approach to the study of the history of built form that also has implications for a theory of such form that I hope to develop elsewhere.

By implication, the present approach might be regarded as contrasting with other approaches that I regard as less valid; more specifically, that of mainstream architectural history and theory. In this connection, note that although a portion of the large literature on architectural history and "theory" was read, I will hardly refer to that, nor will I review it or criticize it *explicitly.* By suggesting a particular approach that should "work better" in achieving certain important but neglected ends, goals, and objectives, which will be more rigorous and which should therefore provide more valid precedents for design, other approaches are *implicitly* being criticized.

Part I

The Argument

Chapter *1*

History of What?

INTRODUCTION

All research and scholarship begins with a question, with the definition of a problem or of a problematic situation (e.g., Northrop 1947). This can only happen within a given domain—the body of data and its interconnections that is to be studied. Domains must, therefore, be defined.

Thus the fundamental question that is typically the bedrock for *any* useful research concerns the nature or definition of the field of study. This is possibly *the* fundamental question in research generally. No science or field of scholarship* can take its object of study as given—it must explicitly define it and give reasons for that choice. The first general step in the process of discovery lies in the definition of a field of study or domain; the second, more specific step is the identification or definition of problems within that field.

In fact the two are related. My approach to domain definition comes from Shapere (1977, 1984). He suggests that in research, certain bodies of information are taken to be the subject for investigation; this applies both to general fields and more specific subfields. Such bodies of related items Shapere calls *domains* that, he argues, are very useful in understanding science generally and theories in particular. Domains consist of both the specific subject matter of a field and the important questions in it. A domain is thus a body of unified subject matter about which there are problems; more specifically, well-defined problems raised on the basis of "good reasons" that are worth tackling and capable of being tackled. In my view, the definition of the subject matter is the more critical. This is the case generally because the questions tend to be suggested by the subject matter. It is also the case more specifically because the field with which I am dealing is still relatively unformed so that there are, as yet, no "well-defined problems." Moreover, the two aspects can be separated. I will thus deal with the subject matter in this chapter and with some of the questions in Chapter 2. Both of these, comprising the domain, will in turn inform the whole book.

---

*It is at least possible to argue against the position that one must make clear distinctions between the natural and human sciences (Freese 1980, Papineau 1978, Wallace 1983, among many others).

It is also my argument that domains can be defined at earlier stages than Shapere assumes in dealing with chemistry and physics. At the very least a domain can be made explicit at the outset (Dunnell 1971, Part 2, Chapter 5). Without necessarily accepting the rest of Dunnell's argument or his method, one can accept the need to define a field early; it cannot be left confused. Once a field is defined, one can begin to think about what needs doing and how.

There is another implicit disagreement between Dunnell and Shapere: Dunnell argues that fields differ not by subject matter but by theory, by how phenomena are viewed. Shapere suggests that theories may often be the same; it is the *subject matter* that is different. This view I tend to accept.

In addition to this most general reason why domain definition is an essential first step, that one cannot study anything at all in any way without defining the subject matter, there are several more specific major reasons for addressing the nature of the subject matter.

1. I suggested in the Introduction that the history of anything is our knowledge of the past of that thing. It follows that one cannot study the history of anything until one knows what that "thing" is. Thus one must consider the body of evidence that forms the domain of the history of built form, that is, the subject matter that is to be studied.

2. The nature of the domain suggests how the items that make it up are to be described and, later, how such descriptions are to be modified and improved. It also suggests various ways of classifying the subject matter, and it is often the case that at early stages in the development of fields, classification plays a very important role (Dunnell 1971, Mayr 1982, Rapoport 1986c, 1988b, in press).

3. The nature of the domain and its components suggests what concepts may be usefully applied to it, and it has been argued that no science can advance without a clear definition of the concepts with which it is concerned (Price 1982; Sanders 1984).

4. The clear definition of a domain has another useful consequence. Such explicit definition may suggest which other disciplines are related to it. Such disciplines dealing with comparable domains, as well as their concepts, theories, and methods may then become mutually relevant, transferable, may offer useful analogs, and so on. In the present case, for example, the specific definition suggested overlaps with archaeology; this, in turn, leads to useful insights discussed in Chapter 5. Other overlaps are also suggested that are discussed in Chapters 3 and 4.

5. The nature of the domain, and the concepts relevant to it, in turn suggest what are *appropriate* questions, as well as what questions can usefully be asked of it at a particular stage of development. In doing so, by defining problems and problematic situations, it also suggests specific lines of research and provides reasons for deciding that some lines are "more promising" than others in resolving these problems about the domain (Shapere's "good reasons"; cf. Medawar 1967). The nature of the domain thus suggests how it is to be tackled because method is inevitably related both to the nature of the evidence and to theory and theoretical presuppositions and assumptions (Kaplan 1964). Thus epistemology

depends at least partly on the nature of the domain because there is a linkage between epistemology and ontology. These linkages will (or may) also suggest theoretical orientations to a given subject matter.

6. Clarifications of the domain of concern of a field also suggests inadequacies; these, in turn, in the usual way suggest further research.

Domains thus guide research. Typically new bodies of data open up new areas and types of research. Therefore, the explicit demarcation of a new set of data as constituting the domain of the historical study of the built environment is an essential first step in any attempt to establish a new type of history. Moreover, if one accepts that history has to do with the nature and use of historical evidence, that is, evidence from the past of any field or subject matter, then the nature of that evidence is an important and natural first question. Yet, in spite of being so important, it is in the present case fairly straightforward. Thus this chapter and Chapter 2 are rather short.

THE SUBJECT MATTER OF ENVIRONMENTAL HISTORY

In order to begin the task of identifying the nature of the product that constitutes our domain of study, I begin by contrasting it with mainstream architectural history.

Traditional mainstream architectural history has had a predominant bias toward "important" works subjectively selected. It has considered substantial religious, institutional and, more recently, commercial and other buildings. When dwellings were considered, these were typically palaces and mansions (Beattie 1984, p. 179). The emphasis has also been on particular (selected) architects, certain hero figures, rather than even the generality of architects. It is clear in reading the architectural magazines that most architects in practice at any given time are unpublished and unstudied (e.g., Prak 1984), although their work may be much more congruent with users. Thus even within the high-style professional tradition, the evidence does *not* include the full range, only a small arbitrary subset selected on grounds that are far from clear and certainly not explicitly stated or argued. Yet to study even that subset, one needs the general trend in architect-designed buildings, which can only be provided by an adequate sample. And even if that were remedied and the work of all designers were to be included, their work (and this always includes only buildings rather than built environments) would still represent a very small part of the built record.

Architectural history, as the history of this small domain, has been shaped by at least two more implicit assumptions about the domain, these *conceptual* rather than *evidential*. The first is the fact that such history *generally* (there are, of course, some exceptions) has emphasized the "hardware," the visible products. But the environment is best conceptualized as the organization of space, time, meaning, and communication, or, alternatively, as the relations between people and people, people and things, and things and things (Rapoport 1977). It follows that the domain must include human behavior, the relation between behavior and the built environment, and the relationships among the components (or elements) of the built environment. Moreover, the latter must go beyond buildings—they need to include systems of settings of which buildings form only a part; they also need to include fixed, semifixed, and nonfixed feature elements (e.g., Rapoport 1977, 1982a, 1988a, 1990). The second concerns the prevailing metaphor of architecture as an art. It then follows that architectural history has been shaped by art history (see Chapter 3). It is the history of major works by major figures, the study of tradition in terms of art. When in the early eighteenth century comparative historical studies appear (with Bayle) the analogy is literary and the subject matter is *monuments* of architecture, and then still mainly in Italy (Heath 1984, p. 31). Even a study such as Banister Fletcher's that is subtitled *comparative* is not; moreover, its subject matter is also restricted to monuments.* The approach to history here being advocated is based on a shift of metaphor to environmental design as a science-based profession, the theory of which will be a science based on EBS (Rapoport 1983b, 1986e). It then follows that a parallel shift is also necessary in the history of the field—to history of the built environment as a science (Chapter 3). It has been argued that when serious theoretical problems arise in the history of architecture, theories are borrowed from outside, such as "rationalism" and "romanticism" (Heath 1984, pp. 30–31). These, however, are *not* theories but rather ideologies, attitudes, or paradigmatic frameworks (Binford 1983a,b); thus when I speak of theory I have something rather different in mind.

A shift to the history of the built environment as a science inevitably entails a need for comparative studies and generalization. In order to develop a theory of Environment–Behavior Relations (EBR) one will need to generalize. This will inevitably involve analyzing a broad body of evidence that provides large amounts of data. The question then becomes: What is that body of evidence (cf. Drake 1974)? The result is that the domain is no longer buildings but the built environment. Seen most broadly, history generally is the study of the record of what people have done. It follows that our domain is the record of what people have designed and built. Our topic is then *the history of the built environment.*

The domain with which we are concerned includes all those artifacts (fixed and semifixed) that comprise the full range of built environments at a variety of scales, as they existed in the past—and the behavior of the people who inhabited these environments (the nonfixed feature elements). The attributes of these environments can be assumed to be the result of human activity: They are all designed in the sense that design is any purposeful modification or change to the physical environment (the face of the earth) by humans (Rapoport 1972, 1980a, 1984b). It is thus the whole cultural landscape that results from the choices made. Its attributes are the result of recurring human activity, of many individual and group decisions, which add up to recognizable wholes. Cultural landscapes can then be treated as manifestations of cognitive schemata, ideas held in common by the makers (and users) of these cultural landscapes, which design as a choice process tries to express (Rapoport 1972, 1984b, 1986b,d). Cultural landscapes are also a subset of *material culture,* that sector of the physical world that is modified by people through culturally determined behavior. The subset relevant to my concerns includes many artifacts such as settlements, buildings, fences, fields and gardens, furnishings of all sorts, and many others, as well as many aspects of human behavior, that is, nonfixed elements (cf. Deetz 1977, p. 24). Although these cannot be studied directly in the case of past environments, we will see in subsequent chapters that inferences about these can be made.

*I have not yet seen the new 1987 edition.

One needs to study the cultural landscape, the sum total of all human modifications to the face of the earth. I said "purposeful" modifications earlier but that is redundant because people only modify the environment, a difficult activity demanding the investment of resources—economic, time, energy, effort, and so on—*for a purpose.*

The cultural landscape is of interest not only because it is the sum total of design and that system of settings in which people live (Rapoport 1986a, 1990) but also because of the way in which the many individual decisions that create it add up to recognizable wholes. This, I have argued, implies the presence of a "template," a shared cognitive schema, as it does in the case of a designed building. It hence allows inferences to be made about cognitive processes—and we shall again see in subsequent chapters that this can, in fact, be done.

The subject matter of our domain thus broadly defined also makes the built environment a much more significant portion of material culture, in some cases almost co-terminous with it. This has major implications in the relationship of such study to other fields that are concerned with material culture (principally archaeology).

History is then the history of the whole built environment, including all types of environments, all cultures, and the full time span about which information can be obtained (see Figure 1.1).

This is not meant to replace traditional architectural history completely in terms of the domain. The intention is rather to add to it, to complement it with the missing bulk of the built environment. I will clearly be concentrating on that part not studied by traditional architectural history, its obverse, as it were. This can be visualized as the relation of the ground to its figure (see Figure 1.2).

This is because such an increase in the *amount* of evidence is essential and also because it greatly increases the *variety* of evidence. Moreover, many environments not designed by architects are arguably better than those designed by architects and can provide different lessons, that is, precedents (Rapoport in press). This applies not only to the traditional vernaculars but also to many spontaneous settlements in developing countries (Rapoport 1988b).

To reiterate: *The domain is the whole environment, all environments in all cultures covering the full time span.* It includes not only all the fixed feature elements and the relationships among them but semifixed feature elements and nonfixed feature elements, that is, people and their behavior, generally and relative to the built environ-

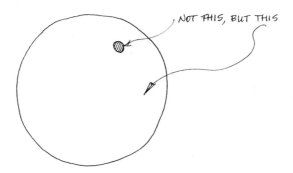

Figure 1.1. The domain of environmental history.

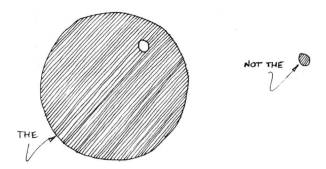

Figure 1.2. The relationship between the domains of environmental and traditional architectural history.

ments. It is thus about people, buildings, objects, and landscapes; the relationships among people and people, people and things, things and things; all the organizations of space, time, meaning, and communications about which we can find out (Rapoport 1977).

I have already suggested that the evidence used must be as varied as possible. It is possible to be more specific. To be effective, the evidence needs to be varied in four ways that, although interrelated, are analytically separable.

1. It must include the whole environment not just arbitrarily isolated parts of it; for example, high-style buildings. As I have long argued, and will elaborate later, one cannot study the Parthenon without looking at it within the context of the Acropolis (Doxiadis 1972). Neither, however, can one study the Acropolis outside the context of contemporaneous Athens (Rapoport 1969a). People live in systems of settings within which activity systems occur (Rapoport 1969a, 1977, 1982d, 1986a, 1990). These need to be *identified* in order to be useful; they can be identified as I have suggested specifically for the case of dwellings (Rapoport 1980c) or urban open space (Rapoport 1986a). More generally this is the process of *progressive contextualization* (Vayda 1983).

2. One must consider all cultures—or as many as possible. Thus one's evidence must be cross-cultural. The great variety of built environments is striking, given that they enclose a much more limited range of human activities (Rapoport 1986c). This variety has tended to lead to an emphasis on differences and cultural relativity—what I have called culture-specific design. This seems correct, but the search for patterns is also a search for regularities behind this variety, especially those based on evolutionary baselines or panhuman characteristics (Bonner 1980; Boyden 1974, 1979; Boyden and Millar 1978; Boyden *et al.* 1981; Dubos 1966, 1972; Eibl-Eibesfeld, 1972, 1979; Fox 1970, 1980; Geist 1978; Hamburg 1975; Lumsden and Wilson 1981, 1983; Rapoport 1975b, 1977; Tiger 1969; Tiger and Fox 1971; and many others; see also later chapters). If any such regularities or invariance can be found, this would be extremely useful because the extent of variability would be reduced (see Figure 3.12).

3. One must consider all *types* of environments, that is, not only high style but preliterate, vernacular, popular, and spontaneous (squatter) settlements. This is not only because the full *range* of environments needs to be used for generalization, but because (related to Point 1) the environment is a system, so that relationships among

elements (which form a system) may be more important, or at least as important as the elements themselves (Rapoport 1969b, 1977, 1982d, 1986a). Also those high-style elements that form the subject matter of traditional architectural history and theory only make sense when seen within their contemporaneous vernacular matrix (as discussed later), they are part of the house settlement system or other systems of settings.

4. Finally, as the point just discussed suggests, one needs to consider evidence across time, that is, look at it historically. In this connection, one must include all of history, the full time span of the built environment. Thus even architectural history à la Banister Fletcher (about 5,000 years) is not adequate. Even less adequate is the emphasis on only the modern period, only the period since the eighteenth century and other arbitrarily selected segments of the past (unless for specific studies). I have recently argued that the origins of architecture can be traced to the first hominid structures in Olduvai Gorge that go back almost 2 million years (Rapoport 1979a; cf. Young, Jope, and Oakley 1981). Even fairly complex settlements go back a very long time—several hundred thousand years (Rapoport 1979b). It is *that* kind of time span that needs to be considered. When it is, it almost inevitably also covers Points 2 and 3; *the evidence to be used in generalization, therefore, includes the full record of built environments by humans and hominids.* It may even prove useful to consider possible roots of built environments in animal architecture (Bonner 1980; von Frisch 1974), in changes from primate to hominid behavior (Isaac 1972, 1983), and recent work in sociobiology and so on to be discussed later (cf. Festinger 1983).

This can be restated in terms of discussing how one might teach the history of the built environment, by listing some of the ways in which it should *not* be taught.

1. Do not confine yourself to high style.
2. Do not confine yourself only to the Western tradition.
3. Do not consider it to be a separate thing, a thing-in-itself.
4. Do not begin with Egypt 5,000 years ago.

The opposites to this are indicated. One must include and even emphasize *knowledge about all past environments.*

This redefinition of the subject matter is equivalent also to a conceptual redefinition. This is what makes it potentially so revolutionary, not least in terms of *precedents for design.* This is because history then becomes the record of accumulated knowledge, ideas and experiments tried, successes and failures, the experience of responses to certain general problems faced by humans in their attempts to build in different places, climates, and sites, and for different cultures and life-styles. It also becomes a record of a range of formal and perceptual experiments. Clearly, then, the more complete that record, the more it covers the *totality* of built environments, the more potentially useful it becomes. In other words, the record is more than the visible (or "sensible") environments that can be identified; it is even more than the inferred behavior (in its broadest sense) of its creation, use, and modification. It is also a potential set of lessons that, if sufficiently generalized in terms of EBR can then become precedents for design in terms of *principles.*

It follows that all evidence available on prehistoric, preliterate, folk, vernacular, and popular environments must be used, both because they comprise the bulk of the built environment but also because they form the setting or matrix within which high-style elements are to be understood. It further follows that one must include all the manifestations of these environments cross-culturally. This not only greatly increases the evidence available but also draws attention to the extraordinary variety of environments and shows other ways of doing things. This is most important because one needs not only a broad body of evidence but a *varied* body of evidence. This is in effect equivalent to a gene pool—a body of environmental experience and achievement based on the full spatial and cultural extent of environments. Similarly it is essential to include the full range of time—prehominid, hominid, and human—almost 2 million years as already mentioned, that is, the full temporal extent of built environments. Finally, history must be considered to be a part of EBR. Hence all this evidence is for the purpose of learning about how people have shaped environments, how environments have affected people, and what mechanisms link people and environments—what I call the three basic questions of EBR (Rapoport 1977, 1986a and later discussion).

It is often argued that the first *known* "architect" was Senmut, the designer of Queen Hatsheput's Tomb at Deir-El-Bahari in Egypt (e.g., Heath 1984, pp. 2–3). Be that as it may, it is clear that people had "designed" before—as already suggested, any built environment is designed in the sense that it represents purposeful action, involving choice. What I call the choice model of design seems universally applicable; what varies are who makes the choices, using what criteria, over how long a period of time, and so on (Rapoport 1972, 1977, 1984b, 1986c).

There were even "known" designers in preliterate societies—they are often *unknown* or *anonymous* only to us. Thus there seemed to be "copyright" in the facade decorations and patterns used in the Sepik River Haus Tambaran (men's houses) in New Guinea, so that when such patterns were used by others, "payment" had to be made (Gardi 1960; Rapoport 1981a). This was also the case elsewhere; so that there were frequently strong sanctions against the alienation of designs (e.g., Mead 1979). In some places in Oceania, there were specialists in the building of Canoe Houses, and some were known to be much better than others and regarded as "artists" (Staeger 1979). Similarly, although all men in the Gilbert Islands could build, specialists built meeting houses, and their primary skill was not carpentry but mastery of the intricate forms of decorative lashings that were employed in the construction of such buildings (Hockings 1984). Generally, in the Eastern Solomons and parts of New Guinea, artists were *very* important and influential; although working within strict patterns, they clearly competed with each other. Mud mosque builders in Northern Nigeria were often celebrated in songs, and these master masons were endowed by popular opinion not only with creativity, intuition, and genius but with supernatural powers: Some of them became folk heroes (Saad 1981). Similarly the creators of canoes, paintings, and songs in aboriginal Australia were also known, celebrated, and held in high esteem. Generally, then, the anonymity of vernacular designers may well have been exaggerated (Rapoport 1986c, in press). This may have major implications for views about design process, the emergence of architects as a recognized group and so on; *note, however, that these questions only begin to emerge when the domain has been redefined—and as soon as it has been redefined.*

In all these cases, as in all others, the high-style elements form a small part of the built environments of those cultures and need to be seen in the matrix of the *contemporaneous* environment; even the high-style monuments typically studied by architectural history only make sense within such a matrix (e.g., Rapoport 1969a). This matrix is the system of settings or the cultural landscape that I have discussed (cf. Rapoport 1974, 1984b). It provides the context for understanding any elements of the built environment and is hence quite critical for the field and for understanding both vernacular *and* high-style elements. Not only is this the environment in which all people at all times have always lived; because relationships may be more important, or at least as important, as the elements (Rapoport 1969b, 1977), it follows that one needs to look at the relationships between high-style and vernacular elements of the built environment.

In the case of high-style and vernacular elements, these relationships can be of two kinds. The most common is the situation where the high-style elements are embedded in the vernacular that is then the *matrix*. This is clearly the case with monumental buildings in all cultures, for example, religious buildings whether the Haus Tambaran, a Gothic cathedral, an Indian temple, or a nineteenth-century church (e.g., Rapoport 1982a). This is also the case with major urban complexes, such as the Acropolis in contemporaneous Athens, the Maidan-i-Shah and its surrounding elements in the urban fabric of Isphahan (Rapoport 1964–1965), or the many examples to be found in Bacon (1967) and other sources. In all these cases, in terms of human behavior and experience it is the transitions and contrasts between the dwellings and spaces of the vernacular urban matrix and the special major elements that are important (Rapoport 1977). The second, and less common, relationship between high-style and vernacular elements is where the former *structures* the vernacular which can then be understood as infil. Among examples are major avenues and other urban spaces in cities (Bacon 1967), urban grids (Rapoport 1977), the ceremonial centers and causeways of Maya cities, the structure of Indian cities (Lannoy 1971), ancient Rome (Rykwert 1976), China and others (Wheatley 1971), or Cambodia (Giteau 1976).

Such examples could be multiplied *ad infinitum*. It seems clear that this is a very different domain to that usually considered in architectural history. This is true not only regarding the body of built forms that typically concerns only the high-style elements. This is also the case regarding relationships as being at least as important as elements—if not more important. The relationship between high-style and vernacular changes our understanding of both. It also makes it mandatory to see the built environment as a cultural landscape composed of systems of settings containing activity systems. All people have built. They have therefore designed—all environments, in all cultures, in all periods. In all cases, these environments have constituted an integral system—cultural landscapes or systems of settings. People have lived in *these,* not in buildings alone. They typically move among settings, link and separate these differently, depending on their activity systems (Rapoport 1977, 1986a, 1987a, in press).

This is important not only with regard to the setting system itself but also in connection with the nature of "design." The most interesting thing about cultural landscapes that I have already mentioned is that many apparently independent decisions, by many people over long periods of time, add up to a highly distinct and, if one knows the cues, easily identifiable landscape (Rapoport 1984b). This is also linked to the important questions about the cultural and place specificity of such cultural landscape; why there should be such a diversity and variability among them given the significantly fewer things that people do (Rapoport 1986c).

Given that built environments form systems as discussed, then the history of the built environment must be about the vernacular matrix or infil as well as the high-style elements, the relationships among the elements and between elements and their contexts; the purposes of environments and their relationships to behavior, and so on. For some questions, high-style elements may be those most usefully studied; for others, vernacular environments; for others yet, the relationship between the two. In some cases high-style elements have a richer vocabulary, a more appropriate—or more easily identified—set of ordering rules or ordering system. Then traditional architectural history may be appropriate, although in its present form it still lacks reference to the vernacular matrix and to human behavior. It can be argued that even the high-style elements cannot be understood or studied outside their context, that is, their contemporaneous setting (Rapoport 1969a, 1977). One does not need to imply that vernacular environments are better—although they may well be; in any case, any individual may prefer such environments. It is more difficult to reject the position that such environments are more important—after all they do comprise most of the built environment— worldwide today less than 5% of all environments is designed by architects (cf. Rapoport 1969a, p. 2). If one considers the full range of past environments that figure goes down to significantly less than 1%. Think of a country like the United States: Suburbia, shopping strips, roadside environments, fields, and the like make up the cultural landscape. The U.S. built environment can neither be understood nor studied nor discussed without those. Most important is probably the nature of the relationships between vernacular and high-style elements, the relationships between monuments and the settlement fabric and whole cultural landscapes (including their "natural" elements). In that context, the problem of defining vernacular or high style, as well as "primitive" and popular that is not as easy as it seems, can be avoided—one needs to deal with the whole system of settings relevant in any given case (Rapoport 1988b, in press).

These relationships can be studied at a moment in time or over time—synchronically or diachronically. In the case of the latter one might pick a given site, for example, Salvador Bahia, Brazil (where this idea was first written down) and document and study the successive built environments there over time—prehistoric, Indian at contact, Portuguese, African, Syncretic—all the way to the present day high-rise hotels and condominia. The implications would be most interesting. Alternatively one could study the relationships between various areas of Portugal and the syncretic environments in different parts of Brazil—the influences of precolumbian traditions and various African and European traditions (for a Caribbean example, cf. Edwards 1979).

Many other interesting questions pose themselves. Note how quickly a variety of questions can be generated—and how different such questions are (see Chapter 2). Note also the way one can begin to use primitive, vernacular, and high-style examples to illustrate EBS concepts. This is a simple and early form of generalization. In this way various topics approached as history are not only related and integrated, but they begin to bear on design. In this way precedents for design can be generated; it is important not to use evidence only as examples and not to use it selectively. The evidence should be selected carefully and explicitly: the case study provides one example. In Chapter 3, I will also discuss the more careful use of data to *test* concepts rather than *illustrate* them (cf. Rapoport 1986c).

It is important to select evidence as explicitly relevant and related to a concept. To use a biological analogy: If one is studying water animals one does not include land animals—unless the question is about the emergence of the latter from the former or the move of the latter back into water. This clearly involved, for example, the modifications of morphological elements and follows from the definition of the problem within the domain, which in the case of evolutionary biology are all living organisms. In our case, that domain is all built environments. From that domain, specific subsets can then be selected that are relevant to particular questions or problems.

SUBDIVIDING THE DOMAIN

In my discussion I have already frequently had occasion to mention particular subsets of the total domain and how these might be related to particular questions. This means not only that given portions of the domain can be isolated for analysis; it means that *this needs to be done*.

This is the case in all fields. For example, it has been argued (King 1982, p. 122) that "progress in medicine comes from *increased discrimination and sharper distinctions*" (my italics). These require knowledge and theory and, in turn, lead to further theory, further distinctions, and so on (Kaplan 1964; Wilkins 1978, p. 125). I would argue that the current clamor for "wholes" and "holism" is mysticism and irrational (if not antirational).

Whole has two senses:

1. The totality of all properties or aspects of a thing and specially all relations among (all) its constituent parts.
2. Certain special properties or aspects of the thing in question that make it an *organized structure* rather than a mere heap.

If 1 is true, one cannot study anything at all; if 2 is true, one can study a whole that can be isolated and that is *relatively* independent of any given environment (Vayda 1983). "Since we cannot meaningfully envision a totality we necessarily *abstract and isolate*" (King 1982, p. 223). As fields grow and develop, single vague concepts become subdivided into much more precise terms each of which then helps to develop explanatory theory (King 1982, p. 64, shows how this occurred in the case of medical diagnosis). In the second sense of wholes, study can occur and can be highly productive. Thus Folan, Kintz, and Fletcher (1983, p. xix) claim their study to be a "holistic contribution to settlement pattern research," but, as we will see in Chapter 5, it first dismantles and dissects the subject matter and then reassembles it using many disciplines and much evidence. This is very different indeed to "holism" as it is often used today and which can be criticized (e.g., Billinge 1977).

It seems clear that greater differentiation, better identification of features, better classification, and, ultimately, better understanding follow from "dismantling" or "decomposition" (Efron 1984; Morris 1983; Nicholson 1983; Phillips 1976). As Jacob (1982) points out: Asking very broad questions often leads to minimal answers— asking specific and apparently narrow questions often leads to very broad answers. Knowing what to ask, and of what subset of evidence to ask it, is crucial: Both require *subdivision*.

Consider the topic of the case study—pedestrian streets. In general one cannot understand the use of any setting without considering other settings in the system. The use or nonuse of streets, for example, will depend on what happens in other outdoor spaces as well as a large range of other settings including indoor ones (Rapoport 1986a). In some cultures, streets are not used, and the activities in question occur in other, often indoor settings (Rapoport 1969a, 1977, 1987a, 1989a). However, if the question is formulated correctly, one can isolate a subset, in the case of the study in Part III pedestrian streets, in such a way that they can validly be studied. More generally, for any *specific* problem one *needs* to select the relevant subset—type(s) of environments, part(s) of environments, period(s), culture(s), activity systems, and so on. Within the domain of all built environments, the problems, the questions, the research tradition and strategy will provide criteria for choice. In any case, as long as the criteria are made explicit, they can be discussed, criticized, altered, and improved. This also applies to the definition of the *extent* of any system (cf. Vayda 1983).

I have argued for quite some time, repeatedly and at length, for the absolute need to dismantle global concepts and to do so carefully; to identify their constituent parts, to study their interrelationships, and then to reassemble them (Rapoport 1976, 1977, 1980a, 1986c, 1990). Although a fuller discussion of this topic will be more relevant to a discussion of theory elsewhere, a few points need to be made very briefly.

I have made this point about the term *environment* (e.g., Rapoport 1977, 1980a) about *housing* (Rapoport 1980c), and about *culture* (Rapoport 1977, 1986c, 1990). This is because otherwise these terms are not helpful. One needs to conceptualize environment—and for different purposes, different conceptualizations are useful; one that I find useful is as the organization of space, time, meaning, and communication. I have conceptualized housing as a system of settings within which certain systems of activities take place; these settings can also be described in terms of their environmental quality that in itself can be dismantled into components and described as a profile (Rapoport 1985b). I have argued that *culture* is both too global and too abstract and can be dismantled along two axes—one leading to more manageable concepts such as life-style and activity systems (both capable of further dismantling) (Rapoport 1976, 1977), the other to social variables such as kinship, institutions, roles, and so on (Rapoport 1990).

Culture and built environments cannot be linked at that level of generality. It is impossible to analyze the relation between culture and environment or to "design for culture." Greater specificity does not help. To "design housing for culture" is indeed more specific but no easier nor more feasible. To consider housing for a given culture is yet more specific but still impossible. To consider how various components of life-style and activities relate to systems of settings and environmental quality profiles is both feasible and not too difficult. This becomes critical in the study of past environments. To attempt to "analyze an environment in terms of the culture of its designers" is an impossible and even meaningless task. To analyze a system of settings, first in terms of activities, then inferred life-styles or roles, kinship, and so forth is feasible—and done in archaeology. It then may become feasible to understand the cognitive domains and meanings involved—as we shall see later. Note that in these various dismantlings it is essential to consider the cross-cultural validity of any units used, particularly when the evidence is broadened in the ways discussed in this book. It is significant that my dismantling of the concept of *house* and its redefinition as a system of settings within which a (specified) system of activities took place was the result of the need to have a definition of dwelling that was valid for cross-cultural analysis (Rapoport 1980c, 1989a). Emic categories are difficult, if not impossible, to derive for most past environments; it then becomes critical that the etic categories used be carefully and explicitly specified for valid comparison. This can only be done if subsets of the domain are used—and these subsets themselves are carefully and explicitly selected.

The point I am making is that if the potential domain is broadened to include everything that has been built, then subsets must be selected. If and when they are selected, however, this is not to be done because one likes them, thinks them beautiful, interesting, or inherently important—as is the case in architectural history. They are to be selected from the much larger domain on the basis of the light they can shed on the specific questions asked, and the reasons for this choice and the criteria used need to be explicitly stated and clearly argued.*

There seems to be an apparent paradox here. On the one hand, I am arguing that selection must be made, and on the other for greatly broadening the domain. Is this not contradictory? It is not, and the reason why has already been given. The selection needs to be made from the domain seen extremely broadly. The definition of the domain as the whole built environment or cultural landscape is critical because to be rigorous and relevant regarding EBR theory, one must generalize. To generalize, evidence must be as broad as possible—that is, include everything people have built (the whole environment, all environments, all cultures, the full time span). Because the only way to learn from historical data is through applying such theory, then within the domain of historical data the same argument applies.

Also, during most of human history, people obtained shelter by "designing" and building it. It is only relatively recently that most environments can at best be chosen and modified (Rapoport 1985b, 1986c). This means that during the very long time since its origins, close to 2 million years ago (Rapoport 1979a; Young, Jope, and Oakley 1981) the built environment *was created for people.* Environments were created as settings for behavior—including its latent aspects such as meaning (Rapoport 1977, 1982a, 1988a, 1990). *This is a constant.* Built environments have tended to be designed so as to be as congruent as possible with people, their behavior, and activities, the latent aspects of which lead to the extraordinary variety of environments (Rapoport 1983b,c,d, 1985b, 1986c). Although this may *not* be the case for some high-style settings, it is *always* the case for other types of environments. Considering the full range of environments, in all their variety, then provides an important way of discovering patterns in which aspects of behavior and activities are important, whether latent aspects are more important than instrumental and under what circumstances. Many hypotheses can be proposed on the basis of patterns (e.g., Rapoport 1969a) and tested against even larger bodies of evidence. In this process, studying the full range of environments is clearly crucial.

The purpose of built environments is therefore to be congruent with and supportive of the life-styles and activity systems of the inhabitants, including their latent aspects. Yet in reading traditional architectural history (and theory), one would think

---

*Renfrew (1982, p. 5) quotes John Disney making that point about archaeology in 1849! Admittedly, he then shows that some regressions occurred later in the nineteenth century (ibid., pp. 5–7). See also discussion in Chapters 3 and 5.

that the buildings discussed (and that is usually all that is discussed) were produced for architects or critics. Yet by asking the basic and apparently trivial questions—Why does one build? What are buildings, settlements, and landscapes for?—one realizes that they are for users. They are also most often *by* users, and even designers need to be seen as essentially surrogates for users (Heath 1984).

Environments are built to be supportive of users' dreams, wants, needs, and activities (instrumental and latent); to satisfy users; to help guide behavior and co-action; to remind people of social rules and situations by acting as a mnemonic (Rapoport 1982a), and to suggest new possibilities by acting as a catalyst (Rapoport 1983a,c). They tend to be much more successful when they are chosen, that is, when wanted (Rapoport 1977, 1978a, 1983c, 1985b). One could say, with others, that built environments enclose behavior if one interprets "behavior" properly; moreover, behavior is active, linking settings into systems (Rapoport, 1990).

This series of apparent assertions are, in fact, hypotheses based on what seem to be patterns revealed by looking at evidence in the domain as I have defined it. As such, they are rather different to the subject matter of traditional architectural history; they have also changed the way EBS has thought about environments. *Both these linked changes followed the shift to the new domain of concern.*

The constancy of the purposes of built environments already mentioned has major implications for both theory and the study of history because of the potential uniformitarianism at the level of *process* (to be discussed in Chapter 3). For theory we can argue that in terms of Aristotle's four causes, in the built environment *final causes* are always the most important: Choices are made for purposes on the basis of ideals and schemata. This means that for built environments teleological explanations always apply (cf. Nicholson 1983). Next in importance are *formal causes* (the nature of the product). *Material causes* tend to be modifying (Rapoport 1969a), and *efficient causes* are even less important as modifying causes. Note that, for example, in Newtonian physics efficient cause is emphasized, and there is much argument about the validity or danger of teleological explanation in the philosophy of science, particularly biology. This clearly has major implications for theory; the foregoing also has implications for the nature of my approach to the past (Chapter 3).

For the study of history, the possible uniformitarianism at the level of process is important. It also has implications for the subsets of the domain chosen for study. The purpose (final cause) in the case of a palace (seen as a ruling machine [Uphill 1972]) will require a different body of evidence than that used to analyze the purposes of settings for family life or commerce (Rapoport 1982a, 1988a). Moreover, even in "palaces" and their equivalents, changes in the purposes can explain changes in form over time, and the latter can suggest changes in purposes—as in a recent study of over 70 U.S. city council chambers (Goodsell 1984, 1988). Again, the specific concern leading to questions will lead to specific subsets of the domain being chosen for study on a basis other than personal preference.

METHODOLOGICAL IMPLICATIONS

The redefinition of the subject matter the history of which is to be studied has major implications for the approach taken: The change is drastic. It tends to change completely the way in which one considers even monuments: Just considering the relationships between monuments and the vernacular matrix will do that, as we have already seen. Redefining the domain *conceptually* as the study of environment–behavior relations changes the evidence used, the questions asked, and hence the approach equally dramatically. Before I discuss that, however, let me pursue some other consequences of the changes in the evidence considered.

As I have been describing it, the history of the built environment is not only closer to science than to art but also closer to prehistory and archaeology than to history. For one thing, in most cases there are no written records or documents, which tend only to be available for high-style buildings over a limited time span and culture range. Written data are not available for preliterate environments (by definition!). They are typically also absent for vernacular or popular environments, for most of the time span and for most cultures. Of course, when documents *are* available, they should (and will) continue to be used although broadened (Rapoport 1970a, 1973a, 1977, 1982a, 1985b,c,d) and should include, for example, the content analysis of newspapers, advertisements, novels, travel descriptions, poetry, myths and sagas, film and TV, photographs, drawings, and so on. Moreover, because behavior cannot be observed, it must be inferred from material culture and its remains or surviving descriptions.

As we shall see, archaeology itself has changed in the direction for which I have been arguing and in ways to be discussed in much more detail later; it also changed by changing its domain in comparable ways by broadening it. Its concern changed from a preoccupation with palaces and temples to settlement patterns and settlements; dwellings; behavioral, economic, social, political, and symbolic systems. This has also tended to occur in other areas of scholarship. Thus in a review of a book* (*Biblical Archaeology Review* 1987), the point is made that the Greek Magical Papyri tell us a great deal about daily life, as well as providing direct evidence about the religious beliefs of the time. The editor of the book makes the point that "modern views of Greek and Roman religions have long suffered from certain deformities because they were unconsciously shaped by the only remaining sources: the literature of the cultural elite and the archaeological remains of the official cult of the states and cities." The point is then made that the material in question serves as a powerful corrective. It is, I would suggest, *essentially equivalent to including vernacular elements.*

There have also been comparable changes in ways of approaching art objects themselves. These also occurred as a result of the redefinition of the domain from "masterpieces" to the corpus of work produced by a group of people or in a given place, as in the case of Bushman or Australian Aboriginal rock paintings, African tribal art, Australian or Oceanian art and the like, discussed in Chapter 3.

We are concerned, of course, essentially with the *history of environment–behavior interaction.* Because, however, the people of that past are no longer around, one needs to make inferences from data *which are in the present,* so that the past is always inferred (cf. Binford 1981, 1983a,b). This can only be done on the basis of the environments themselves—or what is left of them or known of them, or what reconstructions of them have been done either by others or are still to be done by a new breed of historians. I will discuss the matter of inference in much more detail later particularly in Chapter 3 (and

---

*H. D. Betz (Ed.). *The Greek Magical Papyri in Translation, Including the Demotic Spells.* Chicago: University of Chicago Press, 1986.

also in connection with archaeology in Chapter 5). At this point a few preliminary comments are in order. Regarding recent high-style environments, written records are typically used in conjunction with the environments themselves. For other types of settings, one can rely partly on written or graphic documentation of various kinds—by travelers, visitors, and the like (Rapoport 1970a, 1973a, 1977, 1982a). Our reconstruction of data of the past is inevitably based on some form of analogy with the world around us. This can be based on work already available, such as ethnographies. One can also be helped by observing existing environments or equivalent ones still being used by people. One can also refer to specific studies addressing such questions, comparable to "ethnoarchaeology," which is discussed in Chapter 5. Comparative studies are not only useful but quite essential. Generalizations based on empirical studies done in EBS and in other fields can be applied. There are thus a number of ways of making inferences by using contemporary work; these will be discussed in detail later.

What this makes clear at this point is that, because one is interested in environment–behavior interaction, one can begin at both ends, as it were. I mean by this that one can begin with knowledge about people and their behavior and see how this is related to built environments. This is mainly, but not completely, based on empirical work that must be cross-cultural and related to the full range of existing environments. Such knowledge typically comes from anthropology, psychology, the social sciences, physical anthropology, evolutionary studies and sociobiology, EBS. It is essential to check it against history and prehistory—this is, in fact, the reason for studying history in the first case. One can also begin with environments, both existing or from the past. This involves EBS, architecture, urban and landscape design, cultural and historical geography, geology, ecology, climatology, archaeology, and all its associated disciplines. In this way, the full range of built environments and cultural landscapes becomes accessible. These two sets of data about people and environments, from the present and the past, can be seen as a single set of data and can be brought together using a variety of approaches derived from EBS, the social sciences, history, the historical sciences, archaeology, and so on (discussed in Chapters 3–5; see Figure 1.3).

This can be seen as addressing questions about *process*—how environments come to be, what effects they have on people, how people use environments, interact with them, modify or destroy them once they are in place, and what mechanisms relate people and environments. One can also address questions about *product* where again considering the full range of such products is essential. When the questions concern the past, particularly the remote past, products (environments)—or parts of them—are all that is left. Processes, like all behavior, must also be inferred.

As we shall see in Chapter 4, similar views have been expressed within history— for the need to be "scientific," to be rigorous, to use generalizations and explicit sampling, to deal with behavior. This was also influenced by the social sciences, as this book is, via EBS. This also occurred very recently, in the middle of the twentieth century, and was also due to a shift from the "great man" view of history implicit in political and diplomatic histories, or nineteenth-century cultural history, to a concern with social history—the full range of the past. This also is equivalent to a redefinition of the domain in terms of its subject matter in ways analogous to mine—as a much larger set, ideally a universal or at least surviving set of objects and their relationships, producers, and ideas.

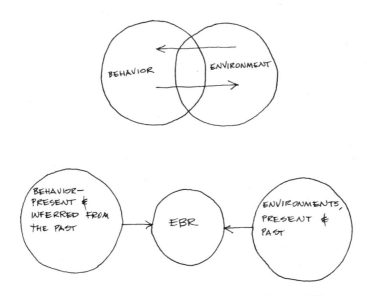

Figure 1.3. Present and past data as a single set relevant to EBS.

Traditional architectural history is still concerned almost exclusively with monuments by a few known designers in the Western tradition for which written records exist. By doing so, it only deals at most with 1% or 2% of the built environment for a small part of the world over a relatively short time span; it also looks at a few selected buildings, doing violence even to those by isolating them from the contemporaneous matrix, the larger system of settings or cultural landscapes within which they are embedded and to which they are inextricably linked, and by ignoring human behavior in all its forms within these environments. It is little wonder that such history has not been able to generalize. It has also not been able to derive precedents in the form of lessons. Copying or mining the past for bits and pieces, and playing games with them, is *not* learning.

By ignoring these aspects even of the selected monuments, it has failed to be duly enlightening, as I will discuss shortly. Even in the historical study of urban design, only monumental complexes and spaces tend to be studied (e.g., Bacon 1967). Yet settlements and urban fabric is mainly about relationships if it is about anything (Rapoport 1969b, 1977, 1982d, 1986a), and we have already seen that relationships to the vernacular matrix are critical for an understanding even of the high-style elements. Moreover, traditional architectural history typically looks at buildings as formal objects; it does that even when attempting to be comparative (e.g., in its recent use of typologies). It ignores people and their behavior, semifixed elements, changes in use over time (as when the same space becomes a series of different settings [Rapoport 1976, 1977, 1982a]). It also ignores changes made to environments. When not ignored, these are often criticized rather than understood (Rapoport 1982a). Typical architectural history is the history of architectural objects and their producers; less frequently of the ideas of the elites incorporated in these objects. Thus arguments for the study of the history of the appreciation of architecture (as in Bonta's [1975] book on the Barcelona Pavillion) still refer to a few selected observers or critics. This, if it were about *users* would mark an advance; it would be a better, or at least different, question. The domain, however, remains the same (cf. Brine 1984).

But even in that case, where the subject matter remains the same, redefining the domain *conceptually,* as the study of environment–behavior relations, would change the approach drastically, even when single monuments are studied. I will conclude this chapter by discussing my review of an outstanding example of traditional architectural history scholarship (Rapoport 1984c). This study of the Plan of St. Gall in Switzerland is traditional in the sense that a specific project is approached through the apparatus of architectural and historical scholarship. The emphasis is on the buildings, neglecting their relation to culture and behavior. Because broader theoretical issues are neglected, the study fails to make some higher level linkages even on its own terms. Thus the buildings are considered as artifacts rather than analyzed in terms of important relationships, for example those between the secular and monastic worlds in terms of domains, barriers and linkages, and cues. The lack of an integrative conceptual framework means that although reference is made to the history of religion, social life, technology, food, and so on, these are neither integrated with each other nor directly and explicitly related to the organization of St. Gall. Neither is the question considered whether *the overall layout* was an invention *de novo* (which seems unlikely) or the culmination of an unknown or undocumented development; yet an attempt has been made to trace the development of specific *buildings,* especially "ordinary" ones, to their Northern European vernacular precedents. In fact, only buildings have been illustrated or studied in any detail. Yet, given that the major significance of the plan is that it was for a *community,* it is surprising that no attention was paid to the overall relationship or hierarchy of the complex in terms of scale, height, or materials; to the exterior spaces, to the three-dimensional nature of the whole, its urban design, and landscape aspects. This is typical of most traditional studies: Elements are studied, whereas relationships among them (and to the larger context) that are often more important, are neglected.

Also lacking as part of a broader framework, even in terms of architectural history, is a comparative approach. This may lead to mistaken implications. Thus the overall layout of the plan is interpreted (although, unfortunately, not diagrammed) as a series of enclosures of ever greater seclusion with the monks' cloister as the most secluded. There seems to be an implication of uniqueness of that organization until one recalls that this is a typical schema in most traditional layouts (ancient Egypt and China, Yorubaland, and some precolumbian settlements, to mention just a few). The commonality of this theme cross-culturally and historically, and any patterns this suggests, can only emerge within a comparative framework; this is essential if it is to be more than just a case study of a single complex. Similarly, the monks' cloister is linked to the notion of paradise; this is also a major theme in Iran, in Islamic design generally, and no doubt elsewhere. Very general questions are thus raised, such as, How, if at all, do these traditions relate? How do they differ? What do they have in common? Are there other examples of "courtyard as paradise"? Such questions go beyond the plan of St. Gall, monasticism, or even the West, although they still belong to the traditional architectural history tradition. Thus, even within its own domain, there is need for a broader theoretical, more comparative framework, even if this framework is not emphasized. Such more general questions about the relation of people and their settings need to be addressed by studies of past built environments. One way this can be done is by considering environments in terms of their relation to culture that, in the case of St. Gall, seems particularly appropriate. For one thing, the plan of St. Gall is ideal and prototypic, that is, it is an exemplar, and as such it is closely related to concepts such as paradigms and schemata. These, in turn, are important and useful concepts in relating environments to culture.

Moreover, St. Gall is an extremely well-preserved version of what appears to be the first design for a community in Western Europe (at least since antiquity); it is also based on a highly structured and well-articulated ideological and behavioral system, the Rule of St. Benedict, which is also extremely well documented. This should make it easier than usual to address one of the most basic questions about design: What objectives is it intended to meet and why (rather than how they are met), that is, on what basis are design choices made. These questions are generally best approached by examining the link between the environment and what is broadly called "culture." Typically, the link goes both ways—culture, through the schemata, life-styles, rules, and behaviors associated with it and social groupings, relations, and institutions shape the environment. The latter, in turn, influence, guide, and constrain behavior and activities, although they do not determine them.

Questions such as these are not even addressed by this fine example of traditional architectural history. Some indications emerge through glimpses, as it were, that allow some inferences to be made with much effort: There is potential information that is mainly about the influence of the particular culture on the environment, partly because the reader is given little detailed information about behavior or about the semifixed elements (the furnishings in the broadest sense) both of which one needs in order to interpret the links between settings and behavior. Moreover, because one is dealing with a monastic community, behavior tends to be constrained more than in other cases by other means through rules (prescriptive and proscriptive), traditions, and the like—although this, again, is the case in all situations, albeit often in a weaker form (however, only comparative studies could reveal this). The environment, however, would still play a role even in this case. First, the plan is designed to be supportive; it is meant to make the desired behaviors easier and more automatic; it helps to reinforce the rules about who does what, where, when, and including or excluding whom—which I take to be a central question about EBR. Also, unlike many ideal plans, this one is very careful to consider behavior and instrumental functions. It pays extraordinary attention to functional zones and relationships and proximities, even the relation of temperature to activities. It tries to integrate ritual and socioeconomic systems with daily life. The plan carefully defines domains and fixes proximities and separations, linkages, and barriers. It does this because at least three very different groups were involved (and clearly distinguished): the monks, those making the community function (workers, gatekeepers, managers, professionals), and visitors/outsiders. In this respect the plan is considerably more sophisticated than most contemporary designs. In this connection, the environment influences behavior in another way: It communicates meanings and provides cues for appropriate behavior to these groups. It also makes potentially important statements about sanctity, religion, the link between religion and state, and social hierarchy during the period in question. It could thus tell us things about both latent and instrumental functions. This makes it potentially a most useful case for such a study, and it is thus even more unfortunate that questions of this type have not been addressed.

More important, however, a single case study is inherently incapable of answering such general questions. Comparative analysis must become the norm—and the comparisons must be among a large and varied body of cases and on the basis of explicit conceptual and theoretical issues related to EBR. But with this we have effectively left the discussion of subject matter and moved into the types of questions to be asked. This is related to the issue of the reasons for which one studies history, or past built environments, and how one uses its lessons as precedents. These form the subject matter of the next chapter.

Chapter 2

History for What?

INTRODUCTION

Having defined the subject matter of the domain, the essential next step is to consider the important questions relative to that subject matter. These questions, in the broadest sense, are equivalent to asking why one wants to study historical data in the first place, that is, defining the problem area. Before *any* research can be done, some questions must be posed; these are critical. What do we ask of the evidence, in this case of past built environments? To obtain answers one must know what questions to ask; to do that one must have clear, hence explicit reasons for studying the evidence. Having defined the subject matter one needs to discuss what questions, in principle, are to be asked of that subject matter. This is particularly the case because the point is made that in research specific results are often of less importance than *how the formulation of a new kind of question led to those results* (Goodfield in Ayala and Dobzhanski 1974, p. 82).

This broad question needs to be asked for several more specific reasons. First, the kinds of questions asked will have a major impact on how the subject is approached and studied—and that is the principal topic of this book (see Chapter 3). Second, the question of why history and historical data should be considered or studied seems to be of great general interest—yet it is rarely discussed or even raised. Moreover, it seems particularly reasonable to ask why history should be studied at the end of the twentieth century, at a time of rapid change, when change is emphasized everywhere and is a major component of the *zeitgeist* (cf. the analogous question in Rapoport 1969a). It is, therefore, useful to ask: Why history or history for what?

It is, of course, also part of the definition of the domain. How history is to be studied follows not only from the redefinition of the subject matter but also from the questions posed; what the purpose of study is taken to be. *What is good history is related to what it is good for* (Fischer 1970).

APPROACHES TO HISTORY

As I have already suggested, four attitudes are possible with regard to the use of historical data or precedents generally. These apply to traditional settlements (Rapoport

1973b, 1983a,d, 1986b), vernacular design (Rapoport 1982b, in press), spontaneous settlements (Rapoport 1983c, 1988b), and so on. They also apply to the full record of historical evidence available, such as actual environments or their archaeological remains; preliterate, vernacular, popular, and high-style design, cross-cultural materials, descriptions and illustrations of past environments and so on, or any subset of them.

The four possible attitudes are the following:

1. The evidence can be ignored.
2. One may acknowledge its existence but reject its value or interest.
3. One can romanticize it and copy various solutions, forms, or design elements; this is equivalent to using past environments as a quarry for mining.
4. One can learn from the evidence by applying various concepts and theories derived from EBS, at a high level of abstraction, derive lessons, and apply these lessons and principles that then become precedents (Rapoport 1982b,c, 1983b,d, 1986b).

The distinction between the last two is of most interest, as I have argued in connection with vernacular design and environments in developing countries. This distinction can usefully be summarized in a diagram (see Figure 2.1).

Part of the reason that this model, developed in connection with studying preliterate and vernacular design, is relevant in the present context is that the principal issue is the notion of generalizing on the basis of the broadest possible body of evidence. Preliterate and vernacular environments are not only a subset of built environments generally and largely historical; as I argued in Chapter 1, the full range of data from the past will, inevitably, be mainly preliterate and vernacular because it comprises the bulk of the material. In my discussion of the subject matter of our domain, I have already implicitly suggested that 1 and 2 are wrong on the face of it because past environments comprise most of the body of evidence regarding all that humans have built and also the bulk of environment–behavior interactions. They are essential for any generalizations: *They constitute, as it were, a laboratory in which generalizations can be tested against empirical evidence.* I have previously argued that 3 seems untenable for vernacular design (Rapoport 1982a, 1983a,d) and developing countries (Rapoport 1983a); it seems equally, if not more, untenable in the context of the history of an even more varied body of evidence. I would, therefore, conclude that the last is the only tenable approach.

These four attitudes are not just hypothetical—all can, in fact, be found. With regard to vernacular, the first is the common view, and examples of the second, although less common, can be found (this is my interpretation of Sordinas 1976; cf. Doumanis and Doumanis 1975). The third approach can be seen in the neovernacular of England, Australia, and other countries, the "style neo-Quebecois" in Quebec (e.g., Despres 1987) and its equivalents in developing countries (e.g., Barnard 1984; Rapoport 1982b). With regard to history generally, the modern movement in architecture, in its more extreme moments, discarded architectural history because it denied continuity. It then followed that it was of no value; it was occasionally also rejected explicitly as of no value. If it was studied (or taught), it was only done because it was "needed" as a liberal art, as part of "being an educated person." This is still partly the case in the revived interest in history. This recent revival of interest in the history of a

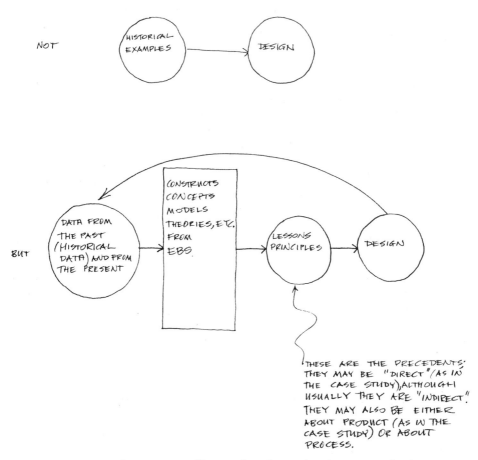

Figure 2.1. Alternative ways of learning from the past (i.e., deriving precedents).

very small part of past high-style environments does emphasize some sense of continuity, but the use of "precedent" seems to be in terms of the third position—as a quarry or mine for forms or elements (as in so-called "postmodernism" and the like).

Even if "copying" seems useful, it becomes stronger and more convincing if more general reasons based on theory support it. For example, a good case has recently been made for using the verandahs traditional to the main streets of Australian country towns (Siksna 1981). This becomes stronger when cross-cultural and other historical examples can be adduced; it gains even more when conceptual reasons for their use can be given on the basis of EBS research (cf. Rapoport 1977, 1987a, and the case study in Part III). It is not then merely the case of copying a device that one researcher likes in Australian country towns. Rather it is a precedent, based on theoretical grounds and supported by a diverse sample of evidence: If certain goals are to be achieved, then the characteristics produced by arcades and/or verandahs are useful, not only in Australia but Singapore, Paris, Bologna, and elsewhere.

But the feeling of continuity needs to be related not only to the full range of environments but also to the idea of *cumulativeness*. One of the characteristics of our society, and increasingly the world, is its extreme awareness of the past, of cultural diversity—what Malraux called *the museum without walls*. This is particularly the case with the professional, intellectual, and academic segments of society. But unless this continuity becomes cumulative, it may become counterproductive. If it does not fit into some conceptual or theoretical framework, it is just more random information—which may, in fact, become noise. It is a *heap* rather than a system. One question, therefore, about why one should study history implies in our case the kinds of connections or linkages that exist between past environments and those of the present and of the *future* (which are currently being designed or are yet to be designed). It has been argued, in principle, that history should be instructive, that it should not entertain but shed light on some of the larger questions—in our case, of EBR theory and problems. Yet historical writing even in general has rarely provided a past that speaks to contemporary concerns (Sarna 1985, p. 76).

Conceptual frameworks and linkages also apply to how the material is being studied. For example, to study historical precedents as isolated pieces of hardware is not useful. Nor do typologies seem very useful at all. But more important, in both cases, it is only shapes and elements that are being considered. I would argue that these must be at a higher level of abstraction (Rapoport 1977, Chapter 1, 1980d, 1982b). There must also be linkages of these environments to human behavior in the broadest sense (Rapoport 1976, pp. 12–17, 1986b).

Thus the approach remaining is the fourth: to use historical evidence for the purpose of learning lessons and principles that then become the precedents. These can be about awareness of process or dynamic forces at work in a field (King 1982, p. 14), about relationships of general interest and so on. It seems very important to emphasize the difference between studying "history" (or its use), as history is understood generally, and the use of historical data. It is the latter that is emphasized in this book. An even better formulation would seem to be *the use of data from the past*. This implies *lateral* rather than *vertical* linkages (see Figure 3.1 and Chapter 5), by which I mean using a variety of evidence related to a topic rather than a chronological sequence.

When the "lateral approach" is taken, this typically changes one's view of current environmental problems. For example, Brooke (1984) reports on a project by Charles Gehring on Manhattan in the 1650s. Great similarities are found with Manhattan in the 1980s. It follows that the problems of New York (and other contemporary cities) are not new. One might also be able to begin to predict under what circumstances they tend to occur, if the sample is large enough to trace patterns and when correlations are made with a variety of other physical, cultural, social, economic, and other conditions or circumstances. In this connection one purpose or value of history is to provide data, as valid and accurate as possible, to avoid the wrong argument; in this case seeing the past as either a paradise or hell—both of which seem common in discussions of cities or other environments (Rapoport 1982c, 1984a; cf. Efron's [1984] similar argument about carcinogens). The neglect of historical data, or the use of wrong data, leads to wrong arguments; for questions like these to be addressed, the lateral approach seems essential.

It is in terms of this formulation—*the use of data from the past*—that the historical sciences, archaeology, "historical sociology," and other historical social sciences, discussed in Chapters 4 and 5, become relevant to this argument. One is not interested in history as such, for itself—as chronology, description, or development—but as an entry point into environmental–behavior relations (cf. Rapoport 1982b, in press). The interest is *extrinsic* rather than, or as well as, *intrinsic* (Rapoport 1983a).

The broadest, most general, question is then: *How can one use historical data (data of the past) for certain (stated) purposes?* In the case of the case study (Part III), for example, the question is how can a sample of pedestrian streets be used to test a specific hypothesis derived from EBS about what makes such streets appealing to pedestrians. This is central in urban design. It becomes essential when designers' objectives are to use the knowledge of the past. Thus in a recent report on an urban design practice, the point is made that the objective is to design public spaces "where people feel welcome and comfortable" and that this should emerge from the study of such spaces (Scardino 1986). Assuming this to be a valid objective (and only EBS theory can help decide that), the issue is then how this is studied. Clearly, the more varied the body of evidence the better. It would also benefit from EBS research, for example, whether *people* is not a gross overgeneralization (e.g., Rapoport 1986a,c, 1987a). Finally, the lessons, that is, precedents, should be in the form of principles and concepts, and potential repertoires for achieving those, rather than imitation. Equally obvious, any designs resulting should then be evaluated to see if the objectives have been met and thus become part of the cumulative body of knowledge. In architecture, urban design, landscape architecture, and the like one cannot in general really perform experiments except under very unreal circumstances. One can, of course, perform some experiments in some areas of EBR and occasionally do "action research" or use "reforms as experiments" (Campbell 1969). Historical data of the kind I am advocating provide one very important way of studying the relation of many different variables, such as different cultures, climates, topographies, and so on to built environments. The data of the past become one's laboratory, as it were. The use of historical and cross-cultural data about the full range of environments is also often the only way in which to broaden the limited range of those experiments that one can use. This applies to EBS experiments, "reforms as experiments," "action research," and "quasi-experiments" (Cook and Campbell 1979) or of postoccupancy evaluations, even in the rare cases of designs based on explicit objectives and thus capable of providing *some* generalizability. In all these cases, the range is broadened in terms of evidence of products and inferred processes, of those contextual variables deemed to influence environments, and in terms of enabling checks of generalizations based on contemporary, culture-specific research that may then be found to represent anomalies rather than patterns. Because both knowing patterns and identifying anomalies are useful (and related!), historical data become most important.

Note that one of the principal questions to be addressed with regard to built environments is their bewildering variety. One species, with a limited repertoire of needs, activities, and behaviors, a limited range of climates, sites, and materials, seems to produce an apparently bewildering variety of settings, built forms, settlements, and cultural landscapes (Rapoport 1980a, 1985b,c,d, 1986c, 1988a). Note an important point: Although any theory of the built environment needs to address this, *the realization itself only dawns once the full range of environments is considered.* It is one of the first patterns revealed. Then questions quickly arise about higher order underlying patterns—the variety behind apparent similarity, similarity behind apparent variety (e.g., Rapoport 1977). This, in turn, raises more conceptual issues about units of analysis, taxonomies, and the level of abstraction used in analysis already briefly discussed and leads to many important theoretical questions; for example, pancultural, diachronic constancies and at what level versus cultural and other variability (see the discussion in Chapters 1 and 3; cf. Rapoport 1975c, 1979a,b, 1986b,d, and references there).

Note that in connection with the variety of environments, biology provides a very useful analogy (see also Chapter 4). First, because biology has always faced this problem, it has greatly emphasized taxonomy, which I increasingly feel is a most important and neglected issue in the study of built environments (Rapoport 1982b, 1988b, in press). This I will also discuss in Chapters 3 and 5. Second, the attempts to explain this variety has been a driving force in the development of the field. Third, it leads to efforts both to discover the true variety (or diversity) of the domain of biology and to preserve it so that it can be studied and its lessons learned (e.g., Eckholm 1986). It is, I think, significant that the arguments for the need to describe, to know, and to study this diversity are almost identical to those I have used. For example, that only on the basis of lessons based on the full range of species can the most recent forms of genetic engineering progress. Also, it is argued that each species is a unique repository of enormous amounts of information. Although each species taken alone throws major light on its relationships with its context, its ecological setting, even more important, the diversity of species and their relationships with their environments considered together may greatly modify many generalizations about ecological principles, about species and their interactions and needs, about the development of life, its evolution, extinctions, spread, and so on. This biodiversity is seen as a fundamental database for the biological sciences, the "library" on which any work in genetic engineering must be based. The point is finally made that biological diversity is one of the key problems of science generally that will not only "resolve unanswered questions . . . [but] will provide answers to questions both large and small, practical and esoteric" (Eckholm 1986, p. 20). Many of these questions cannot be properly addressed until the extent of this diversity is actually known. Similar, indeed identical points can be made about the subject matter of *any* field—and clearly are relevant to EBS—as I argued in October 1984 (Rapoport 1986c). In terms of the topic of this book, this means first identifying, then recording, then using the full range of built environments ever built in answering questions posed; other questions about which one cannot even think will inevitably be raised. An obvious first question is why there should be such diversity. This I have attempted to answer (e.g., Rapoport 1985b, 1986c). Another question is why there seems to be an apparent reduction in diversity—although more diachronic work is needed to check this observation. Many other questions can be raised about this phenomenon (e.g., Rapoport 1985c,d).

The role of history in general is a contentious issue. Can precedents be useful at all? Ever? It brings up the question of the relation of past to the present and future; it also brings up the issue of "historicism" (e.g., Popper 1964). This is often a matter of confusing *historical* sequence with *logical* sequence (i.e., *post hoc ergo propter hoc*). Only if the later development could logically be deduced from earlier general principles *a priori* would such an assertion be true. Hence my emphasis on what one could call lateral rather than vertical linkages. This is because vertical linkages tend to involve developmental historical sequences, whereas using data from the past linked laterally seems to avoid the problem of "historicism." To argue that the purpose of the study of history is to learn from it, by deriving precedents, does not imply "historicism" because it does not imply progress or development—the lateral connections make that untenable. Learning and precedents become possible when the full range of past environments is seen as a treasury, a repertoire of solutions to problems, of ingenious responses to sets of conditions, solutions that contain aesthetic, formal, climatic, siting, cultural and social, meaning and many other lessons. An analogy then becomes possible with the idea of a "gene pool" in biology and ecology and the equivalent danger of losing this diversity of environments (cf. Rapoport 1978c, 1980a, 1983d). This *does* imply a belief in certain unalterable realities and constancies of human nature, based on evolutionary history, and often in terms of process that will be discussed in Chapter 3 (as part of the problem of uniformitarianism). This also makes one skeptical, on the one hand, of all forms of utopianism, in our case those that imply a "new" architecture or a "new" planning" for a "new humanity" for whom all prior environments are irrelevant. On the other hand, it makes one skeptical of any single tradition as the only valid one—which is the danger of "historicism" and against which Popper warned.

In addition to using lateral linkages, it is also useful to be able to specify the processual mechanisms that make later events follow from prior ones; these can be seen as models that make theory operational (Hesse 1966; Morris 1983).* At the very least, one needs to consider mechanisms linking people and environments—the third of the three basic questions of EBR (see later discussion and Chapter 6; cf. Rapoport 1977, Chapter 1, 1986a). These, in turn, can also be more easily derived if a wide body of evidence is used, including data from the past.

In any case, I do not wish to argue this point at length, particularly because my emphasis is not on the use of history *per se* but on the use of data of the past. Note, moreover, that Popper's argument has not gone unchallenged. In fact, a past critique (Wilkins 1978) argues that Popper's extreme views on history are not valid. One of the arguments Popper uses is the lack of the possibility of experiments in history. I have already suggested that the use of historical data is, in effect, a form of experiment. In addition, it has been suggested that experiments are not appropriate in all sciences. History indeed cannot repeat facts, and facts about the past may indeed be collected with a preconceived point of view (another arguable point). But, as we shall see in Chapter 4, there are historical natural sciences such as palaeontology, evolution, geology, cosmology, physical anthropology, and the like. Their existence (and their undoubted success) suggests that Popper's opposition between science and history is not as clear-cut as he thinks—nor is the gap as great. Historians, like historical scientists, do not observe facts but deal with artifacts—whether these be written, built, organic, physical, or whatever. They deal with remains of the past, including information, such as the background radiation bearing on the origin of the universe that survives in the present; the data of history is *always* in the present. What constitutes a "fact" is as difficult and contentious a question in science as in history. Moreover, Wilkins (1978) argues that Popper's emphasis on the role of values in selecting historical problems is logically compatible with claims that although values may determine the *selection of*

---

*This will be discussed in more detail in my book on theory.

problems considered worthy of investigation (as they do in any field including science), they need not determine the content of one's attempts to resolve those problems. For example, the questions or problems studied may change with each generation (although their logic and sequence may be cumulative), but one can still study them in a consensual, objective, and cumulative way. Moreover, many data and "facts" once established do *not* change even if the interpretation of them does (see Chapter 4; cf. Harré 1986).

In any case, I have stated the aim as the use of data or evidence from the past. This puts the emphasis on questions such as how one can most validly use such data, how inferences can be made from them, and consequently how one can learn from them. Moreover, I have already emphasized that the use of historical data is unavoidable and essential, because they provide a crucial, and possibly the only, way to broaden the body of evidence used in EBS. In EBS, as it exists, the evidence is mainly from the West, although this is starting to change; it is based on a very limited range of environments and only from the last 15 to 20 years. This provides a *very inadequate* body of evidence, and broadening the body of evidence used in EBS is essential, first for low-level generalizations and then for generalizations at higher levels. Only in this way will one eventually be able to suggest explanatory theory. This is because, clearly, if one generalizes on the basis of a single piece of evidence—or a small and limited body of evidence—one is not as confident as when one generalizes on the basis of much evidence. One is also more confident if that evidence is very diverse. And, inevitably, such large and diverse bodies of evidence must include the kind of material that I am discussing in this book. The issue is therefore not *whether* to use it but rather *how* and *for what purposes*—the topics of Chapter 3 and 2, respectively.

Once questions are raised, research seen broadly begins with *pattern recognition* (e.g., Judson 1980; Ziman 1984). This often requires the ability to identify essential similarities while ignoring extraneous differences. Although that is less emphasized, it also requires identifying differences behind apparent similarity. More generally then, pattern recognition involves identifying regularities; this is because irregularities and anomalies, in turn, presuppose knowledge of regularities. For all of these one needs both a large and, more important, varied body of evidence—only when those are used can patterns be identified. Patterns, in turn, suggest hypotheses which can themselves best be tested by also using the largest and most varied possible bodies of evidence. These, in turn, can eventually possibly lead to explanatory theory. It is this need for large and varied bodies of evidence which first led me to include preliterate, vernacular, popular, and other environments in their totality, then to consider evidence cross-culturally, and finally to turn to the use of data from the past which provides the rationale for the domain as I have defined it in this book.

A comparable example from another field may clarify this point (see Figure 2.2). This is a hypothetical curve of murders committed in a given place during a century. If one were only to consider single segments of the curve, one would get rather wrong, and often contradictory results; these would be minor fluctuations as opposed to an overall trend. It then follows that one needs to incorporate as large a time span as possible. One may use the analogy of a meandering river—it seems to flow in all directions if seen over a small length; when seen over its whole length, however, it is seen to flow in only one direction (Lopreato 1984, pp. 102–103). This is justification

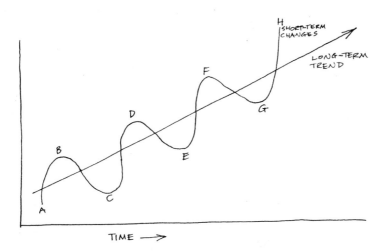

Figure 2.2. The need for large and diverse bodies of evidence (based on Lopreato 1984, Fig. 31, p. 103).

for broadening the body of evidence used temporally—the concern of this book. It has been suggested that the impact of longer time perspectives on various topics can be "revolutionary" (e.g., Szamosi 1986, p. 111). Also, the use of a long time series of almost anything, particularly if quantitative data can be used, is said to throw light on what otherwise might be apparent only to rarely gifted individuals (Price 1982, p. 179). Moreover, such methods, it is claimed, can help with the periodization of topics, in the case in question of science, suggesting, for example, that the Industrial Revolution may be merely an artifact of historiographical convenience. Because the history of science is currently having a major impact on the philosophy of science, it being argued that for the latter to be valid one must have a history of actual scientific practice (e.g., Miller 1984; Nersessian 1984; Shapere 1984, among many others), this is important. It is also similar to my argument about the relation of history to EBS. The point is also made that the history of science is not just data, the memory of science, but also its epis-temological laboratory (cf. Grmek *et al.* 1981)—another point I have been making. For it to become that, however, requires very precise, or at least genuinely accurate histor-ical analyses (Grmek *et al.* 1981, p. 6). For many of the questions in EBS, it is often, or always, also useful to consider the complete spectrum of cultures and the full range of environments (and the whole environment).

If science begins with a search for patterns, that is, regularities, then the purpose of social science is to identify regularities in human affairs and to develop empirically testable theoretical formulations that link together and explain these regularities; in the case of EBS, to do this for environment–behavior relations or interactions. As already pointed out, experiments are by and large impossible, and one must, therefore, examine and *compare* the phenomena in question in a variety of contexts: across cultures and smaller groups, life-styles, locations, climates, and so on. The use of the subject matter as defined provides an opportunity to study EBR in a much wider variety of contexts. The combined use of the past, of different environments, and cross-cultural evidence can then help to formulate hypotheses and test those in the present. At the same time, hypotheses derived from present-day research can be tested against past data. This helps not only systematically to trace trends and changes over time and to study determinants of change; it also allows one to study constancies and to test generalizations as well as to begin to consider the applicability of generalizations, that is, the conditions under which they apply or do not apply, how they need to be modified, and so on (cf. Rapoport 1988a).

An almost identical argument for social science can be found in Clubb and Scheuch (1980, p. 16). The purpose is to use data from the past to construct explanatory EBS theory, not merely to use EBS to understand or explain specific environments or environment–behavior relations of the past.

Thus the use of data from the past in EBS introduces a different model of research in *that* field (Rapoport 1979a,b, 1986b), just as the use of an EBS approach to data of the past changes the model of research in history; *it is a reciprocal relationship.* Thus the case study of pedestrian streets in Part III of this book is not only an example of the use of historical data to test an EBS hypothesis; it is also an example of a model of research that may lead to questioning social science models generally as *the* model for EBS; for example, in terms of empirical work, use of questionnaires and other instruments, and so on. This may lead to a new model of EBR that is not necessarily based only on social science. In this way, it may bridge the apparent gap between "science" and "traditional" scholarly and humanistic study. The latter is not necessarily different in its methods or "softer." It has, in fact, been suggested that rather than thinking of *softer* or *harder* areas of knowledge, it may be more useful and productive to think of better established and less well-established areas in all disciplines and how one moves from the former to the latter (Boulding 1980). Traditional scholarly and humanistic study is more usefully and typically conceived as dealing with the full spectrum of human achievement as studied by a sequence of scholars; the latter base their work on earlier work that then becomes cumulative. For example, a recent summary of scholarship speaks of the laborious, painstaking detailed efforts to make available to other scholars reliable data. Other scholars can then use these data in their own work; for example, to draw inferences in the light of other data known to one scholar or another (*Biblical Archaeology Review* 1987, p. 6) or, of course, in terms of particular concepts, models, and theories. In the passage cited, the data are ancient texts. In our case they may be what have been called *culture histories* in archaeology—reconstructions of past cultural landscapes. These themselves, as will be seen in Chapters 3 and 5, benefit from new approaches. More important, they then become the data for others to use in other ways, which often leads to reinterpreting the data. A major additional shift for which I am arguing is a shift from culture history to generalization and theory. This is a shift from the idiographic approach of traditional historical studies to the more nomothetic approaches of science. For my purposes, the reconstructions of aspects of the past by historians, archaeologists, and others are necessary but not sufficient. The reconstruction or understanding is essential as *data;* the important questions then still remain. First, how can such data be used for generalization or theory building and second, what can the data and the resultant principles tell us or teach us? Regarding the former, one can suggest that traditional historical "explanation" is the equivalent of "proximate" causal explanation, whereas at the theoretical level, one is trying to get at "ultimate" causal explanation (e.g., Mayr 1982; von Schilcher and Tennant 1984; e.g., pp. 130 ff.). In terms of Figure 2.1, it is the task of the history of the built environment to provide the data of the past.

We will see parallel shifts in other disciplines in Chapters 4 and 5. At this point, the major implication is that it clearly raises new kinds of questions—which have not been raised hitherto. The fact that they have not been asked does not mean that they should not—or *cannot*. Thus the use of historical data in the way I am proposing, changing the metaphor to science, has the unintended and paradoxical effect of linking scientific research with scholarly and humanistic research. This I regard as a most important result and a fundamental lesson that goes well beyond the derivation of any specific lessons or precedents, important as these might be.

I have already discussed the essential link between the use of past data in the way I am suggesting and the development of explanatory EBS theory. Because such theory is the only long-term basis on which valid design objectives might be based, this provides a link between this book and the design disciplines, albeit in a very different guise. This approach to EBS also has implications for a potential link with mainstream architectural "theory"—another apparently paradoxical result. This is because historical data, albeit very different in kind and approached very differently indeed, is the only element common to both (see Figure 2.3).

SOME EXAMPLES

The argument so far has been very general. It has concerned the rationale for the use of data from the past in the study of EBR. I have emphasized the value of this to increase the size and variability of any data used. There is, however, another important reason for using historical data. By providing evidence from long time spans, it also makes possible *diachronic study;* temporal patterns of constancy, change, or development become possible. This can be of any given phenomenon: size, elaboration, decoration, siting, how importance is communicated, the role of latent as opposed to instrumental functions, and many others. Both these related uses of past data help in the process of generalization and theory development in EBS. An almost endless number of topics suggest themselves as examples of how the use of such evidence could change the way the subject matter is approached. In addition to the case study in Part III of this book, I will discuss a number of possibilities among the many that could be discussed:

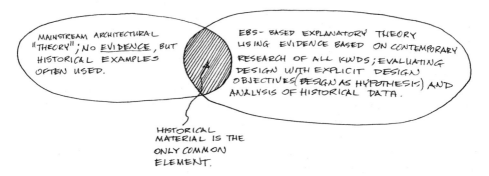

Figure 2.3. History as a possible link between EBS and mainstream architecture.

There is hardly a topic that could (and would) not benefit greatly. These range from the most global to quite specific topics.

An example of the former is provided by what I call the three general questions of EBS in terms of which *any* question in the field can be understood (see also Chapter 6).

1. What characteristics of people, as members of a species and of groups of various kinds, and as individuals influence or, in design, should influence how built environments are shaped?
2. How, and to what extent does the physical environment affect people's behavior, mood, well-being, and so on; that is, how important is the built environment, for whom, under what sets of conditions, and why?
3. A corollary question: Given a mutual interaction between people and environments, there must be mechanisms linking them. What are these mechanisms?

It seems quite clear that all can (and must) be studied over time, that is, using historical and archaeological data (as well as cross-cultural and all types of environments). Only in this way can one discover which human characteristics are important for which environments, under what circumstances, which are invariant and which change, and how; how environments affect people; what groups are most affected and under what conditions; which mechanisms are used in any given case—and what is their full range (because some may not be used in any given case), and many other aspects of these questions would greatly benefit from being tested against the subject matter of our domain. Because these questions are at the base of all EBS, such study, in effect, helps evaluate the validity of social science approaches to design and planning. As we shall see in Chapter 4, very similar questions are being asked in social science (e.g., Laslett and Wall 1972) and in archaeology (e.g., Staski 1982).

One can take a set of formal qualities in design and study these across a wide and diverse body of evidence such as discussed in Chapter 1. For example, one could look at sequences and progression of spaces, entries and transitions, complexity, relation to landscape, relationships among buildings, color, changes in level, textures, massing and fenestration, plazas and how indicated, inside/outside and the interface between them, light and shade, and so on. These could be studied over time, in different cultures, regions, climates, for different types of environments, and so on. In each case, one could study ranges; for example, the range of ways of handling these formal problems. Regularities and patterns could also be traced; for example, the use of color or relation to landscape as a function of type of environment, type of landscape or climate, culture, and so on. The generality of one's conclusions would be greatly increased—as when I argued that the "use of gateways" recommended by an Iranian submission to the Habitat Bill of Rights could be seen more usefully as a range of ways of handling that transition called "entry" (Rapoport 1982a).

One outcome of such study is not only the tracing of patterns and regularities but also the establishing of a repertoire—that is, the range of devices, elements, or means to achieve certain goals. This greatly broadens the range of devices, the repertoire, or vocabulary. One example might be how to communicate various meanings—for example, "sacred" (Rapoport 1968b, 1982a, 1982e; cf. Freidel and Sabloff 1984 in Chapter 5); to achieve complexity (Rapoport 1988b, in press), to communicate the qualities of a detached house (1980b, 1985b), and many others. For example, what kinds of buildings and cultural landscapes tend to attract tourists or retain interest or liking over long periods of time. What are the characteristics of settings that provide pleasure and entertainment, whether cinemas (Thorne 1987), pubs (*Sunday Times* 1968), shopping and entertainment areas, and the like; the case study in Part III is an instance of a study of this type, although approached deductively, from EBS, rather than inductively. Such repertoires could then be used to do simulations based on various combinations of elements—what they might look like, and tested with various groups of potential users. Such simulations can then also serve as hypotheses to be tested against further data. Such patterns of the past can themselves be related to simulation and to contemporary empirical research (see Chapters 3 and 5).

Consider one topic mentioned, the "sacred." On a number of occasions I have considered ways in which the sacred can be communicated through built environments (e.g., Rapoport 1968b, 1969a, 1982a, 1982e, 1988a).

I have also discussed the early beginnings of these expressions, a topic that has also received much attention in archaeology; moreover, it is related to the very beginnings of the built environment and the marking of place (Rapoport 1979a,b). In fact, these studies and the references in them greatly change one's approach to the built environment. This is a direct result of the greatly increased time span and the redefinition of the subject matter. New questions inevitably and immediately follow. These attempts could be broadened by analyzing a large range of sacred environments over time and in different cultures, to see whether they share any characteristics in common or whether they are highly culture specific. From a preliminary sample a set of hypotheses could be derived that could then be tested against a much larger and different body of evidence. In fact, even hypotheses based on introspection, either one's own or of a group of people (such as U.S. architecture students), could be tested in this way. This could be broadened even further in terms of the suggestions that even dwellings and cities in many traditional cultures are sacred (cf. references in Rapoport 1969a, 1977, 1979a,b). This could then be extended to landscapes—among the Navaho, Australian aborigines, and so on (Rapoport 1969d, 1975a), or the Zapotec or Maya (Flannery and Marcus 1983; Marcus 1973, 1976). This could then lead to questions about the relative importance of *building* to the sacred as opposed to other elements of the cultural landscapes. One could also relate this to more traditional history and study it in this way, for example, Roman or Greek temples, Gothic cathedrals, Renaissance churches, and so on (MacDonald, 1976; Rapoport 1982, 1988a; von Simon 1953; Scully 1962; Wittkower 1962). Essentially, hypotheses derived from any source could be tested by looking at sacred environments over time, in many cultures; by using large bodies of evidence one could answer questions, test hypotheses. learn lessons, and derive repertoires. The variety of repertoires could be further refined in terms of *transformations* (Glassie 1975; Gould, S. J. 1980, 1985; Habraken 1983; Jacob 1982; Leach 1966; Thompson 1961; Wolpert 1978; among others). Only large samples could elucidate this (Bakker 1986; Eldredge 1985; Gould, S. J. 1985). Of course, in this, one is essentially looking for patterns, possibly regularities behind the apparent great variability of the repertoire. Many other references, in architecture, history and philosophy of science, biology, and other fields immediately suggest themselves as potentially relevant (e.g., Holton 1972, 1973, 1978; King 1980, p. 29; Tyler 1978, pp. 279–280; among many others). Note that transformations could be studied over time in a single place in different places, for example, different cultures, regions, climates, and the like, different social organizations, in vernacular versus high-style environments; in different building types (e.g., symbolic, commodity, and system buildings [Heath 1984]); in different urban contexts, different economic systems, different resources, and so on. Again, a large and fascinating range of studies quickly suggests itself. If fertility is an important consideration in evaluating approaches, the present one seems to be highly promising.

Moreover, generalization becomes possible, as when assertions about the sacred being indicated by modern materials/forms or alternatively by archaic forms can then be shown to be generalizable to *noticeable differences* (Rapoport 1982a). This could be studied by considering a wide variety of sacred buildings in their respective contexts.

A diachronic study could test arguments about the reduced significance of the sacred and particularly its expression through built form, over time (Rapoport 1986c, 1988a, 1989a). This could then be broadened further to the role settings played in some vision of ideal people leading ideal lives. Again, hypotheses could be tested to see whether this relation in fact does apply to all environments, including the suburban house (e.g., Rapoport 1985b,c,d). Diachronic analysis could show whether there have been systematic shifts in the *type* of ideals involved and the importance of settings (and the *kind* of settings).

Note how many questions can be raised, how quickly and easily; indeed, how they almost thrust themselves on one. Note also how different these are to the usual questions considered in architectural history. Two things seem to be involved. First, that the subject matter of the domain raises new types of questions. Second, that one is using *lateral* rather than linear linkages.

As an example of this use of historical data, consider *meaning*. This is an interesting choice for several reasons. First, because it is my answer to the question about the reason for the variety of environments. Second, because it seems to play a central role in EBR. Third, because due to its centrality in human behavior, it is often used as an excuse to argue that social science cannot be scientific (e.g., see references in Argyris *et al.* 1985 as just one example of a large literature; cf. Hodder 1986). In my own work, I give meaning a central role in EBR (Rapoport 1982a) and give examples to support this argument. These represent the "spread" for which this book is an argument and range from the Sepik River through Gothic and Renaissance churches to France in the 1960s and New Jersey in the 1970s (see also Rapoport 1988a). It can then be shown that this applies to cave paintings, as we shall see in Chapter 3, to tools (Chapter 5), and to almost all human behavior—including Neanderthal burials. Again, it is the long time span and variety of evidence that suggests the centrality of meaning in material culture. Also, given the culture-specific nature of design, such regularities become even more impressive—at a higher level of generalization.

We have, in effect, found a very strong recurring pattern: People seem to shape and interact with built environments/material culture primarily through meaning, and this seems to hold over time, cross-culturally, and in all kinds of environments, contexts, and situations. This tends to give one confidence that it *is* indeed a pattern and that meaning is a central mechanism in EBR. This means that one can generalize with increased confidence: If humans have done something for tens of thousands of years, possibly since their emergence, and in all cultures, *it must be important.*

Identifying this pattern does several more things. First, it suggests that one should look for this in studies done currently. Second, if it is not found this becomes an *anomaly*. We are then constrained, first, to recheck to see whether it exists in other forms, expressed differently, and so on. Second, if it is indeed absent, one is constrained to explain this absence—as I tried to do regarding the "disappearance" of symbolic meaning in built environments (Rapoport 1986c, 1988a). Note, finally, that in addition this generalization then enables us to integrate a very large body of literature from many disciplines (cf. Rapoport 1982a). This again increases our confidence in the generalization—evidence comes from many disciplines, from different lines of evidence, and so on. It also makes available new approaches and new methods that "come with" these disciplines.

In my own work I have always used this approach. For example, my first published piece, on Isphahan, was essentially historical and tried to show the great continuity of the elements (Rapoport 1964–1965). This I developed further in some unpublished lectures I gave at Berkeley in the early 1960s on Iranian architecture where I argued for a close link between vernacular and high style, with major continuities in certain characteristics in spite of major changes in political and religious systems. It is interesting that similar points have recently been made by archaeologists. Thus a 3000-year continuity in settlement structures in the South American Andes has recently been pointed out, which persisted in spite of major discontinuities in social and economic systems (Isbell 1978). Similar constancies of various sorts can be found in Mesoamerica (e.g., Marcus 1973, 1976; Ingham 1971)* and in Europe (e.g., Habraken 1983). To trace such patterns long time series are necessary in the case of specific locales—whether Iran or the Southern Andes (cf. Kramer 1979, 1982, and other examples in Chapter 5). This is the value of history at the level of the study of specific places ("culture history"). They are equally essential for more generalizing and theoretical approaches, where a number of such places might be compared. As we shall see in Chapter 3 and 5, both culture histories and generalizing studies benefit not only from redefining the domain but also from more rigorous and scientific approaches.

In fact, there is no topic in EBS that cannot gain from historical data. Thus in dealing with open-endedness in housing and trying to identify which elements change much and which little, that is, relative rates of change, I relied on some historical data (Hole 1965; Hole and Attenburrow 1966; Rapoport 1968a). The topic would benefit greatly from a much broader sample and a much longer time span. Open-endedness could then be studied in terms of *patterns,* that is, by trying to discover the types of things that change much, those that change little—or not at all; under what conditions they change, for what reasons or purposes, and so forth. One could also try to discover completely new components of the built environments. Some patterns that seem to emerge (for example, that dwellings tend to change less than public areas, or the opposite) can then be tested against longer time series of more varied evidence. Regarding housing also, broadening the sample immediately changes one's view about the importance of site, materials, or climate (Rapoport 1969a). Further evidence cross-

---

*Even larger patterns can then suggest themselves. Thus the symbolic model underlying the Aztec realm has recently been discussed (Broda *et al.* 1987). It matches that of a small town today (Ingham 1971) and links with the Maya (Marcus 1973) and Zapotec (Flannery and Marcus 1983). The use of a single model at various levels (in the Aztec case the universe, city, temple, etc.) also duplicates a similar "secular" pattern in San Cristobal de las Casas (Wood 1909).

culturally and studies over time could help one discover how relatively important climatic variables are as opposed to cultural ones—and under what circumstances. One such attempt was a paper by Donna Wade in one of my classes. She took my hypothesis that cultural variables were more important than climatic (Rapoport 1969a)—itself derived from a large body of evidence, that is, a pattern. She then tested it against a sample of preliterate cultures in the most extreme cold climates, where the contrary was most likely. The hypothesis seemed to be supported. Note that the sample of preliterate cultures is not only cross-cultural (her intention) but *inherently* historical in my sense. Similar tests could be carried out regarding *any* aspect of EBR. Thus a colleague and I once showed that physical standards are culturally variable (Rapoport and Watson 1972, first published 1967–1968). We suggested that standards also vary over time, for example, that domestic temperatures in Britain were going up but had few data. Recently, a time series for domestic temperatures in Britain between 1946 and 1978 has been produced that clearly shows trends toward higher temperatures. This has implications for future planning (Hunt and Steel 1980). This could be done for area or lighting standards—in fact *any* subject gains by being considered over time. One can also study a concept such as "comfort" not only cross-culturally but historically (e.g., Rybczynski 1986), although a stronger, EBS-based theoretical framework, a larger and more diverse body of evidence, and a longer time-frame would help. Regarding climate, studies of past environments of the type defined in Chapter 1 can help one derive precedents for design in terms of energy use of settlements. One such study (Rapoport 1986b) provides an example of how this can be approached, how inferences can be made about behavior and life-style, and how surrogate indicators can be developed for energy efficiency because these data are not available for past environments; clearly some experiments in still existing traditional environments could help. Finally, this study shows the *nature* of precedents that have to do with lessons at high levels of generality rather than with copying forms. In effect, this study is a miniature version of the approach of the case study of this book albeit in a different domain.

As already discussed with reference to the three general questions of EBS and elsewhere, one can test the basic environment–behavior interactions over time, starting with hominids (Isaac 1972, 1983; Rapoport 1979a,b; Young, Jope, and Oakley 1981). Only in this way can patterns of such interactions be revealed—if they are studied using the full range of evidence from the past, based on evolution, sociobiology, primitive and vernacular cultural landscapes, and so on. This might provide evidence for the idea of an evolutionary baseline, extending to animals, already discussed and to be further discussed in Chapters 3 and 4. This could be tested against different cultures and types of environments using historical data. This baseline could, of course, be in terms of transformations as already discussed, or in terms of invariant processes. These, to be discussed in Chapter 3, also need to be tested against historical and cross-cultural evidence. Thus it has been suggested that all organisms seek comfort (Lopreato 1984, p. 256), and this then becomes a generalization relevant to EBR. However, we have already seen that what is "comfort" seems greatly variable cross-culturally and historically and may, moreover, be less important than, for example, cultural variables (e.g., Rapoport 1969a, 1986b). Again, this kind of evidence I am discussing seems essential—as it does for environments that make one "feel good" (e.g., Pulliam and Dunford 1980). In fact, the role of emotions, the affective component, in environmental choice and preference—a constant theme in my work (e.g., Rapoport 1977, 1983c, 1985b)—is again a hypothesis based on patterns revealed by some evidence of the type being discussed and requires further testing against larger and more varied bodies of evidence, more carefully selected (cf. Rapoport 1986c). One could even hypothesize that there should be greater consistency for "natural" landscapes than for buildings because humans evolved in such landscapes. This has been suggested by Kaplan (1982) on theoretical grounds and explicitly argued, using historical evidence (Appleton 1975). In that latter case, a large body of evidence—paintings—is adduced to show that certain landscape preferences, derived on evolutionary grounds, are in fact present. Unfortunately, only the Western tradition is used: Using a wider range of examples and other types of evidence would strengthen this study. This is also an example of an attempt at a "scientific" approach to environmental history—a hypothesis tested against historical data (cf. Salk 1973). When recently tested empirically, the hypothesis of constancy of landscape preference was not supported (Lyons 1983). This study, however, neglected the best developed consistency model (Appleton 1975) and only used contemporary U.S. groups. Moreover, Wohlwill (1983), among others, has found greater consistency in natural landscape preference than in preference for man-made landscapes. Although this consistency/variability issue is clearly very complex and unresolved it begins to point a way to an approach that is not only testable and *demands the kind of evidence this book calls for;* it also raises most important issues and questions in terms of broader theory.

One reason that such questions are extremely important for EBS is that *time* has been neglected in the field—in two senses. In the short-term, synchronic, thin-slice sense, time as an essential component or element of the built environment has been neglected. I have argued for its importance in my work since the 1970s. Time in the longer range, temporal scale sense of the diachronic study of pattern, change, and constancy—with which this book deals—has also been neglected. There is, however, currently a growth of interest in, and emphasis on, temporal aspects of EBS (e.g., Werner *et al.* 1985; Pavlides 1985a,b; and Sarcar 1985, as just three examples). (See also Conan 1987 and Alsayyad *et al.* 1987.) This implies the need to use data from the past. I have argued for its importance but have not systematically used it until fairly recently as in my study of precedents for energy efficiency of settlements (Rapoport 1986b) or, in general (1968c; 1988a, 1989a), as a way of helping distinguish vernacular from high-style environments (Rapoport 1988b, in press) among others. In this latter case, I hypothesized that vernacular changes more slowly over time than high style* and also that the *form* of temporal change is different in the two cases (see Figure 2.4).

Both are hypotheses that require testing against a broad and varied sample that must be historical in nature in addition to the other requirements that I have been emphasizing. In addition, there are some suggestions that vernacular may not be universally as unchanging as has been suggested—for example, U.S. vernacular may be quite different (Jackson 1984). It follows that larger and more varied evidence must be used; it also makes diachronic studies of vernacular itself even more important; such

---

*Cf. Glassie (1968, p. 33) who similarly argues that folk material culture (vernacular) exhibits major variations over space and minor variations over time, whereas products of academic culture (high-style) and popular culture exhibit minor variation over space and major variation through time.

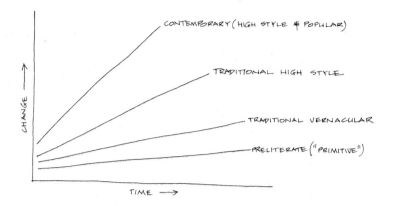

HYPOTHETICAL RATES OF CHANGE OVER TIME.

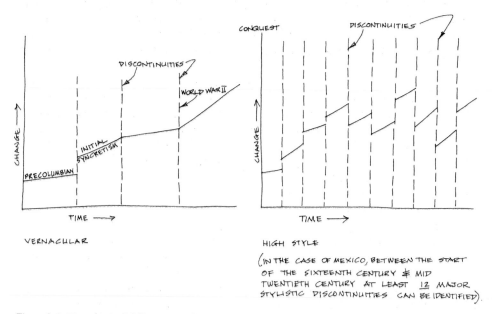

Figure 2.4. Hypothesized differences in changes over time of vernacular and high-style environments (based on Rapoport, in press, Figs. 11 and 12).

studies are, in fact, found. Some of them deal with the United States as one would expect from the foregoing comment and also show the relation between vernacular (or is it popular?) and high style (e.g., Jakle 1978; Jakle and Mattson 1981; Rubin 1979), but not all do. Thus we find a study of Turkish vernacular over a 12-year period (Sancar 1985) and a study of one place in Greece over 130 years (Pavlides 1985a,b). This latter leads to a conclusion at odds with one widely used characteristic of vernacular—that it supposedly does not communicate status. In this case it clearly does so. This then becomes a question to be tested across cultures, in other places, in other environments, over longer time spans. One may discover the conditions under which status is communicated (e.g., Rapoport 1981a, 1982a, 1988a, 1989a). On the other hand, this may become a "universal" statement—that all (most?) environments are used to communicate status (which is, in fact, likely). In that case, the reliance on dwellings and neighborhoods as status indicators in the United States and other Western countries is not an anomaly but a general phenomenon; what has changed is that its importance might have grown and that it has become more generally available. As a result, it is much more widespread and important—as part of the mass culture typical of the latter twentieth century. Further study using broad and varied evidence could then establish the range of devices used to communicate status in vernacular environments: size, height, color, decoration, location relative to some entrance, controlled accessibility, and so on—a *repertoire* once again. The specifics may again possibly be seen as transforms of basic patterns; the specifics will depend on criteria of choice, meanings, schemata, purpose, other devices used or available, life-style, and so on—as well as constraints, such as site, materials, climate, resources, technology, politics, knowledge, or prejudice (such as sumptuary laws or safety) (see the example of the Zoroastrian house, Mazumdar and Mazumdar 1984; see also Chapter 5). Thus, as we shall see in Chapter 5, archaeologists through the use of ethnoarchaeological research derived conclusions about meaning, for example, in contemporary Africa, which are then tested against archaeological remains in the Orkney Islands (Hodder 1982c). These can be seen as transforms of a common basic pattern—the environment as the organization of meaning. This could then be extended and tested against a very much broader and more varied body of evidence.

Third World environments also need to be studied over time. The very change of attitude to these is a diachronic phenomenon—and occurred *because* of studies over time that showed that spontaneous settlements in certain places (e.g., Peru and elsewhere in Latin America) eventually become "normal" neighborhoods (see review in Rapoport 1977). But these conclusions could then be shown to be overgeneralized when tested against a body of cross-culturally more varied evidence that did *not* show that development over time. Thus a purely historical study in one place is not enough; the evidence needs always to be broadened according to the four criteria described in Chapter 1 (cf. Chapters 3, 4, and 5). Note also, that my approach to studying supportive environments in developing countries by identifying the culture core of particular groups was developed by using a much broader sample than is common. Moreover, it depends on a diachronic analysis of a range of environments (Rapoport 1983a). Although the model has been tested in classes against a variety of places and I recently had an opportunity to apply it in the field, it needs to be tested much more systematically against a much broader sample, Many specific questions that arise during the application of this model in a class I teach on design for developing countries in turn require diachronic study, that is, historical data. In fact, knowing whether a particular environment is "traditional" or spontaneous often requires historical data (see examples in Rapoport 1988b). At the same time, one could also test more specific underlying models, such as Siegel's concept of defensive structuring (Siegel 1970). This I have done informally (Rapoport 1978a, 1981a, 1983a); my use of it has also been extended in a historical study of Salzburg, Austria (Rowntree and Conkey 1980). Note that this concept is inherently *diachronic;* it could be studied more systematically against a broad sample of high-stress situations—and compared with nonstress situations. This could not only test the hypothesis but show patterns of the devices used in defensive structuring, that is, a repertoire once again. Once again, as soon as any problem, question, hypothesis, or generalization is tested using such evidence, many new and hitherto unasked questions seem to proliferate: It is a most fertile approach. In each case the fertility is a result of the decision to use evidence "laterally" and to test any and all hypotheses or generalizations against the widest possible body of evidence selected according to the four criteria described in Chapter 1, although explicitly selected to be relevant to the specific question or topic being addressed.

Some time ago, on the basis of both EBS and cross-cultural evidence, I made a series of suggestions about neighborhoods (e.g., Rapoport 1977, 1980/1981). Among these was the suggestion that in general it appeared that neighborhoods were significantly smaller than planning theory had assumed and that *in general* they tended to be homogeneous; this homogeneity was on the basis of perceived characteristics that varied from place to place and from time to time. Whether this was indeed the case, how generally and so on is the type of question well suited to being tested against the type of evidence described in Chapter 1. Moreover, the EBS generalizations underlying this hypothesis would also benefit from such testing. It is also the case, as we shall see in Chapter 3, that at the neighborhood scale there is much greater (in fact, great) constancy and similarity over time; it is at the scale of the city, the metropolis, and the megalopolis that there is variability and change. This hypothesis also requires to be tested against both cross-cultural and historical data, that latter suggestion having been based only on the ancient Greek city (polis). The suggestion about homogeneity is one of these constancies. One of the earliest and best examples of testing this against historical data was a paper by one of my students (Sanders 1975). In it he reviewed my general hypotheses about neighborhoods, specifically discussed the homogeneity argument, and identified possible indicators of homogeneity that which might be found in the archaeological record. He then listed much archaeological evidence that could be tested. In one term he was only able to examine selected examples from Mesoamerica, South America, and ancient China and the Middle East. As far as the study went, the homogeneity hypothesis seemed to be supported. That is less significant, however, than the type of question, the type of subject matter, and the approach.

Many designers have strong views about what makes for good environments. Even as a personal opinion, it can be tested if taken to be a hypothesis and used with a large and varied body of evidence. In my view, one of the major findings of EBS is that environmental quality is highly variable and hence far from self-evident. It follows that one role of past data (as of other types of data) is to study this variability as well as any constancies. Thus one could study the quality of settings, good or bad. For example, one could analyze "delight" or "quality" in urban environments that are either taken as self-evident or not specified at all (e.g., Cresswell 1979); similarly, with hypothetical wishes of urban residents (e.g., Tanghe *et al.* 1984). Such studies would reveal whether they were, in fact, valid (*what* should be done and *why*) and, if so, how they could be achieved—a repertoire. Thus even "how" can be studied in this way. Note that in the books just mentioned, many examples are given but are not used in the ways being suggested. More usefully, one could ask whether there are any patterns—characteristics of environments that are liked or disliked, whether in particular places—or globally; and over short, medium, or long periods of time. If such environments are found and patterns of common characteristics uncovered, that would provide important evidence and precedent. For example, it has recently been suggested that the three-dimensionality of facades is related to preference and liking. This hypothesis, derived from EBS studies in Sweden (Westerman 1976), could be tested against various environments, cross-culturally, over time and so on. If profiles of such characteristics could then be constructed and shifts traced over time in the activities housed, in the populations attracted, by culture and so on (cf. Rapoport 1977; Tuan 1974), this would provide most important knowledge: One could begin to specify what characteristics environments need for given activities, for given population groups, and cultural contexts. Such general, conceptual characteristics combined with the notion of repertoires could begin to provide precedents for design, predictions that could be tested either through simulation or by evaluation once built. Because objectives are set and reasons for them known, such postoccupancy evaluations would also be generalizable—which many so far are not because of the lack of explicit objectives and criteria (Rapoport 1983b,e, 1986e).

Many terms and concepts that are used in architecture and EBS can be redefined when the evidence used is broadened in the ways discussed here, that is, the domain is redefined. This is the case for cities (Rapoport 1977), dwellings (Rapoport 1980c, 1989a), or privacy (Rapoport 1976, 1977, 1980d; cf. Altman 1975). This is most important conceptually because it is equivalent to *deriving* etic categories rather than imposing them. Moreover, it leads to concepts or constructs that are broadly valid and can be applied to many situations: cultures, settings, and so forth. In this sense concepts can be *tested* against much and diverse evidence and adjusted, or they can be *derived* from such evidence and then tested against wider evidence still. Not only concepts and constructs but models and elements could be tested in such ways. Two examples can be mentioned briefly. The first is the idea of space syntax (Hillier and Hanson 1984), itself based on a domain shift—to vernacular design, which could be tested against a large range of settlements over time (e.g., traditional versus modern) in different cultures, in vernacular parts of settlements versus their high-style parts and so on.* The second is the idea that there is a certain spatial unit that is used broadly in assembling environments (Scheflen 1976). Although this was already based on some measure of cross-cultural and historical evidence, it could be tested much more systematically against a much larger and more varied sample. This could be, and needs to be, done for any generalization, concept construct, model, or whatever in EBS.

The number of possible studies and examples could be multiplied endlessly. This is because any topic can be studied in this way. But enough has been said to make several things clear. One is the intimate relation between contemporary and historical work and data: As suggested, *the two together form a single domain.* Second, the types of questions that seem to suggest themselves so easily are not only interesting and fertile; they are quite different to the usual questions addressed in architectural history. They lead to different conclusions and consequences: History is being used for a new and different purpose. Third, all these examples and suggested studies, no matter how

---

*This could also be compared to my hypothesis in the case study that essentially deals with very similar physical evidence at the level of urban *fabric.*

apparently different, have one thing in common. They reveal *patterns*—patterns of the past or over time, of different cultures, different types of environments, and so on; or of various combinations of these. Such patterns, in turn, can be used abductively to suggest hypotheses or develop concepts or can be used to test hypotheses derived from other sources and to check generalizations proposed. In either case, such patterns play an essential role in theory development.

IMPLICATIONS FOR THEORY

As already suggested, in all these, and other examples or cases, the use of historical data about built environments can lead to hypotheses, for example, about human behavior. This involves an interesting paradox. On the one hand, history needs an explicit notion (or model or theory) of human nature or behavior—as we shall see in Chapters 3 and 4; so does social science (cf. Gordon 1978). On the other hand, it is the use of a broad range of evidence, including historical data, that can best provide such a theory of human behavior. One could suggest that housing always expresses status or some identity—or does not always do so or does so in specific cases with specific sets of characteristics (Rapoport 1981a, 1982a, 1988a, 1989a). Alternatively one could hypothesize that it does so at the level of the group in some cases and for individuals in other cases; the nature of groups could also be studied. Other generalizations also become possible that can be tested in the more active, empirical mode typical of EBS. Conversely, hypotheses generated empirically in contemporary EBS can be tested against the data of the past—if one can derive product/built form consequences from these EBS data—to see whether they hold up more generally. This involves or implies a distinction between studying *process* and *product* (Rapoport 1988a, in press), and different lessons might be derived from each. There may be constancies in either—or both; change in either or both—or transformations. One could argue (Rapoport in press) that there may be more lessons at the product level (as in my previous discussion of formal qualities); but that, in itself, is a hypothesis to be tested because the opposite can also be suggested (e.g., Hakim 1986).

This suggestion at a more abstract level is also encountered, often *implicitly;* that *process* may be more unchanged than product and hence may have more lessons: In a way this is implicit in the "evolutionary baseline" argument (see previous discussion and Chapter 3 on uniformitarianism). The point has been made explicitly in the case of anthropology: that there are cultural universals but these are regarding process rather than outcome (Fox 1980, p. 7). In other words, the same process is found; by interacting with specific variables, in our case context, locale, climate, materials, life-style, ideals, and so on, specific outcomes result. I have already briefly discussed the possibility that the outcomes themselves may be transforms of one another. In terms of the environment, this can then provide important lessons: about which wants and needs design tends to try to satisfy (process) or about repertoires or vocabularies of formal qualities or preferred configurations (product). In terms of the use of historical data, the uniformitarianism at the level of process has important implications (see Chapter 3). This point has been made in the case of the history of medicine (King 1982), where one finds that whereas specifics clearly change, certain things remain constant. Thus techniques, conceptual frameworks and theories, knowledge, that is, the quality and quantity of data, have greatly improved—they are progressive and cumulative. Ethics have changed—but less. The problems have remained the same, as have the ways good doctors (and bad ones) think or have thought.* This, of course, relates to another level of my argument about problem identification. Once problems have been identified, good doctors always use the *best available information* and tools (cf. Rapoport 1983b,e, 1986e). It is these latter that have changed and are now better. There is continuity in problems and variation in answers; as the "illumination" has changed, so have the answers.

We shall see in Chapters 4 and 5 that attempts have been made to identify unchanging problems facing people and that these have been used in making inferences about the past. This seems potentially very useful for the history of the built environment. In EBS the problem of teleology that concerns certain historical sciences (e.g., biology) does not arise: Built environments are *always* created for a purpose. One can then try to identify certain problems: What were people always trying to achieve? What problems did they face and which did they, therefore, try to solve? Those that have remained invariant can be distinguished from those that are different, and these can be specified (see Chapters 3 and 4). How did they go about solving problems? Analogously to medicine (King 1982), one can argue that the thought processes were the same.† There were always similar problems with which people coped as logically as now and as best they knew how; their intelligence was the same. In *their* context, designers, like doctors, made very good sense. It is the context, the tools, the knowledge that have changed. The very fact of change and advance implies that something is constant. These are the problems and the attempts to solve them. Although King (1982) only deals with evidence from mainstream medicine, the domain could be extended to include paleomedicine, ethnomedicine, and folk medicine. This would extend the time span beyond 2,500 years and increase the diversity of ways of defining and tackling problems. I am suggesting that we approach EBR in the same way that King (1982, pp. 1–12) deals with medicine: as the history of problems, ways of dealing with them (customs, traditions, behavior), and of knowledge and information available. In our case evidence from the past and the other forms of diverse evidence can help study the heritage of problems, of designing for people, of creating supportive environments; it can help distinguish constant from changeable problems, transformations among problems, and patterns behind apparent diversity. Similarly, one can study the heritage of concepts, knowledge, information, means, and constraints; one can also derive the heritage of behavior of designers—in the broad sense in which I am using this term; the behavior of professional designers can then be compared with that. The value for theory seems evident.

Note how by raising the apparently simple question, "Why does one build?" that some most interesting, potentially fruitful, and different lines of thoughts come up. By changing the domain—in terms of subject matter and questions and problems—many other things change: not only about history and why and how it should be studied but also for theory. Changes to what is to be studied—the ontology or nature of the

---

*I think that there has been some change due to the development of scientific education in medicine.
†See the large literature in anthropology on so-called "primitive" versus "modern" thought.

universe of discourse in question, as is so often the case, lead to fundamental changes in the questions asked and problems tackled—the topic of this chapter. This redefinition of domain leads to changes in epistemology—how things are studied, the methods used; the approach becomes fundamentally different (Chapter 3). This, in turn, leads to changes in theory, although that will not be discussed in this book to any major extent.

All this testing of all sorts of things against the new types of evidence can reveal whether any principles, generalizations, or hypotheses are artifacts of the particular experiments or studies, anomalies, aberrations, and so on. If one finds some things occurring consistently over long time periods, in many environments and across diverse cultures, then one has identified a regularity or a constancy. This is then likely to be an important generalization or may even be a "universal." If it is never found except in the one experiment (or situation), it is not likely to be correct. Alternatively, one can then develop hypotheses about particular forms of change, or transformations, that can be tested in their turn. In this sense *one can use data from the past as a form of replication.*

This has methodological implications for EBS (e.g., Rapoport 1970a, 1973a) because different methods must be used on such data. It also relates intimately to the point already made about pattern recognition, classifications, and the like. Note that in the early stages of *all* research in most disciplines, pattern recognition is essential. This involves classification and taxonomies as important first steps (Gould 1985; Mayr 1982; Northrop 1947; Rapoport 1986c, 1988b, in press). For both purposes—to be able to classify and, hence, see patterns, regularities, and the like, one must have large and varied, that is, representative, bodies of evidence—as argued in Chapter 1. Note that this point is clearly made and widely accepted in the historical sciences that I will discuss in Chapter 4. If the objective of EBS and the use of history are to lead to theory, that is, to obtain understanding and explanation, then this becomes an important criterion. For example, in astronomy (Clark 1985), the point is made (p. 16) that one only began to understand supernovae when one could compare the light curves, spectra, and other characteristics of large numbers of them; over 400 of them have now been detected. At that point they could be classified into two types. One could then begin to understand them and ask questions leading to further advance in both theory and observation. Among such questions were (Clark 1985, p. 10): the nature of supernova explosions; theories of why certain stars explode; the form of the debris ejected (the so-called supernova remnant and its effect on the interstellar environment); what is left (if anything) at the site of the explosion.

The answers to such questions seem to depend on the mass of the star, but there are two evolutionary pathways following explosions. The point is made (Clark 1985, p. 19) that 999 stars out of 1,000 do such and such. This clearly implies a *vast* body of evidence. Of course, astronomy has been collecting evidence for thousands of years.* As it has developed, it has collected evidence systematically, so that one finds systematic surveys of the sky, first optically, then using radio, x-ray, and other telescopes as technology improves. This allows one to identify objects, to classify them, and to begin to see patterns such as regularities, associations between objects—for example, in optical, radio, and x-ray surveys; this also draws attention to anomalies. Thus (Clark 1985, p. 28) there seems to be a uniform distribution of pointlike radio sources all over the sky; when higher concentrations are found in given regions of the sky, this raises

---

*This may partly explain why it developed as a science "unusually" early (Price 1981).

questions. Also an idea generated on the basis of a single case (or, in our case, a very few cases in a very limited domain [temporally, culturally, in types of environments] and extent of environment) is necessarily speculative. With more cases one can evolve theories which may then become widely accepted.*

In biology a similar sequence has occurred: Observing and collecting data (natural history) led to taxonomy and classification (Mayr 1982). The theory of evolution was plausible as a result of the vast body of material that Darwin collected. The current argument about gradual evolution versus punctuated equilibria, that is, about the form and mechanisms of evolution, depends crucially on paleontology. In that field the whole argument is based, and has been since Darwin, on the *completeness of the evidence* (Bakker 1986; Eldredge 1985; Gould 1985; cf. Chapter 4).

If one derives certain EBR principles, in whatever way (e.g., Rapoport 1977; 1982a) as well as ideas about primary and modifying factors in design (e.g., Rapoport 1969a) or the role of constraints of various sorts (e.g., Rapoport 1983a, 1985b, 1988b), the question then arises how these can be studied. The answer, in principle, has already been given. To understand them, and then to use them in design, one must have generalization and theory. In the first instance, such theory will be explanatory; it will explain various aspects and forms of environment–behavior relations. This will then lead to ideas about *what* needs to be done and *why,* although theory (and precedent) can also be useful regarding *how* it is to be achieved—the idea of repertoires or vocabulary already discussed.† For all of these, one needs evaluation or testing, one needs to be able to judge what is good or bad and why (and on the basis of more than personal preference). One also needs to be able to set design objectives on the basis of such theory in order to be able to evaluate, that is, discover whether the objectives have been met. As I have argued, evaluation is a two-stage process—one first asks whether objectives have been met—if so, whether they are valid (Rapoport 1983b, 1986e). This two-stage process is reversed for design. In either case, *design is seen as a hypothesis* that needs to be continually evaluated—before design (on the basis of the best available knowledge and theory), during design (by simulation), and after building (by postoccupancy evaluation). The link between all of these is theory because one must know what a thing is supposed to do before one can judge whether it does it well or badly (cf. Wolff 1982). This means that explanatory theory must precede any normative judgments (or "normative theory").

This link between evaluation and design, through explanatory theory, is intimately related to history in the sense of using data from the past. This should be clear from the discussion so far. *History in that sense can be an evaluation of past successes and failures; current evaluations of specifics of past environments; an evaluation of the success or failure of concepts, principles, and precedents derived from past examples.* History (as well as the other criteria for the subject matter of the domain) is thus *central to a theory of EBR* and of a cumulative body of knowledge. The role of history thus changes: As new questions and problems are asked of the new subject matter, that is, as the domain is redefined, the purpose of history becomes part of theory building and central to it.

---

*I am *not* suggesting that theory emerges from masses of data; I *am* suggesting that one can only develop plausible theories on the basis of adequate data; they must then be tested against further adequate and representative evidence (cf. Bullock 1984).

†This and many other related matters will be developed in a proposed book on theory.

The notion of cumulativeness, that is, of progress in a discipline, can be interpreted in at least two ways. The first, and most common, is the linear form typically taken as characteristic of progress in the natural sciences, for example astronomy or physics (e.g., Ayer 1984, p. 2).* In this, one finds either a succession of systems, each in turn replaced by an improved one, or a variant that has been suggested as more valid for biology where cumulativeness is seen to be the result of *growth and combination* rather than *replacement* (e.g., Mayr 1982). The second type of cumulativeness is related to *the evolution of a set of perennial problems* and is more typical of philosophy (Ayer 1984, pp. 3, 14). None of the issues are resolved, not even provisionally: Progress consists not in the disappearance of any of the age-old problems nor in the increasing dominance of one or other of the conflicting (alternative) views but rather in a change in *the way or fashion in which the problem is posed,* and in an increasing measure of agreement concerning the character of their solution. Although answers have not yet been found, the area in which answers can reside has been narrowed.† There can also be progress in terms of *the greater sophistication of argument* (Ayer 1984, p. 18)—which is what I find characteristic, for example, in reading a large body of work on philosophy of science over the past 8 years. On these views, philosophy has progressed in terms of the study of evidence and the range of that evidence.‡ It is in this sense that I believe that the study of the history of the built environment can progress, or in terms of the "biological" model (Mayr 1982) whereby alternative views are linked into larger systems and often seen as complementary rather than conflicting. This is typical of my work generally (e.g., Rapoport 1977, 1986c).

Recall the point made in Chapter 1 that the use or availability of new bodies of evidence often greatly changes the nature of a given field—a phenomenon often encountered in the history of science. In this chapter the changes being discussed are in the purposes of study and the questions asked and its impact on EBR theory. In Chapter 3 the impact on the nature of the study of past environments itself will be traced.

One purpose of using data from the past and generally broadening the body of evidence is to derive rules or generalizations. These do not have to be universal, but can be statistical or probabilistic, possibly with specified boundary conditions or domains of applicability. These also can only be derived from broad and diverse bodies of evidence covering long time spans. Such generalizations can be about what was, what is, or what *tends* to be, given certain conditions. Then, once theory has been developed to explain these, one can possibly move on to *what should be.*

I do not want to discuss this in any detail here because it belongs in a consideration of theory. However, there is a well-known logical problem, first enunciated by Hume, about moving from descriptive statements ("was," "is," "tends to be," and the like) to prescriptive "ought" statements. The use of a broad body of evidence is critical to derive descriptive statements or rules. The link with prescriptive statements is then through understanding and explanation, that is, theory and the models and mechanisms that

---

*Note that in contemporary philosophy of science the question of cumulativeness in the natural sciences is no longer self-evident; it is in fact questioned and generates much debate. This will not be discussed here but in another book on theory.

†An interesting analogy is my "choice model of design" (Rapoport 1976, 1977, 1983b, in press).

‡Note that Himsworth (1986) questions whether there is progress in philosophy and suggests that many philosophical questions will be solved when they become part of science.

articulate theory and make it work. By knowing that certain things tend to happen and by knowing how particular settings support behaviors, needs, or wants and through which mechanisms, one can then derive objectives or goals. By knowing the repertoires available to achieve these ends, "ought" statements then become possible in the form of: "If you wish to achieve so and so, for such and such reasons, you should do the following." This then is the hypothesis in the form of design that needs to be tested. This is the only way of deriving "ought" statements from "is" statements and depends on the confidence one has in one's generalizations and linking mechanisms; this confidence is related to the use of the types of evidence being discussed.

The rules based on patterns derived from broad and diverse bodies of evidence may indicate that they should apply to all environments, throughout history, to different cultures, possibly to the whole environment, that is, at different scales. Such evidence may also suggest any conditions that depend on various specific circumstances. The conclusions need to be tested against different bodies of evidence. One can test specific aspects of the three basic questions of EBS in this way. I have hypothesized that perceptual aspects will be the most invariant, that cognitive aspects will differ somewhat more in their specific expression, whereas choice and preference will vary most, not least because associational (meaning) aspects are deeply involved and are most variable (Rapoport 1977, 1982a). This hypothesis, based partly on contemporary research and partly on a varied body of evidence can, and should, be tested against a very broad and varied body of evidence. Similarly, with notions of health or supportiveness of environments, one can suggest that the search for them is a universal but with highly culture-specific expressions. On the other hand, the hypothesis that the symbolic importance of built environments diminishes as other symbolic systems become available (Rapoport 1981a, 1986c, 1989a, 1990; cf Goody 1977* for a parallel argument) would require a comparison between a group of literate and a group of preliterate cultures, or of a number of cultures over time. The importance of the availability of the type of evidence discussed in Chapter 1 is critical in both cases. One needs an adequately varied body of evidence and also enough of it to be able to subdivide it into relevant subsets. This also applies to other formulations of the field. Thus instead of my three basic questions, it has been proposed that EBS be conceptualized in terms of settings, user groups, and sociobehavioral phenomena (Moore, Tuttle, and Howell, 1985, pp. 35–36). I have already argued that these need to be considered in a cross-cultural context because culture plays a role in each (Rapoport 1986c). I would now argue that similarly the other criteria for the subject matter of the domain must be met—including reference to the full range of past environments.

The following argument is often heard from designers: Because they work in *their* tradition and at the *present* (which implies *for the future*) and because they are architects, that is, belong to the high-style tradition, why should they bother with all that other stuff, of what benefit is it? The answer to that apparently reasonable argument once again has to do with the *purposes* of such study—the topic of this chapter. If the initial purpose is to *understand* and hence explain EBR, then the answer has already been argued. It has also been suggested that such explanatory theory must precede any normative "theory"—that is, design seen as a science-based profession with EBR as its

---

*I read this first in 1985.

science. If design in a more immediate sense is the purpose, then the answer is that design is for the purpose of producing better environments. Then one needs to know what is better, better for whom, and so on (see list in Rapoport 1983b). That, in turn, as already argued implies a knowledge of environment–behavior relations general enough to be useful: As already pointed out, one needs to know what a thing needs to do before one can judge whether it does it well (cf. Wolff 1982). Similar arguments apply to more specific questions and problems: All require testing, and the data used must be adequate to the task. Moreover, such questions also depend on higher level concepts and generalizations.

The role of the broader body of evidence, including data from the past, is to learn lessons at a higher level of abstraction; to find regularities over time in human wants, needs, and responses; to find commonalities behind apparent differences and differences behind apparent commonalities; to study constancies, change, and transformations, and so on. Only in this way can valid generalizations be derived that, in turn, lead to these lessons at higher levels of abstraction that constitute precedents.

By thus combining EBS and history, one begins to *ask questions of the past.* These questions are in the first instance about what to do and why. They can also be about how to do it, both in terms of formal qualities and product characteristics, but also in terms of process and constraints and also means of design and implementation. Essentially, by approaching history in this way, one naturally and without forcing it integrates both history and EBS into the study of the built environment so as to understand and explain it. Hence, through theory one ultimately also integrates them into the design of the built environment; it is, in fact, the only way so to integrate them.

Even if all this were to be accepted as a programmatic statement, a major question would remain: whether it is *feasible,* whether it is, in fact, possible to approach historical data in this way; whether inferences about behavior, including cognitive and symbolic behaviors, can be made from built environments; whether lessons relevant to the present and the future can be derived from the past.

It is the purpose of the next chapter to argue that all these are feasible. It is the purpose of Part II (Chapters 4 and 5) to show that other disciplines have, in fact, done so and that the ways in which they have done so are very similar to those suggested in Chapter 3. This, it will be argued, is particularly the case for archaeology, the field closest to the one I am discussing in this book. Chapters 4 and 5 also provide precedents for the approach—there is, in fact, no need to invent one *de novo.*

Chapter 3

What History?

INTRODUCTION

The subject matter of study has now been redefined: It is the broadest possible range of environments in type, cross-culturally, and through time. The purpose of study has also been specified: It is to go beyond establishing the nature of past environments, to learn from this material—which, as far as history is concerned, consists primarily of artifacts. On the basis of patterns traced, generalizations can be made, hypotheses tested, and explanatory theory developed about the relation between the built environment (seen broadly) and people (EBR). It now remains to discuss how such study should be carried out because the change of domain must lead to different ways of studying the past. The objective is not to discuss specific methods but rather to establish an approach that would guide study.

In essence the argument of this book moves through the following five steps:

1. The body of evidence of which questions are to be asked
2. The types of questions asked
3. Why such questions should be asked
4. How the questions are to be posed
5. How the questions are to be answered

Whereas 1, 2, and 3 were the subject of Chapters 1 and 2, 4 and 5 are the subject of this chapter.

The kind of history that needs to be done is influenced by the two major factors addressed in Chapters 1 and 2. The change of subject matter has several implications. Including semifixed elements and people, and the whole system of settings as part of the "built environment" means that most of material culture becomes relevant. This redefinition is important not only in making the body of evidence large—the body of evidence also becomes more varied. This is important because, in general, the more variety of types of things considered, the better founded one's conclusions become. It is, of course, essential to argue *explicitly* for the relevance of any variety because relevance is more important than number alone. Relevant variety depends on an explicit consideration of the evidence applicable to any question or hypothesis; this is typically a matter of the subject matter rather than of logic. Redefining the subject matter also leads to a much greater emphasis on prehistoric, preliterate, vernacular, popular, and "spontaneous" environments because they comprise the bulk of what has been built. Most of these environments occur in nonliterate societies or, in any case, are not written about. It then follows that any study of these must take into account the absence of written sources that are typically seen as basic to history *qua* history. This has methodological implications both for history and EBS (cf. Rapoport 1970a, 1973a). Redefining the subject matter is the most important factor in shaping the approach because it tends to suggest new questions. However, the change in the questions asked, the purposes of study themselves, in turn change how such study is done. This is because history, like all research, begins with asking certain questions, *which define the problematic situation* (Northrop 1947). In all research, data and evidence only become available after questions have been asked of phenomena. This can, and often is, based on prior knowledge, earlier theory, and concepts from other disciplines. There is a parallel here to my theory-based approach to design. This argues that considerations of what is to be done and why must precede questions about how one does it (and are also more basic) (Rapoport 1983a,e, 1986e). In research what questions are asked and why (conceptual issues) and of what they are being asked precede how one goes about it, answering them that is, methodological issues (cf. Kaplan 1964; Moore and Keene 1983; Northrop 1947).

These two factors come together to suggest the nature of the kind of study of the past that is necessary, that is, the answer to the question posed in this chapter: What history?

The underlying presupposition of the approach has already been stated: Such study will be more "scientific," *moving from an art history metaphor to a science metaphor.* That, of course, is a totally inadequate statement because the nature of science, whether it can be demarcated, and whether there is any unity among sciences and, if so, of what that consists are all major and highly contentious issues in the philosophy of science. A detailed discussion of these issues would take us too far afield in the present context. It is a topic I hope to discuss when I address the issue of what an explanatory theory of the built environment might be. However, between the overarching global position on the one hand and specific methods on the other, neither of which will be discussed here, there is a broad area which can be discussed. In that area it is possible to outline an approach.

This is the purpose of this chapter. This outline will be elaborated in the next two chapters (Part II) through the use of analogous examples of how other disciplines adopting similar approaches study their respective domains.* The approach will also be clarified and made more concrete by means of the case study in Part III.

BEING "SCIENTIFIC"

It is not an entirely novel idea that the study of the past is of value for the future. After my argument was essentially complete in an early draft of this book, I came across

---

*The discussion of another highly contentious issue: Whether science (whatever it is taken to be) is a valid approach for *any* human or social domain will also be postponed and discussed in the book on theory.

another argument for the "peculiar belief that the value of history is what it teaches us about the future" (Jackson 1984, p. xi). That argument continues by suggesting, as I do, that one must deal with more than the tiny fraction of the cultural landscape studied by traditional architectural history. It is also mentioned, almost in passing, that modern archaeology might be helpful (Jackson 1984, p. xi).

What is new here is the major elaboration of these ideas that Jackson makes in passing, and their detailed and much more explicit development. What is even more novel is the suggestion that the history of the built environment and its role in providing precedents for design (via theory) needs to be studied *scientifically* in the broad meaning of the term. This does not imply that any single specific philosophical position is taken. For example, I do not argue that a rigid hypothetico-deductive program is to be applied (if it ever is). This clearly is not appropriate in the historical science of hominid paleontology (discussed in Chapter 4). There, fragmentary information is accumulated more or less at random and needs then to be interpreted as evidence for or against various equally fragmentary theoretical schemes (Ziman 1984, p. 481). But this makes paleontology no less a science.

This is also the case in art. There also one can search for pattern and developments or "laws" in artistic events but not in terms of "better" (on the basis of aesthetic preference). In one case, a review of a large set of ideas from a range of disciplines leads to the adoption of a hypothesis of cognitive development in Piagetian terms. This is then tested against a large body of art (Gablik 1976). Although it seems highly questionable to apply the Piagetian scheme at the level of humanity, the argument is scientific in the broad sense adopted here. Similarly, in another case (Machotka 1979), a hypothesis derived from clinical psychology and fitting into a larger theoretical framework is applied to the study of the nude in art. The question posed is how the nude becomes an aesthetic object by fulfilling certain psychological functions. Note that both studies deal explicitly with human behavioral processes—cognitive and affective, respectively—and do so "scientifically." Other approaches to the study of cognition in art are also possible (e.g., Crozier and Chapman 1984). Aesthetics generally can also be studied historically and in relation to contemporary psychological research (e.g., Martindale in Gergen and Gergen 1984).

Note that as more data become available, whether in art, paleontology or any other field (e.g. neuroscience, cf. Bullock 1984), that is, as it becomes less fragmentary, more complete, and elaborate, theoretical schemes can be developed. From these, hypotheses can be derived and tested systematically. This shows not only the importance of a "natural history stage" of collecting information (Northrop 1947; Rapoport 1986c) but reinforces the importance of much more complete bodies of evidence (cf. Bakker 1986; Bullock 1984; Eldredge 1985; Morrell 1987). Thus it is pointed out that it is a standard methodological principle that a variety of evidence is better than a narrow spectrum of evidence (e.g., Glymour 1980, pp. 139–140). This emphasizes the importance of taxonomies and classification (Gould 1985; Mayr 1982; Rapoport 1988b, 1989b). More generally, it suggests that very different approaches all play a role in science, sometimes at different stages (Kaplan 1964; Northrop 1947) and sometimes at a given time; such different approaches that in the philosophical literature are often the subject of heated debate work in tandem. This is well illustrated in what has been called an "ethological approach" (Chisholm 1983, pp. 93–94). This is based on a rigorous and often even quantitative natural history approach that makes induction both possible and fruitful (cf. Hinde 1984). The data collection is informed by the research and theory already available; it is thus based on hypotheses, although without too strong a commitment to those, so as to minimize preconceptions; this allows a form of the "natural history" approach. Its goal is to develop further more explicit and testable hypotheses. This approach also encourages casting a wide research net and the use of multiple disciplinary frameworks and methods. This rather eclectic approach not only makes possible the testing of specific hypotheses but also encourages the easier generation of new hypotheses from the data being collected. Although more explicitly developed in ethology, this approach is also used productively in anthropology and other disciplines. It is also *scientific* in the broad sense of the term.

The most contentious aspect of this book (and my current work on theory) will no doubt be the emphasis on "being scientific." In fact, I have been advised repeatedly to say what I am saying without using that term; it needlessly raises people's hackles. The point was made that people would then be much more receptive and likely to consider the case on its merits. I even vaguely considered doing that. But there really seems no way to avoid this term. Nor, as already pointed out, is it possible to argue that topic in all its complexity, particularly given the current pervasive antirational and antiscience attitudes that lead to the formulaic attacks on "positivist" science. Here again it is impossible to deal with this term as it should be to show how meaningless it is. In any case, as Freese (1980, p. 4) has pointed out, the only generic term available to contrast with *positivism* is *negativism.*

I am well aware of the complex discussion on this topic in the recent literature on the philosophy of science (quite apart from sociology of science, psychology of science, science studies, etc.). I will avoid it here. What I will do, instead of arguing the nature of science and whether demarcating it from other fields is possible, is briefly to state what I take to be the characteristics of a "scientific" approach. In other words, I will trace the implications of this position *once it is accepted.* This I have to do because I regard such an approach as a *sine qua non* for any meaningful study of the history of the built environment. I also believe that it is possible to state a series of characteristics, or core attributes of mainstream science (possibly a polythetic set), most of which would be reasonably widely accepted, at least as *claims*—even if their achievement were denied, their desirability questioned, and their applicability to the domain in question rejected.

As already suggested, in arguing the case for being scientific, one needs to take a *broad* rather than narrow interpretation of what that means. First, I do not rule out or question other ways of interacting with the world—art, mysticism, religion, emotion, empathy, or whatever. I do, however, assert that when it comes to *knowing* (at the cognitive level) and being able to communicate and use such knowledge, the thing we call science is by far the best, most elegant, most powerful, and most successful way so far developed to achieve *well-founded, reliable knowledge.* In Boulding's (1980) terms, it is the most reliable and best established knowledge. In that sense one can best contrast science, or *scientific knowledge,* with assertions or guess (or with *wish,* which is an ideological or normative position) rather than with other areas of knowledge, such as history or other forms of scholarship.

In this connection, "scientific" typically means that claims must be tested *empirically*—although not necessarily *directly;* frequently many and complex steps intervene between theory and data (e.g., Leplin 1984; Shapere 1984). This does not mean that it is *empiricist* (like *positivistic,* a popular current pejorative term). This is because scientific understanding is not restricted to observables. In fact, it depends fundamentally on theoretical extension beyond observables. In that sense, science today is both rationalist and empiricist—that is a nonargument. Theory makes known the underlying and otherwise inaccessible linkages among concepts and mechanisms linking concepts to manifest phenomena. This is typically done by constructing analogical models (cf. Boyd 1984; Hesse 1966; Leatherdale 1974; McMullin 1984; Ortony 1979); and through the use of *controlled analogies* (see Chapter 5). But all this activity must always refer back to observables, to empirical data. Moreover, not all models, which are always empirically underdetermined, are equally useful; neither are they purely speculative. Two major constraints always operate. First, the plausibility of the model in terms of background knowledge, other disciplines, the "network of theories" (Hesse 1974). Second, the fact that there are some empirical constraints exerted by phenomena, so that many models are possible but few are likely (Bunge 1983; Jacob 1982).

Note that both these constraints depend on the widest range of evidence and on reference to many disciplines. In these ways, knowledge becomes self-correcting and hence progressive and at least partly cumulative (in spite of early Kuhn, Feyerabend *et al.*). Also, although self-correcting, it is capable of achieving results that are in fact no longer open to debate, that is, the permanent debate about fundamentals ceases, and there is closure (even, if at least in principle, provisional).

Science is rigorous, systematic, and very good at developing sophisticated concepts or networks of concepts (theories) that can and must be tested empirically. It is also objective, in the sense of being intersubjectively valid and empirically verifiable. It moves from consensibility to consensuality (Ziman 1978) and attempts, insofar as possible, to be impersonal and value free. Science is thus to be distinguished from both normative positions and ideology, and human phenomena can be studied scientifically (Freese 1980; Lopreato 1984; Nicholson 1983; Papineau 1979; Wallace 1983). In this process, theory is central; this, in turn, presupposes generalization and hence, in studying the past, one must make *uniformitarian assumptions* (to be discussed later).

The issue of "objectivity" itself is clearly highly complex and contentious, with a vast literature pro and con in philosophy, philosophy of science, sociology of science, sociology of knowledge, and so forth. Without reviewing that literature I will argue that objectivity or relative objectivity is possible. Whatever *general,* basic human subjectivity is inherent has no impact, only *specific* subjectivity (cf. Shapere 1984). In any case, science is the *most objective* human endeavor. In connection with our domain, the separation of the role of historian from the inherently subjective role of "architectural critic," usually confounded (cf. Brine 1986), is likely immediately to lead to much greater objectivity, quite apart from the increased objectivity of the approach being advocated. Most attacks on the objectivity and freedom from values of science are often ideological: It is not so much that these are impossible but rather because they are seen as reprehensible (Passmore 1978, p. 91). The question to ask of any factual statement, provided it is not meaningless, can only be whether it is true or false, not whether it is good or bad, "moral" or "immoral," or whether one likes it or not. A factual statement can be evaluated on only two grounds: whether the facts are correctly perceived and whether any inferences made are valid. To do that, the argument must be explicit. The motives in making such statements, or asking the questions, or their implications are irrelevant (cf. Pipes 1986, p. 36).

Given that, a minimal preliminary set of characteristics of "being scientific" in this broad sense would seem to emerge. This would include:

1. That one begins with clear and explicit statements of the problem situation. From that, specific and rigorous questions can be derived. These must be valid—meaning answerable.
2. Although intuition is central in science, no intuitive, nonrigorous assertions are acceptable without explicit reasons for them being given or without the possibility of empirical testability.
3. Science must concern itself with phenomena rather than solely with ideas, although ideas, concepts, constructs, and the like are essential and central in dealing with phenomena.
4. The phenomena in question must be carefully specified, described, and classified and their relevance for the questions explicitly established.
5. Description alone is not sufficient: It is necessary to proceed through generalization and pattern recognitions to understanding and explanation and, possibly, prediction and control (e.g., in hypotheses to be tested or in setting and satisfying design objectives).

Note, once again, that this broad very briefly stated position does not attempt to identify with any *specific* philosophical position. Yet much of the discussion in the philosophy of science and elsewhere is concerned precisely with such specific positions. Thus even when the desirability of science is admitted, debate continues about what is scientific in terms of specific logical forms of explanations, inferences, and so forth.

Typically, the debate concerns the hypothetico-deductive (h-d), hypothetico-analogical (h-a), statistical-relevance (s-r), and other models of explanation and specific methods. This is premature (Kaplan 1964) and, in any case, multiple methods are preferable (Rapoport 1970a, 1973a). It is also much too prescriptive and, in any case, has much more to do with "reconstructed logic" than with actual "logic in use" (Kaplan 1964). Rather, the explicit statement of the questions posed and the assumptions made; the rationale for the use of the data and its relevant variety; the form of inference and explanation adopted; the use of explicit and repeatable procedures for testing conclusions; the use of systematic, rigorous, and explicit methods; the variety of methods used, and explicit and logical argument make possible mutual informed criticism, increasing verisimilitude and hence progress and cumulativeness.

Note that the use of multiple methods is, in general, as important as the use of the broadest and most varied bodies and lines of evidence and comparative work. In our case, the purpose is first to reconstruct as much of the human past as possible by whatever means are available and then to use such reconstructions in making generalizations and building theory. Because different disciplines use different sources, data, concepts, and methods, the use of multiple disciplines is very useful because complementary. As a brief example (see Chapters 4 and 5 for more), one could use archaeology, architecture, and other fields concerned with material culture; EBS, all the social and behavioral sciences; written and oral history, travel descriptions and illustrations, controlled analogies (ethnography, ethnoarchaeology), experiments, simulation, linguistics, folklore, art, music, genetic data; botany, zoology, and other biological sciences, cultivated plants and other resource studies, paleoclimatology, geology, human ecology and historical ecology, demography, epidemiology, physical anthropology, physics and chemistry, and others.

Although not adopting any specific philosophical position, such a broad interpretation of science does make certain very broad philosophical presuppositions. Any science in the sense of well-founded knowledge begins with a realist position of some kind: that there is a world out there and that it is knowable. It assumes that the human mind is capable of understanding that world and that science provides the best way of doing so. It rests on the belief that one can discover patterns and processes so that generalizations become possible. Such generalizations reveal regularities and provide saliency and legitimacy for knowledge of specific events (Sanders 1984, p. 13); leading to understanding and explanation and suggesting further lines of inquiry. It also provides the ability to predict (although the need for that is not universally accepted) and eventually, in a normative (or applied) sense the ability to control, either in science (as in making electrons or genes do what we want them to [Hacking 1983]) or in designing built environments on the basis of EBS to achieve given objectives.

Much of science begins with pattern seeking—a search for patterns that reveal regularities. These are inevitably based on large and varied bodies of data, that is, observation of some sort related to some problematic situation. It is thus essential, in any field, to go through a "natural history stage" as already suggested (Mayr 1982; Northrop 1947; Rapoport 1986c). The search for pattern also implies the use of sophisticated taxonomies and classifications to order the extraordinary variety of the full body of evidence. This is so as to be able to find commonalities behind apparent differences and to become aware of differences behind apparent similarities.

One looks for persistent patterns—in our case, persistent over time, across cultures, in different environments—or in subsets of these. If they are persistent they are likely to be significant. Persistent patterns and their connections and linkages can become generalizations en route to theory—they express regularities that are taken to be real features of the world.

In connection with science, *knowing* implies *understanding* and *explanation* (avoiding for the moment the debate about whether these are the same or different). This means that one must go beyond description and generalization to concepts and theories (Willer and Willer 1973). One must begin to look for connections and linkages by asking *why such patterns or regularities?* By trying to answer "why" questions, one is well on the way to understanding and explanation, particularly if one can also propose the mechanisms that are involved.*

It has frequently been argued that although generalization is central to science, it may not be possible in history—that whereas science is nomothetic, history is ideographic. Less categorically one finds a discussion in social science, archaeology, history,

---

*While revising the final draft I came across a study that shows this extremely well. Examining cloud pattern reports from many cities over a long period suggested that cloud cover was increasing. It was suspected that the reason for this was a response to the heating up of the earth (the "hothouse effect") due to human activity. The mechanism proposed was the production of certain gases by plankton in the oceans that, through some intermediate steps, led to increased cloudiness (Gleick 1987).

EBS, and so forth, a common persisting central problem: the nature of the appropriate forms for the generalizations that can be properly made about people—now and in the past (e.g., Renfrew 1984). (In our case about human behavior in its interaction with the built environment now and in the past). Note that generalization does not have to be universal as some have argued. It may apply to some subset of one's domain. Whether it is more widely applicable, under what conditions, and so on, become matters for empirical investigation and theory development. It is also the case that if no generalization were possible, no comparative work would be possible and hence no social (or any other) science (Lopreato 1984; Nicholson 1983; Papineau 1979; Wallace 1983; etc.). Even traditional history would become impossible because that always depends on *implicit generalization or laws* (see Chapter 4). To what extent any generalizations hold (or do not) and under what conditions requires precisely the kind of comparative approach based on the broadest body of evidence, for which I am arguing.

In fact, laws are just as necessary for interpretation and understanding as for explanation. They are generally implicit and need to be made explicit, as we shall see later in this chapter and in Chapters 4 and 5.

Thus a scientific approach implies primarily a disciplined and carefully structured search for knowledge, based on explicit questions and research designs and carried out systematically; using all the evidence possible, multiple lines of evidence, methods, and disciplines. Also involved is empirical testing, that is, the possibility of falsification or disproof. The goal is the best explanation at a given time based on the best available data, knowledge, and methods, that is, one is not looking for perfection. There is always the possibility that new and better explanations may exist and need to be sought. This continuous refining and improving of knowledge and explanation is typical of science more than any other human enterprise and makes it self-correcting, progressive, and cumulative. Such knowledge enables specific hypotheses and predictions to be made; these need to be tested against further data. The only way in which this can be done is by relating data to a body of theoretical concepts. These provide a framework and also, as we have already seen, a means to go beyond the data. Otherwise there is only accumulation—as is the case with what has been called "systematic empiricism" (Willer and Willer 1973; cf. Binford 1983a,b; Clarke 1978; Freese 1980; Wallace 1983; etc.). When this process is iterated, when intersubjective agreement is sought and mutual criticism made possible, it becomes possible to arrive at reliable knowledge—in our case about the past.

It is also the case that as such concepts, theories, and models develop one can also reanalyze data in previous publications, museums, the field, and elsewhere from new points of view and with new questions. Thus already known material can provide new answers and understanding (cf. the reanalysis of data in Marshack 1971; Schmandt-Besserat 1983; Hadingham 1975; my case study in Part III).

An approach such as this is helpful in the two types of studies of the past that one finds. The reconstruction of specific sites, periods, environments, and so on improves greatly in validity, confidence, and reliability. Even in this more traditional type of history or archaeology, one can take a broad area (e.g. geographic, but cf. Vayda 1983 on the need to define the boundaries explicitly). One can then review previous research, describe the settlement patterns, settlements, buildings, material culture, establish known chronology, and so forth. All of these can be interpreted in terms of the specific research in question (e.g., Scarre 1983) and predictions and hypotheses made to be tested in further fieldwork, excavation, and so on. This leads to rapid development, progress, and cumulativeness (see Chapter 5 for many more examples).

At the same time, and more importantly, without such an approach the second type of study of the past is impossible; hence, there can be no generalizations and consequently no patterns, regularities, or constancies can be traced. No expectations can then be developed whereby anomalies may be noted; no "why" questions can be asked, and no theory development is possible even in principle; for theory development this approach is a *sine qua non*. As a result, learning from precedents cannot occur except by copying.

It follows that how one approaches the nature of theory, its construction, and its purposes is critical regarding the use of data of the past just as the use of such data is essential for theory development. The link is that both follow from a single approach derived from the substitution of a science metaphor for an art history metaphor.

One can now summarize a somewhat more specific set of core features of what "scientific" is most commonly taken to mean (cf. Argyris *et al.* 1985, pp. 11–14):

1. It involves using data the validity of which can be checked intersubjectively, that is, it must be public (cf. Ziman 1978).
2. It involves making explicit chains of inferences linking data and theory.
3. Its propositions must be capable of empirical testing and subject to public testing. What cannot be shown or communicated cannot be known.
4. It involves constructing theory that organizes such propositions logically and rationally (as well as explicitly).
5. Its theories must propose models linking them to empirical reality and thus making such theories capable of systematic testing.
6. The community of inquiry that is basic to science (cf. Ziman 1978) must engage in rational criticism of each others' claims.

I would argue that although these characteristics are most highly developed in what is called science, they also share characteristics with at least some sections of other academic disciplines, for example, in the humanities and social sciences. These also conduct their inquiry systematically and publicly; they also use explicitly rational, analytical criteria to test and try to disprove claims. In doing so they ask specific questions explicitly formulated of various bodies of evidence: texts, cultures and societies, institutions, environments, artifacts, and so on. In this, they deny special privilege to any of them. In principle this means that one must reject all private, self-validating forms of experience or works, the meaning of which are subjective and can either not be communicated—or can be communicated only to initiates. To the extent that other areas of scholarship share these attitudes they overlap with science seen broadly, and there is an area where the demarcation is far from clear.

Finally, I clearly do not argue that scientific means quantitative, although one cannot object to quantitative work. Nor does it imply the use of any specific set of methods or techniques. This is because methods and techniques follow from theory, or a theoretical position, from the questions asked and from the nature of the domain. Thus the question is not only whether history can be "scientific" in the sense that testable inferences can be made from the evidence available of past environments.

Recall that my purpose in looking at history in the first place is not only for its own sake, interesting though that may be. It is to help with the development of an explanatory theory of EBR. This theory can also serve as the basis for a science-based profession that itself would be much more intimately linked to applied or "clinical" research. This may, in turn, involve historical data when outcomes of such applied or clinical research become subjects for further basic (or pure) research. It is only in this sense that history can serve as precedent for environmental design.

BROAD OUTLINE OF THE APPROACH

The shift is away from the approach of traditional architectural history as neither relevant nor applicable to the domain of study or the purpose of study. The first, and major, shift is of course from the notion of *history* as a narrative sequence of events to that of the *study of past built environments* for the purpose of answering general questions regarding EBR. In any historical study, as already pointed out, one can distinguish between first, the establishing of a chronology, narrative description, or reconstruction of a particular place, person or group, period, event, artifact or work, or group of artifacts or works; and second, the study of the past in a generalizing way, in order to trace patterns, achieve understanding, derive and test hypotheses, and eventually develop explanatory theory.*

I have already suggested that the approach being developed should help in both types of study but that the emphasis should be on generalization and theory development.

Architectural history typically has been concerned only with the former, but it is my argument that the latter is more important because the purpose is to develop an explanatory theory of EBR. Such theory will make EBS into the scientific discipline on which a science-based profession of design could be based. At the same time the approach being developed here also helps with the first by making reconstructions more valid, by posing more useful questions, using more valid bodies of evidence, more powerful methods, explicit inference, and so on. This is essential because the first kind of study is necessary for the second: The past must be reconstructed as validly as possible before it can be used for purposes of a generalizing discipline. As we shall see in Chapter 5, the two have gone hand in hand in archaeology: The attempt to develop theory, the success of which so far is equivocal, has led to much more convincing reconstructions of the past, including inferences about human behavior and cognitive and symbolic behaviors. It has led to what have been called better "cultural histories" (Gibbon 1984), and better history is what is understood as "the achievement of a fuller and more accurate reconstruction of the past" (Goody 1977, p. 90).

One major consequence of a more generalizing approach is a change in the form of relationships among items studied to which I referred briefly in Chapter 2. Typically, traditional history and architectural history have studied things chronologically—they looked at linear relationships (sequences) rather than thematic or *lateral* relationships: They were concerned with chains of events, although that may be changing (see

*Note that some authors use "understanding" and "explanation" as synonymous and interchangeable, whereas others distinguish between them. I will adopt the former position.

Chapter 4). Given a set of phenomena, that is, built environments, traditional history takes a *linear,* sequential route through them; the approach here advocated more typically takes a lateral route. This can be visualized as follows (Erikson 1973; Fogel and Elton 1983; Goody 1977; see Figure 3.1):

This means that the approach should not necessarily be chronological but seeks to use the broadest variety of evidence relative to a given topic. Although this does not, of

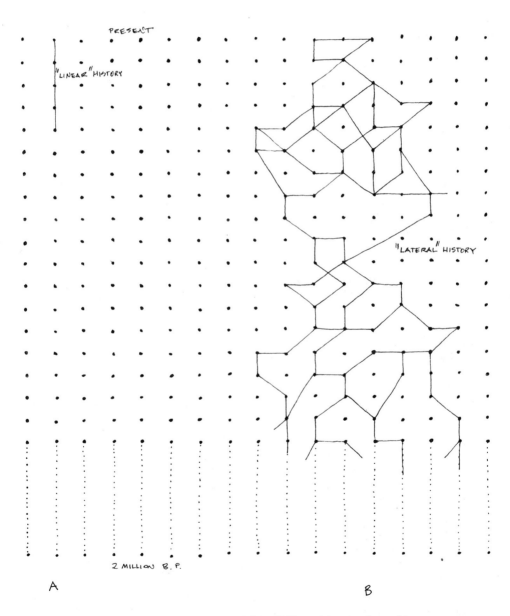

Figure 3.1. "Linear" (sequential) versus "lateral" (thematic) approaches to history.

course, preclude chronological study should the topic require that, the intention is not to create sequences or catalogues of buildings, styles, or the work of "major architects." Rather it is to analyze data that can contribute to one's understanding of EBR theory, environmental problems, and design. These analyses would tend to cut across traditional chronological and geographical divisions and would also concern the fullest range of built environments. Thus, the environments typically considered comprise a small, limited set (Fig. 3.1a), whereas I am advocating the use of a large, varied set. Among examples might be studies of courtyard houses, underground dwellings, or communal dwellings around the world, any of the examples in Chapter 2, the case study in Part III, and so on. A good example in another field is a historical and cross-cultural study dealing with the household (Netting, Wilk, and Arnould 1984).

In fact, a central almost starting point is the need *not* to study past environments chronologically, although, of course, chronologies must be established. Dating is important, for example, for contemporaneity in archaeology or to trace changes, developments, or constancies diachronically: If the question concerns change over time, then a chronological sequence may be required. Rather, such studies should be organized *around a series of topics or questions;* this kind of *lateral connection* is crucial and examples were given in Chapter 2. This, in turn, implies a certain taxonomic structure; taxonomy and classification become most important (Gould 1985; Rapoport 1988b, in press). That, in turn, is related to a theoretical structure, although the latter can be such as to allow an account of a great variety of other taxonomies and theoretical structures (e.g., Chisholm 1983).

The search is always for relevant connections and linkages, not sequences. As I have been arguing since 1969 (Rapoport 1969e), the understanding of any empirical phenomenon is made much more likely when one can trace connections among things that seemed quite separate or disjointed. Theories fulfill that function and provide a broad perspective for ordering and arranging not only isolated facts but also the various regularities derived from these facts—in our case typically through comparative studies. In fact, theories themselves also are typically seen as sets of interrelated connections and linkages among concepts and constructs and between those and empirical data (Figure 3.2).

Thus theories do more than order things; their construction requires not only the development of concepts but also the search for linkages and connections among those. Moreover, increasingly, it seems that one is concerned with networks of theories (Hesse 1974), that is, linkages among theories as well as among concepts. In fact, typically theories in different fields are involved in such networks. The resulting redundancy is important in increasing confidence in one's results. It follows that the more fields are involved, the greater the reliability and the more confidence one can have in results. In that sense, "lateral connections" refers not just to chronology but to different disciplines, hence different bodies of evidence (domains), different sets of concepts and theories, and so forth. All the processes involve isolating and separating issues in complex problems (what one might call "dismantling") that leads to much greater conceptual clarity and also to the study and understanding of the linkages and connections among the components (Rapoport 1977, 1986c, 1989a).

EBS, as a field, has from the outset been interdisciplinary although most research, with the exception of mine and some others, has tended to remain within one disci-

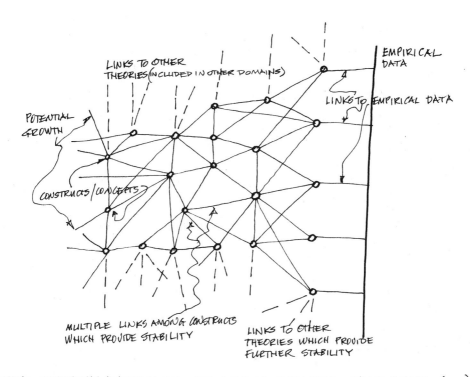

OTHERS GENERALLY CONCEPTUALIZE THEORIES IN RELATED WAYS. PAPINEAU (1979) DESCRIBES A THEORY VERBALLY AS A PYRAMID, LINKED TO EMPIRICAL DATA AT ITS BASE, WHERE CONSTRUCTS FORM NODES & STABILITY COMES FROM "STRUTS" INTERCONNECTING CONSTRUCTS. THIS CAN BE DRAWN, BUT FURTHER GROWTH AT THE FAR END (TOP) AND LINKS TO OTHER THEORIES & DOMAINS ARE DIFFICULT TO SHOW.

Figure 3.2. Theories as networks (based partly on Margenau, in Lesham and Margenau 1982).

pline. It seems clear that no one discipline can deal with the full range of data—nor can any one person. It seems clear that increasingly such study will become an inter-disciplinary team effort, whether the "team" works concurrently or sequentially (i.e., using one another's work). It is, I think, significant that scientific research is typically a team effort—and that this is happening in the various fields discussed in Chapters 4 and 5.

In art history, the paradigm for traditional architectural history, this has not happened, although proposals to that effect were made as early as 1963 when it was suggested that art history must cease to be "wholly" literary humanism, and also must avoid pedantry. The need was for rigorous scholarship, meaning that art history needed to become an interdisciplinary team effort. This was because no one person could combine "the erudition of the historian, the critical abilities of the philosopher, the objectivity of the scientist and the imagination of all three" (Ackerman and Carpenter

1963, p. vii). Neither could one person master the full range of concepts and methods that go with those broad categories of people.

At the very least, then, their argument implies combining art history with archaeology or, more broadly, history, philosophy, and science (see Figure 3.3).

This is related to my redefinition of the domain. In the case of art, it suggests that prehistoric, "primitive," preliterate, folk, popular, and other comparable forms of art should become the subject of concern with major changes in approach, skills, methods used, and so on. Although this will be discussed later in this chapter, an obvious change is that one must then deal with *classes* of objects rather than single works of art; generalizations must become explicit, lateral connections become central, and models must be used, explicitly stated, and tested. Also other disciplines become relevant—not least because archaeology itself incorporates and uses a great range of disciplines; physical sciences for dating and identification of the origins of materials; historical sciences for contemporaneous environmental conditions; geography for plotting distributions in space; travel accounts and ethnography for ethnographic descriptions; experimental archaeology and ethnoarchaeology and many others. The diagram becomes too complex to be useful (see Figure 3.4).

When such changes occur in art history, it becomes possible to study, for example, the symbolism of ancient Chinese Shan bronze ritual vessels by broadening the range of evidence used (Munsterberg 1985). By using inscriptions from a variety of sources, archaeological data, surviving folklore, and popular beliefs, as well as the artifacts themselves, inferences are made about the world view of the Shan. This leads to a proposed iconographic scheme used in studying symbolism, something beyond the reach of traditional art–historical scholarship. In effect, that art is made more accessible by clarifying the cognitive processes of the Shan, something possible even without written sources (cf. Renfrew 1982, and the discussion of rock painting later).

Among existing disciplines the approach proposed is much closer to archaeology, particularly prehistory, than it is to history. This is not only because I also begin with material culture. There is also the fact that the term *prehistory* means before history, that is, before written records. "The emergence of what we call history was linked very closely with the advent of writing as the implicit distinction with prehistory suggests. . . . There is no history without archives" (Goody 1977, p. 148). Written sources,

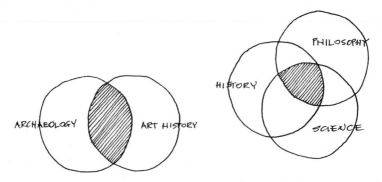

Figure 3.3. The relation between art history and other fields (based on Ackerman and Carpenter 1903).

Figure 3.4. Environmental history as an interdisciplinary endeavor.

of course, never existed in the case of vernacular environments either, as already pointed out. In the case of those, and even in the case of popular and spontaneous environments, this lack of written records continues even after the invention and spread of writing and into the present. Note, on the other hand, that they have been written about; what writings exist about those types of environments are by analysts and scholars (e.g., anthropologists, folklorists, geographers, archaeologists, pre-historians, and the like) not by the designers who are, in any case, unknown. In that sense, these scholars also begin with material culture.

Thus, written documents, although useful and to be used whenever available, cannot be made an *essential* requirement and, in any case, will always be a *minor* part of the evidence.* Note also that the use of documents alone does not make history. It is also how they are used. In addition to the existence of archives and the formalization of information, what makes history possible is the existence of commentaries and the possibility of comparing different versions side by side (Goody 1977, p. 149). This is equivalent to my "lateral connections," and we will see in Chapter 4 that with change in the domain of history comparable to mine, new approaches similar to the one here advocated also began to emerge in history.

At the heart of the endeavor in all such cases is the question whether and how inferences can be made from artifacts to behavior.

Because EBR is my focus, the artifacts are built environments. Because, in the broad sense here used, these include semifixed feature elements, then nonarchitectural artifacts, such as furnishings, tools, signs, vegetation and many others, even art, are included. They are all part of material culture. People, the nonfixed elements, must also be considered.

ART HISTORY AND ARCHITECTURAL HISTORY

I will begin by discussing briefly how the underlying art metaphor of architecture relates architectural history to art history. Both are traditional history in the sense of

*I was once told by an eminent architectural historian that preliterate, vernacular, and popular design could never be considered to be part of architectural history because no written sources or documents existed.

dealing with singular events. Rather than classes of events or generalizations, the singular events of art history are taken to be the works of the great masters. Thus Gombrich (1979, p. 151) argues that in art history one must study the work of masters, although he admits that such study must be broadened both in terms of its time scale and cultural range. This is essential, because one purpose of historical study is to make available the full range of the past to those living in the necessarily limited present. Thus he essentially argues for broadening the database—*but the set are still "master-pieces".*

In architectural history the singular events are the high-style great works of *its* great masters. A critical question is, of course, how one decides who are masters. This necessarily involves subjective aesthetic and value judgments on the part of the scholar in the choice of material for study. That is not justifiable unless the specific question is such that the selection of such a subset can be explicitly and logically justified, as can the selection of members of that subset. For example, it might be worthwhile to ask whether those environments typically judged highly by architectural historians in the past have any characteristics in common and whether they vary from other types of environments in systematic and predictable ways. This means that even masterpieces cannot be selected on the basis of "I like it."

Thus, as I also argued in Chapter 1, although the study of "monuments" may be useful for specific questions (and one can certainly raise those), much more significant, particularly if the objective is to understand EBR and hence to develop explanatory theory in that field, is to consider the bulk of the evidence, the most typical environments and their full range. Moreover, it was suggested that even "monuments" or "great works" only make sense in their contemporaneous setting. They may even be seen as *anomalies,* a departure from the majority of environments. But for that the "normal" or expected pattern must be known. It may be precisely their anomalous nature that may be the most significant thing about them. It may be much less important whether they are "good" or "bad" buildings or who designed them than that they are *different* in some way—"architecture" as opposed to "mere building." Their role or position in the larger system or pattern may be the most interesting thing about them—and the thing to be studied. This may also be the case for works of art.

Yet in traditional architectural history aesthetic, subjective, and typically *arbitrary* judgments are always being made; as we have seen, many architectural historians are also critics (cf. Brine 1986).

Thus common to traditional architectural and art history and criticism is connoisseurship. Even attribution is based on quality and perceived similarities and differences. Yet even "style" can often be discerned objectively (e.g., Lewis-Williams 1981, 1983, and later discussion). Also, attribution can be helped by objective and even quantitative analysis as in the case of a recently discovered poem that is possibly by Shakespeare (Kolata 1986). Moreover, even if selection and attribution are made on the basis of subjective judgments, one can discuss probabilities, levels of agreement, and tests of interjudge reliability that have clear definition and description of attributes and explicit reasons for choice, blind tests, and so on.

This discussion suggests, once again, that a major factor in changing approaches is domain redefinition. In science, for example, hierarchies implicit in an emphasis on "masterpieces" have been given up, and one studies *all* the evidence or data, or representative samples of them; if selection occurs, the relevance of that selection to the

research questions are explicitly established. In our case, of course, this means beginning with the full range of built environments. From that point of view, *history is the
knowledge of the past of a given domain inferred from the remaining traces.*

From this perspective works of art form part of material culture. It is to be
expected, therefore, that if the domain of art history were to be redefined in ways
analogous to those described in Chapter 1, major and analogous changes should occur
in the way the material is studied. If the domain includes prehistoric art, preliterate
artistic traditions, folk art, popular and tourist art, written sources disappear, and what
is available are the artifacts from the past. In addition, the available evidence includes
studies of these artifacts, and relevant to them, by prehistorians, archaeologists, anthropologists, and other scholars, early travel accounts, and a range of disciplines that
can date those works and establish contemporaneous environmental conditions and
available resources. It is likely that the approaches used to study such art would be very
different to traditional art history. *If this were the case, if it can be shown that the changes
in the study of art history, given a redefined domain, are analogous to those I am advocating, that would be highly significant, given the discussion on the use of art history as the
model for traditional architectural history.*

It is striking that one can show just such changes, becoming clearer the further
one moves away from the traditional subject matter of traditional art history—high-
style Western art. Some preliminary indications of some changes in art history have
already been given—in one case, a non-Western domain, that of Shan bronzes; in the
other two, new types of questions were posed involving lateral connections among
items. Even more relevant would be the domain of prehistoric art and that of so-called
"primitive art," that is, the art of preliterate societies.

Almost as soon as the domain is redefined to include such art, one finds that its
study changes: It becomes scientific rather than aesthetic (e.g., Otten 1971). Part of the
reason is, of course, that such art must be studied through its artifacts and as a system of
communication because no written documents exist. Moreover, as a system of communication, it needs to be related to social systems (cf. Layton 1981). It even becomes
possible to link such studies to those on animal communication and to experimental
studies on drawings by apes, a good example of what I have called lateral connections.
They also soon involve a variety of disciplines and come to include experimental work,
quantitative analyses, models of choice, and so on (Greenhalgh and Megaw 1978).
Furthermore, such studies soon tend to become comparative and cross-cultural, to
generalize at the same time that specific groups and bodies of art are studied in new
ways.

Thus one can study the art of a region, for example, Oceania, from the remote past
(relying on archaeological data) through the recent past (including tourist art), relying
on contributions from ethnography, anthropology, and observation; studies like these
require a multiplicity of approaches, which are, in fact, found (e.g., Mead 1979). In that
study the approach reflects the programmatic suggestion that such studies need to be
multidisciplinary and by a "team" (Ackerman and Carpenter 1963). The thrust is
clearly toward becoming "scientific" in the sense of rational, empirical, and objective
rather than being based on one's subjective preferences; comparative and cross-cultural,
and generalizing. It essentially becomes part of anthropology, that is, social science
(e.g., Forge 1973b; cf. Maquet 1979, 1986; Greenhalgh and Megaw 1978, etc.). Note
that, in turn, it is often essential to add diachronic study, that is, a time dimension (i.e.,

history) to social science (e.g., Kashiwazaki 1983; cf. Eidt 1971) as will be discussed in Chapter 4.

The material in such studies of art tends to be studied systematically. This implies using multiple techniques and applying these to large and varied bodies of data. The need for data makes the *descriptive* level important; one needs a great deal more recording and compilation of data, and such data often needs to be broadened even more, for example, to include body decoration (Forge 1973a; cf. Rapoport 1975a). A large and varied body of data, although not sufficient, is necessary for valid work and progress. This, of course, also applies to the natural sciences, for example, neuroscience (Bullock 1984) and biology (Mayr 1982) among many others. Once there is such data, *classification* and *taxonomy* become important; cross-cultural and other comparative work become possible, and patterns can be traced. It also becomes possible to study visual material as an independent system as well as the relationships between that system and other systems. *Classes* of human creations such as visual art or built environments can be analyzed, patterns can be traced, and cultural specifics identified within any such regularities. A series of questions can then be asked and studied using both contemporary materials and those from the past. Predictions can be made as hypotheses and tested.

For example, in a study of Sepik River art (Forge 1973b), a very extensive amount of material was collected, photographed, and tested using native informants. At the same time the researchers analyzed objectively the constituent elements of the material and combinations among them. The two sets of results were then compared and correlated. From this, among other studies, *inferences* can be made about tribal philosophy, social aspects of art, the makers' intentions, how that art communicates, and how it is used as part of the culture. These findings can then be applied to art from the past.

It is significant that the art of the past and in preliterate societies can be studied rigorously, with attempts to generalize, and that it is possible to work comparatively across different cultures, time, periods and the like. Such comparative work is essential in cases where no quantification is possible or where only ordinal scales, but not interval or ratio scales, can be used (cf. Kent 1984). Moreover, in all my own work from the beginning, I have argued for the importance of comparative work. This book, in fact, seeks to extend the body of evidence to be compared from a large variety of types of environments and wide range of cultures (which I have emphasized) to a wider temporal range. Comparative work is essential because EBR, culture, behavior, meaning—or the study of the past—cannot presently, if ever, be reduced to quantitative terms (interval or ratio scales) because they cannot be characterized in absolute terms. It therefore becomes necessary to compare things and place them in a general context, or on an ordinal scale.* Typically different groups, cases, situations, and periods can only be studied comparatively. Such comparisons are often *implicit* and/or *global;* they need to be made *explicit* and *dismantled* (e.g., Rapoport 1980a, 1985b, 1989a, 1990). Only when this has been done can similarities and differences for any given variable or set of variables be identified and, if necessary, located on any continuum. The position of whatever is being plotted on that continuum represents a specific location only in

---

*An example is division of labor: One can speak of more or less, stronger or weaker but not place societies on an absolute scale of division of labor (Smith 1982; cf. Kent 1984). Privacy would be another good example where comparative rankings would be necessary.

relation to other items on the continuum (Kent 1984, Chapter 1). For example, in studying to what extent different societies emphasize boundaries between humans and nature and how these change over time and in relation to epistemological boundaries, only comparative rankings could be used (e.g., Krawetz 1977). It is important that such comparisons, plotting and ranking should be done systematically and explicitly.

The last example and the other studies cited not only relate artifacts and built environments to ethnography but also use contemporary data. This enables fieldwork to be done that in the Sepik River case becomes "experimental" (Forge 1973b; cf. Greenhalgh and Megaw 1978). This process can be taken even further, so that experiments can be carried out on intentions, purposes, production of artifacts, their distribution, discard and preservation processes, and so on (e.g. Washburn 1983*; see the many examples in Chapter 5).

All these studies tend to be very rigorous. There is constant emphasis on the importance of measuring and comparing attributes systematically and objectively through space and time. The relevance of these attributes is explicitly argued, and they are applied to *large bodies of data* (or samples of those). In that way, for example, temporal shifts can be identified. Even if the database is clearly distorted and is not likely to improve (as in the case of prehistoric Southwest U.S. textiles) it should be used; one should do the best one can while explicitly discussing the inadequacies (Washburn 1983). It is, after all, the only data available. In such studies, one also finds the application of multiple lines of evidence and a variety of methods including some from other disciplines: in one case, symmetry analysis from crystallography (Washburn 1983). Again (as in Forge 1973b), art is seen as a system of communication (as we will also find in the next section). So, of course, can the built environment (Rapoport 1982a, 1988a). Moreover, inferences can be made about social, political, and cognitive systems and their relation to art and material culture among the groups studied.

It thus appears that the changes in art history when the domain is redefined to include preliterate art are very similar to the ones suggested in this chapter. We have not yet considered the other extension of the domain mentioned that may be more relevant to this book: prehistoric art. I will now consider, as a case study, a body of prehistoric art: *cave art or rock painting*. Here also the approach is analogous: For one thing no aesthetic judgments are being made, and the full body of evidence is used. These students of past art are *not* also critics; moreover, inferences are made from the art to human cognitive and symbolic behaviors.

CHANGES IN THE STUDY OF ART: THE CASE OF ROCK ART

I will use only a very few studies to show how data from the past that at first glance would seem to be almost impossible to study along the lines sketched out thus far has been so studied. In none of the examples used am I concerned with specific, substantive findings, nor do I take any position on these findings or any controversies about them in the field. The emphasis here, as throughout, is on the approach to the problem of how one makes inferences about the past from material existing in the present and *how*

---

*Note the highly significant point that this book that deals with fieldwork in *contemporary situations* forms part of a series in *archaeology*.

one makes inferences about intentions and behavior when only surviving artifacts are available. The intention is to show how different and, I would argue, better this type of study of art is from traditional art history and, by implication, traditional architectural history based or modeled on it.

Some studies of rock paintings are still not very rigorous and may even superficially resemble traditional approaches (e.g., Godden and Malnic 1982). Yet even in this study of Australian Aboriginal rock paintings we see several differences. First, of course, these are not works of masters in the European tradition. Second, there is an attempt to discover and use the whole set, not just those judged to be "masterpieces." Moreover, there are attempts to plot the spatial distribution of the paintings to develop regional classifications of "style" and to relate these to available ethnographic data. There is also reference to, and use of, other data for dating purposes. But this study hardly goes beyond the data themselves.

In the case of European Paleolithic cave paintings I will consider just one example from a very extensive literature in several languages (Leroi-Gourhan 1982). This study begins by pointing out that there must have been open-air art that has disappeared because people lived mainly in open-air settlements (a useful bit of environmental history!). This emphasizes the inevitably fragmentary and hence "biased" nature of the evidence. This is a problem always encountered and much discussed in archaeology, as we shall see in Chapter 5 (cf. Clarke 1978). As a result of the nature of the surviving material, its *location* becomes important. On the basis of this, a set of EXPLICIT assumptions are then made that, the study emphasizes, are not universally accepted. These are that this art is symbolic, that it had a religious/magic function, and the relative numbers of things represented and their placement were deliberate and systematic. Then follows a *classification* of forms that is deemed necessary in order to study the art and also a classification of the spaces in the caves. On the basis of these an attempt is made to decode the message, that is, to make inferences about the content. Two stages are involved. First, the "actors" represented are identified: animals, humans, monsters, and abstract signs; second, they are subjected to very detailed numerical analyses in terms of assemblages, their locations, and so forth.

The importance of assumptions being *explicit* is, of course, that they can be discussed and criticized (e.g., Ucko 1977, p. 9), for example, on the grounds that they are based on insufficient evidence. Ucko's criticisms are based on extremely detailed studies of various caves that show striking exceptions to the presumed patterns derived primarily from Lascaux. In effect, they are taken as *predictions to be tested*. The more such investigations are carried out the more, it is argued (Ucko 1977), do such exceptions seem to become the norm. The argument is thus about the presence or absence of pattern or regularity. In the long run this discussion is about the nature of this pattern: Eventually, as even more caves are studied, new patterns may emerge. Note also that the criticism and debate become rational and rigorous, and the process both accessible and open to inspection and potentially self-correcting and cumulative. Note also how much depends on the extent and variety of the evidence.*

In essence Leroi-Gourhan's approach is also that used in the remaining studies

---

*In this connection it would be interesting to know, for example, what percentage of the settlements in question show the optimal orientations discussed in Knowles (1974). There, neither the total number of settlements nor the percentage having the "correct" orientation is given. It is at least possible that they

(Lewis-Williams 1981, 1982; Vinnicombe 1976), all of which deal with southern African Bushman paintings. In that case, however, ethnographic data can be used—both early travel accounts and past and present anthropological research. That is, of course, not the case with the Paleolithic cave art. There only the art itself exists, but archaeological data related to the people in question can be used to help infer behavior.* In the case of Australia, there are not only ethnographic data; Aboriginal artists are still working and can be observed, interviewed, and so on; so can the "consumers" of this art. There is still some living (although limited) tradition.

For our purposes, the specifics of the studies of southern African rock paintings are less important than the fact that they can be studied rigorously and inferences made about matters such as intentions, behavior, and cognitive schemata of people in the past.

Vinnicombe (1976) begins with an "objective" numerical analysis of over 1,600 paintings. Her analysis involves the number of sites, the number of individual paintings, their subject matter, color, and technique—and the changes of these over time. She avoids defining "styles" but identifies four phases in a continuous tradition. The result of this objective and quantitative analysis of the large (although still incomplete) body of rock paintings seems to be that they do not imitate nature. They are not a realistic reflection of the daily pursuits or environment of the Bushmen. Rather the paintings are *very selective,* displaying differential emphasis. Certain patterns or basic formulas are selected by the artists for constant use. It is not enough, however, to give numerical data showing selective and repetitive characteristics, that is, to identify the pattern. One needs to ask "why" questions, *to try to understand the rationale for these preferences.* Hypotheses can be made, but these are difficult to test because not only were the painters in the distant past but even contemporary Bushmen seem to have no reliable memory or knowledge of them. Yet, Vinnicombe argues, one must try, and there are clues in the literature that enable one to correlate some aspects of the art with Bushmen's modes of thinking, their cognitive categories, and so on. In this, an important role is played by what I have called *lateral* (rather than linear or sequential) linkages. In this case, comparisons with other groups. There are also interdisciplinary connections—not only the use of the ethnographic materials and early travel accounts already mentioned but also ecological data. Also used are anthropological concepts based on the theory of ritual, which are used to study rituals, music, dance, myths, cosmology, and culture as a system in its relation to the environment, resources, and technology.

If the study had any preconception, it was that the art would be explained in terms of pragmatic and technical (and some aesthetic) rationale. Note two important points. First, these preconceptions are, once again, made *explicit.* Second, as the analysis progressed, *these assumptions changed because of the data:* The emphasis shifted to the communicative ("symbolic") aspects of the paintings. The selectivity of the themes led to search for explanation (not the other way around)—and this selectivity was only revealed because the whole body of work was analyzed and patterns thus identified.

---

constitute so small a proportion of the total set as to be "random." This question also applies to settlements in Cappadocia being studied by Mete Turan. A comparison of two such larger samples, and others, would provide even more evidence (cf. Rapoport 1986b; on the "incorrect" use of examples as illustrations, see Rapoport 1986c).

*In the longer run, when comparative work is done among many and varied bodies of cave art or rock paintings, controlled analogies may be used to bear on Paleolithic art.

The final studies of the Bushman rock paintings (Lewis-Williams 1981, 1983) that date 5 to 7 years later, show clear progress,* partly because they are based on Vinnicombe's analysis of so much data and her identification of a continuous tradition. Thus the study of this particular domain is becoming cumulative. In building on this previous work, these later studies were able to analyze a smaller subsample of her larger set, be even more systematic, less inductive, and more conceptual.

The importance of avoiding subjective, purely intuitive stylistic inferences is emphasized; so is the need for objectivity. The full range of data (or a sample) needs to be used, not paintings selected because they are "best preserved," "more interesting," or "most beautiful." In other words, it is most important to *avoid a subjective selection based on personal preference* (Lewis-Williams 1983, pp. 28–29) which is precisely the basis often emphasized in traditional approaches. An unequivocal definition of style emerges from the analyses.† Regional differences and similarities also emerge—a new pattern that is not yet well understood. Here again subjectivity must be avoided: The classification of rock art (as of *any* data) must have a specific and explicitly stated purpose. This will govern the selection of discriminating features for the creation of classes (Lewis-Williams 1983, p. 36). Note, however, recent suggestive evidence of "natural" classifications (e.g., Rosch 1978; Rosch and Lloyd 1978; Johnson-Laird and Wason 1977; Rapoport 1986c) and also developments in numerical taxonomy, cladistics, polythetic classification, and so on. All these can and should play a role and make taxonomy and classification most important (Gould 1985; Rapoport 1988b, in press).

In these studies (Lewis-Williams 1981, 1983), the main interest is in interpretation. Whereas Vinnicombe (1976) begins with the data and introduces theory at the end, Lewis-Williams begins with a great deal of theory, derives explicit conclusions from it, and tests these against the evidence. He is able to do that partly because of Vinnicombe's previous work.

Once again the approach depends on a detailed numerical analysis of sites, their locations, and orientations; the deposits and artifacts found in these sites; the subject matter of the paintings—animals, humans, other forms (cf. Leroi-Gourhan 1982). Again, *selectivity is avoided*. Lewis-Williams accepts the fact, established by Vinnicombe, that there is a continuous tradition and tests the phases using paleoclimatological and geomorphological data, the degree of patination of the paintings, radiocarbon dating, and a multiplicity of other techniques (Lewis-Williams 1983, pp. 28–29). Given this continuity, he is able to argue that ethnographic data and early travel descriptions can be used: The general uniformity of the art over space and time, that is, the pattern, has important implications for the interpretation. By allowing uniformitarian assumptions, it allows eighteenth-century ethnography to be used (with due care) (Lewis-Williams 1983, pp. 36–37).‡

---

*Note that in this study the term *San* replaces *Bushman*. This is because anthropologists have recently decided that Bushman is a term with pejorative connotations. They now prefer to use the term *San* for the whole group with "tribal" prefixes, for example !Kung San, Basarwa San, and so on.
†Recall that Vinnicombe (1976) avoided any reference to style. This difference may be due to different views held by the two researchers or to the greater maturity of the fields 5 to 7 years later.
‡This later, more popular synthesis of his work is part of a series that also included, in mid-1985, A. Leroi-Gourhan on European cave paintings; A. Beltran on those of the Spanish Levant; and C. Grant on North American Indian rock art.

Lewis-Williams tries to understand the culture through three bodies of eth-nographic material: the older literature already mentioned, more recent anthropological research, and his own fieldwork. (This latter will be seen as increasingly important in archaeology in Chapter 5.) The object is to see whether there is any *convergence of multiple lines of evidence*. In addition, a body of other research on the subject is also reviewed, some of which is criticized and some accepted. Lewis-Williams (1981) agrees with Vinnicombe to a significant extent about the significance of ritual, the centrality of the Eland, and also relies mainly on ritual theory. However, he also discusses alternative models, among them symbolism and semiotics. The result is a partly synthetic model that is *made explicit*. The argument is that if one can identify the relevant ritual context, the rock art can then be "read" and inferences made about how it was used. Having stated this theory and analyzed the data, he then links the two through detailed, quantitative studies. These tend to eliminate the two most popular and widely accepted interpretations of this art. His own interpretation follows.

The point is made (Lewis-Williams 1981) that the process of mutual illumination between paintings and ethnography will continue as new sites and paintings are dis-covered, known sites are reexamined, and ethnography is read more perceptively. One could also add that new theory will play a role, as will more and better ethnography done in southern Africa. A role will also be played by more work done in Australia and elsewhere—that is, as the continuum that I discussed earlier is developed. In this way one can begin to develop a sequence ranging from cases like Australia and others, where most is potentially knowable because the tradition is still partly alive, to that where least is known, for example, Paleolithic art, with southern African rock paintings intermediate. This is an example of the argument that the distinction between "hard" and "soft" sciences is not useful: It is more useful to speak of areas of more or less certain knowledge (Boulding 1980). This is also an example of a more extensive net-work of lateral connections that, at least in principle, may allow larger scale patterns and regularities to be observed and higher order inferences and generalizations to be made.

In this way, one could then relate and link work done in still different areas of the world, an example of one form of lateral linkages, that is, use an even broader sample assuming that the processes are "universal" (see the discussion on the uniformitarian assumption later). In this way, more and better theory can be built—and it, in turn, will help with further comparative studies.

In concluding this case study, one can point to at least one example of a beginning in that direction (Ucko 1977). This explicitly links Australian aboriginal art with prehistoric Europe—although not, unfortunately, with southern Africa. It also incorpo-rates a great deal of theoretical discussion from numerous and diverse disciplines. I will not even try to begin to summarize an extremely dense book of over 480 pages. In addition to cross-cultural and cross-temporal comparisons, however, one also finds a systematic and rigorous ("scientific") approach. The material is carefully and explicitly ordered and classified using systematic taxonomies; there is detailed stylistic and mor-phological analysis and the use of field studies among contemporary populations. The comparative work is precise, with carefully chosen evidence, clear definition of the categories, classes, and concepts used—or at least an explicit discussion of the variety of ways in which they are used (cf. Rapoport 1986c). Cultural contexts are carefully established. There is also an attempt to use clear and agreed-upon terminologies and nomenclatures. One finds the use of large teams from various disciplines. Also important are replication and the restudy of previously studied materials with new questions, concepts, and techniques (cf. Marshack 1971; Schmandt-Besserat 1983; and Chapter 5). Attention is repeatedly drawn to the need to study the full set of data, or a sensible subset, rather than data selected arbitrarily on the basis of "quality," "artistic merit," or whatever. Clearly, to make meaningful statements about differences and similarities one needs patterns that can only be established by analyzing adequate, varied, and relevant bodies of evidence.

It is most significant that all the studies discussed in this section agree with each other about what begins to happen in a field such as the history of art when the domain is redefined in a manner analogous to that in Chapter 1 and when similar types of questions are asked. There is remarkable convergence on an approach quite different from traditional art history. Among the emerging characteristics are objectivity, rigor, explicitness; the use of the full range of evidence; the use of multiple lines of evidence, methods, and disciplines, comparative work; lateral connections. All these characteristics are also very similar to those I am proposing for the study of the history of the built environment; they begin to define a way in which the study of past material culture is to be done. Note also that the goal of such study also converges with mine. It can be understood as the study of a particular person–material culture relationship, an attempt to reconstruct the creators' motives and cognitive systems and what they were trying to communicate—and how their audience used this art (Lewis-Williams 1983, pp. 11–12). One can also begin to see how from such more valid reconstructions, generalizations, and theory development might follow.

FROM THE STUDY OF ART TO THE STUDY OF THE BUILT ENVIRONMENT

The developments discussed are very comparable to those I have been outlining regarding the built environment. The domain, the questions asked, and the approach are as different from traditional art history as mine are from traditional architectural history. In both cases, the intention is not history as a narrative of events but rather the use of data from the past to answer questions raised in a particular discipline and simultaneously to use contemporary data wherever possible.

In my case, the discipline is EBS rather than anthropology (although anthropology plays an important role in my own work in EBS and increasingly in the field as a whole [Rapoport 1986c]). The purpose is to study human interaction with those artifacts that comprise the built environment. The purposes at a higher level of abstraction are the same: to develop generalizations, conceptual frameworks, and theories. In order to do that, however, reconstructions of the past and inferences about it need to be improved. It is a result of this latter fact that there is convergence in the approach. If seen from a research perspective, rather than in terms of the subject matter, the convergence is almost total, as will be seen to be the case with archaeology in Chapter 5.

There is, however, also a difference. This relates to what follows the development of understanding and explanatory theory. In the present case, it is to develop the disciplinary base of a science-based profession of design. In other words, there is an additional possible step of *application*. This is where *precedent* comes in: There is an interest in deriving lessons from the knowledge and understanding obtained that can inform design, that is, a concern with how this knowledge can be used to derive normative statements about *what should be—and why*. But that additional step in no way changes the coincidence of the approaches at the level of research, understanding, and theory.

THE RELATION BETWEEN PAST AND PRESENT

It is a very traditional historical view (e.g., Porter 1981) *that history can only be studied from the point of view of the present*. This position does not have to be accepted because it is at least arguable that history in the sense of reconstructing past events, places, and the like can be done objectively. However, the study of the past to derive lessons for the present and future does mean that the past is indeed studied from the point of view of the present. The specific position taken in the present clearly becomes a critical issue and needs to be made explicit.

I made a very similar point in *House Form and Culture* (1969a) when I argued that vernacular could best be studied from a specific point of view, raising specific questions about topics or themes, rather than chronologically, the way traditional architectural history is studied. In later work (Rapoport 1980d, 1982b, in press), I argued that interest has gradually been shifting from one in vernacular for itself to vernacular for the lessons it can teach us, as an entry point into general questions about EBR. I have also argued that a concern with developing countries has similar extrinsic interest (Rapoport 1983a).

In terms of learning lessons relative to design, my starting point, my *specific position in the present,* is that the work of most contemporary designers is highly unsatisfactory and has been so for quite some time. It is further my position that the differences between movements and "schools" are largely trivial and superficial (Rapoport 1983b); so are the in-group debates among designers, which are often meaningless (Wolfe 1981). It seems to be the case that "primitive," vernacular, and popular environments, and even spontaneous settlements in developing countries are often better than those done by professionals (Rapoport 1987b, 1988b). It also seems to be the case that in many cities it is the new buildings that are unsatisfactory; new towns are notoriously unsatisfactory in many ways; urban renewal and redevelopment have been described as disasters, and housing (in Britain) has recently been convincingly shown to be appalling (Coleman 1985; Stretton 1985), whereas designers' criticisms of suburbia (also in Britain) have seemed to be wrong (Oliver, Davis, and Bentley 1981).

Logically, design is for the purpose of creating better environments. Design also needs to be seen as a set of hypotheses of the form: If such and such is done, the following will happen; these hypotheses need to be based on the best available knowledge. Consequently, the following questions become important: What is better? Better for whom? How does one know? What should one do to achieve better design and why? How does one know *that?* How does one evaluate? How does one improve? and so on (Rapoport 1983b,e).

Given that, the questions in the framework of the present book then become: How does the study of history, that is, of data from the past, change if the foregoing is the position taken in the present? A corollary question is how the study of the past can help answer questions such as these—with which I cannot deal here in detail. The need for such data follows partly from the need for the broadest possible body of evidence already discussed. It also follows partly from changes in the way design is done and how most environments come to be, compared to the way they used to be created. This can be summarized as a transition from *selectionism* to *instructionism* (Rapoport 1986c, pp. 171–172). The former is a process whereby fit or congruence is gradually achieved through selection; the latter refers to short-term design using a model or template and hence achieved through giving instructions. (My use of these terms is based on Lederberg, cited in Jacob 1982, pp. 15–17.) Preliterate and traditional vernacular design worked through selectionist processes*; in current design by professionals the predominant, in fact only, process is instructionism. In order to give instructions about the creation of environments, many of which were not professionally designed in the past, for plural user groups, for "unknown" users who are different from designers as a group, and so on, one needs new kinds of knowledge; designers need to "objectify" the knowledge needed to give valid instructions, hence the need for EBS research. EBS is the start; the full range of data, including historical data is the body of evidence. There is a need to learn to translate lessons about products of the selectionist mode, and inferences about behavior based on them, into data suitable for an intructionist mode. Provided, for example, that one knows why and how EBR worked when the products came about through selectionism, one might be able to achieve comparable results by instructionist means, not by copying but by applying any lessons learned. This can only happen through research leading to theory. Because most selectionist environments are past environments, particularly non-high-style ones, the study of history in the sense being discussed has a major role to play in this process.

Part of the reason is that in developing a theory of environment—behavior interaction, one needs knowledge about both behavior and environment. In general, the former has been much more studied not only in the social sciences but even in EBS. To redress this, one needs much greater emphasis on the built environment and material culture. It just so happens that this is all we typically have from the past, with some exceptions when written history begins that not only tends to neglect most of the built environment but began relatively recently and in relatively few places. Thus this body of data is central. To study it, however, one must rely on controlled analogy with the present to make the necessary inferences about behavior. Only thus can the past be validly reconstructed and understood. Conversely, only thus can this past, consisting of built environments, become more relevant to an understanding of EBR in the present.

In this connection, the focus of the study of the past here being advocated is *the interaction between human behavior and those parts of material culture that comprise the built environment wherever and whenever they occur*. The emphasis then is on the constant processes of EBR. By using the full range of evidence to generalize about EBR, the study of the past is brought into the present and knowledge from the present related to

---

*In my view this is a better term than the operationally very similar *unself-conscious design* (Alexander 1964) because it makes no judgments about the presence or absence of self-consciousness.

the past. We are thus dealing with the study of a single field: EBR. We can use both present-day conceptual, theoretical, and methodological apparatus and data from the past: The issue is how the past relates to the study of ourselves. In this type of study, one obviously needs to use whatever data is available rather than carefully controlled samples. The evidence, however, can be controlled in other ways, for example, by showing its relevance, maximizing its relevant variety, and so on. Moreover, material culture is often more durable than other data and is much more widely distributed, in space and time, than written material. This makes it of critical importance to all social science research and particularly to EBR research. In this way, material culture plays a central role in any diachronic analysis: It provides a time dimension to evidence even when change or development as such are not being studied. One may find, on the contrary, that the pattern reveals constancy and lack of change. The main issue is to compare extensive series of data distributed over time and space. In this connection, it is useful to distinguish two aspects of time. History involves not only the variability and the size of sample, that is, data over time. It also involves diachronic development— that is, patterns of change or constancy in any given "thing," whether size, elaboration, decoration, siting, territorializing, personalization, taking possession (e.g., Conan 1987), and so on and on. In a way one may think of the former as the "lateral" approach, the latter as more "linear." A classic example of the latter is the study of Western ideas about the relation between people and nature over a period of 2,000 years (Glacken 1967).

Thus rather than being concerned with the style of monuments and famous designers, one is concerned with those built environments that will facilitate making inferences about EBR in the past. In this, one is no different from historical geography, prehistory, archaeology, or even social history. To make inferences about EBR one needs, in turn, to make accurate reconstructions of past environments. In this we are not very different from paleo-whatever (ecology, climatology, botany, etc.)—or from cosmology, geology, paleontology, or any other historical science. In all these cases, *the essential issue is making inferences about the past on the basis of evidence, however inadequate, in the present.* The specific subject matter may differ, the data to be used may differ; one thing, however, remains constant: the process about making inferences about the past. Also different are the purposes and hence the kinds of questions asked.

Two related questions then arise. First, the quality of the inferences possible and, second, the validity of any inferences made, given change and the role of change. How these questions have been discussed in other disciplines concerned with them will be the topic of the next two chapters. At this point a few preliminary remarks are in order.

Uncertainty can arise for various reasons (cf. McClelland 1975, p. 226; Clarke 1978; cf. Chapter 5). One type of reason concerns the imperfection of the surviving data, their differential survival, which raises questions about how representative such data are of the whole. Note, however, that they are the *only* data; one can only use what is available. The many existing constraints cannot be removed. The question is essentially about reducing specific doubt: Given all the problems, which approaches work best? This is comparable to the problem of the human mind knowing "reality." Assuming there are built-in problems and limitations, the question becomes which is the best, most reliable (or least unreliable) approach. The answer is what we call science, which has been described as "the paradigm case of knowledge acquisition" (Shapere 1984, p.

200), and regarding science itself a similar point has been made: *General* doubt has no impact on reliability, only *specific* doubts (Shapere 1984).

Regarding the quality of the data one needs *explicitly* to discuss the shortcomings and allow for them—as is typically done in paleonotology and archaeology. There is, of course, a basic assumption central to any inference from artifacts to behavior. This is that material culture gives physical expression to ideas, images, and schemata, that it incorporates information coded in systematic ways so that the materials communicate meaning. What has been encoded by human minds other human minds can decode. *Material culture can, therefore, be decoded in principle.* In this, the historical record of what people have built is like archaeological data—it is *congealed information* (Clarke 1978). So, of course, are other records from the past—in geology, paleontology, and whatever. The question is not whether it can be decoded in principle, that is, inferences made, but how it can best be done.

Another type of uncertainty is caused by the unknown degree of imperfection in the theoretical models that are used. These may be due to the neglect of various factors or use of assumptions that distort reality. The answer is, once again, to be explicit in one's models and assumptions and to test them as rigorously as possible. This explicitness also makes it possible for them to be tested and criticized by others. They can then, and generally do, gradually improve.

A discussion about inference can best be approached by recalling that the built environment has been defined in such a way that one cannot study only buildings. First, they need to be seen as part of a system of settings and the cultural landscape within which people live and behave. Of course, for any given question, specific subsystems not only can but *must* be isolated: "Holism" is a dangerous and pernicious myth (e.g., Phillips 1976; Efron 1984). Second, built environments include all manner of semifixed elements. Third, they mean little unless occupied by people—one needs to know the cognitions, perceptions, choices, preferences, and meanings of these people and also their activity systems, including their latent aspects (Rapoport 1977, 1982a, 1986a, 1990). This applies to contemporary research as well as to studies of past environments (e.g., Rapoport 1984c). This is because the purpose of such study is to develop generalizations about EBR en route to explanatory theory.

It follows that the data necessarily go beyond buildings and include their furnishings as well as plazas, streets, open spaces, settlements, and their landscapes—cultural landscapes. Buildings alone are never sufficient.

Most of these are not available even in present-day material and are only beginning to be considered in EBS. The process has not advanced far (e.g., the questions in Rapoport 1976 produced no results). In the case of past environments, the lacks and gaps are clearly vastly greater. Yet the data needed for generalizations and theory building are reconstructions of past environments "occupied" by people, their activities, and behaviors. We wish to know how they came to be designed, what they meant, how they were used, and so on.

Thus, as already pointed out, the study of past environments involves two steps:

1. Inference about *what was.* One can then ask *what that means* and *why.* This leads to Step 2.
2. Explanation—which is still inferential, but at higher levels. In fact, later I will discuss first-, second-, and third-level inferences.

Also, as already discussed, one can consider Step 1 to be the reconstruction of specific places, sites, groups, periods, and so forth: the more traditional subject matter of history. Step 2 then consists of generalizations and theory based on such reconstructions by asking specific questions of that material in terms of contemporary EBS research. In other words, the first step of the study of past environments is their reconstruction. In that sense, this is not unlike traditional history, although the new approach can help in making the process much more rigorous and credible. The critical step here is inference.

The second step, generalization and theory construction, is, of course, more difficult, and even in archaeology has not gone far although major advances have been made. In the case of EBS it is possible, by taking a position in the present based firmly on current research, which suggests activities, behaviors, cognitions, meanings, and the like to use purely physical attributes of past environments to generalize and theorize: The case study will show how. In Chapter 2 I gave some briefer examples of such studies including meaning and sacred space. Even in that process, however, inference plays a major role—*explicit* inference connecting data to concepts is central in all scientific inquiry including the humanities and social sciences. Moreover, in any process of reconstructing the past or of building explanatory theory, *the use of analogy inevitably plays a central role* because most models are based on analogy (cf. Gibbon 1984; Hesse 1966). That, in turn, depends on an assumption of continuity between the past and the present, at least in certain respects. It thus becomes necessary to discuss first, the process of inference and, second, the question of uniformitarian assumptions.

INFERENCE

We are concerned with inference from past built environments to human behavior—both in their broadest sense. This typically means inferences from organization of space (and artifacts) to organization of communication and meaning (and possibly time). Inference is, first, to past EBR and, second, to contemporary EBR; it is also from behavior in specific settings to generalizations about behavior and, ultimately, to theory about EBR. Because, as we will see, analogy with contemporary behavior is unavoidable, there is also inference from present-day behavior to past built environments.

The process of inference is a highly technical topic about which there is a large and complex literature, much of which concerns specific models, methods, and techniques. These will not concern me here; rather I will discuss the process of inference in very broad, general, conceptual terms.

Making inferences about the past (or anything else) is never easy; it is a lengthy and complex task. Inferences, however, can be less or more difficult and the outcomes more or less valid (or at least convincing). But even the most difficult questions can be answered if certain preconditions are met. Thus inferences must be made systematically, using an explicit, logical, and rigorous argument. Premises must be explicitly stated and, at the very least, a coherent framework or context within which to argue developed and stated explicitly. One then needs explicitly to describe the steps of the inference process, the analogies used, and the rationale for using them. These, and the conclusions, must be related to a body of public knowledge, that is, it cannot be private or esoteric knowledge. This knowledge needs to be criticizable and correctable and, hence, cumulative. It must also include background knowledge from a variety of relevant fields and disciplines. All this allows one to go beyond material culture to behavior patterns of people long since gone—including social, cognitive, and symbolic behaviors. Iteration of this process and rigorous mutual criticism should in principle allow one to do so with greater clarity and increased confidence. Note that all fields in which inferences like those I am discussing are made are equally concerned with making the process more satisfactory. The search is not for certainty but for more convincing, and progressively improving, inferences.

One also needs explicitly to discuss any problems with the evidence. One needs data, many of which are still, in our field, lacking (unlike archaeology, paleontology, neuroscience, etc.). As previously pointed out, we lack the natural history phase (Rapoport 1976, 1986c; cf. the case study in Part III, where data are very inadequate). Not only does one need adequate data, but there must be clear ways of describing what is observed, that is, agreement on criteria for description* as well as for measurement (if any) and of how any characteristics serve as indicators for various constructs or theoretical concepts. This latter point is of crucial importance. In making inferences, or even collecting data, one must decide what are relevant *indicators*. Clearly these depend partly on the research question as far as data are concerned. For example, in studying urban spaces, different indicators would be relevant if concerned with their perceptual qualities (as in the case study in Part III) than in trying to identify different cultural influences on urban morphology (cf. Alsayyad *et al.* 1987).

The effort is for *consensibility,* which is a critical first step to *consensuality* (Ziman 1978). Clear distinctions must be made between description and explanation as well as between empirical generalization and theory. There must be clear and explicit linkages between empirical data and theoretical constructs. Theory must be linked to empirical data by models. Explicit taxonomies must be used to order material and patterns identified and explicitly described.

The centrality of pattern recognition has already been pointed out (Binford 1981, 1983a,b; Judson 1980; Renfrew, Rowlands, and Seagreaves 1982; Salmon 1982; Smith 1977, 1978, 1982; Ziman 1978 among many others). It is also the starting point for inference. This is the case because, first, only from patterns can one derive regularities and constancies, develop expectations, and hence note anomalies. Patterns also allow inferences to be made about parts of built environment and material culture that have not survived, that is, they allow predictions about what *should* be found. Finally, they suggest "why" questions, which lead toward explanation, to the identification of reasons for such patterns (or their absence) or of the agents or processes responsible. They also suggest further questions and lines of research.

The identification of patterns and, even more, of reasons, processes, or agents is inferential. One is, in the first instance, looking for repeated patterns of cooccurrence, whether universal or probabilistic, or for specific groups or periods. The more one finds

---

*In this connection, the lack of a descriptive language in our field is a major problem in spite of efforts by Cullen, Halprin, and Thiel (the last is still trying). Unless built environments and settings, their ambience, character, characteristics, and attributes can be described so that these can be communicated to others, consensibility cannot be achieved. (There can be no essential intersubjective comprehension and agreement). As a result, environments cannot be easily studied or analyzed—consensuality cannot develop before consensibility (Ziman 1978).

such patterns, and the broader the range of evidence to which they apply, the more useful do such patterns become. One may then be able to discover patterns that reflect *constancies in human behavior that may be highly predictable under sets of known and specifiable conditions.*

The longer over time and the more widely cross-culturally and in different environments such repeated patterns occur or concur, the more significant they are likely to be and the more confidence one can have in inferences made from them. If one is able in this way to link material culture and built environments (either in general or specific subsets of them) with generalizable social, cultural, behavioral, evolutionary, or other human characteristics, one is well on the way toward making inferences. In this process one typically adopts a view of how the world works and how one can best come to know that. This has been variously called *metaphysical blueprint, world hypothesis, philosophical presuppositions, orienting strategies,* and so on. *These need to be made explicit* because they influence the inferential process.

In the present case, this presupposes that there will be eventual agreement about such patterns because they actually exist. In history, archaeology, and EBS one is dealing with the mesoscale, the Euclidean and Newtonian world of our experience, the world in which we evolved. This world we perceive veridically and understand clearly and directly (cf. Cook and Campbell 1973, Chapter 1; Gibson 1979; Harré 1986; Ruse 1986; von Schilcher and Tennant 1984). In Northrop's (1947) terms, many of its elements are "intuitional concepts." Even theoretical constructs (like "culture") can fairly directly be linked to this world (Rapoport 1986c, 1989a). In that sense our data are more "realistic" and easier than those of many sciences, above all those dealing with the very small (i.e., particle physics, subatomic physics) and the very large (cosmology) or very fast (relativity). In those, many concepts are often extremely nonintuitive and very difficult (if not impossible) to comprehend in material terms. It follows that because there is a world out there, it will not allow all inferences or models equally (Bunge 1981; Jacob 1982). Many suggestions can be eliminated at this point. Although inferences, models, and theories always go beyond the data, as they must, there must be a genuine possibility of data challenging the claims being made. This implies that the systematic collection of new data is essential as is their analysis. All these provide further empirical constraints. The data are there, and they force the retention or rejection of certain theoretical constructs intended to explain them.

The record of past built environments is empirically constraining in two ways. First, it provides data that will not allow *all* inferences or explanations equally. Second, there is an interaction between data and inferences or explanations. Environments produced in particular ways or for particular purposes can be expected to produce particular patterns. One can then check how the data bear out expectations; one can use simulation, counterfactuals (cf. Leplin 1984), and other techniques.

Both the identification of patterns and the empirical constraints on inference require a great deal of data, a thorough, broad, and varied body of evidence, careful description of relevant characteristics, explicit classifications and taxonomies, and choice of indicators on explicit theoretical grounds. This implies systematic observation of many data and careful and systematic classifications. One must also avoid the use of selective bits of evidence as illustrations, as so much supposedly comparative work, including mine, has done (Rapoport 1986c). Rather, one needs to use broad bodies of evidence selected for explicit reasons, and one needs to ask a series of explicit questions of such evidence. The use of much data, many observations, and the classifications and orderings does, in fact, help to recognize the structure of the data and thus identify the patterns. For example, Boulding (1984, p. 210) reviews a book that by using 2,400 references by 1,400 authors and arranging these not chronologically but around a series of topics derived from a taxonomy based on a theoretical structure produces a new and different kind of history (in this case of a field of study) by revealing certain patterns.

Salmon (1982, p. 182) quotes Bertrand Russell to the effect that philosophy can help knowing and that knowing involves two processes. The first is doubting the familiar; the second, imagining the unfamiliar. Note, however, that what is "familiar" or "unfamiliar" is itself a matter of knowledge. If one only looks at high-style design and the Western tradition of the recent past, then most of the built environment is unfamiliar. If one knows no other disciplines, then useful findings, concepts, taxonomies, methods, and theoretical frameworks may be unfamiliar. In fact, this book is an argument for increasing familiarity with all of these things. But there is also another sense of "familiar" and "unfamiliar": The "normal" or "typical" is usually familiar, the "unique," "anomalous," or extreme usually unfamiliar (cf. Drake 1973, 1974).

In this connection, Boulding (1984, p. 212) makes the point that quantification and statistics alone are only of limited value in dealing with the improbable because one tends to lose the extreme positions of the system—yet it is those extremes that may be useful. It may be useful to emphasize the *comparative study of special cases*. This relates to my long-standing argument for comparative studies (cf. Bullock 1984; Chisholm 1983; Kent 1984) and my argument over the years of starting with extremes. One also needs averages, however; "special cases" and "extremes" only become apparent from patterns that require broad and varied bodies of evidence, multiple disciplines, and so on. The identification of patterns and extremes requires appropriate taxonomies, a topic generally neglected. To know what did, or did not, happen and how frequently requires the broadest possible body of evidence in historical inference of any type. Thus, using *einfühlung* and similar approaches (e.g., Collingwood 1939, 1961), one essentially asks what one would do if one were in someone's shoes. Other approaches are based on what one knows about certain groups or all people ("human nature"). The second seems clearly preferable but requires *knowledge;* of other cultures and environments; of those psychologies, processes, and patterns of behavior, which apply to all people—or to particular subsets (cf. Pipes 1986).

One is concerned with processes, both current processes and data, and data surviving from the past about which processes can only be inferred. The route of inference is from better and more rigorous description and classification, to historical reconstruction, to generalization, to the development of a body of theory that might be applicable to the future. But it all hinges on inference, and *all study of the past relies on inference based on analogies with the present* (or contemporary data and knowledge).

It follows that research on past environments needs to move back and forth between:

1. Recognition of patterns in the record of the past.
2. Asking the important question what that means.
3. Doing contemporary (current) studies (experimental, EBS, ethnoarchaeology, etc.) to help make inferences about the past.
4. Asking why that should be, that is, seeking explanation.

Inference about the past thus depends on establishing a link with present empirical work and involves the use of analogy; hence the inevitable uniformitarian problem. But the linkage also goes the other way, as we have already seen: One can see patterns in present-day research and can then test such patterns to see whether they are anomalies, or examples of larger patterns, by testing them against the record of the past. In other cases, this can also be done by testing against cross-cultural samples. One can also begin with both—as discussed in Chapter 2 and the case study in Part III.

Inferences from pattern involves the question, "What does it mean?" and then "Why did it happen?" (e.g., Binford 1983a,b; Mayr 1982). This question is not unique either to the study of the past or even to science; it is found in all scholarship. One looks for linkages between one set of characteristics or conditions to other sets and then tries to understand their significance.

In making inferences, the critical relationship is between the data and theoretical constructs or concepts—and that is primarily a matter of explicit linkages between indicators in the data to lower and then higher order concepts (cf. Fig 3.2). We could think of these as first-order inferences from observables to concepts such as "privacy," "territoriality," "safety," "status," and so on, and second- and higher order inferences about explanation—what cognitive schemata or meaning are involved, motivations, what role they play in the culture or social system, and so on (cf. Renfrew 1984 about wealth, status, and power; the "Iconography of Power (and Status)"; cf. Rapoport 1982a, 1988a; Uphill 1972; examples in Chapter 5, e.g., Freidel and Sabloff 1984; Flannery 1976, etc.).

Note that inferences about the past, which means all study of the past (or of anything), involves generalization. This is frequently implicit: It must be made highly explicit. This we will see is also clear from the supporting argument, for example, in history (Chapter 4) and archaeology (Chapter 5). This seems very widely agreed to be a major requirement. These explicit generalizations can then be compared with the empirical evidence, for example, a sample of relevant cases and situations, and tested. To do this such generalizations must be capable of being refuted (in Popper's terms—falsified).*

It could be argued that inference means going beyond the data. But it is generally agreed that it is the purpose of theory and hypothesis to speculate, to go beyond the present state of information. Even perception goes "beyond the information given," in Jerome Bruner's phrase, as language relies on extralinguistic knowledge (cf. Johnson-Laird and Wason 1977; Rosch and Lloyd 1978). In that sense, this process is central to *all* research: Theory both points the way and is capable of being falsified. The careful accumulation of tested data, although it does not necessarily lead to theory through induction, can suggest hypotheses and theory through abduction; in any case, such data enable the testing and hence possible revision of at least the provisional validity of the theoretical position. The process can then be repeated and taken even further. In that sense, inference is a "bootstrapping" process, in Dudley Shapere's sense.

There is, of course, much debate about induction versus deduction. Although induction is currently regaining some favor in philosophical writing, in practice it is

*I am aware of the arguments in the philosophy of science literature of some of the weaknesses of Popper's position, which, incidentally, is less extreme in his work than in that of some of his interpreters. It seems generally agreed, however, that the possibility of falsification or refutation is a core criterion of all scientific (and scholarly) inquiry.

frequently used in all science. There have been attempts to develop more sophisticated inductive methods (e.g., Salmon 1976, 1982; Medawar 1969). Another problem, of course, is that in principle there can be an unlimited number of alternative hypotheses that can be generated and need to be limited before testing. This is done by considering how plausible they are on the basis of background knowledge; other theories in the network of theories also play a role because hypotheses cannot violate them. Existing data also limit conjecture as already discussed, and the more varied the disciplines the more limitations that can be derived from such data. In our case, such data can be ethnographic, ethnohistorical, historical, archaeological, ethnoarchaeological, geographic, ecological, EBS, psychological, evolutionary, and from other disciplines and lines of evidence—but always relative and relevant to the domain.

Once hypotheses are derived, their evaluation must also be done explicitly and, once again, one needs a sufficiently large number of sufficiently varied instances, that is, the broadest possible body of evidence. Neither single cases nor data that are too narrowly based or limited arbitrarily (e.g., on the basis of liking) are adequate; they do not provide enough evidence for inference. Selecting cases to "prove" or "illustrate" the argument is also not acceptable: One needs to use either the total evidence or adequate samples as already discussed.

Not only are induction and deduction both used in science, but three processes have been identified: induction, abduction, and deduction (to which one can add measurement) (cf Gibbon 1984). Induction and deduction are both better known than abduction, which is a form of inference (proposed by Charles Peirce) concerned with how one reasons from data to hypotheses or new ideas, that is, the process whereby concepts, hypotheses, and theories are engendered. Induction is then concerned with how one makes inferences about general rules (or patterns) from specific or individual instances, whereas deduction is used to draw consequences from hypotheses or theories that can then be tested in some way. Clearly all these processes, as well as measurement and operationalization (e.g. selection of indicators), play a role in research. Although this is occasionally drawn as a circle linking theory and data (Gibbon 1984) (see Figure 3.5), it is more correctly understood as a "spiral."

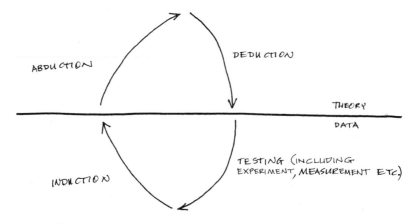

Figure 3.5. Relating theory and data (based partly on Gibbon 1984).

In that sense research can begin anywhere in the spiral, and even design can be conceptualized in such terms (cf. Zeisel 1981). There is a constant interaction between theory and data in the process of inference. One example of an approach, which includes these processes related to archaeology, can be diagrammed as is shown in Figure 3.6 (based in part on Gardin 1980).

A more detailed diagram of the sequence of inference can be produced that can be understood in terms of the increasing distance of different aspects of life from the material data (Mackie 1977). Thus, one moves from (1) material data, to (2) techniques and technology, to (3) social and political institutions (social structures and relations), and finally to (4) ritual, ideology, cognitions, meaning, intentions. In terms of the greater remove of 4 from 3 than 2 from 1 about which some inference is also inevitable, the inferences about them become more difficult. This can be interpreted in terms of greater underdetermination by the data as one goes from Number 1 to Number 4. In effect, given that the built environment can be conceptualized as the organization of space, time, meaning, and communication (e.g., Rapoport 1977, 1982a), this means that typically, regarding the past, all that one can find is *space organization*. One then

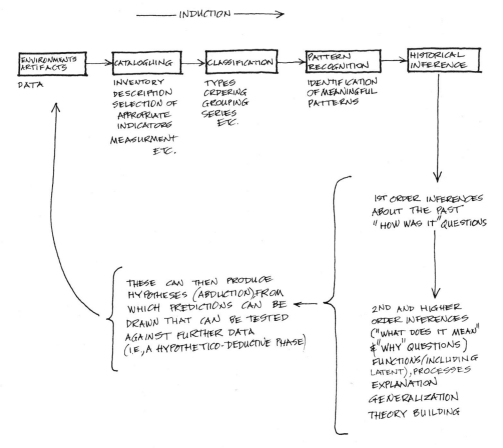

Figure 3.6. A model of inference in archaeology (based partly on Gardin 1980).

needs to make inferences about time, meaning (although one may find evidence in both fixed and semifixed feature elements), and communication. Note that even something as ephemeral as gestures can be studied historically and inferred over a period of 2,000 years on the basis of much varied evidence from several disciplines (Collett 1984). There is a greater degree of speculation and hence a greater reliance on other criteria. But such inferences are not, however, completely speculative. They are always somewhat constrained by the data, by background knowledge, and by other disciplines. Also, over time they tend to become less speculative; see Figure 3.7.

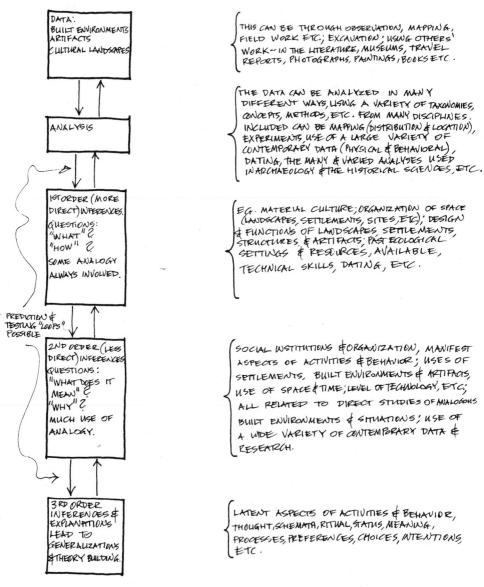

Figure 3.7. A model of inference in environmental history (based in part on Mackie 1977).

This relates well to a sequence that I have long used (based on Gibson 1968; cf. Rapoport 1977, 1982a). This moves from the concrete object through the use object and value object to the symbolic object. I have pointed out the greater variability and difficulty of inferences as one moves from the concrete to the symbolic object, although even in the former inference is involved ("This is a house," "This is a court," "This is a shrine"). There is also a relationship here to my analysis of activities into the activity itself, how carried out, associated activities, and the meaning of the activity (Rapoport 1977, 1982a, 1985b, 1986c, 1988a, 1990). It also relates to Binford's (first used in 1962) sequence of technomic function, socio-technic function, and ideo-technic function. Here also it is increasingly difficult to make inferences. In fact, the sequence of the development of archaeology and other disciplines has corresponded to this order of difficulty: First the "things" are identified and described and their "manufacture" and use inferred; then their social functions and so forth are inferred, and finally inferences are made about their meaning, significance, and the like.

Note, however, that even the data may be difficult to identify without contemporary analogs. Thus, although it is relatively easy to identify a house as such, it is far less easy to identify the extent of a system of settings beyond the house—which *is* the dwelling (Rapoport 1980c). In the case of a single place, the use of the broadest body of evidence and long time series in conjunction with EBS and ethnoarchaeological evidence can enable such configurations to be plotted. Once such evidence is available, broader patterns can be traced (Kent 1987; Rapoport 1990).

This means, of course, that inferences become more difficult, more indirect, and less certain (more hypothetical) as we move from lower order to higher order inferences. Also, analogy is present at all stages, even in identifying a space as a street or court, walls as a building, an artifact as a pot, an axe or a table; its role becomes much more important in higher level inferences and hence so does the use and importance of contemporary data.

Because it is difficult (on some accounts, impossible) to make inferences from material culture traces to behavior—particularly cognitive and symbolic—it may be necessary to use past material culture to test hypotheses derived from the present— EBS, anthropology, general history (based on written sources), and so forth; alternatively one can test inferential hypotheses based on the interpretation of past material data empirically in the present. Historical data about built environments can lead to hypotheses about human behavior that can be tested in an active empirical mode in the present. Conversely, hypotheses generated in contemporary EBS can and should be tested against data from the past to see if they hold up.* The object is to identify interactions between built form in fixed-feature and, as far as possible, semi-fixed-feature elements and human behavior, that is, non-fixed-feature elements, whether the inferences are from built environments to behavior, vice versa, or both ways. The latter is really the most typical, given the cyclic, iterative nature of research generally and the inferential process specifically. This is the case in the study which forms Part III of this book, a topic that has interested me since my student days (both my undergraduate and graduate theses dealt with related topics). My early observations were personal—I found certain past environments particularly suitable for walking; I liked them, found

---

*To avoid repetition, I omit reference to the need for similar testing cross-culturally and across diverse environments (unless the hypotheses are culture- and setting-specific).

them stimulating, interesting, enticing. They made *me* walk, challenged me to do so—often to the point of exhaustion. In this, then, I was like other architectural critics. I then looked for others' confirming or opposing experiences: There seemed to be much anecdotal material and design and travel writing expressing this view. I then turned to some work in contemporary psychology and experimental aesthetics and did some work on complexity in EBR; others did comparable work. This then generated the hypotheses in the case study that are being tested against the collected body of evidence of relevant past environments—and could be tested much more systematically if the data were available. This, in turn, has both theoretical implications for EBR (perception as a linking mechanism; supportive environments, etc.) and implications for design, that is, the future (Rapoport 1977, 1981b, 1986a, 1987a). There is thus a link between the past, present, and future that I would argue can only develop through theory. That link is between history as the data of the past, EBS, and environmental design.

Research in contemporary situations is thus essential: that is, contemporary data and data from the past must be used together. The diversity of data is necessary; not only of many and varied past environments but also contemporary data and cross-cultural diversity (as well as anthropological sensitivity). In that way one guards against both ethnocentric and "tempocentric" assumptions. This latter avoids the error of assuming that the present-day situation is a pattern when it is in fact an anomaly. Thus EBS and ethnoarchaeology (see Chapter 5) can be seen as providing contemporary data to help obtain more accurate reconstructions of the past (i.e., inferences about the past). But I only see this as a first step in three senses. First, these reconstructions are then used to make more generalized inferences en route to theory. Second, these then allow higher level inferences; third, through generalizations and higher level inferences, the better knowledge and understanding of the past can then be used in arriving at a more valid and convincing understanding and explanation of the present. Another "spiral." Thus, the inference is not present → past, or even present ↔ past, but see Figure 3.8.

In linking past and present environments through patterns and regularities of human behavior derived from contemporary research (observation, ethnography, sociology, etc.) experiments *sensu stricto* are possible. Examples are experimental archaeology (see Chapter 5), ethology, ecology, biology, and so on. In another sense, the linking of past and present environments produces the *equivalent* of experiments, as in what I call the "historical sciences" in Chapter 4 (cosmology, astronomy, geology, paleontology, evolutionary science, and the like). *The principles are the same* in all these

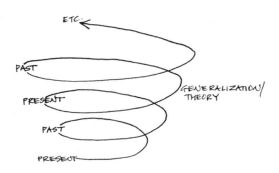

Figure 3.8. Present/past relationships in generalization and theory.

cases: In the case of some sciences, hypotheses are tested by the collection of data in the form of phenomena that have occurred in the past (or are still occurring), either naturally or through human action. In other cases, through human manipulation specifically to test the hypotheses, that is, they are made to occur under experimental conditions. The use of a wide range of data, in the present case, over time, is the equivalent to *using the data of the past as our laboratory.* (In the same way as one can so use cross-cultural data.) This becomes particularly useful if the variability of conditions is explicitly discussed in different cultures, situations, periods, types of environments, and so on.

The hypotheses themselves can be derived from a wide variety of contemporary research, from written history, travel accounts, pictorial materials, and the like and tested against past built environments and material culture. Alternatively, they can be derived abductively from past material culture and checked against contemporary situations (as in Kent 1984, 1987; cf. Chapter 5; see Figure 3.9).

Recall, however, that even when working on the reconstruction of the past of a

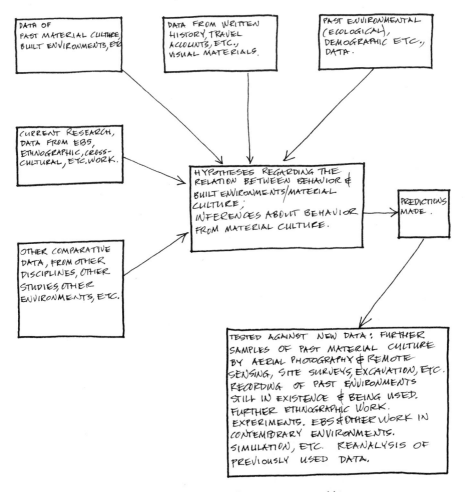

Figure 3.9. Hypotheses and data in environmental history.

single site, period, or group, rather than generalization or theory development, and even in the very identification of "things" as streets, buildings, or whatever, some inference and generalization is inevitable. This not only means that these must be made fully explicit but also that the largest and most varied body of evidence plays a critical role. For example, the widest variety of sources about the given case or the widest variety of elements within the same case, such as cultural landscape, resources, settlements, dwellings, tombs, sacred and other buildings, semifixed elements, and so on. Of course generalizations about human nature and behavior are always present and are inevitably based not only on contemporary work but on cross-cultural comparisons, a large range of data from the past and work on animal behavior, sociobiology, and evolution. These always need to be combined, as they inevitably are, with a variety of disciplinary approaches, concepts, and methods. It is also essential to make assumptions and chains of inference as explicit as possible. All this provides a process of interpretation that leads to more valid and convincing reconstructions of the past, more reasonable higher order inferences about the past, and more credible hypotheses not only about the cultural landscape, built environment, and material culture but also about social, ideological, cognitive, and symbolic behaviors of people long since gone.

Such reconstructions, higher order inferences and hypotheses, then need to be tested against more data from the same site or group and against a broader body of evidence, of cases *explicitly* shown to be relevantly analogous: broad in terms of types of environments, cross-culturally, in different climatic settings, over long time spans, or whatever characteristics are relevant. In this way one can test whether inferences about social, cognitive, and symbolic behaviors apply, whether the inferences and hypotheses are generalizable, whether larger patterns can be discerned or found. This also helps to test the inevitable uniformitarian assumptions involved, to see whether any process constancy holds across the specific expressions, variations, and apparent differences. These generalizations can then lead to further, firmer, and more convincing inferences at the level of specific historical reconstructions, and these, in turn, used for the next round of higher level inferences, generalizations, and the like. This is another "spiral" (see Figure 3.10).

By examining such a wide variety of environments from a wide variety of time periods and cultures, one can even begin to discover to what extent variability exists and in which respects, for example, in settlement forms, dwellings, urban spaces, and

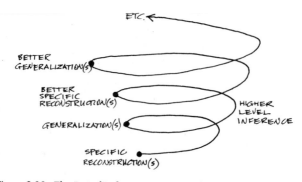

Figure 3.10. The "spiral" of reconstruction, inference, and generalization.

the like. In general, one would always expect variability—variations on a general theme (as one finds in biology [e.g., Gould 1980; Jacob 1982]). The object is to reconstruct a model of the theme (an ideal model or schema) and specify the range of variations and in which domains and under what conditions or circumstances they occur. As I will argue in the section on uniformitarian assumptions, the constant element may be process; the outcomes may be specific and hence variable (e.g., Fox 1980, pp. 470–471).

One could also address the question to what extent this variability may be due to systematic choice (intentions, preferences, and the like) or be "random" (e.g., due to group or individual differences, family size, and so on); to what extent it may be due to constraints and to what extent it is used to express wealth, status, power, roles, group identity, and the like. If the pattern suggests that latter (let us assume that it does), interesting studies could then be done on how size, height, color, newness (of form, materials, or whatever), materials, location, decoration, "furnishings," landscaping and vegetation types, space organization, controlled visual and physical access, and other elements are used to express such various social characteristics and thus to explain the variability. One could derive a *repertoire* (see Rapoport 1982a), a *profile,* or a *polythetic* set (Clarke 1978; Rapoport 1985b, 1988b, in press). This could then allow inferences about social ranking, the presence of different groups, ritual, or whatever. This might also lead to higher order generalizations, for example, that certain elements of the repertoire tend to be used more frequently, either generally or in specific types of environments, cases, and situations, and for expressions of specifics. Or the generalization might turn out to be that the specific elements are less important than *contrast* with the more general pattern (Rapoport 1982a). We will see in Chapter 5 that up to a point this is, in fact, beginning to be done in archaeology (e.g., Flannery 1976; Folan *et al.* 1983; Freidel and Sabloff 1984; Kramer 1979, 1982). Even these studies, however, (except possibly Flannery) are usually of a single site, the typical product being a monograph. For the more general inferences, however, comparative studies of large and varied bodies of evidence are essential.

Even in making inferences about the relation between spatial or formal ("architectural") features or characteristics, with use or instrumental function, one needs to have repertoires of basic elements and knowledge of the choices made among them, to lead to particular arrangements. Such knowledge becomes even more important when dealing with higher level inferences relating form to latent functions or meaning. For that knowledge, one needs first to have analyzed and compared a large number of lines of evidence. It is their convergence that becomes significant. The use of major lines of independent information can lead to hypotheses that can then be tested by other independent lines of evidence. Thus, we will see in Chapter 5 how in archaeological situations background knowledge of the field generally, other work in related areas, ethnohistoric data, surface surveys and analyses, and other sources generate hypotheses that become predictions to be tested by controlled excavation.

To reiterate a point already made, inferences in studying past environments are always, and inevitably, from material culture to human behavior: The people involved are long since gone. The physical artifacts, for example, cultural landscapes, built environments, their furnishings, from which such inferences have to be made are in many ways more similar to the evidence used in the natural sciences (e.g., particularly the "historical sciences" discussed in Chapter 4) than to the data of traditional history (based on written sources) or the social sciences and EBS—where behavior can be observed, people questioned, and experiments done.

The use of analogy is thus inescapable. This depends on some form of uniformitarian assumption. Only in this way can one limit the unlimited possible number of alternative inferences and hypotheses.* Reasoning by analogy is based on the assumption that if one or more things or situations have certain attributes in common, they will also share other attributes; if they agree in some respects, they will also agree in others. In our case, certain attributes of built environments (material culture) are compared with analogical cases where human behavior is known—or can be learned. Inferences are then made about other attributes of the original material—those related to human behavior. For example, if a street excavated in Harappa resembles in many respects many streets found in the contemporary Indian subcontinent, one may infer that behavior in that street was possibly very similar also (Pieper 1980a,b); a similar argument can be made about sacred space.

It has been argued that analogies cannot really be false (Lopreato 1984, pp. 66–67). Rather they vary along a number of dimensions, for example, informativeness, detail of comparison, explicitness, fruitfulness, heuristic value, and so on. The latter is particularly important no matter how gross the similarities between the analogies. This is because it corresponds to the crucial question of how one can learn about an unknown phenomenon or event from those known. This always involves positing systematic relationships among things that seem unrelated. The ability to do so effectively depends on being able to relate relevant attributes.

The use of analogy is more likely to be correct if certain rules are met. These are ones we have already encountered repeatedly since Chapter 1.

1. The attributes in question should be shared by the largest possible number of the most varied situations.
2. The shared attributes should explicitly be shown to be significant on the basis of background knowledge, other theories, data from the widest possible variety of disciplines and sources, that is, on the basis of the convergence of multiple lines of evidence. The more starting points, approaches, and disciplines, the more potential analogies become available.†
3. The number of shared attributes should be as large as possible and should be as specific and as explicit as possible.

---

*While I will discuss this when I deal with theory elsewhere, it is worthwhile to note the growing acceptability of analogy and even metaphor in science generally (Boyd, 1984; Gibbon, 1984; Hesse 1966; Leatherdale 1974; McMullin 1984; Ortony 1979; Schön 1969, etc.).

†It follows that one should use a variety of analogies. After having completed the penultimate draft of this book I came across an almost identical and well-stated position in a totally different domain—the study of strategy (Cohen 1985/1986). The point is made that some historical analogies must inform decisions—and be used in research even in the most present-minded social science. No problems can be considered without considering similar events in the past. Because historical data must be used, the question is how. The choice is not between using or not using historical analogies—but between using them well or badly. Looking at single cases is very bad; the first step is to "examine a variety of episodes, rather than building our theories on the narrow and treacherous foundations of one case, no matter how attractive . . ." (Cohen 1985/1986 p. 11). The same is the case in using analogies from the present for inferences about the past.

Given such conditions being met, the likelihood of inferences being plausible will also depend on how logically and explicitly the attributes in the various situations used in the analogy are linked and related (cf. the case study in Part III); see Figure 3.11.

As already repeatedly pointed out, both the rules and this process require:

1. Accurate and detailed collection and description of data, detailed analysis of their attributes, and classification in terms of explicit taxonomies based on some theory. It also needs explicit specification of which attributes are relevant and why. The lack of relevant data and attributes in the domain under discussion is a major problem.
2. Similarly accurate, detailed, and complete data about the analogous situations from which attributes of human behavior and EBR can be obtained, that is, contemporary empirical work of all sorts related to theory development.

The quality of inferences depends not only on the explicitness and logic of the two links between the cases. It is also essential to state any further assumptions, auxiliary hypotheses, and the like made. Ideally, alternative hypotheses would be explicitly listed, although frequently just one is. Thus I only test one in the case study although another one, in fact, exists (Hillier and Hanson 1984); the two are not mutually exclusive. From these, predictions can be made and tested against further observational data. The object in such inferences is never to "prove": to demonstrate "absolute truth" or "total accuracy." It is rather to show that the hypothesis is supported by the data or in the case of multiple hypotheses, which is best supported. This is usually a matter of comparison, of

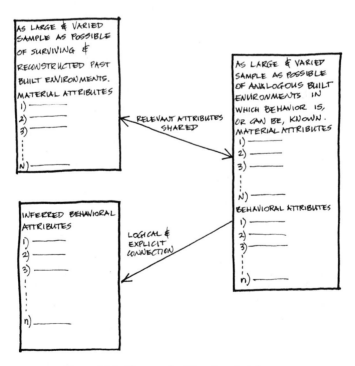

Figure 3.11. The use of valid analogy in inference.

ordinal rather than interval or ratio scales. In any case, there are frequently interim inferences and hypotheses and better ones may emerge.

Note that because every step is explicit, it becomes much easier for other (and future) researchers validly and rigorously to criticize and to build upon what has already been achieved, improving on the inferences and leading to a progressive and cumulative body of knowledge (or discipline).

It is difficult enough to make inferences about human behavior from material culture even when a complete set of such data is available. In fact, as already mentioned, and as we shall see in Chapter 5, the very possibility of doing so has been denied. In dealing with past environments and material culture, the data that survived, that have been recovered, and that have been adequately recorded, described, and studied are extremely incomplete. This applies particularly to the types of environments that I am emphasizing and to those attributes that this type of study requires. This makes the inference process even more difficult. As already pointed out, however, all one can do is to use the best available data; one needs only to be concerned with specific inadequacies and not be influenced by general inadequacies (Shapere 1984). Thus the general inadequacy of the data in any given research is of concern, and it is, of course, part of the research endeavor to improve data. One result of the case study in Part III is, in fact, to suggest a large set of attributes of streets which can be recorded and studied but which has not been. Thus one product of research should be an improvement of the quality of the data, in addition to their quantity and variety. This is common in research: It is possible progressively to improve the quality of the data as a byproduct of research.

This improvement typically occurs as a byproduct of the improvement of processes of pattern recognition and inference. Also, by using the insights, concepts, methods, models, and theories from other disciplines that provide a form of redundancy of networks of theories, data can be improved. It is important, however, to *adapt* such borrowings, not *adopt* them. It is also essential not to borrow specific models or theories, that is, not to borrow literally but to borrow approaches (Moore 1983). Such borrowings, however, clearly multiply the lines of evidence usable and help progressively to improve the data as well as reconstructions and higher level inferences. They also help in the process of generalization and theory construction.

I have referred several times to the inevitable use of analogies. Because models typically are analogical, this means that models are being used in making inferences. Note that the inevitable simplification implicit in models is no reason to criticize them, as is so often the case. On the contrary, the *raison d'être* of models is precisely to simplify the world (Gibbon 1984; Harré 1983). Models also provide ideas about the mechanisms operating, that is, they help to introduce the dynamics to the present record of the past that is inevitably static. They thus help in inferring process from product. They also help by suggesting the types of data and attributes necessary, thus improving the quality of data. Finally, models are also increasingly seen as providing an essential link between theory and empirical data (Hesse 1966; Morris 1983).

The simplifying assumptions must, however, be *explicitly* stated, discussed, and evaluated. It is also useful to estimate the extent of simplification; this is because the choice of a wrong simplification can be disastrous (Moore and Keene 1983; Ulmer 1984).

The choice of analogous situations, attributes, models, concepts, lines of evidence, bodies of evidence, methods, and so forth used in making inferences depends on the questions asked, that is, the purpose of the research (and ultimately, of the domain). Thus the explicit posing of questions is also essential. This, of course, relates to Chapter 2 where the general purpose of the study of past environments was described as being for the purpose of discovering patterns that reflect possible constancies in human behavior as it relates to environment–behavior interaction. From this specific questions follow. It is *these* that will inform the specific choices made in any given research and the types of inferences made. For example, in the case study, the question concerns how particular perceptual qualities of streets are supportive of walking, the link being *interest* in the environment. Given a question of what makes other urban spaces attractive to people in other situations, for example, for sitting, the questions might be in terms of *liking* (e.g., Rapoport 1986a). Very different attributes would be used, different analogous situations, possibly different evidence, different situations, because streets are more ubiquitous than urban spaces for sitting. (See also my examples in Chapter 2.)

The assumption is that inferences can be made from material culture to behavior partly because encoded in artifacts is intelligent, purposeful behavior and choices made by the makers of these artifacts. It is also possible partly because such artifacts were used by other people equally intelligently and purposefully. It has already been suggested that in some ways our data resemble that of the natural sciences (particularly the "historical" sciences) more than those of traditional history or the social sciences. Inferences are made through interpretation. Thus much depends on the interpreter and how the interpretation is approached. The outcome of research thus depends crucially on:

1. The nature of the phenomena as defined by the domain. The substantive criteria derived from that immediately constrain the inferences being made and control the strength of the inferences; they also constrain possible models, explanations, and so on.
2. The questions asked, which depend partly on the objective and partly on "philosophical presuppositions" or "orienting strategies."
3. The use of large and appropriately varied bodies of evidence, that is, the broadest possible range of accessible evidence. This further helps limit the number of possible inferences, models, and so on.
4. The rigor and logic of the argument, certain formal criteria related to rules of logic, rules of inference, criteria for explanation and confirmation, and the like, many coming from the philosophy of science.
5. The use of explicit procedures, so that the inferential and other processes can be followed, tested, and criticized: They must be intersubjective.
6. The use of all possible linkages to the present to develop the requisite analogies. This requires making uniformitarian assumptions, that is, rejecting any radical discontinuities in the attributes in question.
7. The controlled use of analogies and comparisons. These go from the known to the unknown, from the more certain to the less certain. Here the use of multiple disciplines is often extremely helpful.
8. The development (or at least attempts to develop) of a coherent and explicit

body of explanatory theory and of models linking theory to data. Models are constructed analogically, drawing on subjects different from the domain in question. Model construction needs to be explicitly related to one's philosophical presuppositions or orienting strategies.

9. The establishing of explicit connections between theory and empirical data, for example, by specifying mechanisms that suggest how and why the patterns found might have been produced.

10. The use of multiple disciplines, their concepts, models, and methods; these also help to provide the critical background knowledge on which any initial inferences depend.

Most of these points have already been discussed in this section. The last one has not. I have long argued that an essential requirement in EBS generally is to establish relationships among many disciplines with similar concerns. EBS has, from its inception, been an interdisciplinary field although still inadequately so. Although researchers in the field come from many disciplines, relatively little work is in itself interdisciplinary. Yet such interdisciplinary linkages are essential. In all fields, the most interesting work has been at overlaps. These hybrid disciplines have shown hybrid vigor, for example, bionics, cybernetics, biochemistry, molecular biology, biomedical engineering, communication theory, information theory, mathematical biology, evolutionary science, functional morphology, and many others.

The study of past environments must not only relate all the disciplines involved in EBS (design fields, psychology, sociology, anthropology, geography, ethology, etc.). Also involved must be history, the social sciences, historical sciences, and archaeology discussed in Part II (Chapters 4 and 5). Through these, more "indirect" links are established; for example, through the "new" history to econometrics and social history, through the historical sciences with science, philosophy of science, and mathematics. Archaeology, in its new form as part of a "science of man," interacts with linguistics, ethnology, anthropology (cultural and physical), toponymy, paleoanthropology (itself a mix of archaeology, paleontology, paleocology, geology, evolutionary science, animal behavior, systematic zoology, and others), economics, geography, and so on. In this way the multiple disciplines and their concepts and models, the networks of theories, the background knowledge when applied to the phenomena in question (the domain) all provide the "redundancy" that gives one greater confidence in patterns identified, the reconstructions, and in the inferences being made. Moreover, the inferential process can improve over time, as already pointed out several times; this is an example of intellectual bootstrap operations that apply to all science (e.g., Kaplan 1964; Shapere 1984). In that sense, the process of inference being discussed is at least as valid as that used in any science.

INFERENCE IN SCIENCE AND HISTORY

Before proceeding to discuss the uniformitarian assumptions inevitable in the inference process, it may be useful to pause briefly to consider more generally the relation between inferences in science and history.* The point is that there seems to be much more agreement than one would expect to find given the commonly accepted attitudes. These include the view that science is nomothetic, whereas history is idiographic. This implies that science deals with generalizations losing the richness of particulars. In contradistinction to traditional history which tends to deal with single and unique occurrences, whether people, events, or buildings,† science deals with classes, for example, hominids, *Homo sapiens,* or cultural groups; the recurrent creation of sacred settings, delimitation of domains, or the tendency of grids to disappear when central authority disappears (with the U.S. as an anomaly; cf. Rapoport 1977); or built environments. Yet history is also capable of dealing with classes, as we shall see in Chapter 4. Although traditional history may deal with individuals, the "new" history deals with classes of events. For example, instead of dealing with the poet John Keats dying of tuberculosis, social history like science may examine tuberculosis deaths in the early nineteenth century (King 1982, pp. 257–258; cf. Fogel and Elton 1983). It may even go further and study even longer time series (e.g., LeGoff and Nora 1985), changing rates of tuberculosis deaths (or other diseases) over long periods of time, or changing patterns of mortality, thus becoming historical demography or epidemiology.

We will also see that in many "historical" sciences, single and unique cases are generalized. Thus, any mountain range is unique in some way—it is part of earth history. But one can generalize to the class "mountain ranges" and consider the processes involved. Thus geologists deal with classes of events each of which may be unique but that share certain attributes. Similarly, "species" in biology describes classes of unique individuals (note the importance of taxonomy and classification in all this). Moreover, cosmology studies the origin and development of the universe, surely a unique event—because there is only one of those!

In many sciences, certain observed sequences of events can be artificially isolated and studied in order to discover the conditions under which they do and do not occur. In the historical sciences, as in history, such experimental isolation or manipulation is not possible: Inference must be used. This is the point at issue: inference in science and history.

To consider the possible relationship between historical and scientific inference, it is useful to distinguish *events* from *facts* (King 1982, p. 255).‡ An event is to be understood as any past happening; a fact as a true descriptive statement about such an event (Fischer 1970, p. xv, note; cf. McCullagh 1984). Thus that Senmut designed the tomb of Queen Hatsheput at Deir-el-Bahari (in Egypt) around 1500 B.C.E. is an event; the statement that he did so is a fact.

Now, it is clear from what I have already said that such statements are not derived from historical events themselves; those are beyond our direct knowledge. Statements are based on *traces* that events have left. It then follows that in history a fact is an

---

*This section has been greatly influenced by some of the arguments in King (1982). I read this after I had completed most of this book in drafts one and two.

†We have already seen that generalizations, usually implicit, are inevitable in all history—a point to be elaborated in Chapter 4.

‡Note that this distinction is used in a very different sense in French historiography (LeGoff and Nora 1985, pp. 21–26).

interpretation of some traces, an inference allowing statements to be made about the original event. This is clearly the case in archaeology, for example, where this is a major tenet and the "study of traces" has been greatly elaborated—hence my emphasis on that discipline in Chapter 5.

Traces must, therefore, be evaluated, studied, and interpreted in order to make inferences about them that become statements. In traditional history, traces are documents, letters, diaries, inscriptions, records, and the like. But statues, monuments, coins, utensils, tools, furnishings, clothing—all artifacts, including buildings, settlements, and cultural landscapes, are similarly traces—as the new history has realized (Kammen 1980; LeGoff and Nora 1985). In that sense, as I have already suggested, they are closer to the natural sciences than to history.

The traces of material culture are very different from those of traditional history, that is, written records. They are therefore often better handled by methods of the natural sciences because these do not expect data to have self-evident meaning. The traces there have to be interpreted in ways very similar to those of material culture. The traces of material culture are no more biased or distorted than those of history: They are biased and distorted in *different ways*. The traces of material culture are in some ways easier to deal with *because the distortions are not deliberate*—as they often are in the case of written sources. Social-science-oriented study, such as EBS, ethnoarchaeology, or experimental archaeology is, in turn, different because it can *directly observe events*, in our case, social and behavioral data as well as material ones.

It follows that all these are complementary. As usual, the more data and kinds of data, the more approaches and methods, the better. Contemporary studies and experiments, historical documents of all sorts (travel, popular, official, descriptions, diaries, advertisements, drawings, photographs, and paintings) and the traces of material culture should all be used whenever possible. The type of study with which we are concerned is eclectic and considers all data and aspects of human activity that has relevance to understanding EBR. They are all traces from which inferences can be made so that statements can be made about events.

In many, if not most, sciences traces are also what one studies—particle decay tracks in film, bubble chambers, and the like in particle physics; marks on photographic plates of various sorts (visible, infrared, ultraviolet, x-ray, etc.) in various fields; radio tapes or whatever in astronomy from which inferences are made about events many billions of years in the past; photomicrographs in microbiology; spectrographic or chromatographic traces in chemistry; x-ray diffraction patterns in crystallography or early molecular biology, fossils in paleontology, x-rays or other data in medicine (cf. Abercrombie 1969) are all *traces* that need to be interpreted and inferences made about events (cf. Shapere 1984 on the chain of inferences used in studying solar neutrinos). Background radiation is a trace of the Big Bang—which is an event that happened 15 billion years ago—or whenever. The statement that it did is a fact. Thus the inferential process in science and history is very similar indeed.

Both also rely on analogy. In the case of studying the past, this is quite crucial because the data are in the present; but this is also the case in geology, paleontology, astronomy, and the like. Thus most models in science tend to be based on analogies (Gibbon 1984), and the role of analogy in all science is increasingly recognized (Boyd 1984; Hesse 1966; Leatherdale 1974; McMullin 1984; Ortony 1979; Schön 1969; etc.).

It is the controlled use of analogy that avoids many of the problems against which one is often cautioned and warned. But, as we have seen, because the use of analogy is inevitable, even to identify buildings, streets, courts, settlements, tables, or chairs, one can only try to use them as rigorously and explicitly as possible.

One thus begins with Step 2 of a three-step sequence, as much in history and archaeology as in the natural sciences (cf. King 1982, pp. 256ff.).

1. The event itself that is now past, either just past or a long time ago, and about which direct knowledge is not possible.
2. Traces of the event that persist into the present. These, if suitably interpreted,* can provide inferences and reveal much about the event.
3. Step 2 has the logical consequence that there must be someone to examine and interpret these traces and make inferences about the events.

Facts, then, are the statements about the inferences (or interpretations) of historians, archaeologists, or scientists from traces of events in their respective domains. It is important that facts only take on significance when linked into larger frameworks—theories. Witness the well-known caution expressed by Eddington that one should not trust experimental facts until confirmed by theory.† One gives meaning to facts through theory and then evaluates how useful the theories are. It is the combination and interaction of facts and frameworks (theories) that one tends to believe even though they often have a subjective component. This is much greater in some disciplines than others; it can be made explicit, and it is then minimized by open mutual criticism leading to intersubjective agreement. As a result, one can trust these inferential structures not only in science but also in history (e.g., McCullagh 1984).

It is often argued that every generation rewrites history—and even science. However, this is too strong a claim: There are limits to that. What these limits are is a hotly debated topic in the philosophy of science, for example whether science is cumulative, whether successive theories are commensurable, and so on. But even in history, certain disputes are limited, although they are more prevalent than in science and take longer to settle. In science, intersubjective agreement, consensuality, and closure tend to be more common and occur much more quickly. This is one reason science makes such a useful metaphor for knowing. Certain things are facts not because they are true beyond doubt (they are often not even that in science) but because there is sufficient agreement about their truth to "dry up controversy" (King 1982, p. 257)—even if controversy may resume later. Even in history there is often enough available evidence supporting certain facts, so that one believes, on good grounds, that future research would neither increase nor decrease confidence in them. There are historical statements "so strongly supported by evidence . . . that to deny them would be equivalent to making nonsense of large portions of history" (Gardiner 1952, p. 80). In the case of history, because the past is both fixed and real, one can pick out what is genuine historic reality by the way that it bears the same appearance in different kinds of historical narratives; this is equivalent to invariance under coordinate transformations in, for example, physics

---

*The meaning of "suitably interpreted" is the topic of this book.
†The same point has been made about meaning (Leach 1976), and culture has recently been defined as that framework which gives meaning to particulars (see discussion in Rapoport 1986c).

(Harré 1986, p. 257). Thus the more sources one uses and the more evidence, the more invariance or certainty and consequently the more confidence in any facts. Admittedly, these are more frequently at the level of reconstruction than explanation, of lower rather than higher level inferences, but even these latter tend to show convergence and closure if inferences are strong and well supported. It has thus long been beyond argument that between the eleventh and fourteenth centuries, many religious structures we call "Gothic" were erected in Europe, for example, France. One of these was Chartres. Recently views about how Chartres was built, how it came to be, have been greatly changed by new inferences from various traces, mainly in the building fabric itself (James 1979, 1981, 1982). These new inferences are being increasingly accepted because they are highly convincing. Generalizations are also being made about other comparable structures in this general area of France (James 1988). As a result certain higher level inferences about intentions, ideology, economics, institutions, even symbolism, are beginning to change. They will change even more as work is done, as new questions are asked, and as inferences are tested against new data. It is possible, indeed likely, that these inferences will then be accepted as facts—correct statements about events. A comparable argument can be made about Renaissance churches. That they existed no one has doubted for a long time. In the 1950s the explanation about their form was transformed by new inferences (Wittkower 1962). This set of inferences was so convincing that it was accepted widely and quickly—and, as far as I know, not questioned since.

The reliability of statements about events, that is, facts, in any given discipline depends on the level of general agreement about them in that discipline.* Assertions will stand up as facts if there is no good evidence to the contrary—although this can change. Because it is only specific, not general doubt that is relevant (Shapere 1984), it follows that in that respect historical facts are no different *in principle* to facts in science that are also based on intersubjective agreement (consensuality).

Like scientists, historians study diverse kinds of events. They go through similar steps starting with relevant traces. These depend on the definition of the subject matter and questions posed. Studies, replications, and evaluations lead to pattern recognition and statements made that are called facts. The total event is then reconstructed through a process of inference. In this respect biography, archaeology, and historical geology (or the other disciplines listed before and discussed in Chapter 4) are quite similar, in spite of the obvious differences among them.

I have already referred to the typical distinction made: History is idiographic, science nomothetic. The former deals with individual people, places, or events, the latter with classes. Of course, "classes" involves questions of classification and taxonomy that are indeed very important in science (Clarke 1978; Crease and Mann 1986; Dunnell 1971; Gould 1985; Johnson-Laird and Wason 1977; King 1982, pp. 90–120; Mayr 1982; Rapoport 1988b, in press; Rosch and Lloyd 1978; Sokal and Sneath 1963; Sokal 1977). It follows that "insofar as history deals with classes and generalizations it partakes of science" (King 1982, p. 258), and it increasingly indeed deals with both classes and generalizations (Kammen 1980; LeGoff and Nora 1985; see also Chapter 4).

*"Disciplines" in which there is no agreement and argument about basics and everything else continues are not disciplines.

On the other hand, to the extent that science deals with unique events it partakes of history—and it does indeed deal with unique instances as we have already seen and will discuss in Chapter 4. In fact, King (1982, p. 258) makes the intriguing suggestions that experiments can be seen as unique historical events, as unique as any in history. They are as unique as Michelangelo designing the Medici Chapel, certain architects designing churches during the Renaissance, and early Britons designing megalithic tombs (cf. Rapoport 1988a).

The difference between science and history, in this connection, lies not so much in the subject matter of the domain as in the questions asked and the purposes of studying traces of past events. In science, one is not interested in reconstructing an individual past historical event, such as an experiment, or in how the experimental animal or experimenter felt. The purpose is to generalize, and from these generalizations develop explanatory theory that can be applied in the future. The purpose is to construct generalization and explanatory theory, not reconstruct a particular event in the past. To the extent, therefore, that historical study has that goal (as in Chapter 2 I argue it does) it becomes scientific, even if, en route, it needs to reconstruct past events. What is thus essential is to define clearly and explicitly what one is trying to know from the traces. Questions must be explicitly and clearly posed before research can be done—a general point, germane to all research. Thus the redefinition of the purposes of research in Chapter 2 tends to lead to a scientific mode of historical study. Note that my purpose was to use it to develop more valid generalizations about EBS so that explanatory theory could be constructed. The purpose is to provide a disciplinary base for design, which would become more predictable. Note that in science the test of generalizations and theory—ultimately based on inference (which is an inductive leap, a reconstruction)—is that it works (Hacking 1983). This is, in fact, the point King (1982, pp. 258–259) makes about the outcome of Villemin's experiments on tuberculosis: That disease has now almost been eradicated.

Thus generalizations in science are constructs based on inferences from traces of past events. The interest is on abstraction of certain important features and attributes of the inferential reconstruction of events. *These become stepping stones toward explanatory theory.* The accumulation of descriptive or narrative facts is not useful; for example, lists of human behaviors even when organized by classes (e.g., Berelson and Steiner 1964) are no more useful than books of records, almanacs, or telephone books (at least for some purposes). Generalizations alone are also not enough (Willer and Willer 1973). One must take the next step of conceptual development: The goal is to develop explanatory theory. Because most history, no matter how narrative and traditional, is much more than a telephone book or list of records, it must, at the very least, certainly contain lower level generalizations. In most cases implicit higher level generalizations are to be found and, in most cases, implicit theory of various sorts (see Chapter 4). These need to be made explicit.

Note that the relation of history to science, of unique events as opposed to generalizations, has been a topic of discussion in a number of fields, including general history, and will be reviewed in Chapter 4. At this juncture a different point needs to be made.

Inferences about the past, as repeatedly pointed out, hinge critically on the assumption of continuity between the past and present. In science, this is *relatively* uncontroversial. Moreover, this assumption of continuity is also central to the idea of learning from the past, that is, of deriving generalizations leading to theory from which can be derived lessons applicable to the future, precedents. Issues of change, as opposed to continuity (which I tend to emphasize) and even of drastic breaks and discontinuities, whether in evolution, between humans and animals in human development, or in the physical sciences greatly complicated matters and have already been addressed briefly in Chapter 1 (cf. Eldredge and Tattersall 1982; Eldredge 1985; Gould 1985; Prigogine and Stengers 1984; Stanley 1981). Yet implicit in the very possibility of making any inferences at all about the past is the assumption that there is at least some uniformity in human behavior and also in material culture. Only thus can inferences be made about past behavior and only thus can formal features of built environments and material culture be related even to the most instrumental functions. Yet our interest is precisely in the more difficult latent functions, in cognitive and symbolic behaviors. In our discussion of southern African rock paintings, explicit arguments were made about the continuity between the past and present that alone made possible the use of ethnographies, travel accounts, and fieldwork (Lewis-Williams 1981, 1983; Vinnicombe 1976).

The central role of such uniformitarian assumptions, the fact that they present undeniable problems and that their use is frequently questioned and criticized make it imperative that they be discussed, particularly given my emphasis on the need to be explicit in one's arguments.

UNIFORMITARIAN ASSUMPTIONS

We have seen that any study of the past, particularly when moving from material culture to behavior, involves inference and that this is based on the controlled use of analogy. The use of analogy, in turn, inevitably involves making uniformitarian assumptions; some form of uniformitarianism is assumed. These need to be carefully and explicitly stated so that similarly explicit and logical series of inferential steps can be made.

Thus the study of the past and arguments that it has anything to teach us need to justify the use of uniformitarian assumptions.

As is well known, the modern, explicit discussion of this issue stems from that in historical geology by Lyall who, in turn, influenced Darwin. It is now suggested that such assumptions are of two kinds (Binford 1983a, pp. 418–420; Salmon 1982, pp. 80–81):

1. *Methodological or epistemological uniformitarianism,* assumptions about scientific procedure, of which there are two:
   a. That the world is governed by natural laws that are invariant in space and time and essential for any analysis of the past, any generalizations, any inferences, and even any reconstructions. At least some commonalities must be assumed.*

---

*Even traditional history (à la Collingwood) makes an implicit assumption about being able to think like people in the past.

    b. That processes and mechanisms now operating can be used to explain the past: One assumes uniformity or process over time. This is essential because, we have seen, one can only observe present processes directly. This is, however, more controversial and may need to be qualified or argued.

2. *Substantive or ontological uniformitarianism,* that is, assumptions about the world.

    a. That change has been largely uniform in rate—slow, gradual, and steady rather than cataclysmic or catastrophic. This is essentially an *empirical* question that has largely been sustained by research in some fields but is being hotly debated in others; for example, the "punctuated equilibrium" debate—or that about catastrophic extinctions, to be reviewed in Chapter 4. Thus it seems that one can allow for sudden breaks and discontinuities if there is evidence for them. In archaeology, for example, one can allow for major changes and shifts as from nomadism to sedentism, tribal societies to state societies (Renfrew 1984), and others. This is also found in the other historical sciences (e.g., cosmology and physics from the Big Bang on). This is then an *open* and *empirical* question.

    b. That any domain has remained unchanged through time or, at least, has been fundamentally the same since the beginning. This has been demonstrated to be quite false in geology, biology, climatology, physics, cosmology, and so on; it clearly is also false regarding built environments (hence the problem with Naroll's [1970] floor area norms; e.g., Binford 1983a, p. 82; cf. Kent 1984; Kramer 1982; Rapoport 1980c, 1986c). It is an open question regarding human characteristics about which there is also much debate (see later discussion). There are thus questions to what extent uniformities can be traced between humans and extinct hominids; to what extent these apply to rapidly evolving *Homo sapiens* and so on.

Although debate continues, there is certainly much suggestive evidence for such continuity. It is, however, an important issue, potentially knowable, and therefore to be part of research in EBS and on history and precedent in environmental design (cf. von Schilcher and Tennant 1984, p. 57). In an example, Betzig (1986) explicitly applies a combination of biological theory and hypotheses and ethnographic data, often dated and, in any case, of past societies, to the study of history in the broadest sense. In effect this is a sociobiological analysis of history that combines biology, history, and ethnography.

It is thus possible, by assuming methodological uniformitarianism, to *discover* whether the two assumptions about substantive uniformitarianism are correct or false. These become *empirical questions.* The use of methodological assumptions is, however, inescapable. *Some* uniformitarian assumptions must be made if *any* understanding of the past, or learning from it, is to be possible. These act as "intellectual anchors" even when adjustments to them must be made (cf. Tversky and Kahneman 1974). They provide the known from which one can move to the unknown characteristics of the past. It is an example of moving from the better known to the less known, from the more securely established to the less securely established (Boulding 1980). One moves from parts of the domain where inference is more direct and easier to those where it is less direct and more difficult. This also implies that there may be areas of greater and lesser uniformity.

Among critical uniformitarian assumptions in the present case are those about the species, about "human nature." This assumes and needs to make explicit, that the anatomical, physiological, neurological, perceptual, cognitive, behavioral, symbolic, and other human characteristics that apply to humans today also operated in ancient humans, that at the level of process they had the same abilities, needs, desires; that they made choices, marked settings, defined the sacred; that they were goal oriented. This could be interpreted as a form of "goal-directed" theory (Nicholson 1983). One can than argue that people always have built in order to achieve a particular environmental quality profile that is congruent with their activity systems (including latent aspects), life-style, schemata, images, values, ideals, perceived needs, and, ultimately, culture. Because we know and accept that cultures vary, this in no way complicates the matter. If people have ideas, schemata, symbols, and the like, which are cognitive templates for material culture, which is an expression of those (Rapoport 1976, 1980a, 1986c, etc.)— then traces of past built environments are congealed information (Clarke 1978). One can also argue that people disposed of a repertoire of means with which to achieve their intentions. The specifics of all those and the resultant environmental quality profile are highly variable; but because the *process* is constant, it can be understood and the specifics reconstructed and inferred. This is equivalent to the argument (Lopreato 1984, p. 176) that if one is to be scientific in social science, it is most fruitful to explain specifics (in place, time, or whatever):

1. By reference to fundamental processes that are universal in time and place, as well as with the aid of historically and spatially particular circumstances. This provides specifics with which the general process works.
2. By taking into account any evolutionary constancies.

This not only makes social science more scientific, it also makes social science historical and comparative in its approach—a highly desirable outcome.

An example of Lopreato's Point 1 above is King's (1982) study of the history of medical thought, to which I have already referred. People's intentions (to cure) and abilities were always the same (an example of a goal-directed theory [cf. Nicholson 1983] mentioned before). What differed were the means available—conceptual, technological, observational, and so on. As these changed, and improved, over time (and this is, in itself, a pattern identified through studying history!), medicine changed accordingly. A major impact was also the introduction of new disciplines. This not only introduced new data resulting from new scientific research, for example, from physiology, pathology, chemistry, physics, molecular biology, genetics, and so on. It also introduced new concepts, one early example being Sydenham's application of botanical taxonomy to the study of diseases and their classification (nosology) (King 1982, p. 11)—a typical example of cross-fertilization among disciplines as well as an argument by analogy.

One can also argue that these different skills and means were applied to comparable sets of constraints and problems that have existed from prehistory through historical periods and those groups studied by anthropologists. These constraints can be listed (see Chapter 4), and their solutions are incorporated into built environments and material culture. Given that people in the past have had the same abilities and same ways of coping with problems, although the problems may have differed from case to case, as did the means available and what was deemed to be acceptable, adequate, or satisfactory, the outcomes or solutions will vary, but there will be uniformity at the level of goals and processes; at the very least all of them are *comprehensible*.

More correctly, therefore, one assumes that the cultural variability among ancient humans was similar to ours as already discussed (cf. Rapoport 1986c). Thus because it can be shown, as I have argued for years, that the use of space is culturally variable today (cf. Kent 1984), one can infer that it *always* was. This then throws doubt on assumptions that the use of space is uniform and constant for all humans (as Naroll 1970 assumes and as, e.g., Binford 1983a, p. 420, argues). If one can show that built environments are much more influenced by latent than by instrumental functions, then utilitarian interpretations of past environments become suspect.

It is an urgent task for EBS to pursue research into the past and present along as many lines as needed to discover and elaborate a list of those spheres, realms, or situations where constancy and uniformity apply and where not, to what extent, and so forth. This is, of course, one of the purposes of this book.

Such research could then provide empirical evidence for or against, and clarify the view, that underlying the striking variability of built environments there could be a measure of panhuman constancy. If so, that could be important because the degree of variability would then be greatly reduced. The constancy can be identified with an evolutionary baseline, related both to the physical and social environment in which *Homo sapiens* evolved. In other words, there may be certain critical requirements in both the physical and social aspects of built environments that all people require, and the ignoring of which may lead to phylogenetic maladaptation as in other species (see Figure 3.12).

The existence of such an evolutionary baseline is, as already pointed out, an aspect of uniformitarianism open to empirical testing. It involves empirical claims about uniformities between past and present. It has been discussed in many different forms all related to a more empirically based notion of "human nature" and its evolution. Even

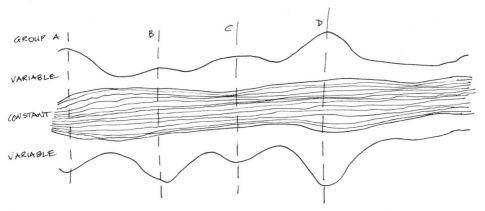

Figure 3.12. Evolutionary baselines, constancy, and variability (redrawn from Rapoport 1975b, Fig. 1, p. 146).

the possession of culture, which makes humans human, has been discussed in such terms; so has the constancy of human responses to and perceptions of the world in which they evolved, the tendency to attach meaning, to categorize and so forth, the tendency to reinvent the same kinds of things; all bearing on the continuity of human behavior in its broadest sense (Barash 1979; Betzig 1986; Bonner 1980; Boyden 1974, 1979; Boyden and Millar 1978; Boyden *et al.* 1981; Eibl-Eibesfeldt 1972; Fox 1970, 1980; Geist 1978; Hamburg 1975; Konner 1982; Lopreato 1984; Lumsden and Wilson 1981, 1983; Maxwell 1984; Rossi 1977; Ruse 1979, 1986; Swanson 1983; Tiger 1969; Tiger and Fox 1971; Tiger and Shepher 1975; von Schilcher and Tennant 1984; Wilson 1975, 1978; among many others). There is considerable evidence beginning to accumulate in favor of such uniformitarian claims, although they are clearly not settled; they can, however, be evaluated.

It also seems likely that there will be less variability in some areas of life than others, for example, in neurology and anatomy than in values and schemata, in instrumental activities than in symbolic ones. We have already seen that variability goes up as we move from instrumental activities to latent ones, from the concrete object to the symbolic object, from technomic to ideotechnic functions. This needs to be specified before inferences are made (see the case study). It is also very possible that, specifically regarding physical forms and built environments, and certain social and economic forms, institutions, and the like, there may be some that are essentially unchanged, some that are somewhat changed and others yet that are totally new (e.g., Laslett and Wall 1972 and Chapter 4). I have already referred to these categories in settlements— neighborhoods relatively unchanged, metropolis and megalopolis totally new; pedestrian paths unchanged, freeways—new, but based on perceptually constant human needs.

Regarding our particular domain, built environments, there may also be scale effects. One can suggest that at smaller scales, particularly the mesoscale, there seem to be greater constancies than in larger environments (Rapoport 1977, 1981c, 1986b, etc.). One can also suggest that those specific elements that change or remain constant may themselves vary and are a matter of empirical investigation (e.g., Rapoport 1968a, 1969a, 1983a). All of these are uniformitarian assumptions about the world to be empirically tested using methodological uniformitarianism.

Also, as already suggested, there may be much more invariance and uniformity at the level of *principles* or *processes* (or mechanisms) than regarding specifics or specific outcomes or results. Thus, for example, Fox (1980, p. 7) argues that "cultural universals are at the level of process not outcome" (cf. Lopreato 1984, p. 176 discussed before). It follows that any analysis frequently needs to be at higher levels of abstraction, going beyond shape, materials, and so on (Rapoport 1977, 1980d, 1982b, 1986b).

The argument that the reconstruction of past lifeways and the study of process are often related (Deagan 1982) also supports the point being made that process may well be more constant than outcomes. The point that universals may be regarding process rather than outcome is also supported by Jochim (1976), who proposes a model based on *choice*—another example of goal-oriented theory. He argues that choice is a universal—what one chooses varies. This is, of course, central in my own work (Rapoport 1977, 1983c, 1985b,c,d, 1986c, 1988b, 1989a; cf. Kearney 1972; Wilk 1984, 1989, who also argue for the importance of choice).

The study of process and the idea of repertoires as aspects of constancy are related to uniformitarianism. The emphasis on them is based on the notion that if certain things happen over and over again, throughout history and in many different cultures, they are certain to be important; they may be the "norm"—any may reflect or represent human nature (e.g., Gordon 1978). For example, one constant already discussed may be the human predilection to mark the sacred, however different the *means* that are employed (Rapoport 1968b, 1982a,e). In fact, a possible repertoire can be suggested (Rapoport 1982a) as already discussed. By studying a large body of examples, cross-culturally and over time, a large and diverse body of evidence, the widest range of independent lines of evidence and multiple methods, one might derive a hypothesis: There is a general strategy used, that of making the sacred noticeably different, that is, the EBS concept of noticeable differences becomes applicable. This can then be tested against other samples of data, for example an extreme case such as Australian Aborigines (Rapoport 1975a). One might also begin to test higher level uniformities. One could ask whether all sacred spaces, however defined, have any constancies at the level of their attributes, character, ambience, or whatever—as already discussed briefly in Chapter 2.

Not only history and archaeology but all modern science—whether biological, geological, medical, physical, or anthropological—is predicated, explicitly or implicitly, on the idea that the various laws of nature and various processes now in operation—or at least some of them—have operated in the past and will continue to do so in the future. Certain rates and intensities may vary, but the principles governing the actions are fixed. This is both clearer, less problematic—and certainly less controversial—in the case of physical, chemical, and geological processes (at least since the universe and planet reached their present forms) than in the case of human ones. However, as we have already seen, some minimal level of uniformitarianism must be accepted if the study of the past is to be meaningful or even possible: At the very least we must assume that *we* can understand what *they* were doing or building in the past.

One can also raise other questions, for example, about what kinds of uniformitarianism apply. It is quite clear that certain values, social arrangements, economic forms, technological capabilities, and the like have changed dramatically, radically, and abruptly. It is also clear that human ideas and ideals can change dramatically. Evolutionary change in humans is in the realm of ideas. But that implies some continuity by which we can assert that claim. Also, there may be invariance in the role that values, social arrangements, economic forms, technological capabilities, and the like play in given domains or areas of interest. There may be also certain limited repertoires regarding values, social arrangements, and economic forms, the full range of which may be observed by using large bodies of evidence (cf. Dalton 1981 to be discussed later). Technological capabilities—in comparative terms—may also be much more invariant than one thinks and, in fact, it may be that human uniformities need to be used to *evaluate* technological innovation (Rapoport 1982c).

In all this, there is still a direct link with the first true humans and, possibly, even hominids, the first dwellings (Rapoport 1979a), first settlements (Rapoport 1979b) from the first city to present-day conurbation, certainly at the mesoscale (Rapoport 1977, 1986b). In each case, humans are projecting their ideas, ideals, schemata, and so forth into the environment; moreover, these environments rarely, if ever, start *de novo*. Typically, existing forms are used, combined differently and manipulated analogously to biological evolution (Gould 1980; Jacob 1982); this was certainly the case with early cars, electric lights, trains, and so on. Research is needed on buildings, neighborhoods (but see Rapoport 1977; Athens Center of Ekistics 1980), and settlements. I have also argued before that process may be constant even though outcomes and results may vary. Also environments are always used by people—and we have already seen that at least some arguments exist that, in principle, humans are the same anatomically, physiologically, neurologically, and perceptually as they have always been. Hence, all human beings have this degree of uniformity (cf. Leach in Spriggs 1977). The question is whether certain patterns and categories of thought are also uniform and *whether one can infer, and thus reconstruct, the categories of thought that dominated the creators of past environments. This can be done because they were people like us.* One can also specify the differences, such as the development of writing and other symbolic systems (e.g., Goody 1977), and trace their implications for built environments (Rapoport 1986c, 1988a). We can be sure that they had *categories* of thought. Again, we find a common process with different specifics, although even those, some recent work suggests, may be much more constant, at least in some respects, than had been thought, as already discussed (cf. Rapoport 1986c).

I have already suggested that a most significant characteristic of built environments that needs to be explained is their extraordinary variety that, however, seems to be the result of a much more limited set of needs and activities (cf. Rapoport 1986c, 1989a). Thus the processes, the ways in which values, social arrangements, economic forms, technological means available, and the like link with built form may show regularities.

In terms of the three questions of EBS, one could suggest that the first shows great continuity as discussed. The effects of environment on people is a combination of constant and variable effects, whereas the mechanisms vary from extremely constant and uniform at the level of perception, through those, like meaning and the role of schemata, where their centrality is constant but specifics may vary, to preference and choice, which is the most variable although even there some suggestions of constancy can be found (e.g. Purcell 1984, 1986; Smets 1973).

Note, however, that these once again become empirical questions. They become hypotheses that can only be tested against the broad body of evidence urged in this book: long time series, diverse cultures, and situations. The stating of these hypotheses explicitly and their testing against such data are, in themselves, part of the central argument of this chapter and this book.

There is some ambiguity between regularities and patterns as opposed to uniformitarianism. Such patterns and regularities can be understood as *high-level uniformities.* "We believe that there are laws that govern human development and that these laws can only be discovered by the systematic testing of hypotheses within and across cultures" (Munroe, Munroe, and Whiting 1979, p. xiii) and, in our case, I would add, over long time spans and looking at the full range of built environments. But this assumption itself is not only part of the approach but becomes a hypothesis to be tested.

This assumption is, of course, as basic in historical research as it is in cross-cultural research. The comparative aspect, which requires these large and varied bodies of evidence, becomes central and most significant. Interestingly, although it is and has been very basic in anthropology, it is less common in archaeology (but see Kent 1984) and very new indeed in history (see Chapter 4). It is also of increasing importance in paleontology, astronomy, neuroscience, and so on—all of which emphasize comparisons of large and varied bodies of evidence.

The uniformitarian issue as it applies to making inferences about the past and even more to learning from the past is the following. Our own culture is clearly different from many now existing and even more different from the prehistoric past. The question then becomes: How can the two possibly be related? As already pointed out, they can be so related if they are used together to try to answer "why" questions about the links between built environments and social and behavioral variables. As long as contextual variables are researched and explicitly specified, and comparisons made among the largest and most varied bodies of evidence, then one can say something about the ways in which people and environments interact or discover the range of such interactions. In this way insights relevant to both the reconstruction of very different pasts and lessons for the present (which in design means the future) can be obtained. Again, the uniformity is at the level of constant processes, even though these may be regarding product. For example, it has been suggested that the way vernacular design is produced is now of little use because it cannot be produced in this way in the future (Doumanis and Doumanis 1975; Sordinas 1976) because we have moved from selectionism to instructionism. This is an open question, but it is, however, the case that vernacular environments tend to be liked by people and to evoke certain reactions. One can then ask the question "why" and by looking at a large variety of such environments, discern patterns of certain perceptual qualities at the product level as already discussed in Chapter 2. Recall also that the lessons (or precedents) are at some level of abstraction (see Figure 2.1).

Any study of the past must use the present as a model. At the same time, the present and future can best be studied and understood by reference to the past and using evidence from the past. This may possibly be even more relevant to the future—it can stop us violating long-standing "universal" processes and patterns related to human nature and discovered by studying the past (e.g., Rossi 1977; Tiger and Shepher 1975). One must assume that the past is dead and only knowable through the present that itself needs to be better known (hence the need for EBS as a way of studying the present adequately and properly). One must also assume that accurate knowledge of the past is essential to understanding the present—the past must be studied in order to be able to generalize about EBR now. As a result, uniformitarian assumptions cannot be avoided.

Because the use of some form of current data in studying the past is inevitable, one must make the best of it and make the best use of both types of data—current and past. Recall that general doubt is never the issue—only specific doubt (Shapere 1984). It is, therefore, futile to argue about using uniformitarian assumptions in principle, as it is about whether data from the past, being incomplete and biased, are adequate for making inferences. The solution is not to argue the general point to but argue the specifics of any case—how, in a given problem, best to use both kinds of data. This use of analogy is both unavoidable and essential in science as a whole, as we have already seen. In all cases, it is important that any implicit assumptions, analogies, connections, and so on be made explicit. One can then ask the critical question about what are adequate criteria (and why) and also make explicit the chain of argument and reasoning. This can help settle specific doubt.

Note a problem with inference not yet discussed that bears on the uniformitarian problem. There are clearly cases of groups with very complex behaviors whose material culture remains (traces) would never reveal that (for example, South American Indian groups; Australian Aborigines—cf. Rapoport 1969a). Note, however, that *we only know that from the present*—in fact, that is the *only* way we can know of that possibility; we cannot possibly know that from the past itself. It is ethnographic work that has shown that this possibility in fact occurs. What we have here is, in Dalton's (1981) terms, a genus: It is one of the range of possibilities, a type of human society. There is thus, again, an essential link between studies done at the present comparatively for diverse environments, cultures, and the like revealing possible types and ranges of EBR and the equivalent study of the past.

In fact, it is important to realize that the use of the broadest and most varied evidence as well as of approaches and methods is in itself helpful in coping with the problem of uniformitarianism. At the same time, as the effort of studying and using the past makes it necessary to make these assumptions, it makes their use less problematic. The use of the largest range of both contemporary and past data greatly increases the sample of whatever one is studying for purposes of comparison. Because one must link actual behavior to past environments, the discussion cannot be general and polemical; it needs to be specific and analytical. Rather than arguing about relevance in general, it is more useful to discover mutual relevance or to assume specific forms of relevance and to test them. This is best done by posing questions that require answers from both past and contemporary data, as two different but complementary sources of information about the relation between human behavior and the built environment. The fact that some questions may never be answered does not mean that one should not ask others; in any case, one cannot know which are answerable and which are not until one has tried—repeatedly. The past should not be used as background, nor contemporary material in straight and simple, uncontrolled, analogy. Rather, questions should be so formulated that answers are needed from both kinds of evidence. By using such joint bodies of data and by involving varied disciplines, methods, and approaches that share interest, problems, and questions, specific doubts can be resolved. The question is what approaches, concepts, and methods can make the analysis of sets of data relevant and appropriate to the study of EBR independent of particular space and time locus or particular disciplines.

A particular question is posed. One then examines how given cultures operate today to produce and use cultural landscapes, built environments, space, or semifixed elements. One also looks at a large variety of material data from the past—historical, ethnohistoric, archaeological, and others. One uses all these together to try and answer the question: *The past and present bear together on that question.*

We have seen that material data of the past are in some ways more like the data of natural science than they are like those of history. This is generally ignored. They are also different to natural science—a difference greatly emphasized: They reflect human conception, cognition, perception, social usages, and so on: Because material culture data reflect culture, *meaning is central* (Rapoport 1982a, 1986c, 1988a, 1989a). The question is then raised whether this requires particular epistemologies. That issue is at the heart of arguments that scientific approaches are inappropriate (see Hodder 1986 as just one example). *Material objects must be seen as concrete expressions and embodiments of human thoughts, images, and schemata,* that is, *as congealed information.* They can only be decoded if there is some uniformity at the level of these human processes as already discussed. Also, once again, there is an essential link between studies done today, comparatively for diverse cultures, environments, and so on and for equivalent study of the fullest possible record of the past. One additional reason for this type of study is to become aware of and establish the extent of environmental diversity that, by analogy, parallels species diversity—as already discussed.

In making uniformitarian assumptions, one is often cautioned not to take it "too far" (e.g., DeBoer and Lathrop 1979). The question is, of course: What is too far? What is the permissible extent? How can one know and judge the permissible limits? At the same time one is also told that one *must* make such assumptions, that one cannot give up and only describe or measure built environments and other artifacts (although, as we have seen, even identifying and describing artifacts involves analogy and hence some measure of uniformitarianism). It appears that the extent of permissible uniformitarianism is a matter of judgment, of reasoned argument, and explicit statement. "It is the familiar quandary between significant research based on possibly faulty method or using sound methods for work which is trivial" (DeBoer and Lathrop 1979, p. 103; cf. Crease and Mann 1986 on theoreticians and their wild ideas!).

The puzzle of uniformitarianism is that although the past can only be understood through a knowledge of the present, the past probably differed in many significant ways from the present (Watson 1979). Valid critical questions have been raised whether present-day behavior is relevant to the distant past, whether one can assume that contemporary hunter–gatherers are similar to early humans. This can only be answered in ways very similar to those suggested to the related problem of analogy already discussed. One makes as many and as varied as possible hypotheses and tests them in as many ways as possible. Present-day and past data must be used concurrently as a single body of data against which to test hypotheses that can be derived either from studying the past or the present. The former can be specifically tested against the present, the latter against the past; both can be tested against the total evidence. These hypotheses should be derived from settings sharing as many attributes as possible—the more continuity the better (as in the case of analogy—see Figure 3.11). For example, villages and cities in the Middle East were, until very recently, very like their equivalents thousands of years ago (Fagan 1983, pp. 136, 190, 193; cf. Kramer 1979, 1982; my comments on Jansen and Harappa [Jansen 1980] and my discussion of ethnoarchaeology, both in Chapter 5). Even today many continuities exist—and the differences and changes can be identified from the pattern (Rapoport 1983a). As in the case of any analogy, it is necessary to specify explicitly the attributes and characteristics that one believes makes the situations alike. Although one can argue that different situations are too different to be compared, they can be grouped. One can then say that some things are more alike than others, and the differences among things in a cluster become trivial compared to the similarities (cf. Dalton 1981).

All such arguments hinge essentially on analogy and hence on some form of ethnographic parallel. Even though I have argued that general doubts about such parallels should not inhibit their use, only specific doubts, arguments have been proposed against even general doubts. It has been suggested that for such doubts to have merit, an entire *set of societies* (a genus) would have had to exist that were markedly different from any single society known either to historians or to anthropologists (Dalton 1981). This is deemed to be implausible particularly if one adopts Clarke's (1978) suggestion for the use of polythetic classifications (cf. Rapoport 1988b, in press). Using such classification, any known society or community can be assigned to a set, group, or cluster containing other members. Dalton (1981) then asks how one can be certain that there are no "extinct societies" that have no equivalent in the ethnographic record. This of course, depends on what one means by "extinct society." Populations have disappeared, and other environmental, social, political, and religious, and sociocultural systems have been either lost or radically transformed. In each case, he argues, this has been an individual or a species—*not a genus or genus cluster*—a set of societies that have more in common than those in other such clusters classified polythetically. Examples would be hunting bands, lineage, clan, tribe, state, and so on. Each society that has disappeared had *genus mates* in the historical record, the *core attributes* (cf. Rapoport 1979c, 1983a) of which were sufficiently similar to be clustered together with those that have disappeared or been radically transformed. In fact, *all* early historical, let alone prehistoric, societies are extinct. It is not enough to say that a species is extinct (although even that can be studied, as in paleontology) because there are still genus mates, the basic structures of which are enough to reconstruct the core attributes from whatever partial evidence is available—but only if one knows that what one is attempting thus to reconstruct is a member of a particular genus. (Note once again that for this one needs to use the widest and most varied body of evidence possible.)

It is thus the case that Dalton's (1981) concept of the *genus set* is of central importance to the argument that one can infer social, behavioral, cognitive, and symbolic behaviors from material objects if one can use historical and ethnographic analogies; one can identify the genus set to which any society belongs (p. 42). If models of such genus sets can be shown to have explanatory power and one can give convincing reasons to justify parallels and analogs, then they can be used because it is unlikely that a whole genus is missing or had never existed.

Note that this depends, once again, on both a wide and varied body of evidence and on the use of comparative work. Researchers (historians, archaeologists, anthropologists, and in EBS) need to invent conceptual models and pose questions that will enable them to extract what is important and not obvious from the diversity of evidence, that is, patterns. One needs to be able to justify that these patterns are not spurious and to give persuasive reasons to explain *explicitly* why what is extracted is important. That cannot be done either without classification and grouping or without constructing models (Dalton 1981, p. 43).

In order to understand the relation between people and material culture, models must be constructed that allow the use of the full range of evidence. In the case of evidence from the past, which is the primary concern of this book, one can try to generate models relating human behavior in its broadest sense to the variability of built environments and material culture over the widest range. This range encompasses the whole environment, all environments, all cultures, and the full-time span from existing societies where both behavior and material variability can be tested to past traces of material variability. All such models and their testing involve inference, analogy, and, ultimately, some form of uniformitarian assumptions.

CONCLUSION

In concluding this chapter, I will summarize the answer to the question, what history? This will be done in a series of attributes or themes that characterize the approach. These will be found implicitly in the supporting argument in Chapter 4 and will be used (although in a more "collapsed" form with fewer categories) explicitly to structure the supporting argument in Chapter 5, in order to show how closely current archaeology fits my proposals.

1. A conscious application of issues and developments from the philosophy of science (theory, models, inference, explanation, etc.) as part of an effort to become a science.
2. An emphasis on an *explicit* discussion of ontological, epistemological, and other philosophical issues. Both are intended to help develop a philosophical base. This is a crucial step but will not be pursued or developed here but in a book on theory. Here I just wish to draw attention to the importance and centrality of this issue.
3. An explicit definition (or redefinition) of the domain; this can be seen as related to 1 and 2 as a concern with ontological issues.
4. A clear definition/statement of the questions to be asked of the materials from the past, whether for the purposes of reconstructing that past or in generalizing and developing more general theory.
5. The use of reasoned argument and a concern with logic, rigor, and making all arguments and their steps as explicit as possible.
6. An attempt to be as objective as possible. As part of that, the unequivocal, explicit, and careful statement of presuppositions, assumptions, criteria, and so forth (e.g., uniformitarianism). Careful definition of terms, isolation of useful conceptual entities, and so on.
7. A reliance on explicit, rigorous, mutual criticism, based on public (hence reliable) knowledge; the use of previous work; cumulativeness (cf. 18).
8. A commitment to an empirical content of any generalizations, inferences, hypotheses, models, and the like. Hence, a commitment to, and concern with, empirical data and empirical testing.
9. Careful attention to the nature and availability of data, their completeness, effects of time, and differential preservation; sampling. Efficient and rigorous methods of identification and collection of data, description, measurement, hence an emphasis on taxonomy and classification and use of such data.
10. Reliance on multiple bodies of evidence, both within material culture and the use of oral traditions, ethnography, ethnohistory; the use of contemporary data to be used in controlled analogy.
11. An emphasis on methodological sophistication and use of multiple methods.
12. The maximum possible connection with other disciplines (and their theories, data, and methods) that share similar or related concerns, that is, inter-disciplinary and multidisciplinary approaches. Points 6 through 12 are all for the purpose of identifying regularities—a search for pattern.
13. An emphasis on attempts to generalize, to seek explanation, and to develop

theory; a conscious and explicit commitment to the development of concepts and theory, as the only way to link data from the past and the present and to derive lessons applicable to design.

14. A careful and controlled use of analogies.

15. The making of predictions with indications of how they are to be tested.

16. The development of careful operational linkages between concepts and data (i.e., identification of *indicators* for theoretical constructs); the explicit building of models to link theory with empirical evidence.

17. The testing of such models in a variety of ways, including simulation.

18. The development of a sense of continuity of the work of all members of the discipline and related disciplines. Hence, an emphasis on the cumulative nature of the field; the use of others' work to do one's own and, therefore, a critical knowledge of the various literatures, new developments, and so forth.

19. As a result of all these and others, the careful, sophisticated and hence convincing (if, like in all sciences, frequently provisional) process of inference about the past. These inferences become a tested rather than untested "leap of faith" and never mere assertion.

Part II

The Supporting Arguments

Chapter 4

Supporting Argument 1

INTRODUCTION

I have developed the approach discussed in the first three chapters over a number of years and fairly independently. However, this position is rather controversial and very much a minority position not only with regard to architectural history but even in EBS, where there is much and increasing criticism of "positivistic" (read—*scientific*) approaches, analysis, rationality, and objectivity. This is also the case in the social sciences (e.g., Brown and Lyman 1978; Saarinen, Seamon, and Sell, 1984; cf. Rapoport 1985a; Argyris, Putnam, and Smith 1985; Reason and Rowan 1981; among many others).

Without discussing these matters, in this book it seemed useful to see whether one could find similar views or parallel developments in related disciplines. In other words, I was looking for what one might call supporting arguments.

There was another reason for this search. My conclusions seemed so self-evident that I could not believe that they had not been reached elsewhere before. Moreover, if they had been, then one could take advantage of what they had concluded and what they had done rather than to reinvent the wheel *de novo*. One of the characteristics of real disciplines is their cumulativeness—an absence of what Sorokin called the "New Columbus Syndrome." Consequently it is important to look at any precedents that might be available in other disciplines addressing similar topics. Not only would that provide relevant ideas, insights, concepts, approaches, and methods that would not need to be invented; the network of theories works across disciplines also and helps limit possible inferences, hypotheses, and so on. I will, therefore, review a number of fields the relationship of which to my argument is itself part of that argument.

Chapters 1, 2, and 3 and the case study were in various drafts before I reviewed some of the literature in art, social science, history, the historical sciences, and archaeology. I felt that these disciplines were "historical" in the sense of this book but had not been related among themselves nor to the history of the built environment. My intentions were to discover, first, whether similar or analogous arguments or positions were to be found in disciplines concerned with making inferences about the past of their given domain. Second, if so, whether there would be convergence among the various fields about the characteristics necessary for any useful study of the past.

I quickly found a great deal of support for my position: Shifts similar to those I was advocating could be found in other disciplines. There was convergence even if these positions sometimes represented an equally minority position. The question is not whether this position is generally accepted, or whether there is opposition to it or debate about it. It is rather to establish that this position *exists* even as an alternative. Once this is established, the concern is to identify the rationale for the proposed changes and the characteristics and implications of this approach.

The intention of the discussion that follows is not to make readers familiar with the literature of the different fields; the review is neither systematic nor even-handed enough, nor complete. Neither is it to take positions on any of the substantive issues or questions. It is rather to introduce the reader to what are essentially other comparable forms of studying the past or relating it to the present, which are possibly unfamiliar in themselves, likely unfamiliar in their mutual interrelationships, and almost certainly unfamiliar in relation to the history of the built environment. The intention is to find supporting arguments for my proposed approach to historical data by emphasizing that some precedents do exist in a number of other fields.

I asked the question of how other disciplines that face the problem of analyzing data of the past have approached this rather difficult task in ways different from those based on the metaphor of traditional art history. By posing the question in this way, it became possible to relate history, *as the study of the past of ANY domain,* to a large number of disciplines. Some of these are obvious; others, however, are far from obvious and are not usually regarded as relevant to my topic. It is precisely by also considering those latter that major support would come if they are found to bear directly on my argument. This is, of course, because their diversity becomes much greater.

More generally, when reexamining basic issues, one's conclusions tend to be made more convincing if, by systematically looking at the literature of other disciplines, one finds comparable, similar, analogous, or identical arguments that have been arrived at independently, in different domains, from different starting points and using different concepts. The greater the range of disciplines and the greater the agreement or convergence, the greater the confidence in these supporting arguments.

In Chapter 3 I dealt briefly with the supporting argument from the history of art. In this chapter I review, in significantly more detail, the supporting argument from a number of other fields. In the next chapter, I consider archaeology as the discipline dealing with the domain closest to that of this book. By presenting these one by one, I am allowing the reader to develop the sensation I had of how it all adds up.

Postponing the discussion of archaeology and prehistory to Chapter 5, this chapter will review supporting arguments from three broad areas. First, *the social sciences,* including sociology, anthropology, and historical geography. Second, *history proper,* especially historiography and philosophy of history. Third, *the "historical sciences,"* a number of sciences that deal with "unique" events such as our universe, our solar system, our planet, life on earth, and so on, in which explanation is often more important than prediction and, in any case, the data used are essentially historical.

THE SUPPORTING ARGUMENT FROM THE SOCIAL SCIENCES*

There is a body of writing in the social sciences that deals with the past in the sense of emphasizing the need for the use of historical data. In social science these arguments are essentially analogous to my argument about EBS, that is, the rationale for the study of the past is its value for the field in question. There is also discussion of the converse, of the value of social science for the study of the past (which one also finds in history). There is thus convergence in the emphasis on the "mutual benefits" of looking at both past and present, and also in the implications in terms of approaches of such closer links between social science and history, and this even applies to books that are explicitly antiscience (e.g., Gergen and Gergen 1984). Generally, the social sciences, like EBS, have tended to ignore the temporal dimension even in the short term, that is, at the synchronic level. As early as 1976 I argued for the importance of dealing with *time* as an essential component of the environment. Even more underemphasized has been the long-term temporal dimension: Diachronic analysis related to history and historical evidence. This is changing, however (see references in Rapoport 1986c). A brief, far from complete, review of a small set of such social science argument seems useful.

One emerging theme closely related to my concern with the nature of the domain is a growing interest in new types of data. Among these are artifacts, travel descriptions, catalogues, advertisements, novels, films, TV, and so on (Rapoport 1970a, 1973a, 1980a, 1982a, 1985b,c,d). Examples come from political science: The use of city halls to study politics (e.g., Goodsell 1984, 1988), the use of folklore to study vernacular design (Bunkśe 1978), the use of artifacts in sociology and anthropology (Bourdieu 1968; Csikszentmihalyi and Rochberg-Halton 1981; Douglas and Isherwood 1979; see also references in Rapoport 1982a). More generally, the domain definition is changing as artifacts and other aspects of built environments become part of the evidence (e.g., Ginsberg 1975; Jackson 1985; Kamau 1978–1979; King 1980; Moore 1981; etc.). This emphasis on both fixed and semifixed elements of the built environment also leads to inevitable methodological changes; for example, inference from material culture or the use of new types of written materials such as novels, advertisements, and catalogues. Also, in dealing with material culture from the past, one cannot use survey research, interviews, questionnaires, observation, or many other traditional social science methods.

Among changes in the nature of the domain is a growing interest in oral data (Harms 1979; Thompson 1978; Vansina 1965, 1985) that marks a change in the view of what is admissible evidence even in history. Reference to this work here is due to the fact that some of the preceding references are from the *Journal of Interdisciplinary History*. This itself marks a change not only in the domain definition and evidence used but also a link or bridge not only between social science and history but is also analogous to my argument about the need for multiple disciplines. Such attempts to link and relate different disciplines are becoming more common. For example, cross-

*Note that this review only deals with these in relation to the past. The more general question about whether social science can be "scientific" will be discussed in a second book.

cultural and comparative studies, history, archaeology, and others are applied to past built environments and the past, more generally, to provide an insight into current issues. To give just one example: the special issue of *The American Behavioral Scientist* (1982), *The Archaeology of the Household—Building a Prehistory of Domestic Life.*"

Linguistics began as a historical science (Rapoport 1982a)—as did many, if not most social and natural sciences (e.g., Gergen and Gergen 1984). The point was made in the nineteenth century that "there is no linguistics but historical linguistics" (H. Paul cited in Weinreich 1980, p. 26). Note that some recent criticisms of Chomskian (or structuralist) linguistics (Moore and Carling 1982; cf. references in Rapoport 1982a) can be interpreted as a criticism of the neglect of the use of the broadest possible body of evidence, including "behavior" and, implicitly, of history. Weinreich (1980) also argues that after a hiatus, due to the structuralist notion that systems can only be studied synchronically, there is now a realization that historical knowledge is essential for linguistics. There is a need to preserve the distinction between synchronics and diachronics. One can then identify patterns such as the stability of structural characteristics while other aspects change—an argument I make in this book and have made before. This may be a useful notion regarding environments (e.g., Downing and Flenning 1981; Edwards 1979; Glassie 1975; Hillier and Hanson 1984; Isbell 1978; Seligmann 1975; etc.). This may also have interesting implications for my discussion in Chapter 3 on uniformitarianism of process, not outcome. Note that Weinreich (1980) emphasizes that linguistic periods are a retrospective and relative, not an absolute, concept. That periodization is best that considers the largest number of simultaneous changes—linguistic, historical, sociocultural, and so on. Breaks are best located where a number of changes coincide (p. 720). Not only does this correspond to my arguments about redundancy, for example, in neighborhood definition (Rapoport 1977, 1980/1981) or meaning (Rapoport 1982a); more importantly, in terms of this argument, it emphasizes the need to search for patterns, the importance of taxonomy in this, possibly the use of polythetic classification, as well as the need for multiple lines of evidence and various disciplines (which Weinreich 1980 actually uses, e.g., pp. 178–179). Their convergence enables inferences to be made about something as ephemeral as phonetics, for example vowel sounds, 1,000 years ago in the absence of obvious data such as written documents. Clearly such reconstructions may be wrong, but the approach, based on multiple lines of evidence, explicit assumptions, certain uniformities based on much comparative work, and the like, seems most relevant.

Historical linguistics tries to go further: There have been attempts to reconstruct the very beginnings of language among the hominid precursors of *Homo sapiens*. More generally, by comparing widely scattered languages and thus tracing patterns it is possible to find great similarities at the level of process and structure.* This has been done (e.g., Bickerton, 1983). What is important for my argument is the attempt to use historical data to make linguistic inferences. On the basis of a large body of evidence, with a clear temporal dimension, an explicit hypothesis is derived: That the grammar of Creole languages is innate and easier to learn because children learn language by first constructing an abstract form of Creole. The specific conclusions are not universally

---

*A comparable example from cognitive anthropology and cognitive science is the patterned development of color nomenclatures and taxonomies based on many cross-cultural instances.

accepted; in fact, some linguists in the field reject them. But that is not important for my purpose because the hypothesis can be discussed, reviewed critically, tested empirically, and so on. This seems very close indeed to my view about the potential relation between historical and contemporary data discussed in Chapter 3.

A variant of that very point is also a major theme among a number of recent attempts to link social science and history in various ways. It is suggested that *the past in effect constitutes (or provides) a body of data against which to test hypotheses that are needed to develop systematic theory in social science that is essential—and lacking.* In this connection, an implicit distinction is made between history as the reconstruction of specific events, periods, places, or whatever and the use of historical data for generalization and theory construction. This distinction I have made explicitly in Chapters 2 and 3; we will also encounter it in Chapter 5.*

One specific version of this position parallels my argument so completely as almost to reproduce it (Clubb and Scheuch, 1980, pp. 16, 19). Analogously to my argument about EBS and history in Chapter 2, the purpose of social science is taken to be to identify regularities in human social behavior and to develop empirically refutable theoretical formulations that link and explain these regularities. Because experiments are impossible, social phenomena must be compared in a variety of situational contexts: across nations, regions, cultures, smaller groups, and so forth. The past provides an opportunity to examine a wider variety of human behavior in a wider variety of contexts. In this way hypotheses derived from the past can be tested in the present and vice versa; trends, changes, and patterns can be traced systematically and reasons for them investigated. The purpose is to use the past to construct empirical social theory, not merely to use social theory to explain specific events in the past (Clubb and Scheuch, 1980, p. 19). *Note that if one substitutes references to our domain this statement becomes identical to my formulation; the convergence is complete.*

As I do in Chapter 3, they further argue that in order to do this, it is necessary to consider carefully the epistemological bases of one's work. Among those they emphasize the need to consider classification in terms of one's particular theoretical formulation, conceptualization, explicit inference, careful research design, measurement (and choice of indicators, i.e., operationalization of concepts), and interdisciplinary cooperation. Although their main emphasis is on the methodological implications (which I tend to ignore), the overlap with Chapter 3, other sections in this chapter, and Chapter 5 is quite remarkable (cf. Murphey 1973, Chapter 6).

The essential point in this, and much of the social science literature on history generally, is that the interest shifts from an intrinsic interest in particular phenomena to an interest in their relevance for theoretical concerns. This I have called extrinsic interest (Rapoport 1983a); I have also identified comparable shifts in approaches to vernacular design (Rapoport 1982b, in press). In this case, the concern is less with history *per se* and more with the use of historical data, the relation of such data to any given topic.

One immediate consequence of such a shift is the shift from linear to lateral use of

---

*This position is explicitly criticized by Guelke (1982) from an idealist position related to Collingwood's empathy model. That position is, in turn, criticized in Green, Haselgrove, and Spriggs (1978) and by Nicholson (1983).

historical data (Chapter 3), although comparability and relevance of such data must be explicitly and carefully considered (cf. Rapoport 1986c). One is interested in testing questions against historical data—whatever these questions might be. For example, if the concern is ecology, one result of using historical data is the development of paleoecology (see "historical sciences" later) or historical ecology (e.g., Bilsky 1980).

In studying human ecology, the topic is essentially how human society and nature interact. It is useful to identify problems and to study them in the past (e.g., Rapoport 1986b). This then helps to understand contemporary problems. For example, the use of such data may disprove or greatly modify beliefs such as that traditional tribal societies nurtured their environment (e.g., Flannery 1982), that pollution, environmental destruction, or species extinction are to be identified only with technologically advanced societies (i.e., not even traditional complex societies). This correction is now so well established that it has even reached the general press (e.g., Browne, 1987). Another view, that cities of technologically advanced societies are far worse than those of the past rather than as is the case, much better, can also then be modified (Rapoport, 1982c, 1984a). In fact, in a recent analysis of the so-called environmental movements, their extreme positions (and consequent policies) are shown to be untenable on the basis of historical data and patterns (Efron 1984). It may also help to understand contemporary problems by emphasizing that the interactions between societies and nature remain as givens—a parallel to my discussion of process as constant (cf. King, 1982). It can be argued that all earlier societies experienced institutional problems related to the environment. This provides a broad, conceptual framework for examining human (or cultural) ecology. One can study the impact of societies on the natural environment. These can be of the ancient world and the European Middle Ages of the modern world (Bilsky, 1980). One could, and should, broaden the coverage even more cross-culturally to include non-Western traditions such as the Maya, Khmer, Harappan, and so on (Hamilton 1982). It should also be broadened temporally to include prehistory, and the role of early humans in animal extinctions and also preliterate or tribal societies (e.g., Australian Aborigines, Maori, Polynesians, Anasazi, etc.). This expanded coverage provides not only a perspective over time but also over space, being cross-cultural. One could also easily ask the obverse question and look at shifts in environments (using archaeology, paleoecology, paleoclimatology, etc.) and study their impact on societies; one can study their *mutual* interaction (as one does in EBS) and examine the resulting cultural landscapes. Note that such studies tend to be inherently interdisciplinary and eclectic—another theme in my own work over the past 20 years and one that comes up repeatedly not only in Bilsky (1980) but in other disciplines we will review later (e.g., Fogel in Fogel and Elton [1983] and Berkhofer [1971] in history). In the case of ecology that would involve biologists, geographers, and anthropologists and, I would add, archaeologists (and all their associated multidisciplinary specialists), historians, and a range of natural sciences (evolutionary science, geology, paleoclimatology, and many others).*

*Johnson (1980) using archaeological data covering several centuries from southwestern Iran, southern Iraq, the Aztec Valley of Mexico, China, India, and the United States is able to show that rank-size distribution of settlements only holds when system integration is strong (e.g., political unification, effective administration, intersystem trade). When this weakens, the size distribution of settlements sharply deviates from the rank-size slope.

It is similarly possible to study health and disease by combining epidemiology, geography, and history; for example, health in Britain since pre-Norman times (Merlin Jones, 1976). It is then possible to examine changes by period, locality, climate, water, housing, diet, and so on. In effect, one can do medical history. This can reveal, for example, whether health is improving or getting worse; they may help avoid wrong, alarmist, or "apocalyptic" claims or statements (Efron 1984; cf. Rapoport 1982c, 1984a). Major contributions to the study of disease in the past are also made by paleopathology (as will be seen later in this chapter) and by prehistory. This is done by analyzing mummies, bones, teeth, coprolites, and the like.

A well-known example of a linkage between history and social science is the "historical sociology" of the Cambridge population group. In essence, this highly inter-disciplinary group examines, and often disproves, widely held views about populations and the family by using historical data and rigorous methodologies.* An important aspect of their argument is that *all historians to some extent engage in simulation* (Wachter *et al.* 1978). They simulate past situations, past processes, past plans, inten-tions, and the like that might (or must) have taken place in the minds of people. One might add that they do so on the basis of traces—the data available, which may be inadequate. The Cambridge group, however, makes this explicit and the data is then used explicitly (in this case using computers; cf. Wachter *et al.* Chapter 5). They also use counterfactuals (cf. McClelland, 1975; Todd, 1972) to judge what would (or could) have happened if what did happen had not happened. Of importance is the *explicit and systematic treatment of hypotheticals*. In their view this is the vocation of statistics and is one reason that this group emphasizes statistics.

An important point is the distinction made between implicit and explicit quan-tification. It is argued that most history is in fact *implicitly* quantitative but needs to use *explicit* quantification (Laslett and Wall, 1972). Once again, much more important is the general point, made in Chapter 3 and to recur in this chapter and the next, that an important aspect of becoming scientific in history, archaeology, or whatever is *making explicit what is usually implicit,* whether this be quantification, generalization, models of human nature, or anything else.

As do historians who address the relation between history and social science (e.g., Erikson 1973; Fogel in Fogel and Elton 1983; Pitt 1972; etc.), Laslett and his colleagues argue that historians dislike social science because it is the opposite of traditional history that is literary and narrative, humanistic (?), and concerned with unique events, unique personalities, and, hence, unrepeatable episodes. The Cambridge group, how-ever, argues that historians need social science that makes history more objective. At the same time, social science needs history, not least because it needs many data so that hypotheses can be tested against large bodies of evidence. Both these points closely parallel my argument. It then follows that historical social science can help in discover-ing in what ways the present differs from the past and in what ways they are the same; what I would call patterns of constancy and change. The suggestion is made that in some areas or domains things may be so new that appropriate social forms must be

---

*The specific methodologies are of less interest in the present context. It is of interest to note, however, that examples of methodologies are not only available here and elsewhere but tend to be emphasized (e.g., Clubb and Scheuch 1980; Murphey 1973, Chapter 6).

invented (Laslett and Wall, 1972). Because the past is not all useful, it is an issue relevant to uniformitarianism and in itself one way in which studies of the past can help the future. Recall also the similar point made in Chapter 3 about scale effects—the metropolis versus the neighborhood (cf. Rapoport, 1977)—that can also be made about new types of settings (e.g., Rapoport 1986c, 1987b). The argument can be made more general by stating a hypothesis: There are some elements that are unchanged: There historical precedents may be directly relevant (albeit at a certain level of abstraction). Some elements are changed: There transformations of the past, general lessons or principles derived from the evidence may be useful; it is also a possibility that these changes may still be a result of constant processes. Finally, there are new elements: There the past may not be useful, although there is the possibility that these new elements still result from constant processes or are related to certain contexts that show constancy. In that case the past is still of use; it is, of course, also of great use in identifying these new elements. Moreover, even here, both cross-cultural and historical data are still useful; lessons can still be derived. For example, the aging of the population is said to be a new phenomenon for which there are no historical precedents (Laslett and Wall 1972). I would argue that although the specifics are new, both cross-cultural and historical data are still useful both in understanding aging, distinguishing biological from social and cultural effects, and in understanding supportive settings in both social and physical terms, such as the family and other mediating institutions (Rapoport 1974, 1983a).

Attempts (not always successful) have also been made to study human communication historically (e.g., Williams 1981). Once again it is the general point that is significant: To study the impact of new communication technologies on people one needs to know the history of such communication; what is unchanged, what is changed, what is new, and what impact previous major innovations have had. One can also criticize new developments or forms as inappropriate, unlikely to work and likely to have negative consequences on the basis of strong and persistent patterns revealed over long periods of time and across many cultures. This is, of course, the case with such criticisms based on perceptual constancy, sociobiology, "human nature," evolutionary baselines, and the like, already discussed.

This point is of course part of my argument for the role of historical data; that is, only part of a much more general argument for the use of much larger bodies of evidence. In fact, I began by looking at non-high-style environments and cross-cultural evidence, then at the whole environment before I turned *explicitly* to historical data (I have always used it) in this book (cf. Rapoport 1986b). The same argument about this role of historical data is also found in social science (e.g., Bradfield, 1973) where, in emphasizing the need for comparative study generally and cross-culturally, four points are made that I have been emphasizing for some time:

1. The need to use different disciplines (for example, ethology and, I would add, sociobiology) to link humans to primates, protohominids, and hominids (Bradfield, 1973, Vol. 2, p. 442). This corresponds to what I have called *using the full range of history*.
2. The use of materials already available in various fields (e.g., ethnographies, ethology, archaeology, etc.).

3. The need to make inferences about behavior, thought, and so on.
4. The need for theory rather than just data.

The point is made that thinking and theory are as essential as observation, data collection, and experiment (see Figure 4.1). It follows that to find out what happens, one needs not only observation and experiment but data already collected. Thought is thus needed not only in the original question but on the kinds of observations to be made on *others'* observations (cf. Chapter 3 and the case study in Part III; reanalysis of existing data by Eldredge 1985; Gould 1985; Lewis-Williams 1981, 1983; Marshack 1971; Schmandt-Besserat 1983; Vinnicombe 1976; and many others; cf. also Chapter 5). As in all scientific and scholarly work, thought and data need constant interplay: The issue is about what are the data in question. In general, there seems to be agreement with my argument that they need to be as broad as possible.

If social science restricts itself to empirical work, then things can only be seen at a given point, highly restricted in space and time. This limited observed segment, however, forms part of a temporal continuity and is also related to a broader context (Pitt 1972). Although cross-cultural comparisons have been used in anthropology (although not sociology), the historical record provides important insights and data not accessible to empirical observation. It follows that the present is too restrictive; what is significant is often that which lasts and shows continuity or constancy (Rapoport 1970a, 1979a,b, 1983a, etc.). Thus the temporal dimension, that is, historical data, is essential in order to understand the constant, and even more necessary to understand change (Pitt 1972, p. 4). The present tends to appear stable; change can only be seen through historical knowledge. It is even more necessary to understand the interplay of constancy and change and, most generally, to reveal patterns.

The point has already been made (cf. Pitt 1972) that originally anthropology and sociology contained a great deal of history. This then declined due to an emphasis on empirical work. In effect, synchronic studies won out over diachronic (although both are essential). At the same time, at least in anthropology, diachronic work never completely disappeared because of a concern with cultural dynamics and culture change. To give just one example, hypotheses about the relationship between population and changes in technology are tested against historical data (e.g., Boserup, 1965, 1981; cf. Renfrew, 1983).

The loss of historical depth in social science was also due to the nature of theories being considered. In reality theory, and specifically explanatory theory, must use historical data in order to identify patterns (e.g., Fox 1980; Leach 1976). The loss of

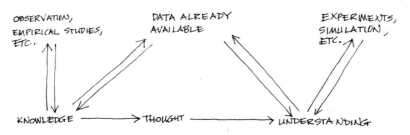

Figure 4.1. Developing understanding (based partly on Bradfield 1973 p. vi).

historical concern in social science was also due to traditional history that, as already pointed out, tended to deal with unique situations; it was narrative and literary and hence emphasized writing style; it was also concerned only with written documents. The rapprochement began relatively early and, due to problems with historical data, the emphasis came to be on the use of inference with the goal of establishing probabilities (Pitt 1972, pp. 12, 46ff., 58ff.; cf. Chapter 5).

This point about problems of communication between social science and history (Laslett and Wall, 1972; Pitt, 1972) can be made slightly differently. Although history is needed in sociology (read "social science"), it has hardly been used because of professional filters and disciplinary boundaries (Erikson, 1973, p. 28). This is, of course, why one must be interdisciplinary. In the case of history and social science there are major discrepancies among the sets of ideas shared between these two communities of scholars about the proper standards of performance, the proper order of evidence, and standards of criticism. They can be resolved either by social science becoming more like traditional history or, as I argue, also by history becoming more scientific.

Recall the point already made about the different way in which traditional history and social science look at the connections among variables. Whereas history looks at *sequential relationships,* the social sciences and EBS look at *lateral connections* (Erikson 1973; Fogel in Fogel and Elton 1983; Goody 1977) (see Figure 3.1).

There are other problems that make links between social science and history difficult. Among them are differences in conceptualization, terminology, and methodology. Regarding the latter, it is useful to think of the possibility of using social science methods in history and historical methods in social science (Erikson, 1973, p. 26). *Method* may, however, be the wrong word. It may be better to consider *ways in which one can systematically study both the past and the present in order to answer certain questions.*

Although, as we have just seen, by the early 1970s people were writing about a reconciliation, the relationship between social science and history was still being described as one of conflict almost a decade later—although a convergence is said to have begun in the 1950s and to have been clear by the 1960s (e.g., Burke, 1980). The break is dated to the 1920s; just as social scientists lost interest in the past some historians began to work on the "natural history of society" as, for example, the *Annales* school in France (cf. LeGoff and Nora 1985). A great deal of social history has now been written that social science could have used. Although history and social science are complementary, the conflict continues. Moreover, it is due not only to professional differences. It has become "subcultural": The two fields have different values, languages, and styles of thought—a common theme in the literature (cf. Erikson 1973; Fogel in Fogel and Elton 1983; Laslett and Wall 1972; Pitt 1972, etc.). Burke (1980) argues for the need of a *synthesis* between the sense of structure and function in the social sciences and of change in history. The former can contribute a variety of approaches, the use of concepts and models, and methods (for example, comparative studies—strangely neglected in most history, see next section). History can contribute models dealing with change, how they can fit particular societies, and so on. In any event, the argument seems to hinge on the explicit use of models, a greater emphasis on concepts, greater concern with methods and rigor, and the like.

Although I will be reviewing some supporting arguments from history proper in the next section, it seems useful briefly to consider here the same essential argument found among some historians. This is occasionally called "applied history" or "applied historical studies" (Drake 1973, 1974) as opposed to the sociological name: "historical sociology." Drake (1973, p. 1) defines applied history as the exploration of the past undertaken with the *explicit* purpose of advancing social science inquiries by *testing models of social science by running them through the dimension of time* (cf. Quale 1988). This is not only essentially my argument in this book but also very similar to the motivation of ethnoarchaeology as discussed in Chapter 5. In all these cases, the objective is to help in discovering and developing generalizations about human behavior relevant to the particular domain in question. This is because by testing the propositions of social science through time, against historical data, they are made more verifiable; this is also the case when they are tested cross-culturally and in other ways. In all these cases, the goal is to broaden the evidence in important ways. Note that, as in my argument, the time dimension is often seen as being long term, covering large temporal spans paralleling the breadth of evidence, in our case cross-culturally, types of environments, and components of environments.

This process of testing social science propositions through time is said to involve three tasks (Drake 1973). First, searching for appropriate historical data, which are readily available, contrary to the belief of historians that they have few data (p. 2). Second, devising ways to use these data. This means approaching historical data in ways very different to those used in traditional history—and more like those discussed before. Finally, presenting any findings in forms usable by social science or, in the present case, EBS.

One field that has always linked these two approaches is, of course, *historical geography*; that, moreover, is also concerned with cultural landscapes. It has also benefitted from developments in geography analogous to those advocated here (and very similar to those to be discussed in Chapter 5).* Neither in geography generally, nor in historical geography is there agreement on this. In the latter, there is much debate in the literature and, as usual, I will not review the whole literature nor be even-handed in considering the different positions. My purpose is merely to show that there is at least some agreement with my position within a given field. However, consider at least a few contrary views.

Extreme rejections are to be found (e.g., Gregory, 1978) and are typical of current phenomenological, structuralist, "hermeneutic," and other "antipositivist" approaches to geography and social science, which are quite common. This was briefly discussed at the beginning of this chapter. Although I intend to discuss this issue with respect to EBS elsewhere, this position has been explicitly criticized in historical geography (Billinge 1977). Because that critique deals mainly with the views of Leonard Guelke, it may be useful briefly to consider his more recent work (Guelke 1982).

In this, both natural science and social science approaches to historical geography are criticized and rejected. Instead, the approach is based on Collingwood's (1939, 1961) theory of historical knowledge based on imagination and empathy, as well as a

---

*Note that even in geography proper, for example urban geography, history can play a major role and help illuminate particular developments. They gain interest and importance when seen within patterns, as anomalies vis-à-vis other places but part of the pattern in a given context (e.g., Holzner 1970).

critical attitude. There are obvious problems with such an approach (for example, the emic–etic distinction, culture, etc.), but I will not deal with those (cf. Martin 1977, for a recent philosophical analysis of Collingwood; see also Pipes 1986). On that view history is an autonomous field of knowledge, requiring an idealist rather than a realist position. (Note that in archaeology also, Collingwood's approach is being revived, e.g., Hodder 1986). Guelke (1982) further argues that *all* human geography is historical geography because human geography studies human activity on the land as a creation of human ideas. This has, of course, been my argument about the cultural landscape, but it does not necessarily lead to an idealist position. The question is, of course: Given that, how is it to be studied?

Guelke (1982) accepts the view that history is cumulative not circular (as discussed earlier, like natural science, it is a "spiral"). He rejects the concept of "scientific" history, *while accepting the use of scientific methods and techniques!* He argues for the use of *any* methods that help to reconstruct past situations and evaluate evidence. In this the gap between the two positions seems minimal. It becomes bigger when he explicitly criticizes the theoretical model-testing approach, where the past is seen as a reservoir of data for testing such models, akin to a laboratory. This is precisely the view that I adopt, and have discussed before, which we will encounter in Chapter 5 and which other historical geographers explicitly accept or adopt.

As I do, some begin with attempts to define the nature of the domain. On one view (Moodie and Lehr 1976), historical geography is essentially geography rather than history; however, it deals with the past. In that sense its unique perspective is that it deals with time. It should be added, however, that it deals with time in ways different from other recent geographical approaches that incorporate temporal variables, such as "chronogeography" (e.g., Parkes and Thrift 1980; Carlstein, Parkes, and Thrift 1978a, b, c). The difference is that *the observer and the observed are not contemporaneous.* It is this temporal distance that is critical; in this, history is like archaeology in that the data are in the present, but inferences are made about the past. The data are also static, but inferences are about dynamics; also, as in all forms of the study of the past, the observations are of what has survived; the data are thus incomplete. They are also frequently second-hand and made by others, although they do not have to be. It follows that they are not fully empirical or, at any rate (given some positions in current philosophy of science), less empirical than others. It follows further that one must have both theory and data—again, from some points of view a questionable distinction and from others far from unique to historical geography. The point is, however, that the theory is today's, the events in the past* (cf. Berkhofer 1971 in the section on history). On this argument, what distinguishes historical geography is a form of interaction between theory and data that makes it impossible to advocate the primacy of either. In this, Moodie and Lehr (1976) criticize the often cited distinction between "'real,' 'imagined' and 'abstract' worlds of the past" (Prince 1971) but again do not really establish the uniqueness of historical geography.

Historical geography has also engaged in explicit reexamination of its basics because, it is argued, the orthodox views are not convincing (e.g., Baker 1972)—very much like my criticism of traditional, orthodox, architectural history. The crucial shift recommended is essentially identical with mine: To use the data from the past to test

---

*Using King's (1982) distinction, the facts may be either in the past or the present.

concepts and theories. The emphasis is on the role of historical data in *clarifying patterns* and leading to more reliable *generalizations*. The more concepts, generalizations, models, and theories are tested across time (and, of course, different cultural and other contexts), the more confidence can one have in them. *The purpose is, therefore, to illuminate the concepts and theories rather than to illuminate the past* (Baker 1972). This is, of course, a behavioral geography approach to historical geographic data, a close analog of an EBS approach to historical built-environment data. Moreover, links then become essential, as I argue, to other disciplines, for example, archaeology, history, sociology, and anthropology. Interestingly, an anthropologist (Leach in Spriggs 1977) argues for links to geography, as well as archaeology and history. The argument extensively quotes Berkhofer (1972), Gottschalk (1963), and others whom I will discuss in the next section.

Paralleling my argument, Baker (1972, pp. 21–23) emphasizes the importance of theoretical approaches in framing questions, establishing hypotheses, and collecting data. Theoretical positions should be formulated on the basis of the best available knowledge from geography and other social sciences and should be formalized to spell out the implications and tested against new data, in this case historical. That is not only identical to my argument, and to my case study, but will also be seen to be identical to many arguments in archaeology. At the same time, of course, data are essential, as are empirical studies (i.e., in the present). In dealing with epistemology, Baker argues for the importance of induction, not only in history, but more generally. Again, as pointed out in Chapter 3, inductive and deductive approaches are inseparable in scientific inquiry. Also, again echoing my argument, Baker argues that historical geography is *eclectic;* it has adopted the philosophical underpinnings, concepts, analytical techniques, and so on from related disciplines (even Guelke 1982 accepts this position). These disciplines may include not only other social sciences but even physical geography and, I would add, other sciences. On this view, and paralleling my argument, historical geography has become more generalizing in the last 30 years, moving from being idiographic to being more nomothetic (cf. Hall in Barker and Billinge 1982, p. 274).

It is a widely held view in studies of science, and central to my argument, that new approaches and methods open up new areas of inquiry and generate excitement. At the same time one should not neglect or lose those traditional skills that may still be useful (Baker and Billinge 1982). At such periods of renovation, as I argue in Chapter 1, the domain often needs to be redefined as does the purpose of research. We have already seen one such attempt in historical geography (Moodie and Lehr 1976). In another (Baker and Billinge 1982), the purpose of historical geography is said to be essentially to understand and explain the man-made (i.e., cultural) landscape. This is, of course, the objective defined in Chapter 2. This cannot be done by looking at a single period, a temporal cross-section; one needs both synchronic and diachronic analyses. This is particularly the case if the emphasis is on dynamics and change, which requires a temporal perspective although, as already suggested, so does constancy.

Before one begins to generalize, understand, and explain, one must know, as clearly and dispassionately as possible, what actually happened. The starting point must, therefore, be data (but this evidence needs to be inferred). The data are the landscape, as it exists, as it is described in oral traditions, written documents, visual materials, and a wide range of archaeological data (including climatological, botanical, zoological, and so on). This corresponds to my argument for the use of the widest variety of lines of evidence. In order to do this, one needs to consider not only such facts but theory, various actors, observers (including contemporary analysts). This implies the use of explicit models of human behavior because, in order to understand the meaning of past human landscapes, one needs a theory of collective (or individual) behavior, culture, environment–behavior interaction appropriate to their context (see next section and Chapter 5). Baker and Billinge (1982) argue that historical and cultural geography are weak in all these.

There is general agreement in history, archaeology, and other disciplines with the argument in Chapter 3: that *the critical problem is that of inference.* This is also the case in historical geography (Baker, 1976). The conclusion reached after reviewing over 40 publications on the purpose and practice of historical geography is that although one must always accept an irreducible minimum of uncertainty, that is also the case with the geography of contemporary environments. The gap between the two is imaginary—in the sense that inference is also used in contemporary work. Certainty is impossible because, in all cases, one is testing theories and models against data (p. 177).* All inferences, whether drawn directly from empirical evidence or indirectly (whether from theory or from the past), are probabilistic (cf. Salmon 1982). One also needs to distinguish *explicitly* among different generalizations (e.g., first order from, say, sixth order) (cf. also Chapters 3 and 5). Baker suggests a specific approach, but the general point is that in historical geography one needs to frame alternative hypotheses and test them; the argument is very similar to what we will encounter in archaeology. Such more rigorous approaches also make possible simulation studies in historical geography (Widgren, 1978) as they do in history, archaeology, and the other fields being considered. There is agreement on the need for greater rigor, comparative approaches, explicit argument, the testing of explicit theories, hypotheses, and models, often based on various contemporary studies, against historical data, the need for interdisciplinary work, and so on. These can all contribute both to social science (or EBS) and to the study of the past.

Although many of the studies discussed contain interesting case studies, I would like to conclude this section by discussing one example that shows both the promise and difficulties of an interdisciplinary approach to a problem *within a domain as I have defined it in this book* (Green, Haselgrove, and Spriggs, 1978). On the one hand it shows the difficulty of such work, for example, the difficulty of communication across boundaries of disciplines, or the possible lack of convergence leading to incoherence. This incoherence is attributed not so much to any inherent problems of an interdisciplinary approach but rather to the uncertainty among the authors as to what should constitute their object of study: That must be defined in a uniform fashion (Lewis, 1978, pp. 513–515). In other words, Lewis argues that the domain has not been defined so that, in essence, incoherence is attributed to ignoring the conditions stated as being essential in the Introduction and Chapter 1 of this book. In this respect, Lewis and the editors of this book agree in principle and they also support my argument; in fact, even this difficulty (or rather its resolution) confirms my argument.

---

*Although I will not argue this here, the point is made in at least some writings on science that there are cases where certainty is, in fact, possible and has been achieved.

If there is, in fact, incoherence, it results not only from the fact that there is no clear definition of the domain. It can also be argued that it follows from the inadequate definition of fundamental questions. Without clearly defined questions, one cannot compare "answers" or assess their adequacy. This, again, confirms my argument in Chapter 2 about explicitly defining the problem with the domain; this allows more specific questions to be formulated.

In spite of these criticisms, the book can also be taken as showing how, at least in principle, past environments can be studied by stating a problem, however inadequately, discussing epistemological bases, and then explicitly bringing to bear on it a series of different points of view from a variety of disciplines. In that sense it also strongly reinforces my argument. It also allows many further links; for example, it appears from historical data that innovation is typically based on the old, using existing forms. This suggests interesting links with other archaeological work (e.g., Renfrew 1984) and particularly biology and evolution; for example, Jacob (1982) and Gould (1980) on the use by evolution of existing forms and "tinkering" and hence with d'Arcy Thompson (1961) and Leach (1966) on transformations and many others. (Some examples of built environments were discussed before.)

Moreover, this study also makes a series of useful supporting suggestions that we will encounter frequently in the remainder of this chapter and the next chapter.

1. Inferences are better if data are collected with greater rigor and interpreted more rigorously.
2. It is useful to bring different orientations (in this case, evolutionary and biological data, history, archaeology, anthropology, geography, etc.) to bear on a single topic particularly if that topic is clearly and explicitly defined. (In this case, the topic is the concept of territorial boundaries.)
3. The time dimension reveals the constancy of the *problem* (cf. Hodgen 1974; King, 1982) that is, that of humans in ecosystems; the responses and outcomes differ. One can then *explicitly* review various models, ranging from extreme ecological determinism to ignoring the ecosystem (e.g., mentalist, idealist). The conclusion that is based on the large body of varied evidence is that both humans and the ecosystem must be considered.
4. More detailed theoretical models are developed, for example on carrying capacity, resource exploitation, spatial patterning, and the like, which are then applied to data from the past including the Paleolithic. These are tested in different ways, which include experimental archaeology and could include simulation or ethnoarchaeology (see Chapter 5).
5. The development of models based on specific current empirical research to illuminate the past. This is equivalent to developments in archaeology to be discussed in Chapter 5 and to the central point of this book: to use empirical EBS research to study past environments. In fact, this study uses known or explicitly hypothesized behavior of a given population rather than not specifying it at all (when it is essentially implicit—see next section on history).
6. There are explicit definitions of concepts such as "peasant." These are then explicitly discussed and criteria established whereby they can be applied.
7. There is a corresponding, explicit definition of focus; for example, "village"

(Figure 1, p. 293) and an emphasis on taxonomies. It is suggested that those have heuristic value and are for the purpose of imposing order so that generalizations become possible (see Chapters 3 and 5). As a result, the discussion of built environments becomes far more sophisticated, and a broader range of issues is considered than in traditional architectural history.

8. There is a clear emphasis on the cumulative nature of scientific data that becomes extremely helpful in historical study. One example is the accurate identification and dating of environmental change (Chisholm in Green, Haselgrove, and Spriggs 1978, pp. 213–220).

Although many of the specifics are related to the specific topic addressed, the general conclusions seem relevant to any study of past environments. In this case, it becomes possible to study human habitations (our domain) in terms of their *contemporaneous* topographic, climatic, and resource conditions. Although this means that one needs to know a vast body of data provided by a great variety of disciplines, it is not a one-way street. Those disciplines gain by the study of human artifacts. The uniformitarian problem is explicitly defined. Although there is never any "proof," rigorous levels or degrees of certainty are able to be specified; inference becomes more convincing. There are also attempts to provide explanation. In doing so, three "cardinal rules" of explanation are proposed; it is suggested that if all three are applied, the probability of having found the correct explanation is high (Green, Haselgrove, and Spriggs 1978, p. 219). These rules are, once again, consistent with Chapter 3 and with the other fields discussed in this chapter. They require, first, internal consistency of any explanatory model; second, external consistency, that is with other results (from the broadest range of other disciplines possible); third, parsimony of explanation (to which one could add other, more general, criteria for good explanatory theory that are more appropriately discussed in a book on theory).

The train of reasoning is from the study of ruins, earthworks, and the like, to isotope dating, the ecological aspects of botany, zoology (more correctly paleobotany and paleozoology), the nature of environmental conditions (climate, soil, etc.) and so on. It then moves to asking which events are endogenously and which exogenously determining. This involves a vast breadth of knowledge and an exciting range of work. Even so there are gaps: The emphasis tends to be "materialist," emphasizing ecology and resources; less attention is paid to aspects of human culture such as symbols, meanings, and schemata, to the role of cognized environments and resources, and to latent aspects of behavior and human settlements. Be that as it may, however, the range of work shows two important general things: The artificiality of disciplinary boundaries and also of the division of knowledge into "contemporary" and "historical."

At an even more general level, both the subject matter and the goals of the study, as stated and as I interpret them, as well as other material reviewed in this section, strongly support my argument. The need is said to be generalization in order to build explanatory theory. This generalization must be on the basis of a wider body of evidence than used hitherto; it is suggested that even in anthropology and archaeology the evidence used has been too limited. The search is for regularities, that is, patterns: These enable the essential step of establishing generalities. One cannot begin to explain except through *comparison,* as I have done since *House Form and Culture;* this book emphasizes the need for broadly comparative study (cf. Kent 1984). Such comparative work must transcend spatial, that is, cultural and temporal boundaries, and, I would add, arbitrary limitations of types of environments. This corresponds to what I have called *lateral,* rather than *linear,* relationships among variables. General statements need a comparative base that is as broad as possible so as to explain variation in human behavior—and in environments, where variation is even greater and which is such a puzzling phenomenon (e.g., Rapoport 1985b,c,d, 1986c). This study (like others reviewed) attempts to do what I suggest needs to be done: It takes specific variables that, at least in principle, are precisely definable. These are studied across the widest possible spatial (i.e., cross-cultural, ecological, etc.) and temporal ranges, drawing on material from a considerable variety of disciplines (ethnography, archaeology, history, geography, and a variety of sciences). This produces patterns of association but not of explanation, although hypothetical generating processes can be suggested for these patterns. These regularities and the hypothesized processes that underlie them are then tested using a variety of techniques. Although this is not easy, it is feasible and can be done. In doing so, one even uses contemporary data to test regularities derived from the past; the reverse is also possible.

These conclusions, derived from a variety of social sciences, strongly support my argument; they will also be seen to be congruent with comparable thinking in archaeology and other fields.

THE SUPPORTING ARGUMENT FROM HISTORY

Introduction

Before I briefly review a very limited portion of the literature, four points should be noted. First, my concern is not with history *qua* history but with the use of data from the past, historical information for particular purposes and to answer certain questions concerned with constructing EBR theory as the disciplinary base for design as a science-based profession. This implies a distinction already introduced between history being concerned with "vertical" relationships and my concern with "lateral" connections. One immediate implication is that the essentially "narrative" nature of history, which many traditional historians emphasize, is no longer terribly interesting whether or not it is valid. Second, even when in the historical literature the relation between social science and history is discussed, it is always in terms of how social science can help history (social science→history). In the previous section, the interest was primarily the reverse, that is, history→social science. My interest here is their *mutual interaction,* that is, EBS↔history.

Third, not only is there a very large literature, pro and con, as to whether history can be objective, can establish "truth" or "facts," can be scientific, and so on, but the use of the term *scientific* varies tremendously and is often used very differently to the sense of this term in this book (e.g., Fogel and Elton 1983; cf. Chapter 3, 5). Thus, for example, Collingwood (1961) calls for a scientific view of history but in such a different sense as almost to be contradictory (cf. Martin 1977). Finally, although one finds occasional reference to physics or the philosophy of science more generally and, even less frequently, to biology, there is rarely a more detailed examination either of what "being scientific" might mean or of the implications of certain sciences for the topic (cf. McClelland 1975); this I will do in the next section.

Even a fairly cursory review of the literature on history and philosophy of history reveals an extraordinarily wide range of positions. As one example, Lord Acton's suggestion that historians study problems not periods is denied by Oakeshott (cited in Wilkins 1978, p. 128). Clearly my concern is with problems rather than periods. Whichever is studied, *how* that is done ranges from extreme antiscience positions that accept history as rhetoric, narrative, or storytelling (e.g., Hexter 1971; Porter 1981) through intermediate positions (e.g., Aydelotte 1981; Bullock 1977; Gottschalk 1963), to the other extreme where one finds strong arguments for history as a science (e.g., Berkhofer 1971; Fogel in Fogel and Elton 1983; Todd 1972—although the latter uses *scientific* in a sense different from mine, as noted before). One even finds explicit debates (e.g., Fogel and Elton, 1983).

I will not try to review all these (and other) points of view in any detail; nor will I attempt to resolve the arguments. This would be extremely foolhardy. For one thing, I am clearly not competent to do so (nor have those more competent than I succeeded!). For another, it would take us too far afield. Doing so is not directly within my problem area. As usual, I only wish to draw attention to the fact that this view can be found represented in recent historical thinking. Moreover, once that position *is* accepted, the arguments used and conclusions drawn are very similar to mine: The logic of the situation produces a supporting argument.

I begin with two general reviews, one dealing with history in France (LeGoff and Nora 1985),* the other in the United States (Kammen 1980).†

LeGoff and Nora (1985) point out that this "new" type of history began in France (with the establishment of *Annales* in 1929 and the work of Febvre and Bloch). By the 1970s it had spread to other countries and is no longer new. In fact, several "generations" can be identified in France. The first two generations began with redefining the subject matter of history (my domain redefinition!). First, the subject matter was greatly expanded to include many more and new topics. This changed the notion of what were proper subjects of historians' concern. There was more diversity of subject matter and new fields: material culture, climate, demography, ideology, "mentalities" (i.e., reconstructions of beliefs, thought, imagination—even the unconscious). Second, an attempt was made to get away from history as the study of high politics and diplomatic relations among nation-states. The concern shifted to "total history" or "global history," the history of human experience. An important consequence was the shift to *social history,* from studying elites to an inclusion of common people—a close parallel to a shift from only high style to vernacular and the whole environment! This, in turn, led to an attack on narrative history and a new way of thinking (the new history was a mode of thought rather than a "school"). This involved a shift from description to analysis and explanation, and from monocausal explanation to multidimensional expla-

---

*I read this review in November 1986, after draft three of Part I, and especially Chapter 3, was complete. Note also that this is a translation of a French text published in 1974, that is, it is rather earlier than appears.
†This review, which I also read in 1986, was kindly drawn to my attention by Professor Reg Horsman, Department of History, University of Wisconsin–Milwaukee.

nation. There was a different view of what writing history was: It was to be a human science (*science humaine*) among the social sciences; there was to be a two-way interaction between social science and history. In this connection "science" was understood as well-founded knowledge, not necessarily a copy of the natural sciences.

There were new questions posed and an emphasis that questions *needed* to be posed, explicitly and "tightly"; the result was a new rigor and the use of new methods and techniques with a strong initial emphasis on quantification. This also meant new approaches to earlier fields, the reanalysis of previous work. The new history was eclectic, drawing not only on new methods but on new kinds of data from many diverse fields and also concepts and models from such fields. It was deliberately interdisciplinary, in an attempt to forge a new historical discipline that would bring together history with geography, economics, sociology, psychology, anthropology, linguistics, and any other relevant human or natural sciences—very much like my view of EBS. The objective was to produce a total picture of past societies, to develop "serial history." By this was meant a search for patterns in order to generalize; this meant the use of wide samples from long periods of time (*longue durée*). As I have argued, the point was made that only this can reveal patterns and regularities; moreover, also as I have already argued, only thus can different rhythms or rates of change be identified for different phenomena—short, middle, and long (cf. Laslett and Wall 1972; Rappaport 1979). Serial history also involves using a much broader and more varied range of evidence and, therefore, also of sources (not just archives). "Events" were seen as only relevant within series: Individual events or facts were only significant when fitted into a more general context, a series (or framework) of comparable facts or events. Hence, an "event-based" history was not acceptable.

There was a vision of a relationship between the study of humanity in a historical dimension to other social sciences and, in some situations, natural sciences. The epistemological, conceptual, and scientific validity of the field was explicitly addressed and discussed. It was also suggested that topics previously seen as being without history or ahistorical had a temporal dimension. This reinforced the trend toward a greater diversity of subject matter and their study through conceptual and methodological approaches from other disciplines, and an emphasis on definitions, validation, and so on. In this historiography, history becomes essentially *the diachronic study of human phenomena*.

The third generation narrowed the scope, away from "total" or "global" history. It became more limited in its ambitions and hence, it is argued, more successful. This is a more general point: Often, asking broad questions leads to narrow answers, whereas limited questions give very general answers (Jacob, 1982). The emphasis shifts to definable problems in history. Taken together, all these changes very closely approximate those I have proposed.

Various criticisms have been raised against this approach. Among them, an obsession with quantification and a neglect of what is not quantifiable, a neglect of change, an overemphasis on data gathering, doubt as to whether one can be value free and objective, and whether one can have "true history" (*une histoire vraie*). Many of these are always debated in the philosophy of history (and elsewhere). Others do not seem to be fair and can be refuted; some are no doubt correct. But in spite of these criticisms, many of the specific studies that have resulted have been highly positive. The general impact of what is analogous to my proposal has also been most significant because not only has it become the orthodoxy in France but it has influenced historiography in other countries including the United States. This becomes clear in a comparable review of American history in the 1970s (from the 1960s to 1980) (Kammen 1980).

History in the United States was transformed between the 1960s and 1980 by a "historiographical whirlwind" (p. 21). The general characteristics of these changes is what interests us here. Among them is an emphasis not only on quantitative methods (made possible by computers) but on methodology generally, including a self-conscious choice of methods and a concern with precision also in the use of qualitative data. There was the "discovery" of new areas of inquiry, such as "lower" groups and popular culture—which corresponds to vernacular in my case. There were also many new topics and the development of new subdisciplines, as well as an interdisciplinary approach linking with economics, political science, geography, archaeology, sociology, anthropology, and other disciplines. With this came both borrowing and adaptation of methods, concepts, and theories from other disciplines. Not only are all these similar to what happened in France—so was the result: a redefinition of history as social history. With this came a great expansion of topics, such as the history of groups, social structure, mobility, the family, sexuality, institutions, the private domain of the household and house, cities, and so on.

There was also a self-conscious concern with, and discussion of, assumptions, procedures, epistemology, and the content of the field (the domain). There was an emphasis on explicitness, regarding frames of reference, concepts, theoretical constructs, and theory generally; the relations between theory and data; choice of methodology; the posing of questions. Although attempts were made to pose questions and generate hypotheses before data were collected, more explicit and rigorous questions were also posed of data collected first or already available. In either case, more carefully specified topics of study, explicit questions, and models were emphasized. There were attempts to generate models that were more persuasive because they were not only internally consistent and capable of explaining particular cases but because they were tested against a variety of situations (geographic, social, cultural, temporal, and so on). Their external validity, that is, congruence with findings of other disciplines, also tended to be considered. Only then could one begin to extrapolate any conclusions, discuss longer range validity, and begin to generalize; generalization often proceeded through the use of analogies and comparative studies. There was thus the beginning of explicit comparative studies and even a small body of work with the principal goal of the systematic comparison of the same processes, institutions, and so on in varied settings. Comparison became the core of the enterprise for at least some, although still few, historians.

New types of questions were asked about all the new topics: material culture, cities, epidemiology, technology, transport, and many others. The many changes that one could list are essentially the development of a new, "scientific" approach leading to a generalized social science, with hypotheses to be tested, an explicit interest in the relationship between the present and the past, and the testing of contemporary work against data of the past (and, possibly, vice versa). This implies a two-way interaction between history and a range of social sciences, a mutual impact of anthropology, sociology, and so on, and history seen as diachronic study, which are deemed to be complementary. Historical research and the behavioral and social sciences become interdependent. Although becoming a generalizing discipline with a trend away from narrative, specifics were not lost because local case studies continued: There was an interaction of the general and the specific, as of theory and data.

Many specific achievements of these 20 years could be listed. Among the major, general achievements were a range of new topics studied, the realization that old topics can be studied in new ways; that old evidence can be reanalyzed and can also be supplemented by new material that becomes usable through new methods; that one can generalize and study topics rather than events. Moreover, there was explicit and rigorous debate, always based on more rigorous theory and methods, explicit reasoning, and data. As a result, even the most heated debates clarified issues and arguments by being able to retrace reasoning and to refer to the same data. There were signs of convergence, and at least a hope that closure was possible. Mutual criticism improved both social science and history.

All this is not only very similar to the developments in France but fully congruent with my argument. Moreover, the meaning of *scientific* in this case is also fully comparable to that used in Chapter 3. It does not necessarily mean the discovery of universal laws but rather establishing a widespread consensus on some set of important, if initially narrow, facts; attempts at generalization; having replicable findings; engaging in systematic research that builds on previous firmly grounded results; the use of explicit assumptions, methods, and rules of inference (Kammen 1980, fn. 30, p. 446).

Note that we come up against the issue of inference, which has been a constant throughout much of the discussion, because it plays a role in all science and is central to any study of the past. One can find a number of discussions and philosophical analyses of the logic of historical methods of inference, and their relation to the philosophy of science. At least some conclude that *valid historical inference is possible* (e.g., McCullagh 1984; cf. Murphey 1973).

Some Specific, Implicit Supportive Arguments

There have also been some recent developments, for example, of comparative history, the basic argument of which is very similar indeed to mine. The point is made that although there is little apparent agreement about what comparative history is, its main object seems clear; it is the systematic comparison of the same process, institution (or anything else) in more than one situation (society, culture, place, period, and so on). This can reveal regularities as well as peculiarities, hence patterns and anomalies, and enlarge history, bringing it closer to generalization and scientific theory (e.g., Frederickson, 1980). Not only is this very recent, not very developed, and not yet at all widely accepted, there also seems to be a remarkable lack of knowledge of, and reference to, other fields where comparative studies are advanced: anthropology, comparative anatomy, and so on.

For rather longer than that, however, at least some historians have given a positive answer to the question whether history can become a social science. One important position in this regard, which is part of the more general issue of the need to be explicit, involves the argument that to become scientific history needs to make explicit statements about a theory of human nature rather than "sneaking" it in implicitly (Berkhofer, 1971; Fogel in Fogel and Elton 1983; cf. Laslett and Wall 1972 on explicit vs. implicit quantification). This point will be developed later; the essential argument, however, is that *there is ALWAYS some use of a theory of human behavior, or human nature,* of human thought or emotions that needs to be made explicit and clarified. Although Collingwood's theory explicitly rejects that (e.g., Martin, 1977), in fact this is required by Collingwood's position, if one is to be able to "think oneself" into past ways, although one, of course, also needs specific knowledge of the group, its culture, the context, and the like (cf. Pipes 1986). An explicit model of human nature is also required of any form of uniformitarianism, essential if one is to link past to present, as all historical study must do. At the same time, it is made more feasible by recent work in sociobiology, cultural evolution, and evolutionary baselines. It is for social science what other elements of so-called *metaphysical blueprints* are in the natural sciences— and forms part of them.

Underlying that argument in turn is the view that history needs to become more rigorous and explanatory, and possibly even causal. Even those opposed to a "scientific" approach often accept this, even while pointing out some of the pitfalls. One, for example, is that a historian cannot do what a scientist might do; that is, repeat the whole process after having abstracted one or more of the factors to see whether in such cases the outcome would be the same (Porter 1981, p. 167, citing G. Kitson Clark 1967, pp. 22–23). This position may be questioned. Its prevalence, however, is precisely why I will examine the historical sciences in the next section. Most of them cannot repeat the process either. On the other hand, the use of comparative approaches of many varied instances, as well as the use of simulations, experimental, and observational work in contemporary situation may, in fact, offer possibilities not admitted by those like Clark and Porter.

What I propose to do is to accept the fact of the existence of at least a minority "scientific" position (in my sense) among historians as significant evidence that it is at least a possibility. I will then, as before and later, attempt to summarize and synthesize their positions and try to show that they converge with mine and hence constitute a supportive argument.

I begin with what I called the "intermediate" position on that question. One such position argues that although the application of social science methods to history is acceptable, and the search for patterns and regularities useful, this is different from turning history into a social science (Bullock, 1977). It seems that the first three points support my argument; the conclusion seems to involve partly a jurisdictional dispute and, mainly, disagreement between what I called a *narrow* and *broad* definition of science, and history could still meet the characteristics I have listed. Thus, for example, Aydelotte (1981) accepts the need for rigor, for avoiding "intellectual sloppiness," and for systematic research. He argues for the need to consider theory, test hypotheses, and to look at the relation of history to science. However, he cautions against excess, suggesting that behavioral historians are afraid to admit the need also to interpret. However, given my argument for the centrality of inference in any study of the past and its essential role in much (if not all) science, this position is quite in keeping with mine.

As we move closer to those positions that support my position, we find the view that by applying scientific approaches, old topics can be seen in new ways. Moreover, such approaches can introduce new substantive topics and the use of models of various sorts—as already seen (cf. Berkhofer 1971; Fogel in Fogel and Elton 1983; Frederickson 1980; Kammen 1980; LeGoff and Nora 1985; Mandelbaum 1977; McClelland 1975; Todd 1972; etc.).

An important additional point needs to be made. As is the case in architecture, the discussion often ignores the distinction between what has been referred to as public science and private science. What I mean by this is the difference between how a discipline as a social organization evaluates the validates claims and admits them into its corpus in consensual, open, objective (because intersubjective), and explicit ways, and the much more idiosyncratic ways in which individual researchers operate— although they, unlike architects, typically try to test and validate their own work according to the public criteria (e.g., Rapoport 1983b; cf. Ziman 1978, 1984). This public dimension, group consensibility, the use of agreed-upon taxonomies, definitions, rules for testing, and mutual criticism all leads to consensuality, convergence, and cumulativeness. Yet even when this is not explicitly discussed, it is often implicit, for example in the call for the use of models, for making models explicit, and so on (McClelland 1975, p. 229). At the same time it is emphasized that methodology in history involves advancing hypotheses and showing that the surviving evidence (which is all one has) is consistent with these hypotheses (McClelland 1975, p. 241). He also points out that all causal explanation or generalization is uncertain but essential. Generalizations are a gamble, but they must be used; the key to the best possible success is reliance on the "best possible" generalizations. In a history that is scientific, generalization needs to be done more self-consciously, that is, explicitly. There needs to be more clarification of variables and of the mass of causal generalizations, consistency between hypotheses and data, and so on. Again, the argument parallels mine. In fact, there is agreement about the conclusion that generalization is not only inevitable but always present and must be made explicit and moreover that contemporary questions need to be asked of the past, questions often surprisingly like those I ask in EBS (e.g., Gottschalk 1963, p. 125).

These positions provide potential linkages with the historical sciences, archaeology, and the social sciences. In the last section, I reviewed suggestions coming primarily from social science about closer links to history. Similar suggestions come from history. Thus, for example, the point is made quite early in this redefinition that history and anthropology can draw strength from each other. Ethnographies are as particularistic as histories, but anthropology has had a more developed and explicit comparative and generalizing tradition and a greater emphasis on theory. As a result, it has been able to go further than history. On the other hand, the interest of anthropology has been more limited, being concerned mainly with cultural change, although that in itself is historical. By joining with history the more general purpose of developing time depth would be served; both would thus benefit (Lewis 1968).

Somewhat later, it is reemphasized that anthropological attitudes, approaches, and methods could greatly enrich history. At the same time history could help anthropology. Links between them, a "blending" between traditional history, philosophy of history, anthropology, geology, paleontology, biology, archaeology and so forth would be most useful (Hodgen 1974). This is the kind of interdisciplinary approach I have repeatedly discussed. In this connection history is defined as *the total dated and documented record of the human past.* This is much broader than traditional history and, if "documented," is interpreted broadly—virtually the same as my view of history as everything we know about the past. Again, in line with my argument, the available corpus of data about the human past is seen as a resource, a vast body of evidence, as yet unarranged, about possibly repeated attempts of people trying to cope with various situations and trying to alter or change their condition (or the environment). As I do, Hodgen (1974) then sees this material, about many groups and many cultural and geographical settings over long time-spans, as providing much richer evidence than naturalists have, although the latter have handled it better than historians—or social scientists, who have ignored it. If it were ordered, classified, arranged, and correlated with other events, patterns and regularities could be identified. This often inductive process is, as we have already argued, an essential first step in much research. Taken *overall, this argument is almost identical to mine.*

The examples used, for instance, the spread of printing presses in Europe 1450–1500 C.E., or the spread of Christian churches between 1–400 C.E. illustrate how patterns can be identified: These spreads are neither random nor formless. At the same time the point is made, once again, that they need not be universal either (cf. Murphey 1973). I would make the point that such data could be used to test models or theory that is relevant today: for example, the diffusion of innovation in developing countries. Moreover, this has potentially fruitful links with other diffusion studies such as, for example, the many in geography. The time depth provided by historical data would help to reveal further patterns and possible regularities. In order to explain, description or classification is not enough; one needs hypotheses or theory. Such theory would often tend to be related to process as I argued earlier. Hodgen also makes this point, suggesting that to explain one should study processes. She posits the existence of repetitive actions and operations over time that are involved in the accomplishment of an end—whether natural, artificial, or cultural. This is related to the notion of people trying to solve the same problems (e.g., King, 1982; see also Chapter 5). In this connection I would argue that the inherent teleology of the built environment is useful, that is, the implication that the final cause is more important than the formal, with material and efficient causes as modifying (Rapoport 1969a; cf. Collingwood's view of history as the study of purposes; cf. King 1982; cf. Nicholson 1983 on "goal-directed theory"). The problem in such process-oriented explanation is no greater than in geology—the field that introduced the historical dimension into science. Neither is it greater than in botany, zoology, evolution, paleontology, or the other sciences discussed in the next section. Finally Hodgen agrees with my broad view of science when she suggests that the scientific attitude is distinguished from the nonscientific not by the failure of the latter to get results but *by the willingness of the former to recognize that when acceptable results cannot be gotten one way, others are available for trial.*

The same theme, of trying to answer questions *by means of all relevant evidence,* which is the essence of my argument, is also emphasized in a consideration of the different approaches to the past of archaeology and history (Dymond 1974). Given these differences, it follows that history and archaeology (and, I would add, all the other disciplines concerned with the study of the past) should get together. It is possible to combine written and documentary evidence (typical of traditional history) with material culture and artifacts studied by archaeology. Again, one encounters the common and sensible argument that this will lead to new types of studies—a "total archaeology" that would then enable studies of, for example, a section of the cultural landscape as fully as possible. This is, of course, also what historical geography does. One could also add that this might lead to old topics being seen in new ways, particularly if the many other disciplines that study the past are included. Note also that this topic of the "cultural landscape" is the domain as I have defined it in Chapter 1 (cf. Rapoport 1972, 1977, 1982d, 1986a, etc.).

It is interesting to note that although most such arguments refer to the mutual help history and social science can give each other, the emphasis tends to be on the use of social science methods in history. We have already seen some reference to the help history can give social science; that tends to receive much less attention. Yet it seems clear that social sciences, for example, anthropology, can benefit by using oral and literary sources dealing with the past (e.g., Jain 1977). This is equivalent to the suggestion that cultures be studied historically, using both historical and anthropological methods, that is both past and present data. This, of course, supports my argument.

In essence, all such arguments are supportive in the sense that they agree that "historians" are people who study the past; that such study needs to use diverse evidence and hence must become interdisciplinary; that historians need to be interested in the present because that is the point of departure for studying the past. This does not, however, necessarily involve subjectivism (in the sense that each generation rewrites its history); it can mean studying the past as objectively as possible and even achieving convergence and establish "truth." (There is, of course, much debate as to whether either of these is possible.) What changes are the *questions* posed: *These* relate to the present and need to be posed *explicitly*. Frequently there seems agreement that such questions, in one way or another, have to do with processes; for example, with how people have coped with certain problems or challenges. The need to be explicit also comes up in connection with many other matters, for example, about models of human nature and behavior (e.g., Berkhofer 1971; Fogel in Fogel and Elton 1983; Todd 1972).

There is also an emphasis on the need to reflect more on the philosophical and methodological bases of history. This involves a concern with generalization (or with uniformities and differences), with explanation, with theories (e.g., Finley 1975; Gottschalk 1963; McClelland 1975, etc.) and also with the logic of the inferences and argument. There have, in fact, been a number of fundamental reexaminations of such matters, for example, the logic of history. In one such study, the argument is made that all historical disciplines have neglected to identify, and hence to develop, their logic. It is suggested that, first, there is a *tacit* logic of historical thought; second, that this can (and should) be made *explicit;* and, third, that historical thinking can be refined by applying logic intelligently and purposefully (Fischer 1970, pp. xv–xix).

As we will find in archaeology, although there is agreement on the general need, the specifics recommended tend to vary. In the arguments about these specifics, which often dominate the literature, the more important underlying general agreement is often difficult to see. However, even the specifics often make points that are germane to my argument. In this particular case (Fisher 1970), the tactic is to get at logic by considering illogic, that is, fallacies. This, however, is of less interest than the suggestion that the logic is neither that of deductive inference nor inductive, but *adductive*. This involves adducing answers to *specific questions* so that a satisfactory explanatory "fit" is obtained. Note not only how closely this supports my argument but also the often found emphasis on the importance of asking questions as the critical step (my Chapter 2, cf. Northrop, 1947). The answers may be general or particular as the questions may require; this seems to imply that history is not *inherently* either idiographic or nomothetic.

To use an example already used, a question requiring a particular answer may be: What were the Acropolis and contemporaneous Athens like? The purpose here is to see how the particular high-style environment fitted into its context, how the two worked together. A question requiring a general answer might be: In what ways do high-style environments relate to their vernacular ones? A large sample might then be studied and, for example, two major types of relationships discovered. High-style elements might fit in a certain number of ways within a vernacular matrix or, alternatively, the vernacular environment may occur within a variety of high-style frameworks. In each case, if the evidence is large and diverse enough, the variety and range of specific relationships might be specifiable. The case study later in this book is another type of question requiring a general answer.

Fischer (1970) relates this adductive logic to Popper's (and others') *conjectures and refutations* model. (Parenthetically, increasingly that is also seen to be the model of design as opposed to the analysis/synthesis model.*) Historians ask open-ended questions about the past. The answer is in terms of selected facts arranged in an explanatory paradigm. These questions and answers are fitted to each other by a complex process of mutual adjustment. The resultant explanatory paradigm may take many forms—statistical generalizations, causal models, narrative, motivational models, analogy, and many others—as they do in archaeology. Most commonly it is a combination of these. The important requirement is that it be articulated in the form of a *reasoned argument* that, I would add, needs to be made explicit. The point is also made (Fischer 1970, p. 307) that supports my argument in Chapter 2, that any serious attempt to answer the question of what is good history leads to the question of what history is good for.

Explicit Supporting Arguments

I have now discussed some broad overviews of recent changes in history very similar to those I am suggesting and some specific examples that implicitly support portions of my argument. I will conclude with a review of a series of works that most closely approximate my position and provide explicit supporting arguments. I will arrange these chronologically (1963–1983) to show three things. First, that they all agree in certain essential features; second, that they show development and become more definite over time; and third, that they support my argument.

As early as 1963, in a report of a special committee, the point was made that even narrative history contains generalization because it is essential (as in analysis); such generalization must be made explicit (Gottschalk 1963, p. 186). Arguments advanced against generalization are reviewed; the conclusion is that they are not valid (Gottschalk 1963, pp. 145–146). The point is made that *significant general statements do not need to be universal laws and that the problems of verification are not insuperable.*

---

*I am ignoring the many recent criticisms of Popper's model.

Generalization advances, from simple to increasingly complex, so that events originally viewed as unique may eventually be discovered not to be so: Narrative then tends to give way to ever more complex analysis. This view is not unique, and there is also agreement that how one forms generalizations consonant with evidence is the critical issue (Aydelotte, 1981). Data do not automatically reveal pattern—that must be sought, which is not easy. How one gets ideas is not the important issue—how one *tests* them is (Rapoport 1969e, 1983b). Such tests of explicit ideas, assumptions, and hypotheses can then be against current research, research in other relevant fields, new data can be sought, and so on.

The fact that generalization and verification are both possible means that the scope and reliability of historical statements can be extended. Although it is impossible to be completely sure about anything, one can come closer to making a convincing case for some things than others. Any disagreements or biases (partly inevitable in all research, including the natural sciences) are solved by intersubjective agreement (Gottschalk 1963, p. 160; cf. Ziman 1978 and a large literature in the philosophy and sociology of science). Also, once things are made explicit, so can controversy be made explicit, and points of disagreement can be identified (cf. the comparison between my views and Sennett's about neighborhoods in Rapoport 1977). They can then often be resolved, the extent of agreement identified, and so on. Arguing parts is more useful than arguing wholes (cf. Efron 1984; King 1982; Morris 1983, Phillips 1976; Rapoport 1977, 1986c, 1989a; Wilkins 1978). Scholarly discussion also depends on total, unqualified, and unconditional agreement on the ineluctable and binding quality of the data (Handlin 1979, p. 20). This is the first form of intersubjective agreement necessary. In this process, theory is essential and is inescapable in history because facts do not "speak for themselves"; if one wants to know why, one must theorize. The choice is, therefore, not between "factual" and "theoretical" but between theoretical assumptions that are *half hidden, unordered, and chaotic* and those that, as far as possible, are *recognized, made rational, and explicit* (Handlin 1979, p. 187). The question for history is then whether its common working assumptions can be made explicit, refined, and systematized. This and the emphasis on testing the validation are very close to my argument, as is the need for theory and the need to make theorizing explicit (e.g., McClelland 1975, pp. 217, 223).

This need for theory, or theoretical frameworks in history, is also emphasized in one of the most explicit supporting arguments I have found (Berkhofer 1971). Theories, as we have already discussed, are taken to be integrated sets of concepts, generalizations, and hypotheses. The point is made that theory is more developed in the social sciences such as psychology, anthropology, or sociology—although not everyone would agree (regarding sociology, see contrary views in Freese 1980; Wallace 1983; Willer and Willer 1973, etc; on anthropology, see Betzig 1986). One could, however, agree that when Berkhofer wrote his book (first published in 1969) those fields were more aware of the philosophy of science, more self-conscious about theory (and the methods that follow from it), and more rigorous than history. In any case, this leads Berkhofer to argue for *a behavioral approach to history based on social science models.*

The crux of the argument follows from this point (Berkhofer 1971). Because history is the study of the past of humans, it must begin with human behavior. Behavior, however, can only be inferred—from the remaining evidence. Hence in dealing with the past it is necessary to infer behavior—which social scientists can study or observe. One must, therefore, have some theory about how physical objects relate to behavior and about "human nature," as it were, and about how such evidence can be interpreted and synthesized. It is remarkable how close this is to the way I have defined the problem; it is also almost identical to the key arguments in archaeology, as we will see in Chapter 5.

It follows that regardless of the specifics of any investigation, historians (i.e., those studying the past) must make assumptions about how the things studied fit together: what the relation is between parts and wholes. These assumptions (or theory), although typically implicit must be made explicit. Two points need to be added. First, that this is at odds with Collingwood's argument (1939, 1961), although, in my view, his approach also demands that (cf. Martin 1977, who, however, only addresses philosophy and ignores the scientific literature on human behavior). The second point follows. Since Berkhofer wrote (first published in 1969), one can say much more about human nature and behavior as already discussed (see also sociobiology and biocultural evolution in the next section). In that literature, one even begins to find lists of components of such behavior (*repertoires* in my terms) as well as discussions of the possibility that there may be constant causes of variable expressions in time and place. For example, social stratification suggests assumptions about human nature in the evolutionary character of society that are promising for a general theory of social stratification. As the behavioral sciences effect their rapprochement with evolution, there is greater promise still for cooperative pursuit. It is likely that one will be able to begin to generalize about human nature, human biograms, and the like, and the interaction of these constancies with contexts leading to specific expressions.

After this digression, let us return to Berkhofer (1971), who continues by making the by now well-known point that all data, whether in science or history, are approached within some framework but that *this framework needs to be made explicit and its choice justified*. He further argues that comparative history is essential, that hypotheses are to be tested by comparison of similarities and differences of units deemed to be comparable. Uniqueness can only be studied by comparing it to other, more typical things that show pattern and regularities, that is, create expectations—a point I have already made. His emphasis on comparative work is, of course, central to my argument; it also reappears in Chapter 5. The choice of units can be a difficult task and needs to be based on an emic analysis (i.e., be a *derived* rather than an imposed etic) (Rapoport 1973a, 1975c, 1980a,c 1987a, and case study). In any case, the comparability needs to be argued and made explicit (Rapoport 1986c). The emphasis on comparison, of course, leads one inevitably, I would suggest, to my argument about the characteristics of the evidence required.

Berkhofer denies the distinction made between history, as being particular, concrete, and unique and science (including social sciences) as being general, abstract, and repetitive. Like science, history seeks explanation; but in history an eclectic approach to explanation is necessary. This may be a "hideous hybrid" from the point of view of the logician, but it is the only possible approach—as all my own work shows and as I have long argued. It is, therefore, sensible to use an eclectic approach in reconstructing the past; it is eclectic because "analytically disparate" not because it is "miscellaneously organized" (Berkhofer 1971).

In spite of this eclecticism (or possibly *because* of it), Berkhofer concludes that there is a need for the use of explicit models. He even proposes a specific general model. However, Berkhofer's specific model is far less relevant for my discussion than his general argument. Moreover, what is in this case a programmatic statement is becoming almost a reality (e.g., Betzig 1986; Ruse 1986; von Schilcher and Tennant 1984).

The point is also made that, in effect, "history" is overgeneralized because there are many different kinds of history; one incomplete count lists 13 (Todd, 1972). However, in spite of this, these different kinds of history have some commonalities and some general conclusions are possible. For example, historians use laws rather than discover them. Their task is very similar to that of empirical scientists in that the same reasoning applies to history as to science: the forming and testing of hypotheses. One difference is that most historians are primarily interested in special cases of general laws rather than the laws themselves.* Todd defines the subject matter of history as *human activity directed toward solving problems*—a view we have already encountered in the previous section and before that (see also Rapoport 1969a; cf. Hodgen 1974; King 1982; Chapter 5 on historical human ecology). In our case, the activity is directed toward solving problems concerned with the shaping and using of physical environments in ways congruent with culture, values, life-style, ideals, meanings, and so on. The models (or laws) used are thus implicitly about human behavior. One needs to replace such implicit models, often based on oneself (or one's own culture) as the yardstick, by explicit models based on derived etic categories.

Like Berkhofer (1971), Todd proposes a specific model, in this case based on operations research that, he argues, is more applicable to the past than the present or future. This, through dynamic simulation and the like allows conditions and rules to be changed, counterfactuals to be used, and so on.† He argues that counterfactuals are very important in all 13 types of history which he considers. Others have also argued that causal explanation in history is related to model building and the use of counterfactual speculation (e.g., McClelland 1975, p. 170), although this view is also rejected and there is much debate. Again, the specific model is less relevant than the general argument that concludes that the purpose of history is explanation and understanding; in fact, historians should understand the historical situation (the past) better than the participants. Todd thus takes an extremely optimistic view of the possibilities of historical inference if it is approached as a science.

The emphasis on the need for self-consciousness and explicitness is almost universal in this literature. *Historians generalize all the time*—the more conscious they are of this fact, the more control can they have over such generalizations (Finley 1975, pp. 60ff.). Because generalization is inherent in their work, it follows that it is both absurd and self-contradictory to refuse to generalize. On the other hand, Finley is more skeptical about being "scientific," arguing that one cannot make explicit and define every term, concept, and interrelation. Two comments can be made. First, his argument for self-consciousness implies the maximum possible explicitness. Second, the lack of the

*The fact that this is their primary interest (or was in 1972) does not necessarily mean that it should be—or that one type of history could not deal with establishing laws.
†See Chapter 5 on simulation in archaeology and the next section; in simulation even chance and randomness can be introduced, and one can create "different worlds" (e.g., Cornell and Lightman 1982; Hodder 1978b, 1981; Renfrew 1984, etc.).

concept of "public science" again plays a role. Taking that into account, the cumulativeness and growth of intersubjective agreement, and eventual convergence and agreement means that one need not define over and over, and the explicitness grows and also does not need to be repeated at fundamental levels.

As one continues to review this chronologically arranged material, the recurrence and repetition of themes seem to be equivalent to convergence; they also tend to become specific and, particularly in the latest work, less programmatic statements and more actual achievement. At the middle of the range, for example, we find agreement about the need for an eclectic approach in history because no single method is sufficient (Mandelbaum 1977; cf. Berkhofer 1971; Kaplan 1964; Rapoport 1970a, 1973a). Although this is also the case in science, it is even more important in history because the latter is less unitary; thus different approaches, methods, and modes of explanation are necessary. The aim of all these is objectivity. Although this is a term used in different ways, all such uses concern the accuracy or reliability of knowledge.

Mandelbaum (1977, p. 150) suggests that knowledge is objective if, and only if, it is the case that when two people make contradictory statements about the same subject matter, at least one of them must be mistaken. For any area of discourse, therefore, it is necessary to specify how one establishes which statement is mistaken or gives reasons to reject both. The test must be applied directly to what is being affirmed or denied: It cannot involve the individual's experience; in this, history is like science.

I would add another important proviso. The individual's preferences are equally irrelevant; this is why the close conjunction of architectural history with architectural criticism, and the reliance of both on preference, has such negative consequences. The result of this is a critical additional requirement. Statements purporting to be true are factual statements. Providing that these are not meaningless, they "can only be either correct or false"; not "good" or "bad," "desirable" or "undesirable," neither do their consequences matter. "Factual statements can only be faulted on the grounds of the wrong perception of the facts or faulty inferences from them" (Pipes 1986, p. 36).

The cumulative results of history can, in most cases, be regarded as establishing knowledge that is objective in that sense (Mandelbaum 1977, p. 150). In some cases, general theory is needed in order to decide among alternatives; it is therefore necessary to consider theory. Mandelbaum argues that there is no universally agreed meaning of this term but suggests that it can be seen as a widely applicable hypothesis that serves as an explanatory framework through which a variety of observations and, ideally, a variety of laws, can be connected with one another. The unifying function of a theory depends on the theorist's ability to show that the basic concepts and assumptions of the theory can be applied to a wide variety of phenomena and can usefully connect a diverse set of apparently independent laws (Mandelbaum 1977, pp. 156–157),* and, I would add, phenomena, laws, and theories from other fields. In spite of statements to the contrary, one cannot have history without at least implicit theory, although it is far

---

*Darwin provides a very useful example (see next section on evolution). At a much more modest level this has always been central in my work. Cf. my recent work on nomads, children, elderly, streets, plazas, or Third world environments. These all are applications of a single theory and seem to work. Moreover, they agree with much other theoretical and empirical work from many other disciplines (see second book on theory).

better to make it, and generalization, explicit; in any case, historians use theory and generalization even if they deny doing so. To confirm theory, it is necessary to look at *a sufficiently broad range of the available data.* In both respects, the need for theory and the use of broad bodies of evidence to test it, the argument supports mine.

Finally, in a debate between an advocate of the new economic history (cliometrics) and a more traditional historian, many of the themes discussed are most clearly brought out, as well as areas of agreement and disagreement (Fogel and Elton 1983). This also makes the point about the value of explicit and reasoned mutual criticism that I have emphasized. Note also that it is possible to generalize about history in general, while starting with a specific and specialized field within it.

The distinction is made (Fogel in Fogel and Elton 1983) between traditional history concerned with the specific, whereas scientific history is concerned with the general. The example used is that the former is concerned with the death of John Keats, the latter with why deaths from tuberculosis were so common in the first half of the nineteenth century.* In that sense Fogel agrees with the common view that the singular event is the domain of history, classes and generalizations that of science. The disagreement lies in the further argument that this traditional preoccupation of history with the singular is neither inevitable nor desirable. On the contrary, it *can* and *must* change because, in fact, generalization is always present but only implicitly—a point we have already encountered throughout this review.

My central point about a shift from an art to a science metaphor is implicitly echoed by Fogel (pp. 54–55) when he contrasts the "artistic" nature of traditional history with the "scientific" nature of the new history. The distinction can be represented thus:

Traditional History	*Scientific History*
This is more artistic, in the sense that it is seen as the perfect creation of a single person in a short period of intense activity, like a novel, painting, concerto or poem (or building). It engages in heated controversy about wholes.	Like other scientific creations, it is the result of long periods of work, involving many people, which approaches "perfection" gradually. It is essentially a collective effort (and I would add, interdisciplinary). Criticism is more of parts than wholes. I would add that there is *organized* controversy (e.g., Ziman 1978, 1984).

Convergence with my argument (and the other studies reviewed) continues with the emphasis that the study of history always involves models of human behavior (Fogel in Fogel and Elton 1983, pp. 25–26). These are usually *implicit,* hidden, vague, incomplete, and internally inconsistent as well as, I would argue, externally inconsistent, that is, at odds with findings of other disciplines. *They need to be made explicit.* The relevant assumptions must be clearly stated and formulated so that they can be tested rigorously and empirically. This then leads to *new types of evidence,* although, as I have argued, redefining the domain to include new types of evidence can lead to the other changes: the interaction is two-way. Fogel argues strongly for quantitative approaches;

---

*There is a prior question omitted: *How* common were deaths from tuberculosis in the first half of the nineteenth century?

although these can be very useful, this is not, in my view, central to the argument. *Empirical testing and rigor can be qualitative.*

There is also agreement with the view already encountered that universality is not essential. Thus Fogel (p. 67) contends that models of human behavior must be place-and-time-specific to be useful in history (cf. Martin 1977; Murphey 1973). I would suggest that it is possible to distinguish between general and specific behaviors and that time and place specificity is also essential in cross-cultural analysis and other forms of comparative work (e.g., Rapoport 1986c). This means that *all* models and units of comparison must be carefully and explicitly specified and derived rather than imposed.

Fogel and Elton (1983, pp. 41ff.) agree that the differences between traditional and "scientific" history can be discussed under six categories. They also agree that not all characteristics are the same in each group: There may be a wide range in each group, and there may be overlaps. For example, at times scientific history looks at singular occurrences, whereas traditional history may look at general forces. Fogel and Elton thus call these categories *statistical.* I would suggest that more correctly they form a polythetic set (Clarke, 1978). This, I have recently argued, is essential in dealing with built environments and, probably, other matters (Rapoport 1982b, 1988b, in press).

Fogel and Elton agree that the first difference concerns *subject matter;* what I called the *domain.* This is fundamental: Science deals with collectivities of people, recurring events, patterns, and the like; traditional history with individuals.* This leads to a most important difference regarding *research agendas.* In *traditional history,* the focus is on individuals, particular institutions, particular ideas, and nonrepetitive occurrences. There is a reliance on literary evidence and the use of implicit behavioral models. This seems very close to traditional architectural history, for example, the high-style individual building (masterpiece) by a famous architect with no reference to users or their behavior. In *scientific history,* on the other hand, the focus is on repetitive occurrences and categories of individuals, groups, and institutions. Evidence tends to be quantitative, although that is not essential. Explicit behavioral models are used. The evidence used is different; it is studied differently and authenticated differently. (This seems very close to the domain as I have defined it and to the approach I have recommended: all environments, cross-culturally and over time, and the whole environment and behavioral models based on EBR, etc.).

This leads to the second difference that concerns *preferred types of evidence.* Both types of histories use surviving evidence because that is all there is (cf. Murphey 1973; and discussion in Chapter 5). One needs to use all the evidence available—which supports my argument. In this, Elton agrees with Fogel and goes even further (e.g., p. 80). One must not eliminate any question about the past and use the broadest range of evidence as well as the greatest possible variety of techniques and methods. Essentially, the argument is for being *eclectic regarding both data and methods,* and they agree with the point already made repeatedly that new kinds of data and new methods open up

---

*Note that I have argued that neither in EBS research nor in design is one interested in individuals but in aggregates or groups, although one certainly wants to know the variability within groups (e.g., Rapoport 1990). Hence my interest in culture, life-style, and so forth. The question of what are relevant groups in any given case then becomes critical (e.g., Rapoport 1980a, 1983a, 1985c,d, 1986c). This, in turn, involves questions of taxonomy and classification.

new areas of research. They also agree with my point that there is a vast amount of data already available, and constantly growing, so there is no need to start from scratch.

Elton most strongly disagrees with Fogel about the third point, dealing with *standards of proof.* In scientific history, it is essential to make explicit the assumptions and then to search for empirical evidence to support or refute hypotheses. For this also large amounts of data are essential.

The fourth point concerns the *role of controversy* that is different in the two types of histories. In traditional history, controversy tends to be much more heated and to be about fundamentals and wholes. In scientific history, it is more about parts and specifics, less heated, and closure is more rapid (cf. Kammen 1980). Note that this point has also been made regarding the natural sciences vis-à-vis the humanities and even the social sciences. The former are constituted to have organized (productive) controversy leading to closure and consensuality.

Elton and Fogel agree about the fifth point, that there are major differences in *attitudes to collaboration.* Scientific history, like science, typically involves many people and is interdisciplinary (like EBS and, increasingly, archaeology and social science). The Cambridge Population Group that I discussed earlier (e.g., Laslett and Wall 1972; Laslett 1977), is used as an example: It includes programmers, statisticians, mathematicians, demographers, sociologists, anthropologists, economists, physiologists, geneticists, nutritionists, epidemiologists, as well as historians.* I would add that this means not only that one inevitably uses more, and much more varied evidence, and a great variety of approaches, methods, and techniques; by increasing internal and external validity and rigor and because one needs explicitly to discuss the varied approaches and their assumptions, convergence is more likely and so is the chance of becoming cumulative.

Finally, Elton and Fogel also agree about differences in *communication with the history-reading public.* Traditional history aims for, and reaches a wider public than scientific history that is aimed at specialists. However, I believe that as it develops and grows, it also will begin to reach a wider public—as does the new archaeology and scientific research generally.

The different approaches discussed and contrasted by Fogel and Elton are useful in several respects. First, they show the extent of agreement and disagreement, making argument both rational and useful. Moreover, they well characterize the differences between the two approaches. These, however, are not mutually exclusive, nor need they be antagonistic—although they often are. I would argue that they can be complementary—as I tend to argue, for example, about different definitions of culture (Rapoport 1980a, 1986c). This point has been made by others. For example, it is suggested (McClelland 1975, p. 242) that both scientific history and traditional history are useful. One needs both the use of theory and models, of all the available data, rigor, and so on, as well as scholarship, an emphasis on the unique and distinct expressions of even "universal propositions" (assuming there are any). Context remains important. Both modes of research are needed to deal with all the questions: They would enrich each other, particularly if the more traditional approach adopted the general characteristics for which I am arguing.

---

*Other examples are given in Note 50, Fogel and Elton (1983).

Significantly, the supposedly "traditional" historian in this debate (Elton) agrees that one must be honest, look at all the evidence, and test the probabilities. About this all historians agree; they tend to disagree about what this means and how one goes about it. Among those historians sympathetic to a scientific approach, spanning 20 years (and longer in the case of France), there is much agreement: about the importance of generalization, models, and theory; about the need to be explicit and the use of explicit models of human behavior; about the need to be eclectic regarding method; to ask specific questions and formulate hypotheses; about using as broad and varied an array of data as possible. All these points are highly supportive of my argument.

THE SUPPORTING ARGUMENT FROM THE "HISTORICAL" SCIENCES

Introduction

Traditionally, as we have already seen several times, history has been distinguished from science on the basis of one major criterion—that whereas science deals with generalizations, classes, recurring phenomena, and even with "universals," history deals with singular, unique, and nonrepeatable phenomena.* We have already seen that there is a view in history that generalizations are possible and even inevitable, that patterns can be traced, particularly at the level of human problem solving, behavior, processes, and so on. There are, on the other hand, also sciences that deal with unique events, for example, our universe, solar system, the planet Earth, the origins and evolution of life on earth, and so on. These also tend to deal more with explanation than with prediction and also deal with data that is fragmentary and incomplete, for example, the paleontological record.

These sciences include evolutionary science, paleontology, geology, cosmology, the various "paleo" sciences (paleobotany, paleoecology, to name just two), and they can be referred to as "historical" sciences.† They are of interest in this context because at the same time as they deal with what are essentially historical phenomena and data, they are also accepted as being sciences. For example, Ziman (1984, p. 44) specifically refers to them as perfectly sound sciences, although the phenomena with which they are concerned cannot "strictly be predicted since they all occurred in past." However, theories about these phenomena may be satisfactorily validated by the discovery of "predicted" evidence after these predictions were made. This is the same type of prediction made in archaeology that I will discuss in Chapter 5. These sciences are clearly aware of their historical character (e.g., Wilford 1985) and it is significant that many titles can be found like *History of the Earth* or *The Earth: Its Origin, Structure and Evolution; The Origin and Evolution of the Earth's Continental Crust* (S. Moorbath and B. F. Windley (Eds.); London, Royal Society, 1981). These are clearly historical, yet part of the earth sciences. They are cumulative and show rapid progress; for example, at the Milwaukee Public Museum, the Geology Hall became out of date within 10 years.

---

*Like all single criteria for distinguishing between things (monothetic) this is inherently inadequate: One needs a polythetic approach (Clarke 1978, pp. 35–37; cf. Rapoport 1982b, 1988b, in press).

†I have found since I wrote this that others have used this term (e.g., Gould 1985 among others, who also makes some of the points I make in this section).

Moreover, they also increasingly show predictiveness, for example, where oil might be found and, eventually, about earthquakes.

The relationship between prediction and explanation is a major topic in the philosophy of science, and I will not discuss it here. It is often accepted that the relationship between them is neither symmetrical nor transitive. One can explain the mechanism of hurricanes or earthquakes without being able to predict them, although that is now beginning to change. Frequently, once explanation is achieved, prediction follows. However, genuine predictive power may be too distant an ideal and too narrow an indicator of progress for science in general (Ziman, 1984, p. 48). Also, once again, reliance on a single criterion for distinguishing science from nonscience is questionable (and questioned); a polythetic approach (or at least a multiple variable monothetic approach) is likely to be much more useful.

For example, neither paleontology in general, nor hominid paleontology in particular, is appropriate to the application of a rigid hypothetico-deductive program. There, fragmentary information is accumulated more or less at random and has then to be interpreted as evidence for or against various equally fragmentary theoretical schemas.* Yet such disciplines are as much part of the scientific enterprise as are theoretical physics or molecular biology (Ziman 1984, p. 48).

Note that, as I have already argued, historical data become a laboratory, in my case for the study of EBR under different conditions not achievable in the laboratory, by observation or interviews, questionnaires or tests. In this sense, these data are just like historical data in paleontology, evolution, geology, or cosmology. For example, in studying the sun, data from stellar evolution are used. Moreover, I would argue that physics (and possibly chemistry) far from being the models for science may be the exception. Thus even molecular biology is earth-bound, at least at the moment; there is also no evidence, and some doubt, that its generalizations and predictions will apply to extraterrestrial life, if any (e.g., Gould 1985). Finally, there is also increasing emphasis on the temporal dimension in physics and chemistry. Physics has come together with cosmology, and there seem to be arguments that there are differences in physical laws depending on the time after the origin of the universe. Regarding chemistry and science generally, it has been suggested that increasingly the dimension of time is important: Science is being "historicized" (e.g., Prigogine and Stengers 1984).

It is also of interest to note that some of these sciences, for example, astronomy, geology, and biology (in addition to meteorology and ecology), were used as analogs by Roger Barker and Herbert Wright in their development of ecological psychology; this, in turn, has had a considerable impact on EBS—for example, through the concept of "settings."

Although these sciences have been discussed before, they have not been related to one another quite as broadly as this, not applied to the subject matter of this book (the built environment), nor yet related to arguments about history as the study of past phenomena.

Even less than in other sections of this chapter is my interest in any specific conclusions, findings, and the like; even less than in other sections can I review many

---

*As will be seen later in discussing paleontology, information is no longer accumulated quite as randomly: Systematic searches are often carried out.

studies in any one field. Neither can I review all the possible fields. I can just discuss a sufficient range of them to give their "flavor," as it were, and briefly review certain limited features of them, one by one. I wish to draw attention to the *process* whereby problems are tackled; problems that, by the very nature of these fields, are historical. The issue is not the conclusions or results reached—from the point of view of this book, they are not important. Of interest is how they go about doing what is essentially history (e.g., Schwartz 1986; Wilford 1985). My purpose, as in the discussion of art in Chapter 3 and the other fields, is not even to suggest that the position discussed is the only one; in fact, it is often controversial. Because even experts disagree, I will not argue the merits of any position vis-à-vis any others. My purpose is rather to show that positions analogous to mine are at least tenable in disciplines that I regard as related in that they deal with the past even if those seem unrelated on the face of it, as do the historical sciences. In reading this section of the supporting argument, it is thus essential to bear in mind links to others reviewed and to note that there is a major core of similarity across all the disciplines reviewed and between them and my argument.

Science in General

Before briefly reviewing some specific historical sciences, it may be useful to elaborate on a point already mentioned: the existence of a more general argument about the increasingly historical nature of science as a whole (Prigogine and Stengers 1984). This view is, as yet, singular but even if accepted only partially, it further reinforces my argument by showing that the rapprochement is two way. Not only is history becoming more scientific, but science is becoming more historical. Given that, it follows that the importance of historical data increases in theory construction, generally, and in EBS.

The central idea in this "historization" of science generally is that time plays a role in all science. It is suggested that history began with the study of human society. Then temporal considerations entered geology and biology, then cosmology, and only now are they entering chemistry and physics (Prigogine and Stengers 1984, pp. 208–209). This means that in science, history has moved from the middle scale (biology and geology), where it has long been accepted, to the cosmic scale (cosmology) and the subatomic scale (physics), where it has hitherto been excluded. It is worth noting the important point that EBR also occurs in the middle scale, which has major theoretical implications, some briefly discussed here, others that I hope to explore in my book on theory.

Thus cosmology, which I included as one of the historical sciences several years before reading Prigogine, deals with the birth, development, and death of stars, galaxies, and even the whole universe. The universe itself has a history. This is the final stage of the *progressive reinsertion of history into the social and natural sciences* (Prigogine and Stengers 1984). This historization of science has several implications. One is the discovery of temporal heterogeneity, for example, irreversibility, and the importance of time and questions about its nature (regarding our field, see Rapoport 1977; cf. Carlstein, Parkes, and Thrift 1978a,b,c; Parkes and Thrift 1980). With the dynamics of time entering the natural sciences, the opposition (or gap) between the "two cultures," for example, science and history, is no longer valid (if it ever was). Of course, not only is science becoming more historical on this view; in turn, as we have seen, history can be seen as part of science and hence more scientific.

Furthermore, if the universe itself has a history, this means that nature is irreversible and not symmetrical relative to time; it also means the end of universality (Prigogine and Stengers 1984, p. 217). For example, one now sees discussions of dimensions that were different immediately after the Big Bang, and of forces, fields, and particles that have changed over time (e.g., Crease and Mann 1986; Feinberg 1985). If accepted, this has several implications. First, it has a major impact on the role of "covering-law" models of explanation that, for example, have been a major bone of contention in archaeology. Second, it means that some physical phenomena are no longer accessible to laboratory (i.e., experimental) manipulation. More important for the present discussion, it does not mean the end of science, nor does it mean that physics becomes "subjective," the result of human convictions, preferences, and beliefs: There can be an objective theory of irreversible processes (Prigogine and Stengers, 1984, pp. 218, 252). Science remains subject to constraints (e.g., Bunge, 1983; Jacob, 1982; see Chapter 5). One implication is, however, that one needs to study things "statistically," that is, deal with large populations. This clearly reinforces my argument about using the widest possible body of evidence, all that people have ever built.

I will not argue the pros and cons of Prigogine's view, which is not accepted universally, nor his interpretation of the philosophical implications, which I find rather unconvincing and which, in any case, is not germane to the present discussion. What seems significant is that even these "non-Newtonian" phenomena, self-organizing processes, and the like can be studied "scientifically"—rigorously (even quantitatively) with generalizations, using both experiments and computer simulations. It is also interesting that Prigogine and Stengers (1984) move from thermodynamics, chemistry, molecular biology, and the like, through the role of "randomness," fluctuations, and chance, to the self-organization of termites' nests without a "mastermind" (Prigogine and Stengers 1984, p. 205). This has implications for cultural landscapes (Rapoport 1972, 1980a) and vernacular design (Rapoport, in press); thus I have recently argued, using biological concepts, that vernacular and other such design is selectionist, whereas professional design is instructionist (Rapoport 1986c; see previous discussion). If the processes underlying this selectionist nature of self-organization and evolution of order found in termites' nests can be understood and if they can be shown to apply to vernacular, then they may be translatable into an instructionist mode. It also provides a linkage with sociocultural evolution, the evolution of human culture from animals (Bonner 1980), and animal architecture (von Frisch, 1974). Prigogine's analysis eventually includes the evolution of urban systems, and the mathematical computer simulation includes history that earlier work (for example, Christaller) neglected; this makes it more sophisticated. It is significant that in archaeology also, very similar simulations are being made (e.g., Renfrew 1984; Sabloff 1981; and others discussed in Chapter 5).

Cosmology

The popular literature on cosmology, the history of the universe, is very extensive (e.g., Crease and Mann 1986; Gibbons *et al.* 1983; Pagels 1985). I will review just one book that seems to show how what is essentially a historical subject can be studied "scientifically" (Cornell and Lightman 1982). This book also demonstrates well a topic we have already briefly seen as relevant—the relationship between theory and empirical data. Only a small part that seems particularly relevant to history is considered.

The interaction between theory and evidence is complex. Because science (like all scholarship) is a human activity, a verb not a noun (Cornell and Lightman 1982, p. 1), this relationship is dynamic and varies in different sciences, in the ways in which they are combined and to what degree. It is also the case, however, that whereas history is also data based, it deals with what no longer exists and what is inferred from evidence surviving in the present. Evidence is obtained by some form of observation and is critical; without it, theory can have nothing to say about what actually does exist. Theory itself can only suggest all the possible things that *might* exist: observation shows what actually exists (cf. Bunge 1983; Jacob 1982). But observations (or other ways of obtaining data) need theory that provides the conceptual framework.

It follows that there is not a neat, systematic progression from theory to observation to testing to solution (Cornell and Lightman 1982). It is a much more complex process where theory and assumptions may lead to blind alleys, observational evidence may challenge theory, new data or phenomena may require new theory. On the one hand, fresh data often lead to unexpected ways of answering questions. On the other hand, such fresh data are often the result of new questions or new conceptual frameworks. This is reminiscent of the arguments we saw in history, but the point is also made that the human mind, *if it goes about it correctly,* is able to perceive, sort, and select from the wealth of data and variety of explanations the precise information necessary to understand a given topic or problem.

Astronomy and cosmology, although historical, are cumulative, showing increasing sophistication. They are, of course, closely related to physics and use well-tested theory from that discipline (Cornell and Lightman 1982, p. 12). This we have also discussed repeatedly: That concepts or theory from one domain are used in another domain, for example, social science in history, ethnography in the study of art, a variety of sources in archaeology, and so on.

Typically, theory leads to prediction. It can then be asked whether one can predict the past. The answer Cornell and Lightman (1982) give is: *YES!* For example, Chapter 7 deals with the evolution of the universe and its age. This is difficult because it is a *unique event.* Yet prediction can take place for example by using simulation, doing cosmology in the computer (what they call an "alphabet soup"). They use data going back to 1720 C.E. to extend the range of observational data. At the same time, the ability exists to represent models of reality that go beyond all current observational knowledge, that is, to *predict observations.* I would add that one can also make predictions of consequences of given theories that should still be observable, for example, background radiation.

In such studies it is also possible to use different lines of reasoning, from different methods and even different disciplines, a point I have already made and that is much used in archaeology. If, when one looks at changes in such lines of evidence over time, one can see them converge, with greater agreement and greatly reduced change, then one is probably correct. This has happened when the age of the earth is considered but not yet concerning the age of the universe or stellar ages. These are still going up and down like a yo-yo (Cornell and Lightman 1982, Chapter 7). Although it is still relatively unclear, there does seem to be some convergence, and this has probably increased since 1982 (cf. Creese and Mann 1986). There is, of course, also a lower limit constraint—the age of the earth. Although these questions are distant, abstruse, difficult—and historical, answers are emerging: The approach works.

Similar points can be made about the evolution of the solar system (Cornell and Lightman 1982, Chapter 3). Once again, this is a unique, historical event. Yet convergence is high, and the field is clearly cumulative, showing a steady increase in understanding. Once again, different variables are studied by simulation of the beginnings. As in archaeology, predictions can be made that are testable in reality, that is, the solar system as it is must fit views about how it has evolved. As in history and archaeology, the data are in the present, and inferences have to be made from it to the past, using models and theory. Note also that the use of very divergent bodies of data can be helpful. For example, the recent discovery of solar cycles in the geological record has had a major impact on the study of solar physics (Gleick 1986).

Even apparently ahistorical topics can be shown to be relevant to my argument, for example, the puzzle of the sun's hot corona (Cornell and Lightman 1982, Chapter 4). In examining how a set of questions has gradually been resolved over time, it illustrates the role of evidence; theories, even if *beautiful* (to use Dirac's term) give way if not supported by data and observation. Moreover, in this case, there are problems of connecting theory to data: This connection can only be indirect (which is increasingly the case in all science). It is this highly indirect connection that is very similar to, and highly relevant for, history. The indirectness of the connection is due to two reasons. First, the physical system is so complex that a rigorous complete theory that accounts directly for the data cannot be constructed. Second, laboratory approaches and experimental studies are not possible either in cosmology or in astrophysics. It should also be noted that in the case of many solar (and other) phenomena, even observations are highly indirect and inferential, for example studies of the sun's interior that rely on neutrinos (Shapere 1984). All these circumstances make the topic very relevant to the study of the past.

It is suggested that these indirect connections between data and theory can be made by the use of *scenarios,* plausible stories connecting the two. Note two things. The first is that scenarios and their equivalents are playing an increasing role in cognitive science (e.g., Minsky 1977; Schank and Abelson 1977); I have used their equivalents in studying meaning (Rapoport 1979d, 1982a). More important, their equivalents are implicit in traditional history, for example in Collingwood's views (1939, 1961) and explicitly used in more recent work (e.g., Murphey 1973).

In the present case (Cornell and Lightman 1982, Chapter 4), a scenario is a metatheory that provides both tools and a framework for comparing theory and observation—an approach similar to EBS and this book (see Figure 4.2).

From the scenario, one then tries to develop more detailed theories. The scenario itself provides an interpretive framework within which the theoretical consequences are then developed as observables and ultimately compared with observation. The data are then broadened further. In the present case, for example, data are used not just from the sun but from a variety of other stars. This, once again, is very similar to history and archaeology.

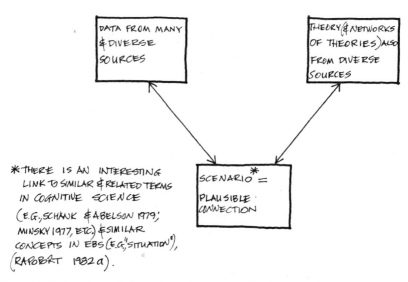

Figure 4.2. Linking data and theory through "scenarios" (based on Cornell and Lightman 1982).

Life on Earth—The Biological and Evolutionary Sciences

The origins of life on earth (and its subsequent development) are, once again, a unique event and hence historical even in a traditional sense.* Although there are many diverse theories about these origins, and limited agreement so far, the significant point is that this highly historical subject is being studied in a scientific context (e.g., Glassner 1984). Even when "extreme" (that is, nonparadigmatic) approaches are advocated, such as the "mineral origins of life," the approach is still scientific, the process of reasoning and inference explicit and logical, and hypotheses are closely related to data (e.g., Cairns-Smith 1982).

The study of the origins of life involves the use of a vast amount of data from diverse sources that need to be reviewed, fitted together, and explained before a theory can be proposed. This is equivalent to my argument for a broad and diverse body of evidence and the search for pattern. Moreover, one seeks convergence among many and diverse lines of reasoning and evidence, internal consistency, and, finally, external consistency: Any proposals must be consistent with existing theories in many different fields. The point has also been made (e.g., Cairns-Smith 1982) that any proposed theory needs to clear the "logjam of objections," that is, it needs to take into account existing theories that, in this field, are essentially chemical, based on various chemical "soups" and the action of electric sparks (lightning), and the like. The new theory leads to a need to reinterpret experiments, and it is also constrained by its need to be congruent with subsequent events, such as evolution. Finally, there is a need to suggest which experiments and observations can best help test the theory. It seems to me that

---

*This applies even if one accepts life elsewhere (e.g., Gould 1985) or that life arrived from elsewhere (e.g., Crick's Panspermia hypothesis).

the parallels with possible studies of the origins and development of the built environment are quite striking (e.g., Rapoport 1979a,b).

Similarly historical is the study of the evolution of life, which includes all biology, because the origins of species and their interrelationships, their DNA, and all other topics are related to evolution. The literature in these fields is immense, yet a few general points can be made that seem highly relevant to my argument.

At the outset, note that one can speak of "the growth of biological thought" (Mayr 1982). In other words, there is cumulative development from the ancients to the present. Aristotle is discussed at the outset—but not as *the* theoretician, as Vitruvius still is in architecture! In fact, Mayr argues that there have been no "revolutions" or paradigm shifts in biology but rather steady growth and synthesis. This does not, however, imply the absence of argument. On the contrary, it is through reasoned debate, question and answer, and mutual criticism that biology, like most disciplines, progresses.

As an example, there is much debate at the moment between orthodox neo-Darwinians (who themselves represent a development of the Darwinian view, the "new synthesis" of the 1940s) and the punctuationists (e.g., Eldredge and Stanley 1984; Eldredge and Tattersall 1982; Gould 1985; Stanley 1981; etc.). Evolution is as well verified a hypothesis as can be found in science; it has occurred, but its form must be explained by theory. This involves questions of how it works—a different matter entirely. This, of course, brings up the *importance of mechanisms*. In this case, the argument is once again about punctuated equilibria versus gradualism. This argument, like others in science, is concerned with understanding the distribution of properties in nature, that is, patterns, and the man-made environment is part of it. Even with regard to fossils, one can make predictions and test them against the fossil record. For example, gradualism predicts a systematic pattern of gradual, progressive change; punctuated equilibrium predicts quite a different pattern in the fossil record: changes in distribution and extinction (not adaptation) (Eldredge and Tattersall 1982, p. 53). There should be long periods of stasis and periodic, rapid bursts of change. These contrasting predictions can then be checked against the evidence of the fossil record. According to Eldredge and Tattersall (1982), that record supports the predictions of the punctuated equilibrium model (cf. Gould 1985; Stanley 1981; Eldredge and Stanley 1984). More generally, data destroy theories, ideas, and stories (Eldredge and Tattersall 1982, p. 122; cf. Bunge 1983; Jacob 1982). *Note that this all involves a search for pattern in the historical record and that the patterns depend on theory/data interaction* and even prediction. In fact, they explicitly turn to *patterns in history* and try to show that in history proper, one finds the same patterns of punctuated equilibria as in evolution (Eldredge and Tattersall 1982, pp. 161ff.). They conclude that *the scientific study of history is possible*—clearly a supporting argument.

Punctuationism is a response to three questions raised by Darwin. There have been different answers to these questions that themselves have remained essentially unchanged.* These three questions are, first, where are the links among species; second, what is the mechanism of evolution, and how can new species come into being; third

---

*Cf. King (1982) on the history of medicine as changing answers to the same problems—a view we have also encountered in history, historical human ecology, and so on.

(and the most challenging theoretically), how can one explain how small, random changes can result in the development of new, highly organized organs integrated with the rest of the creature's anatomy. Answers to Question 2 and, particularly, the third question only became answerable in the mid-twentieth century following the availability of new knowledge, a point we have already repeatedly encountered. It also shows the cumulative nature of the study.

Note how similar these questions are to those often posed in anthropology and archaeology, for example, about culture change. Note, moreover, that evolutionary theory is being approached as a science, yet as Mayr (1982) emphasizes, it is *essentially historical in nature,* which is clear from the literature (e.g., Colbert, 1980; Szalay and Delson, 1979). Also, in studying evolution one cannot do experiments, one cannot revive earlier species, one cannot create data other than those which have survived— which is also the case in history, archaeology, and so on. What can one do? According to Mayr (1982), one can try to connect observations with logical reasoning; try to devise tests for hypotheses; try to keep testing the whole structure for consistency internally (logic) and externally, that is, against empirical data and other branches of science. These criteria we have already repeatedly encountered; they are quite similar to those commonly found elsewhere. It follows that in spite of being historical, evolutionary biology is like the other sciences.

Mayr shows how in the 120-odd years since *Origin of Species,* new techniques and new methods, new disciplines (such as molecular biology), and developments and new data in biology, genetics, and other fields have changed the inferences being made—but in a consistent way. This is similar to cosmology; there the inferences are also historical; changes have also been cumulative (although for shorter periods than in biology), and progress has also been due to growth in observational capacity, the impact of new disciplines (for example, radio astronomy and the like), developments in physics, and so on.

Biology began as natural history, essentially observation. This was followed by classification. The latter (and the science of taxonomy) is typical of biology and is most developed there; many recent changes and developments, such as cladistics, numerical taxonomy, polythetic classification, and so on began there. These latter have now influenced archaeology (Clarke 1978), studies of primitive art (Ucko 1977), and some of my recent work (Rapoport 1982b, 1988b, in press). These changes are also conceptually relevant, because evolutionary biology is concerned with the history of life and that involves systematics (classification and taxonomy) that defines the structure of the evolutionary pattern. Evolutionary science is concerned with trying to reveal and explain the features of this pattern: the history of structural diversity and the history of taxonomic diversity. Note that diversity is also a major puzzle in environmental history (Rapoport 1985b,c,d, 1986c) and that the role of taxonomy is also emphasized in archaeology, as we will see in the next chapter.

These issues of diversity are approached through *pattern analysis* because "it is only by having some aspect of pattern that science has something to explain" (Cracraft 1983, p. 273). Also, only by having *general* patterns, can one recognize unique historical events that, Mayr (1982) argues, are "noise" obscuring general historical patterns. Recall my argument on the importance of pattern recognition and that only historical (and cross-cultural) data can reveal pattern so that comparative work using the widest possible body of evidence is essential (cf. Bullock, 1984). Note again the highly scientific nature of the argument regarding what is explicitly seen as history.

It can be argued that the history of living beings has many similarities to the built environment as I have described it. The latter is also characterized by extreme diversity, is complex and hence difficult to study: There is a great variety of forms, many of which can be seen as "transformations" of others as in biology (e.g., Thompson, 1961) or that seem to use and modify certain previously used elements, again as in biology (e.g., Gould 1980; Jacob 1982). It is, therefore, likely that there is some unity behind the diversity. Again, as in biology, built environments tend to show new (emergent) qualities that cannot simply be deduced from existing forms (or they would not be new) but that arise just from these forms (cf. Ayala and Dobzhansky 1974). It is also orderly and can be understood analogously to biology where this order is due to the interaction of subjects (organisms) and objects (the outside world, the environment) (e.g., Beurton 1981; Bonner 1988).

This is also the case with history that, it is often argued, is written by humans and therefore subjective. But, I have already argued, it cannot be made as one pleases because there are certain constraints both internal (human behavior and nature, evolutionary baselines, processes, etc.) and external (previous history, ecology, culture). The interaction of these produces the empirical evidence that constrains history in ways analogous to those discussed by Beurton (1981).

In spite of some attacks on attempts to use evolutionary analogies in architecture (e.g., Steadman 1979), the parallels between my description of the built environment and its history and the subject matter of biology and the "history of life" are very striking and highly suggestive. Partly, these different views relate to the use of rather different bodies of literature and are also based on a very different reading of the nature of the resemblance.

Some of the other implications of certain discussions in the philosophy of biology are also most useful, particularly, because the problem of teleology can be avoided; built environments, indeed all human artifacts, are made for a purpose that, of course, includes latent aspects that are frequently the most important. Most of these implications are more relevant to a discussion of a theory of the built environment so that here only a very few of these will be discussed.

For example, in the philosophy of biology, there is much discussion of explanation in terms of function. It is suggested that the scientific explanation of some phenomenon in terms of its function involves four steps (Bhaskar 1981, p. 199). First, a *functional claim,* for example: "a cow's tail is useful for swishing away flies." Second, a *consequence claim:* because of 1 cows come to possess long tails. Third, a *theoretical elaboration* of the mode of connection presupposed in 2, that is, between the functional fact and consequences or of adaptation and functionality (for example, Darwin's theory of natural selection). Finally, a detailed *natural history* of a specific case presupposing 3: for example, *evidence* that short-tailed cows perished. In biology this apparently causes many philosophical problems absent in the case of the built environment. The latter is always created for a purpose and for a function (including latent functions) whether in a selectionist mode close to natural selection or in an instructionist mode. Thus this kind of analysis should be easier in the study of the history of built environments than in evolution.

Inference is central in evolutionary biology as it is in all historical sciences and science generally. Such inferences are often described as having the fundamental characteristic *in principle* that they can be changed on good grounds. Thus, even when wrong they will change, that is, they are *self-correcting*. This can be seen in operation over short time spans. Thus it was recently claimed that the very different views about the time of divergence of hominids from other primates based on the analysis of DNA, so-called "molecular anthropology" (the work of Vincent Sarich, Allan Wilson, and others) have been rejected by evolutionary science (Gribbin and Cherfas 1982). In fact, these views have been carefully evaluated and, where shown to be correct, have been accepted. As early as 1981 (Young, Jope, and Oakley 1981), there was much use of DNA data, and by 1984 this had already reached popular science writing in spite of time lags (e.g., Pilbeam 1984; cf. Wilson 1985). This shows how these new analyses have led to a position quite different to that in the 1970s. By 1986, the new timetable was very widely accepted as being closer to the 5 million years according to molecular biology than the 25 million years according to paleontology (e.g., Schwartz 1986). This acceptance was helped by the convergence of a variety of other lines of evidence. This shows that even a historical field can be self-correcting when scientific. Even in their rather extreme criticism of the supposed refusal of evolutionary science to accept the new data (which they attribute to ideology rather than to sensible caution), Gribbin and Cherfas (1982) actually show how scientific evolutionary biology really is, how well structured and logical, how it develops. They also trace some of the other lines of evidence that have now convinced most scholars, such as findings from geology, the study of ice ages, skeletal studies, animal communication studies among primates, and so on.

This also becomes clear from other critics of the traditional view—the punctuationists (e.g., Eldredge 1985; Eldredge and Stanley 1984; Eldredge and Tattersall, 1982; Gould 1985; Stanley 1981). In criticizing the traditional view, the essentially historical nature of the discipline is made clear, as is the great progress it has made and its cumulative nature. The debate is not about fundamentals but about specifics: mechanisms, rates, and forms of change (cf. Gould 1985). In fact, the point is emphasized that the changes being proposed in Darwin's gradualism are *forced by the data*. For a start, the "new synthesis" of the 1940s was itself due to new data and new disciplines, for example, genetics, a field of research unknown to Darwin, who also had no knowledge of genes (although he posited something like them) and the process or mechanism of inheritance. More recently, the challenge by people like Eldredge, Stanley, Gould, and others is due to new techniques and to new data from the fossil record (i.e., *historical data*)—a topic to be discussed later. Moreover, in arguing the case, Stanley (1981) and others discuss the nature of scientific proof and disproof, that is, they self-consciously discuss the epistemological bases of the field.

Fossil research, according to this argument, has got so much better and more highly developed that it forces changes in the theoretical position; for one thing, Darwin had very limited fossil and geological data. The new proposals are thus not arbitrary: The shift or reinterpretation is *based on new and better evidence,* on new research, new disciplines, new techniques, new data. This is also the case with the redating of the divergence of hominids from other primates. In that sense, the field is progressive through research, better data, cumulativeness, and so on. There is, of course, also a *different interpretation of the data, but the new punctuationist model must also account for the past data.* The point is made that Darwin was also forced to change his views against his will by glaring biological and fossil evidence; he set out to prove the orthodox view and proved the opposite.

One problem with Darwin's use of fossils was also due to the use of a wrong conceptual model, which was too "typological."* For him one good specimen served to define a species; there was no formal study of the variability of fossils. This is, of course, an aspect of taxonomy; one can thus argue that the revision of the model is due to a shift in classification to a more polythetic mode—another instance of the importance of classification (e.g., Gould 1985). Moreover, this argument is very close to mine, about the history of built environments and the study of art discussed in Chapter 3. This is because the ability to deal with variability, or even be aware of it, requires large bodies of evidence (cf. Eldredge and Stanley 1984).

The argument continues to support mine in other respects. Stanley (1981, p. 71) also argues that the evidence of paleontology (which he calls paleobiology), that is, of the fossil record, was never in accord with gradualism. He claims that this evidence was rejected and suppressed. If that is correct, and I take no position, one again finds an instance of the self-correcting nature of science in a historical field. The point is further made that the fossil record *must* be used, however inadequate it is, because it *constitutes the only direct source of information of evolution.* I would add that there is also evidence on time scales, although now supplemented by DNA and other analyses. Such data are, of course, historical data and their inadequacies are matters of concern also in history and archaeology. Recall also the argument about the essentially data-bound nature of history and the impact of new kinds of data—the two fields are not that different; if one is a science so can the other be.

To take advantage of the fossil record, one must look in the right places and ask the appropriate questions (Stanley 1981, p. 78)—as I argued in Chapter 1. There is general agreement that "the secret of science is the secret of asking the right questions" (Stanley 1981, p. 80). The key to success lies in choosing from the many possibilities those questions that are both important and answerable. However, the spectrum of appropriate questions will change over time as data, techniques, and theories expand—which is precisely my argument in this book on new questions for the history of the built environment with the development of EBS. New questions also arise as new disciplines are seen to be relevant, or develop. This is important because sciences and other fields are often at their most fruitful when they exploit analogies and borrow conceptual systems and modify them in the process (Harré 1981, cf. Koestler 1964; Rapoport 1969e; discussion in Chapter 3 on analogy; Hesse 1966; Leatherdale 1974; Ortony 1979; Rapoport 1986c).

The study of fossils has often been disparaged because one cannot use the experimental approach. We have already seen that this view is incorrect (Ziman 1984). The fact that the data are historical does not mean unbridled speculation. *One can be a historical scientist reconstructing the past from whatever scraps of evidence are available* (Stanley 1981, p. 80). One can do thought experiments, do simulation, and use scenarios, and many other approaches are available (see Chapter 5). One can make hypoth-

---

*In this connection, the current interest in "typology" in architectural history and theory is ironic. Note that the use of "typology" is also being criticized in archaeology.

eses and deduce consequences and then see whether they are found in the fossil record (Eldredge 1985). *In effect, one is using critical experiments of the past involving large and complex systems over long time spans; this, in effect, constitutes a laboratory unmatched in scale and sophistication.* The parallel of this with my argument earlier that the historical data about built environments are *our* laboratories, and splendid ones at that, is evident and striking. The supportiveness of the argument seems clear. Moreover, Stanley (1981, p. 108) argues that paleontology and other historical data are given high credibility by the history of science. Although its evidence is historical, it is, in fact, the best evidence.

Also, the debate between the punctuationist and gradualist models is both rational and resolvable. Each predicts a different shape of the tree of life. These are clear-cut differences that can then be tested both against the fossil record and in other ways, both specific and broad and basic (Eldredge and Tattersall 1982; Eldredge 1985; Stanley 1981, p. 82). The specifics of these tests are not important for the argument; moreover, they are not relevant in our domain. The principle is important, that *historical data can be studied scientifically.*

In such an approach, any interpretations or inferences must be testable; that is a *sine qua non* of science. The gradualist model was not testable until the evidence accumulated slowly over time (cf. Bullock 1984). Thus collecting information about the full range of environments is essential. Even 15 years ago, it would have been difficult to fashion the published fossil record into a picture of long-lasting species or the rapid formation of distinctive species from small populations—two essential points of the punctuationist model (Stanley 1981, p. 110). Now that information *is* available. Moreover, whereas in the past genetics tended to favor gradualism, new discoveries in genetics and molecular biology and so on are also said to favor the punctuationist view. This supports my argument that *new data from other disciplines* are also highly significant.

Clearly I cannot judge or resolve the debate between gradualism and punctuationism. In any case the debate is still ongoing; moreover, the conflict itself is sometimes questioned, the point being made that the synthetic theory of evolution is evolving rather than being overthrown (e.g., Stebbins and Ayala 1985; cf. Bendall 1983; Dawkins 1986, especially Chapter 9). What is important is the larger picture, the metamodel, of how one historical science studies its evidence that is clearly historical. In essence fossils are no different than past environments, descriptions, or depictions of such environments, archaeological, or any other historical remains, although all these latter are admittedly man made. In my view, however, if the former can be studied scientifically, so can the latter.

Paleontology

Because we have been discussing historical evidence in the form of fossils and its link to contemporary biological studies, it may be worthwhile to consider how fossils are studied and analyzed. One can also study how changes in paleontology itself have occurred over time, either over long periods (Behrensmeyer and Hill 1980; Rudwick 1972) or shorter time spans (Laporte 1978; between 1949–1978). These show the usual growth and development found in science, in what is clearly a historical field. The study of this type of historical material has reached a rather sophisticated stage (e.g., Paul

1980). It is suggested that even the absolute (as opposed to relative) use of fossil evidence is possible and that even one bone may be enough, although it is admitted that this does present problems. For example, one needs data such as complete fossil sequences to be able to deal with commonality, diversity, distribution, and the like (cf. Bakker 1986; Eldredge 1985; Eldredge and Stanley 1984; Gould 1985). It is suggested, however, that even the incompleteness of the fossil record can be addressed, although few have tested it (Paul 1980, pp. 197–198). The one test described suggests that it is reliable (p. 199). The question is: What proportion of species that have even lived are known as fossils? This question of whether one is looking at the 10% of the iceberg above water or the 90% below water is answerable. It varies. For jellyfish, insects, and worms, it is nearer 10% (although increasing as techniques and theory improve); for brachiopods, trilobites (and vertebrates?), it is nearer 90%. Gaps can be identified and so can the number of families involved. Survivorship analysis can be done, the types of curves depending on randomness. Three methods are possible: cohort, census, and Lyell's methods (pp. 199).

For specific studies, one could consult this and other methodological discussions (e.g., Schopf 1972). This might prove very useful.

The latter deals with strategies of research in paleontology other than description, which is, of course, very common. In addition, in a series of papers on specific topics, it urges the use of *analytical inductive inference,* which is fast and strong inference. The need is to move to *interpretation of the great amount of data* already available—a point quite important to my argument and a number of the other supporting arguments. Schopf (1972) also argues that some of the basic theses are also already available. *The need is to organize paleontology around ideas not objects.* This again is part of my argument. This implies the development of models. Models, of course, involve inevitable simplification, but that is precisely their value. Such simplification involves some costs: either in the degree of generality, degree of realism, or degree of precision. Typically, one can only satisfactorily consider two of these three properties in any one model (cf. Gibbon 1984). Also, models can be either equilibrium (steady state) or historical (non steady state, where the system is being perturbed) (Schopf 1972, pp. 421–431). The process involves analysis (or re-analysis) and resynthesis.

Simple induction is not enough because the data cannot be manipulated. One needs deductively to make predictions from sets of data and test the consequences of such predictions. Models (and theories?) test ideas for major internal inconsistencies before one turns to large amounts of data. Before models in a given field can be understood, however, one *must have a reasonable idea of the nature of the things being studied.* That I take to be equivalent to the definition of the subject matter of the domain and to involve the use of classification/taxonomy and the need to be explicit.

The notion of absolute versus relative evidence and the possible usefulness of a single bone (Paul 1980) come together. For some purposes, one specimen may be enough, for example, for absolute evidence of existence. For other purposes a modest sample may be enough, independent of the size of the population (p. 204). For others yet, for example, relative evidence, one may need samples of thousands (cf. Eldredge and Stanley 1984; Gould 1985). One can also study curves of new knowledge of fossils: If they flatten out, this suggests that there are not many more to discover; if not, that there may still be undiscovered and undescribed data. Recall a similar suggestion about the study of convergence and agreement we discussed earlier (Cornell and Lightman 1982). This method may also be useful regarding built environments.

The point is made that the fossil record is perfectly adequate for most purposes (Paul 1980, p. 207). I would add that this is because there has been vigorous and systematic search; regarding built environments this is not yet the case because it has never been studied in the way it needs to be. Moreover, fossils, however inadequate, must be considered and used because, for tracing the history of life, that is, constructing phylogenies, it is both the only test* and also a rigorous and valid test. This is, of course, a point we have encountered repeatedly—one must use the best available evidence. Also, any theory developed may then be modified if more or new evidence becomes available. But *to ignore this (historical) evidence is infinitely worse*—which is, of course, the crux of my argument.

Paul (1980, p. 213) agrees with Stanley (1981; Eldredge and Stanley 1984) and others that the fossil record is much less incomplete than one is led to believe and that the missing portion is small. At the same time, when new or previously unknown fossils are found, they fit easily into what is already known although; first, argument may ensue for some time and, second, the new evidence may change the existing framework (cf. Johanson and Edey 1981).

For example, in 1983 the finding of a skull halfway between a whale and tapirlike animal was described (*The Times*, 1983). This shows a form intermediate between a land-dwelling and water-dwelling animal. This conclusion is based on its earbones. These are useful because they tend to differ greatly in land- and water-living animals due to the different ways in which sound waves are transmitted in the two media. This illustrates several points: There is an explicit rationale for the valid use of particular evidence; this is based on theory from another discipline (physics) as well as biology; it provides a good example of the role of indirect evidence and the close link between theory and data, which has already been discussed generally and in cosmology and astrophysics (Cornell and Lightman 1982; Shapere 1984) and regarding history (Murphey 1973). In the process of making inferences from fossils, one also comes across the use of contemporary data. For example, in Jerison's work on brain evolution based on plaster casts from skulls, the inferences are checked by making casts from skulls of *existing* animals, which can then be compared with their actual brains. In this way the reliability of inferences is established (Dawkins 1986, p. 190; cf. Johanson and Edey 1981).

Overall, I would suggest that the consideration of the literature on paleontology (paleobiology) suggests that its task is more difficult than that facing us with regard to the built environment. Yet that task is handled with much greater sophistication and rigor. Thus it seems to be possible to make inferences about what dinosaurs looked like, how they lived, what they ate, their physiology, how they raised their young. There are, of course, debates and disagreements, but they typically show the characteristics that I have been emphasizing (e.g., Bakker 1986; Wilford 1985; cf. Morrell 1987). By this I mean that even the most heterodox proposals and arguments, whether about the form of evolution, dinosaurs, the origins of life or of humans, or other topics reviewed in this section are made and debated in the same way.

---

*That is *not* correct: For some questions there is DNA analysis and an increasing number of other methods, already briefly discussed.

This is also the case regarding an event even more unique than evolution and thus even more inherently historical in the traditional sense of the term. I refer to the "great extinction" of dinosaurs and other organisms 65 million years ago. The debate continues and is far from settled, and it is becoming more heated and even politicized (due to debates about "nuclear winter"). The point, however, is that it is being studied in what can only be described as a scientific way even though science supposedly only deals with general matters unlike history that is restricted to unique events. Consider just one example (Allaby and Lovelock 1983) that shows that the *process* of debate is scientific in the sense that *explicit* hypotheses and mechanisms based on the best available evidence are proposed to explain an "anomalous" event.* One is thus ready, at least in principle, to abandon the proposed models should they prove to be inadequate to account for the evidence. They would be given up if they can be kept from being entangled in politics and ideology. The data may even be challenged—there are questions being raised as to whether the extinction was quite as sudden as believed (e.g., Bakker 1986; Wilford 1985).

Disregarding the political/ideological irrationality (cf. Pipes 1986), what is of interest in the debate at the factual level is how much can be inferred and reconstructed about a singular event so long ago. Very many different disciplines are involved (Allaby and Lovelock, 1983, pp. 27–28), as is so often the case. An important point is also made regarding the continuity between animals and humans (see later discussion). This is that all living things change the environment—humans only do it more; in other words, the difference is quantitative rather than qualitative. Reference is made to beavers which, if they were as numerous as humans, might change the environment as much. I have personally seen what excessive numbers of elephants can do. This also applies to prairie dogs, insects, and microorganisms.† Thus only the scale of our modification of the physical environment, which is how I have defined the environment, is different; I have already discussed the continuities between traditional societies and ourselves in this regard.

Human Evolution

This last point brings me to the next subject, which links evolutionary science and paleobiology with human evolution—physical and cultural—a topic of central importance to any study of the history of the built environment given its antiquity (close to 2 million years: Rapoport, 1979a). Once again only a very limited number of publications from what is a vast, most exciting, and rapidly growing field will be discussed.

This topic ranges from questions as to when hominids split off from other primates (already briefly discussed as involving both paleontology and DNA and other analysis), through the evolution of bipedalism and various skeletal characteristics to the evolution of the brain, behavior, culture, and values. This, of course, raises questions about possible evolutionary baselines and regularities in human behavior and hence in the origins and use of built environments and is critical as part of any uniformitarian assumptions—as already briefly discussed in Chapter 3. Again, this is very much a

---

*This is now also being challenged in the sense that such extinctions are said to occur periodically and even at regular intervals—a good example of the relation of pattern and anomaly.
†Note that 4.4 billion-year-old (and older) fossils of microorganisms have now been found.

matter of history and one, I would suggest, that presents a very much more complex and difficult task than the history of the built environment. Yet its progress in recent years has been spectacular because it has been tackled in much more productive and sensible ways—because scientific; the rate of progress is such that there are almost daily changes (e.g., Wilford 1987).

Many of the recent changes in views about human evolution have come about, at least partly, as a result of the explosive growth in the amount of data. As we have already seen repeatedly in this discussion, new and more plentiful data frequently challenge and change theory. At the same time, such new data are incomprehensible unless they are fitted into a scheme of hominid evolution that comprises both theory and previous data; in this it is no different than any other field, including history. This scheme or framework of hominid evolution also includes scientific logic and has been laboriously pieced together over more than a century, all over the world, by numerous specialists from disciplines such as botany, nuclear physics, geology, microbiology, and so on in addition to paleontology itself, which help to establish the dates, environmental conditions, and even what animals ate (Johanson and Edey 1981, pp. 24, 375; cf. Bakker 1986). We have also seen that more recently, molecular biology and other disciplines have played a major role. All these data have been cumulative and are beginning to make sense, filling in what had previously only been sketched.

To understand anything, that thing needs to be described, measured, and named. This involves one in taxonomies, criteria for description and classification, identifying ranges of variability, and so on. Basic to all scientific inquiry is the organization of data in sense-making arrangements (Johanson and Edey 1981). Note how closely this resembles the studies of rock art already discussed (e.g., Lewis-Williams 1981, 1983; Vinnicombe 1976) and other discussions of the importance of classification, pattern recognition, and the like. This, in turn, implies using *all* the data that are mapped, counted, classified, analyzed, and so forth—not just a few selected pieces of data. It is impossible either to trace or to understand evolution without a sequence of fossils, without working out relationships through careful description, analysis of differences— the tracing of patterns.

In this particular case, the dating of the Hadar fossils took 7 years and synchronized five techniques: geology, potassium-argon dating, fission-track dating, paleomagnetism, and biostratigraphy (Johanson and Edey, 1981, pp. 202–203). This is an example of using multiple disciplines and techniques and of the convergence of multiple lines of evidence. Such use of multiple techniques is most important: If they substantiate each other (i.e., converge), one can have much more confidence than in a single technique (Johanson and Edey 1981, p. 206; cf. Rapoport 1970, 1973a; Chapters 3 and 5 of this book).

After the dating was completed, debate started about the nature of the fossils, their location in the evolutionary tree, and the shape of that tree. This debate was explicit and referred back to the evidence. Additional disciplines were involved: Thus regarding the development of bipedalism, one had to leave paleoanthropology and involve a specialist on locomotion (Johanson and Edey 1981). Comparisons were also made with primates other than chimpanzees, so as to identify any behavioral differences between hominids and other primates. One can even make predictions about what one can expect to find, given particular models (Johanson and Edey 1981, p. 347). It seems clear that the more relevant data are used, and the more disciplines and methods, the better.

The debate is far from settled and is not the issue; it is how it proceeds that is of interest. Also significant are some of the specific problems explicitly discussed. For example, first the need to enlarge one's knowledge of what is already available by going back to find more and better evidence and data and completing interpretation of it. Second, having identified major gaps (on either side of the "new species" that they have identified—and that others, for example, Leakey, deny). These gaps need to be clarified by concentrated attempts to fill them.

The search, in fact, goes on. One can hardly read the scientific literature or popular scientific literature without constant reports of new data and attempts to fill and clarify gaps (e.g., Schwartz 1986; Wilford 1987). One of the important gaps has been the species that might be the common ancestors of apes and humans. There has been an active search for this. *Proconsul africanus* has long been a candidate, but only a single skull and a few other bones have been available. Recently thousands of bones have been found near Lake Victoria in Kenya (*New York Times,* 1984a). The new abundance of specimens, that is data, means that "there is every hope that we shall know the anatomy of *Proconsul,* their brain size, how they moved and so on, together with their growth patterns, *almost as well as we know it for some living species*" (*New York Times* 1984a, p. 18, my italics). Many fossils of other animals, including new species, were also discovered. Together with sediments, and many other data, these should also clarify the environment in which *Proconsul* lived, what they ate, and many other things. In other words, an extraordinary amount of inference becomes possible, and it fits into a gap already identified, although, of course, what the outcome of such studies will be is, as yet, an open question. What is important is how one goes about studying what is clearly a historical subject.

One of the differences between traditional history and science that we have already encountered is, reputedly, that history is narrative, whereas science is not. Yet there is support for the view that science is also storytelling, even "myth making" but of a special kind. This kind of storytelling or myth making is the invention of explanations about what things are (or were) like, how they work, and how they come to be (Eldredge and Tattersall 1982, p. 1). It is widely accepted that there are rules for such a statement to be scientific. For example, that one must be able to go to nature and to assess how well the explanation fits the observed universe. Statements not susceptible to disproof are essentially unscientific. Some theories are better than others; some have been tested more severely than others and over longer periods of time. Theories that remain unexamined for a long time become myths (Eldredge and Tattersall 1982, p. 2; cf. Rapoport 1986c), and, more generally data destroy theories, ideas, and stories (Eldredge and Tattersall 1982, p. 122; cf. Bunge 1983; Jacob 1982).

Clearly one of the differences between humans and other species is their behavior. In fact, the subject of EBS and the built environment is predicated on this, on the role of culture as distinguishing between the human species vis-à-vis other species whereas, at the same time, culture, as the properties of human groups (pseudospecies) leads to the extraordinary variability of built environments (Rapoport 1986c). It follows that the study of human evolution must include the evolution of behavior and even of culture (e.g., Bonner 1980). This has become a rather charged subject, involving politics and ideology, not only resulting in very heated and vituperative argument but even leading to physical attacks on some sociobiologists. The issue is, of course, not whether one likes the conclusions, or whether they are "good" or politically or socially acceptable, but merely whether they are correct (Pipes 1986).

But disregarding some of the more extreme and irrational attacks, one can argue that even some partly ideological discussions can still be firmly within the approach being advocated (e.g., Tanner 1981). There, a clear and explicit model is being proposed of the evolution of human behavior, action, social organization, kinship, and language. Because *none* of these are fossilized, they are all based on inference, as is all historical study (and much science) and the study of EBR in the past. Tanner suggests that one needs to develop a comprehensive model that tries to bridge the gap between biology and culture. Such a model will need to incorporate much highly specialized information from many disciplines—cultural, social, and physical anthropology, primate behavior, paleontology, molecular biology, geology, the study of paleoenvironments, archaeology, and others. These are all necessary for the process of inference that Tanner calls an "exercise in biocultural anthropology," an attempt to reconstruct behaviors critical to the transition to hominids. The task is made more possible than previously because *there is much new data and information,* such as fossil sequences, dating, detailed studies of primate behavior, paleoenvironmental data, social and economic behavior of contemporary hunter–gatherers, and so on. There are also many new methods and techniques, such as the DNA analysis discussed earlier. The question then becomes how one can reason about this wealth of data, and one answer is that comparative research (cross-cultural and other) is essential to help with generalization. Note the remarkably close parallels between this and my argument.

There is another requirement. Any model developed needs to have two characteristics. Both of these are encountered repeatedly, both in my argument and in the supporting arguments in this and the next chapter. First, the model needs to be *explicit;* second, it needs to be *process oriented* (Tanner 1981).

Only if a model is explicit is it useful. This is because it can then be evaluated, debated, examined for logical coherence, that is, consistency and interconnectedness between components, and inspected for its fit with data. In this way such explicit models can be further elaborated or modified. The requirement for a process orientation of the *model itself* is necessary to deal with questions about how things happened, in what sequence, and so on. Too many models, Tanner argues, are static. This is, of course, a general point: There is increasingly emphasis on the importance of mechanisms (and models) in linking theory and data.

"Facts" alone mean little; they are only intelligible when viewed in a theoretical framework—a point emphasized in almost every work cited. Facts are like pieces in a puzzle; the model is analogous to the completed picture. The key is to visualize how the final picture might look in order to figure out how to put the pieces together (Tanner 1981, pp. 17–18; cf. Eldredge and Tattersall, 1982 cited before). Effective models do a number of things. They tie data together; help make sense of these data; generate novel questions and lines of research; provide a framework for testing hypotheses against existing data and new data as they appear; suggests new data to look for. Note a common problem of confusion in terms between *theory* and *model:* In other formulations similar characteristics are said to describe theories.

Tanner then proceeds to construct an explicit model (or theory) of human evolution. She argues that it is only as good as its constituent parts so that each stage of the evolutionary sequence must be expertly reconstructed.* In doing so, one finds an example of how *Ramapithecus* can be reconstructed: What it looked like, what it ate, how it moved about, its environment, and so on—all on the basis of fragmentary and minimal data (Tanner 1981, p. 52; cf. *Proconsul* previously mentioned). In some cases she uses experiments, for example on chimpanzees, if one begins with them as she does, *giving reasons for this decision* (cf. Schwartz 1986 on starting with orangutans).

A good example of inference of the evolution of behavior, which is not fossilized, is Tanner's reconstruction of the evolution of communication, on the basis of pattern seeking (cf. Ruse 1986; von Schilcher and Tennant 1984). The importance is emphasized of being as rigorous as possible and of integrating multiple (and diverse) types of disparate information or evidence into a model that is reasonably consistent and makes some predictions possible. Although being "scientific," Tanner also tries to be "broadly humanistic"; that involves the use of primate behavior, cross-cultural research, contemporary hunter–gatherers, and so on; the congruence with my approach and my argument is again virtually complete (cf. Chisholm 1983; Hinde 1984; and even early studies on birds that can be relevant, e.g., LeGay Brereton cited in Rapoport 1977).

The evolution of behavior, culture, communication (including language), and so on is also discussed in a large literature on sociobiology (e.g., Bonner 1980; Dawkins 1976; Hinde 1974; Konner 1982; Lumsden and Wilson 1981, 1983; Maxwell 1984; Wilson 1975, 1978; and many others). This literature, as already mentioned, tends to be highly contentious; some recent reviews try to assess its merits (e.g., Lopreato 1984; Ruse 1979, 1986). There are also popular books that include much interesting material, for example, a greater amount than usual of human data and that explicitly discuss the relation of sociobiology to sociology and anthropology (e.g., Barash 1979).

Even if one disagrees with some specific attempts to deny the implications of sociobiology, all such material, from the point of view of the present book, is useful not only to show how an essentially historical science goes about its work and makes inferences about behavior. It is also very useful (at least potentially) in addressing the fundamental issue of uniformitarianism, that is, how much constancy there is. In this connection, there are also substantive contributions this literature can make, for example in providing links with the idea of an evolutionary baseline in EBR already discussed. It also provides a more direct link to history, where it is emphasized (as we have seen) that some explicit model of human nature is essential (e.g., Berkhofer 1971; Betzig 1986; Fogel in Fogel and Elton 1983; for Collingwood's contrary view, see Martin 1977).

Sociobiology deals with precisely that question: human nature, its limits and ranges. It has been defined as the knowledge of evolution, genetics, and ecology applied to understanding behavior; human sociobiology deals with human behavior in an analogous way (Barash 1979, p. 1). Generally, the effort is to establish patterns, that is, that which is most typical or common. This clearly involves large bodies of evidence and is central to my argument as to why one should use historical, cross-cultural, and

---

*This is analogous to my argument in Chapter 3 that en route to generalization or theory, the past needs to be expertly reconstructed. Thus both better culture history *and* theory construction are necessary.

the other kinds of data. An example of such a pattern is the finding that *there is no human society in which women did not have the primary responsibility for child care* (i.e., that it is sex-linked) (Barash 1979, p. 108; cf. Rossi 1977; Tiger and Shepher 1975; and a large literature on this topic generally). The question can be asked about "new" behaviors among women now. These may be selected out against; they may also cause much damage and undesirable consequences in the long run. These might be predicted and checked in the future; other predictions based on models from sociobiology (and elsewhere) can also be tested against different cultures, past and present.

Assuming that one can avoid the heat and irrationality that the ideological and political reactions to these topics seem to engender, it then becomes possible to concentrate on the factual validity of such claims and predictions. This means that one can discuss them in the usual scientific fashion, because they are based on explicit models and processual mechanisms are posited, data cited, and so on (e.g., Lopreato, 1984; Ruse 1979, 1986). Many examples of such predictions can be found in the literature (e.g., Barash 1979); others related to our topic of EBR could be derived.

Once again, in such work, the convergence of diverse lines of evidence from different fields plays a role. There is a vast amount of relevant material on the brain and nervous system, anatomy, body chemistry, relation to genes, relation to animal behavior. In fact, the subject can be called *behavioral biology* and involves not just sociobiology (Konner 1982, p. 16). Included are ethology, neuroethology, physiological psychology, neuropsychology, behavioral neurology, behavioral endocrinology, biological psychiatry, comparative psychology, biological anthropology, behavioral genetics, psychopharmacology, and many other disciplines in addition to anthropology, culture study, and others that one could add.

The similarity to my argument about EBS and the need to use multiple disciplines is again striking. It is also striking how many data are available, and how varied, that can be brought to bear on the topic once all these disciplines are included—and yet the topic is essentially historical. This approach also makes another point: The problems are with splintered, disparate, and piecemeal data that need to be put together; once they are, they begin to show synergistic effects. What is known must be assembled, but this requires the development of a conceptual framework; that requires the first step of identifying the domain, whether as "sociobiology," "behavioral biology," or whatever.

Behavioral biology essentially sets limits to error rather than reveals truth (Konner 1982). It is, however, the case that the progressive tightening of such limits may lead to convergence and hence increasing verisimilitude. The field looks at genetic influences, or limits, on human behavior so that essentially it is to be understood as *the science of human nature* (Konner 1982; cf. Lopreato 1984; Maxwell 1984) that is an essential topic generally and, as we have seen, central to any historical study. Strictly speaking, the method of extrapolating backward in the realm of behavior is no different to the same method applied to soft body parts in paleontology (Konner 1982, p. 33; cf. chapter in Johanson and Edey [1981] on the brain among many others). One begins with comparative studies of living forms of known degrees of relatedness and generalizes from principles derived from the patterns revealed by these comparisons. Also involved are some elementary assumptions about the *uniformity of processes* of evolution—a topic already discussed in Chapter 3). It then becomes possible to make inferences about certain characteristics to be expected and to check these against the evidence: fossils, primates, a variety of cultural groups, and so on. There is agreement about the importance of large bodies of evidence, careful description and classification, for example, of the nervous system; *what* questions have been answered for some time; *why* questions can be answered by *adaptive teleology; how* questions (i.e., *mechanisms*) have only recently begun to be answered (cf. Bullock 1984).

To understand human nature, it is necessary to discover the human condition when evolving (Konner 1982)—what I have called the "evolutionary baseline." This involves not only evolution, ethology, adaptation, and the like in terms of both physical and social environments but also contemporary hunter–gatherer and other groups. The purpose is to understand and to learn. This is very similar to my argument in Chapter 2 and generally on using the record of human environment for understanding rather than imitating, learning rather than copying.

It is often possible to draw contradictory conclusions from the same body of theory and data. This is useful because it identifies the point of disagreement but only if the argument is explicit and based on evidence. I have previously made this point about my views and Richard Sennett's regarding urban neighborhoods (Rapoport 1977, p. 336). We agree about their nature, what characteristics they possess, what people prefer, and what effects such characteristics have on their inhabitants. We disagree about the policy implications for quite evident reasons having to do with values. Similarly I draw very different conclusions from the research they cite than do Barash (1979), the later Wilson (1978), and Konner (1982); these have to do with certain values and ideological positions (which we do not share); these also underly the wholesale attacks on sociobiology generally as inherently "conservative." But even their effort is an attempt to derive knowledge about human nature and its development. This not only illustrates the process of inference about past behavior and provides links with history, through explicit models of human nature and evolutionary constancies, but also make possible the tracing of areas of agreement and the point where one's conclusions or policy implications begin to differ. Once made explicit, even argument can provide more light and less heat.

In attempting to link such work to history generally, to anthropology, and to built environments, a question arises that is also frequently raised by critics of sociobiology and behavioral biology: Why should there be such a variety of cultures, societies, and environments (Rapoport 1985b,c,d, 1986c, 1990, in press). That is one of the central tasks of the human sciences and humanities (including history) and a frequent argument against their being scientific. But this does *not* invalidate sociobiology, behavioral biology, comparative ethology nor their relevance to history; it does not follow that "history is a thing in itself; independent of human nature" (Bock, 1980, p. 6). On the contrary, it is a combination of all these disciplines that will provide answers to such questions. It is precisely the type of study that I am advocating that can begin to answer the question about the puzzling variability of built environments, which increasingly seems to me to be a central question. In fact, all these disciplines have contributions to make and are complementary.

There are useful examples of bringing together students of genetics, biology, the brain, as well as those of culture, explicitly to address common issues (in this case, human evolution) even when it is explicitly argued (p. 4) that cultural aspects of humans *cannot* be studied scientifically (Young, Jope, and Oakley 1981). In this book the relevant time scales are established through the use of multiple methods—fourteen in one case! Temporal, morphological, and ecological data are used to study the immediate forerunners of humans. Data from molecular biology are also considered, although they are questioned. This is additional evidence that (*contra* Gribbin and Cherfas, 1982, previously mentioned) they are not ignored.* Many and varied points of view, disciplines, and methods are used to discuss the question of human emergence and specific aspects of it, such as locomotion, bipedalism, the use of tools, and so on. For example, it was possible to determine at the time of the conference that this book describes (1980), and probably even more now, the diet of extinct hominids using different methods. Just in the case of fossil evidence, the convergence among eight different methods was used (Young, Jope, and Oakley 1981, p. 56).

Throughout this study there is an emphasis on careful and explicit hypotheses, on scenarios (cf. Cornell and Lightman 1982) and on their relation to data. Although it is not possible to deny subjective and analogical thinking, it is *possible to reduce subjectivity to the lowest possible level* (Young, Jope, and Oakley 1981, p. 76). There is also the use of prediction—in a study dealing with the remote past. For example, given a certain hypothesis, and taking plate tectonics into account, where certain fossils should be found (Young, Jope, and Oakley 1981, pp. 102–106). To study the origins of human behavioral patterns immunological data, DNA analysis, studies of the brain, and endocasts of skulls are used, allowing inferences about, for example, vocal skills (cf. Ruse 1986; von Schilcher and Tennant 1984). Alternative models of early hominid behavior can also be used through the use of archaeological data. In effect, excavation and experiment are combined. One makes (or takes) rival hypotheses, for example about tool making, increases in hunting, food sharing (cf. *New York Times,* 1984b, to be discussed later) and tests these against the archaeological record. Such tests need to be clear and explicit. Experiments can also be done on bone breakage, tool making and tool use, site formation, and the like. These can also be tested through ethnoarchaeology—not used in the present study. That method and experimental archaeology will be discussed in Chapter 5. What is most significant is the approach to the study of the most ancient human prehistory.

The point is made that studies of human evolution and cultural development are always based on implicit uniformitarian assumptions (Young, Jope, and Oakley 1981, p. 177). Clearly, it would seem, they need to be made explicit. Only then can one become aware of their limitations. For example, early hominids may be very different from any living counterparts such as hunters and gatherers.† The essential point is that it is possible to develop a theory, make hypotheses, and test them by experiment, simulation, and against the archaeological data. Not only does this book contribute to studies of the origins of the built environment—its great antiquity and early development (Tobias in Young, Jope, and Oakley 1981, p. 48, Figure 2), it is also possible to study symbolic abilities, higher thought, and cognitive processes in the remote past (see Chapters 3 and 5; Lewis-Williams 1981, 1983; Marshack 1971; Renfrew 1984; Sch-

---

*As we have seen, they have since been accepted much more as skeptics have been convinced (cf. Pilbeam 1984; Schwartz 1986).

†Cf. *New York Times* (1984b), later discussion; see also Chapter 5 on how the use of such analogies has changed, showing the cumulative and progressive nature of the work.

mandt-Besserat 1983; Ucko 1977; Vinnicombe 1976). There are also specific studies on nonutilitarian activities in early humans (e.g., Edwards 1978) and applications of ethology to the study of the origins of human art production (e.g., Dissanayake 1980).

One can, in fact, even study human values and their evolution in connection with biology (Pugh 1977; cf. Bonner 1980; Lopreato 1984; Pulliam and Dunford 1980). Again the issue is not whether the arguments are correct or whether any specific model is accepted but the fact that the topic, the evolution of values in humans, is studied the way it is. Thus Pugh (1977) introduces a decision science model of conscious behavior that is based on the design of artificial decision systems. This attempts to show how genetically inherited behavioral tendencies can be linked to actual behavior in higher vertebrates. According to this view, human values are a manifestation of a built-in value system that is an essential part of a basic "design concept" for a biological "decision system." Fundamental human values are not an accident nor a mystery but a natural almost inevitable consequence of the basic "design concept" of the brain (cf. Ruse 1986; von Schilcher and Tennant 1984). Value systems are needed in order to make judgments, to compare alternatives as being "best." Interestingly, this is also an issue in design as I have long argued in developing the "choice model of design" (Rapoport 1976, 1977, 1983b, in press). Human values are central in the human brain and guide all human decisions and, as such, *they are the driving force of human history.* They must, however, and can be studied scientifically (Pugh 1977, p. 9). It must be admitted, however, that this book is more programmatic than substantive in that regard.

There is also much work on human evolution as a result of information from both biology and culture—on the coevolution of organic and cultural systems (e.g., Durham 1986; Lumsden and Wilson 1981, 1983; Wilson 1975, 1978). There is also, of course, disagreement with this view (e.g., Boyd and Richerson 1985; Brown 1986).* Once again, however, the nature of the debate itself is useful and an example of an approach that can only be described as "scientific." Moreover, the topic itself, the suggestions made, and the way in which such an inherently historical subject is tackled are useful in this argument.

In such studies there is an explicit concern with research and how it should be done on a topic of this nature. Typically a great deal of data from diverse sources and disciplines are closely integrated (e.g., Swanson 1983). There is much about culture and the implications of the increasing complexity of culture for various coping or adaptive mechanisms. This, in turn, illuminates problems caused by high and continuous levels of change over protracted periods of time. These not only have major implications for contemporary planning and design (Rapoport 1977, 1978a, 1983a), they can also generate hypotheses testable against historical and cross-cultural data; these data are essential in testing such hypotheses and, in turn, can provide a framework that helps make sense of such data. Without reviewing this large and complex literature, I would, therefore, merely argue that the implications of *the way in which such studies are done* is highly significant in developing approaches to the study of historical data.

The evolution of culture itself can be studied in ways different to (and complementary to) those used in traditional anthropology, archaeology, and history all of which, in

---

*In October 1989, I came across a study of human territoriality in an EBS context that explicitly uses sociobiology and, more specifically, the Boyd and Richerson (1980) approach (Taylor 1988).

one way or another, address this issue (e.g., Bonner 1980; Pulliam and Dunford 1980). The former, similar to the coevolution studies already mentioned (e.g., Lumsden and Wilson 1981, 1983), argues that it is important, indeed essential, to develop theories even if only to disprove them. At a minimum they are tools that help arrange and order an otherwise unorganized (or disorganized) mass of facts; I would add that they can also help rearrange and reorder such facts. It is important to group and arrange ideas and facts in new ways—*even if these facts and ideas are themselves not new.* By putting things in a new way, or in a new arrangement, *one can learn new things from old facts* (Bonner 1980, p. 3). This is not only in itself a theme supporting my argument; so is the explicit discussion of epistemological issues.

The fundamental question in all such studies is *why humans have culture at all.* Bonner (1980, p. 13) considers this question by examining what aspects of culture are selectively advantageous, that is, why there should have been selection for organisms capable of culture. Because culture did not arise full blown, there must have been a whole series of precursive steps, each adaptive. It is therefore possible to study the evolution of culture in animals not by lumping together cultural and genetic evolution but, on the contrary, by clearly distinguishing among them and explicitly specifying the differences (cf. Brown, 1986).* In doing so, Bonner begins at a surprisingly low level— with motile bacteria! He distinguishes between what he calls closed versus open programs. He then argues that the open, that is, flexible-response system becomes more effective as the environment becomes more unpredictable—a view also currently much used, and much debated, in human ecology (e.g., Fratkin 1986; cf. Swanson 1983). He then shows how such flexible response systems could have evolved. One can have "flexibility" in the sense of behavioral strategies that are alternative ways to achieve stability; one can then speak of "alternative stable strategies," that is, *different cultural devices for the same purpose.* This suggests a very close link to the great variety of built environments and the possible underlying regularities behind apparently different forms or different cues (Rapoport 1977, 1982a). This is also *extremely important and relevant regarding the study of history,* because it reinforces the possibility of *different solutions to constant problems*—a suggestion that has already occurred several times in this book.

Culture and customs can then be considered as alternative steady states, and one can even devise (i.e., simulate) customs or cultural behaviors that may be disastrous, that is, *postulate biological limits* (Bonner 1980, p. 197), an idea we have already encountered. Because built environments are an expression and also part of culture and customs† and also enclose and support life-style and behavior, there is a twofold link here between the evolution of culture and environments and the possibility of biological (evolutionary) limits on environments that, if violated, may lead to phylogenetic maladaptations (e.g., Boyden 1974, 1979, 1987; Boyden and Millar 1978; Boyden *et al.* 1981; Geist 1978 among others).

---

*Programmatically, one could ask why there should be built environments. One could then examine the selective advantage conferred by them and trace the development of built environments, first among animals (e.g., von Frisch 1974), then among hominids (e.g., Tobias in Young, Jope, and Oakley 1981, Figure 2, p. 48; Isaac 1972, 1983; Rapoport 1979a,b), and then among humans.

†In my formulation, culture is ideational and customs (as part of life-style and/or social variables) concrete (Rapoport, 1990).

In the same way that we are interested in the variability of environments, social scientists are interested in the variability of cultures; so, implicitly, are historians. Social scientists (and historians), however, refuse to look at biology that might provide uniformity (Bonner 1980, p. 198)—better, suggest underlying uniformity, possibly at the level of process. Although social science and biology deal with, or consider, different causes, they need to look at both because they are complementary. Bonner concludes that the origins of culture are very early and that it is essentially the development of flexible response systems. The origin of culture is the consequence of natural selection for bigger and better brains rather than for more elaborate genetic signals.

The interest for our purposes of approaches such as these is threefold. First, they show ways in which highly varied data can make sense within a new theoretical framework, as long as the very different data are carefully and explicitly shown to be capable of being related. Second, such studies show how sensible (or at least plausible) inferences can be made about nonfossilized processes in the remote past. Third, there is not only a clear indirect link with history because the subject matter is historical, but also a direct link with the history of the built environment. There are two characteristics of such environments that I have emphasized. The first is their great apparent variability and hence the need to discover underlying regularities or constancies (Rapoport 1977, 1986c). Second are the very early origins of such environments. I have already discussed the very early origins of built environments found in Olduvai Gorge about 1.8 million years ago and even fairly complex settlements that seem to begin very early (Rapoport 1979a,b). In fact, stone walls (we do not know about attempts in more perishable materials!) come immediately after stone tools and before the use of red ochre and fire (Tobias in Young, Jope, and Oakley, 1981, Figure 2, p. 48).

I argued (1979a) that this was a way of marking a special place, basing it on the suggestion by Glynn Isaac (1972, 1983) and others (e.g., Swanson, 1983, pp. 62–63) that food gathering and foraging *and bringing it back to a home base for sharing* was a crucial step in human evolution, distinguishing hominids from pongids and other primates. This was because it would have had major implications for social interaction, culture, greater knowledge, neurological changes, and so on. Such views about that behavior have recently been challenged (*New York Times,* 1984b).

On that new view, based on analyses of fossils at Olduvai Gorge, Tanzania, the hunter–gatherer life-style and its associated social patterns developed later than had been thought. This tends to challenge the hypothesis about the very early origin of campsites/home bases for food sharing. The problem is said to have been due to the use of extrapolations backward of a human hunter–gatherer model that is seen as basic. The current argument is that the past was *different* from the present and that one needs to ask what animals those were rather than what these ancestors can tell us about ourselves.

The attempt is then to reconstruct life at that point on the basis of archaeological data. Hominids were indeed different from other apelike animals in that they stood upright, made tools, and ate meat. Also, *Homo habilis* carried both stones and portions of meat to the same places. The linking of the transport of tools and food is a new development, but one cannot clearly answer the question whether foraging and sleeping occurred in groups or alone, or whether families shared food and so on. The argument is that the walled places were not camps but places for storing tools used in butchering; they would have been too dangerous for habitation. New techniques of electron microscope study of tooth marking on bones reveal competition between hominids and other carnivores. Computer simulation also suggests that the use of numerous long-term sites throughout foraging zones is most efficient in time and energy.* Finally, it is suggested that campsites are impossible before fire. At the same time, the point is made that even if these are not yet campsites, they are antecedents of home bases. Others disagree with these arguments and still accept the home base hypothesis even while admitting that the new questions raised are important; there is less certainty about home bases—it has become a more open question. There is little doubt, however, that hominid patterns were different (see Isaac 1983 for his response to such criticisms, especially pp. 533–535; see also Taylor 1988, Part I).

The question remains, however, why stone walls should have been used; they still seem to suggest the marking of especially significant sites. Moreover, it strengthens the argument that the origins, as well as the shaping and use of built environments, can only be understood in terms of behavior, that is, in terms of environment–behavior relations. It also shows, once again, how new data and new techniques raise new questions and how the ceaseless questioning of theories and their testing against the evidence leads to scientific advance. All one can ask is that similar advances occur in studies of built environments as settings for hominids and humans.

Even more important, all this discussion shows that the past, even the remote past of behavior, can be studied rigorously and scientifically. There are in fact numerous studies of the remote past from a range of what one could call the "paleosciences." These show that settings in the past—climate, vegetation, and the like—can be studied in such ways (e.g., Berger 1981; Connah 1981; Evans 1975; Miggs 1983; Scarre 1983; Simmons and Toole 1981; Wigley, Ingram, and Farmer 1981). One can also study ancient neurology (Kochetkova, 1978) and diseases (paleopathology). There is, in fact, a paleopathology association that studies infections, bone diseases, biochemistry, serology, parasitology, and the like in the past (e.g., Cockburn and Cockburn 1980, as just a single example).

I have already referred to studies of paleoecology. There also the process of debate and advance continues (Colchester 1984; Levin 1983). In such studies, ethnographic analogies are often used as well as generalized models, for example, of optimal hunting (but see Moore and Keene 1983). In one case, such a model is applied to help simulate Pleistocene extinction (Webster and Webster, 1984). In doing so, the model is explicitly specified; the argument is rigorous and explicit, so that it can be followed and criticized. Moreover, the type of data necessary is specified so that archaeologists can look for it; that is, there is prediction. Although the complexity and difficulty of the problem are not minimized, it is *assumed that it can be done*.

The list of studies and disciplines and this chapter have already grown too long. They could all be greatly extended. All these studies in many fields show the sophistication with which one can approach what are essentially historical data, many of them more difficult and more remote in time than those facing us. The rigorous approach that in all cases involves being explicit, framing hypotheses, testing these, explicitly choos-

---

*Note that foraging models and optimization models like these have been criticized and questioned (e.g., Moore and Keene 1983).

ing models, using large and varied bodies of evidence, engaging in interdisciplinary work, and so on is in all cases very similar, in principle, to what I am suggesting. These arguments are highly supportive. These supporting arguments from the various disciplines that I have reviewed so far have been greatly developed in one discipline, the domain of which is closest to that being discussed in this book. I refer to archaeology and now turn to an examination of that field in Chapter 5.

Chapter 5

Supporting Argument 2
From Archaeology/Prehistory

INTRODUCTION

Of all the fields to be reviewed, archaeology comes closest to my topic. Much of the more recent work reads almost like EBS (e.g., Binford 1981; Hodder 1982a; Kent 1984). This is because it is essentially history but by the nature of the evidence approaches it through material culture. In most cases, particularly in prehistory, material culture is all it can study. Thus, although it shares with history a concern with the past, chronology, and change, it is inevitably preoccupied with material culture (Hodder 1982b). In fact, of all the social sciences other than EBS it is the most sensitive to the important role of material culture in human behavior. It has also been described both as a bridge between the arts and sciences and as seeking knowledge in both humanistic and scientific ways.

Archaeology, and particularly *recent* archaeology, thus becomes a paradigm case or exemplar for my argument about using all the data, about rigorous inference about the past and an emphasis on generalization and theory. It provides a particularly useful parallel of how to do history in a scientific manner that results not only in better histories of specific groups or places but in the development of a generalizing, theoretical, explanatory discipline.

In this sense, recent archaeology (e.g., "behavioral archaeology") is closely related to parts of paleontology, another discipline dealing with past remains recovered from the ground from which inferences are made about behavior. Archaeology has also moved from a descriptive, largely historical (in the traditional sense) field to a nomothetic discipline, generalizing and theoretical. Moreover, when recent developments in archaeology, such as ethnoarchaeology, are applied to existing human groups, it essentially becomes EBS (Binford 1981, Part II; Hodder 1982a,b, Chapter 9; Gould and Schiffer 1981; Kent 1984; Kramer 1982, etc.). Thus Moore (1983) for example, argues that archaeology should analyze not only flows of energy and matter but of information and recommends the use of geographical studies on the diffusion of information. This is very close to my own work on the built environment as the organization of space, time, meaning, and communication and the argument that environments both reflect and guide communication, one form of which are information flows. This is also the case with views of privacy as the control of information flows (e.g., Altman 1975; Rapoport 1976, 1977, 1980a).

At the same time, this new archaeological literature seems unknown both to EBS researchers and to architectural historians. It thus seems important to bring it to their attention and to those concerned with the history of the built environment. Archaeologists, on the other hand, with a few exceptions, are not in general aware of the EBS literature (Rapoport 1988a, 1990).

The literature on these recent developments in archaeology not only provides the most highly elaborated supporting argument, which is also congruent with the supporting arguments from other fields. Many of its programmatic statements (e.g., Binford 1981, 1983a,b; Clarke 1978) show extraordinary parallels to my arguments in this book. This literature is also very extensive; this creates problems in summarizing and making clear the extraordinary overlaps that I found with EBS, without this material dominating this book. As it is, I struggled to reduce the length of this chapter, and it may still appear that I devote an inordinate amount of space to it. However, the relation to EBS is extremely close conceptually so that Staski (1982) defines the new field of urban archaeology as the relation between human behavior, human cognition, and material culture in urban settings, describing it in terms of five categories that are almost identical to what I call the three basic questions of EBR (Rapoport 1977, 1982, 1986a).

Moreover, also important is the possible mutual interaction between EBS and archaeology (Rapoport 1988a, 1990). The developments in archaeology that can be identified with my approach to the history of the built environment also become central to *any* discipline that must test any of its theories against data of the past if its generalizations are to be valid. Seen in this way, the archaeological approach can make available the full span of the past. Thus, the case study in Part III of this book uses much archaeological data—as it must do to extend the temporal scale.

Archaeology is only between 100 and 150 years old as a discipline—much younger than architectural history. Yet in that time it has achieved much in dealing with a domain virtually identical to that of this book. Clearly its concerns and the questions asked of the material have been rather different, but even that is changing (e.g., the overlap between Spriggs 1977; and parts of Chapter 3; cf. also Kent 1984; Kramer 1982; Staski 1982; etc.). *Particularly striking has been the development in the past 20 years based on adopting a position and program remarkably similar to the one that I advocate in this book.* Although the roots of these recent developments clearly go back earlier, and many "beginnings" of this recent approach could be identified, I will take, in Britain, David Clarke's *Analytical Archaeology* (1968, 2nd edition, 1978) as a seminal critique of the then state of the discipline and as a programmatic statement. At roughly the same time the so-called "new archaeology" began in the United States (e.g., Binford and Binford 1968; Watson, Leblanc, and Redman 1971).

The point is not that development followed directly. In fact, these two approaches (and many others) are highly critical of one another. It is rather that these approaches marked a watershed: Developments either followed, with increasing sophistication, or particular lines suggested or resulted from criticisms of, and reactions to, these same positions. Important was the process of continuous argument and counterargument, criticism, evaluation, and so on—the essence of scholarship and of science.

This still continues. There is, in fact, no *apparent* consensus among practitioners of the type of archaeology I am considering, whether "new archaeology," "processual archaeology," "social archaeology," "behavioral archaeology," "archaeological anthropology," "ethnoarchaeology," or whatever. There is even heated debate about whether a scientific archaeology is either possible or desirable.

In fact, even books denying that possibility can provide supporting arguments. One striking example is provided by a book that explicitly rejects a scientific approach to archaeology (Hodder 1986). It is interesting to note that it does so on the basis of what are claimed to be new insights. Yet the emphasis in all my work since the very beginning has been exactly the same: the critical role of cultural variables, the centrality of what I call latent aspects and hence context, meaning, schemata, and "perceived" environments as leading to human actions, including the design and active use of built environments and material culture. It is, I think, fair to say that I have been the leading exponent of these views in EBS: that material aspects are not primary, that cognitive and symbolic approaches are essential, and that meaning is more important than instrumental function; in fact, I go further and argue that meaning is the main function (e.g., Rapoport 1982a, 1988a). Given this, I fail to see that it precludes the approach I am advocating. On the contrary, the only way these, or any, matters can validly be studied is as proposed in this book; in fact, I cite much of Hodder's previous work in support!

I would also argue that Hodder's book in fact supports my argument. First, taken as a whole, it is an example of what I have called explicit debate about epistemology. It is also essentially about specific approaches; the disagreement is partly due to his use of what I called a "narrow" definition of science (Chapter 3). In terms of what I call a broad view of science, there is almost complete agreement with my argument. An informal content analysis shows almost complete congruence. Thus explanation is the goal (much of the argument being about what constitutes better explanation). In this, theory is seen as central and methodological sophistication and rigor essential; explicit criteria must be used to choose among alternative hypotheses (which need to be used) and theories. The absence of something may be as important as its presence (which implies expectations and pattern). Inference is central, logic and coherence in argument crucial, and there must be correspondence with data. It follows that sampling, research design, and explicitness are emphasized; so is the need for long time periods and much, and varied, data that need to be carefully described; all of this helps to identify patterns: core elements, commonalities at different levels of generality. Convergence from multiple lines of evidence provided by different disciplines also plays a critical role. There is a need to reestablish links between history and archaeology and to use concepts and approaches from a variety of disciplines. It is accepted that the past can be reconstructed and that our knowledge of it can be cumulative; in this, mutual criticism is important.

I could go on, but the point has been made: The argument in Hodder's book is, in fact, highly supportive. At first, the reasons for the apparent disagreement are puzzling. As suggested, they are partly due to the difference between a broad and a narrow view of science. They are also due to another reason that becomes significant when identified.

It is clear that the reasons for the argument are political (e.g., Hodder 1986, pp. 101–102, 106, 157ff.). It is almost a perfect illustration of Passmore's (1978) statement that it is not so much that one cannot be rational and objective, it is rather that one does not wish to be. Like most attacks on "positivism" and "empiricism" (read *science*), it is ideological.*

But this debate itself has had a major impact on the discipline, not least by greatly raising the level of discourse. *In the process the field has become scientific.* Also, behind these apparent differences there are many common characteristics and themes; these are so accepted and shared that they do not even elicit argument. I will be using these to structure this chapter. Briefly, it is a concern with *process;* an emphasis on asking a series of *explicit questions* about the human past and increasingly relating these to the present and future. There is a realization that the data remaining from the past require *better and more rigorous description;* that one must then move through *generalization* to the development of a body of *theory.* There is an emphasis on the process of *inference* because the data are in the present and conclusions are about the past; and agreement that this process needs to be made explicitly and rigorous, and rules developed for making inferences. These characteristics that structure this chapter are virtually identical to those listed at the end of Chapter 3.

As Salmon (1982, p. x) among many others (e.g., Renfrew, Rowlands, and Segreaves, 1982) points out, one can find a philosophy to support almost any position on any topic. Much vehement argument can be found in archaeology regarding *specific* positions (e.g., Kelley and Hanen 1988), ranging from deductive-nomological (covering law ["positivistic"]) through probabilistic (deductive-statistical or statistical-relevance based on Bayes's theorem), inductive-statistical, hypothetico-analogical, structuralist, historiographical, Marxian, materialist, system models, and many others. These do not concern us here. Of importance are the (possibly unintended) consequences of this process. Even if the specific philosophical model initially selected (e.g., by Clarke or Binford) was wrong (and I am not convinced that it was), it did the following:

1. Engendered a great deal of debate that increasingly became more sophisticated.
2. This forced the discipline to become much more self-conscious and explicit about what it was trying to do, why and how.
3. This, in turn, led to what one can only describe as a flowering of the field in the last 20 years in the way in which it goes about studying the past.

This is, of course, the reason for the emphasis that I give it here. It seems to be a paradigm case of what is likely to happen to the field of the history of the built environment once a science rather than art metaphor is adopted. It is also the reason why, for the moment, I will *not* be recommending any particular philosophical position; what seems important are the consequences of debating the philosophical bases of a discipline.

The recent developments in archaeology can be understood as of two kinds. First,

---

*One example is provided by the proceedings of the World Archaeology Congress held in Britain in September 1986, the title of which places the word *objectivity* in quotes (fortunately, what I have so far read of the contents seems more immune). ("Archaeological 'Objectivity' in Interpretation," three volumes, The Congress and Allen & Unwin).

the more traditional concern with writing what has been called "culture history" (Gibbon 1984), that is, reconstructions of particular groups, places, or even sites based on remains of material culture, has been made more convincing because it is rigorous, explicit, methodologically sophisticated, related to human behavior, and linked to theory, models, and explanation (e.g., Canby *et al.* 1986; Flannery 1976; Flannery and Marcus 1983; Folan, Kintz, and Fletcher 1983; Freidel and Sabloff 1984 among many others). Second, there has been a new concern: to make archaeology into a generalizing discipline concerned with models, with explanatory and predictive theory based on cross-cultural comparative work, controlled work with contemporary people, and so forth. This has meant seeing archaeology as a social science, or even modeling it on the natural sciences (quite apart from applying natural science methods to it, which has also had important consequences). This, in turn, has led to taking further the developments mentioned before: considering the full range of evidence or sampling; generalization; prediction; model building; clear operationalization; theory development; explicit, rigorous argument, and the like. The specifics both of the new writing of culture history and theoretical development, that is, the substantive and programmatic concerns of archaeology, frequently differ significantly from those of this book and are not directly relevant, interesting as they may be. The questions that I think EBS should ask will in many cases be different from those archaeologists ask; for example, we will not tend to deal with human origins, the use of resources, the origins of civilizations, the development of technology, and the like—although we may *use* such findings that are of great significance. Some concerns may overlap; for example, culture change (Rapoport 1983a, 1986c, 1988a, 1989a). The main importance of archaeology in the present context is its elaboration of an approach to a comparable body of data that has produced results incomparably better than architectural history. Moreover, as already mentioned, some of its very recent developments, such as ethnoarchaeology and related approaches such as archaeological anthropology and ethnography (Kent 1987), are almost identical to EBS.

What is relevant is the success of the epistemological and methodological developments in the approach common to both aspects of archaeology: the reconstruction of specific groups, places and sites, and theory development. These show an extremely close correspondence to my argument. For example, there is programmatic agreement on the need to be scientific, to accept the way science does things that is the most elegant method so far devised for augmenting human knowledge (Neill, 1978, p. ix). Many of the arguments that appear to be about this are typically rather about *what* "being scientific" means. In other words, they are discussions about what epistemology (if any) is characteristic of science. Thus much of the recent literature on the ways of making inferences, on methods, on the philosophical bases of archaeology are *essentially arguments about the meaning of "scientific,"* a topic that I hope to discuss in a second book (cf. Rapoport 1986e).

What is common to all the specific positions taken is really a concern with one central issue: *how to make more valid inferences about human behavior from material culture.* This implies that such inference is *possible* and that the archaeological record and material culture generally provide "congealed information" (Clarke 1978) that can be read or decoded (Rapoport 1982a). Thus material culture communicates status, identity, and the like and is used for such purposes (Hodder 1981, 1982a,b; Moore

1983; Rapoport 1981a; Wobst 1977; etc.). Many archaeologists would argue that the purpose of such inferences is to describe and explain the growth, development, and diversity of human societies (e.g., Renfrew 1982). The goal of my approach to the history of the built environment is different: to describe, understand, and, eventually, to explain the diversity and commonalities in the mutual interactions of people and environments and the mechanisms that link them. In other words, to give a temporal dimension to the three basic questions of EBS and thus further to broaden the body of evidence on the basis of which generalizations about EBR are made. Yet, even here, there are parallels with the goals of archaeology; for example, as the provision of a coherent set of reasonably high-level generalizations relating material culture and human behavior (Salmon 1982, p. 6). Moreover, even learning, that is, the generation of precedents, can be found. Thus, a study of Maya subsistence begins with a theoretical question of how traditional farmers adapted to the environment (Flannery 1982). It uses folk knowledge, observation, detailed recording (for example, a 12-km-long transect), data on food collection and storage—that is, it combines archaeological, ethnographic, and experimental data. Answers derived challenge certain common assumptions, for example, that traditional farmers are conservationists, have an ecological wisdom, and effectively regulate environmental conditions over the long term. The contrary is the case: Most of their effort is to resolve short-term goals (cf. Browne 1985). In spite of that, however, findings from the past are used to derive principles that can be applied to the present, in order to develop certain neglected forms of agriculture and to use unskilled labor and little machinery; there is learning but not by imitation.

In general, I would argue that all the studies cited in this chapter (and the many not cited) provide an almost perfect match to my argument in Chapters 1 through 3. The explosive growth of archaeology over the past two decades is a result of the implementation of a program very much like the one I propose and is strong support for the argument of this book; the history of the built environment would show equally rapid progress.

It has been suggested that the major changes in how archaeologists work and behave and in how they invest their time can be considered in terms of three contexts of explanation (Plog 1982, pp. 29–32).

1. The *formal component* refers to the process of constructing clear, or at least testable, arguments. This involves an emphasis on laws, clear statements of alternatives—hypotheses, arguments, and so on, and the conscious use of alternative methodologies. *The "idle speculation" model is no longer acceptable* (Plog 1982, p. 30).
2. The *substantive component* that involves accounting for the observed patterns and variations. This involves familiarity with these patterns, more rigor, use of models, and *explanation as articulated diversity* (Plog 1982, p. 31).
3. The *operational component* that concerns the activity of designing research and attempting to ensure that the patterns under study are at least observed. There is great attention to research design.

All such pronouncements—and there are many—share certain themes. One sees a growing sophistication, an emphasis on a variety of approaches and methods, on explicitness of the argument, the use of multiple disciplines, and so on (cf. Binford

1981, 1983a,b; Hodder, Isaac, and Hammond 1981; etc.). These are an actual expression of programmatic suggestions made in Chapters 1 through 3.

There is also agreement with one basic argument of this book: that the history of the built environment is a way of extending research into the past, both using data from the past to help generalize about EBR and using data from the present to help illuminate the past. Similarly, in archaeology, we find statements to the effect that *archaeology can be seen as a way of extending ethnology into the remote past* (Longacre 1981). This also helps illuminate the present because patterns are revealed. Through this type of interplay of past and present, which is increasingly typical of much archaeological research, both gain, and the sophistication of research grows, studies become more empirical, and theory and methods are refined. The result is a greater understanding of how human behavior and cognition are both reflected in, and affected by, material culture. In doing so, as already pointed out, recent studies not only become close to EBS, *they become virtually identical with it.* Thus, in one book (Gould and Schiffer 1981), topics include Williamsburg, architectural change among Mennonites, the archaeology of ethnic relations in Hawaii through the use of graffiti and racial insults (on graffiti, see citations to parallel EBS studies in Rapoport 1977, 1982a), analysis of herbalist's shops, studying the National Air and Space museum to understand the relation of ideology and material culture, analysis of cemeteries, study of fences in Mormon culture, analysis of supermarkets based on missing items and anomalies in the placement of items, which can be used to predict neighborhood composition, the community store in rural America, a study of front/back in modern U.S. dwellings, studying settlements by analyzing building materials, and so on.

These could all be EBS papers. This virtual identity with EBS emphasizes the point that the recent archaeology, under whatever label, has from its inception essentially adopted the goals of social science, especially anthropology: to understand and explain regularities and differences in cultural behavior. It has had to do that by looking primarily at material culture. In both these respects it greatly resembles the goals of EBS: to understand and explain regularities and differences in human cultural (and other) behavior as related to material culture (the physical built environment and its "furnishings"). Also, in adopting this goal, archaeology has implicitly (and increasingly explicitly) looked at the past and present as a *single body of data,* useful and relevant to help answer specific questions. In this it, once again, fully coincides with the argument of this book.

This overlap between my argument and archaeology is so close that it is possible to organize the subject matter of the remainder of this chapter using the characteristics developed in Chapter 3 (with the addition of "examples" that could easily have been subsumed under the other categories). In some cases these categories have, in this chapter, been "collapsed," that is, combined. This is the case with the first two as well as others.

A CONSCIOUS AND EXPLICIT CONCERN
WITH EPISTEMOLOGICAL ISSUES

Recent archaeology exhibits, first, a conscious application of issues and developments from the philosophy of science as part of the effort to become a science and, second, an emphasis on an explicit discussion of various philosophical issues. Both are intended to develop a philosophical base for the discipline. This is a crucial step, also underway in a number of social sciences. This will not be developed here; my intention is rather to draw attention to the importance accorded to these issues in archaeology.

I have already mentioned that it is often pointed out that one can find a philosophy to support almost any position on any topic (Johnston 1983; Morgan 1973, p. 250; Salmon 1982, p. x) and that the significance is in the debate itself, and although that debate is essentially about specific positions, even it has occasionally been acrimonious. The emphasis in the large literature on this topic in archaeology is primarily on epistemological issues, although there is also a less extensive body of work on ontology, that is, concerned with the definition of the subject matter of the domain (to be discussed later). This self-conscious and deliberate effort to analyze the philosophical underpinnings of the field is an attempt to clarify what is meant by becoming "scientific" that, as we have seen, was the stated goal of the proponents of changes in archaeology 20 years ago. One result has been that the level of debate itself has improved; also, the involvement of philosophers who often ask penetrating questions (even if their answers do not always convince) has been useful even if, as some argue, there has been excessive emphasis on philosophy (Plog 1982; Renfrew 1984; Renfrew, Rowlands, and Segreaves 1982; etc.). There is also much, often unstated, agreement that is the subject matter of this chapter.

Much of the literature has to do with the rationale for accepting knowledge claims. It has been pointed out (Watson 1979, p. 287) that this always involves a "leap of faith," but that there is a crucial difference between *tested* and *untested* leaps of faith—a topic central to an understanding of science (cf. Bunge 1983; Jacob 1982). This topic is essentially concerned with how one can best make inferences from past material culture to behavior and how well founded such inferences can be. To improve such inferences, the process must receive systematic attention; this is exactly what has happened in recent archaeology.

It is difficult to review some of these attempts for several reasons. First, the literature is extensive; second, it is highly technical; third, it is not unequivocal; fourth, much of it is philosophical so that it addresses issues more properly developed elsewhere. On the other hand, the discussion of the question I pose in Chapter 3 is explicit and, in general, the answer is like mine: Well-founded inferences depend upon archaeology becoming more scientific. Although there is much discussion about what that means specifically, there is much less on whether it is appropriate.

It should be noted that this kind of discussion is often normative, concerned with what archaeologists *should* do, rather than what they in fact do; in Kaplan's (1964) terms, with *reconstructed logic* rather than *logic in use*. This separation is not, of course, either neat or complete. First, the very discussion of such issues changes what is done. Second, my argument is precisely that there have been major effects.

The issue is not one of data but primarily about what happens once data are available: how one can then draw valid conclusions and "interpret it in ways related to the data in coherent and justifiable ways" (Binford 1983b, p. 7), that is, how inference can be made sounder. The point is frequently made that the data (the archaeological record) *are always in the present;* one is not dealing with *past events* but with *present events* from which inferences are made about the past (e.g., Binford 1981). Although the possibility of such inference depends on a large body of information, it also depends on theory that enables the decoding of the data. This is difficult because it involves inferences from observations on contemporary static material artifacts about the dynamics of past life-styles, cognition, and symbolic behavior (Binford 1983b). The emphasis on process that we will encounter in this chapter is part of the response to this problem (cf. Sabloff 1981, p. 121). It is significant that there is, generally, increasing emphasis on mechanisms and models as providing dynamic processes linking theory and empirical data (Hesse 1966; Morris 1983; etc.).

As discussed in Chapter 3, inferences about what I call "manifest functions" (cf. Binford 1962, "technomic function"; McKie 1977, "techniques and technology") are easier to make than inferences about what I call "latent functions" (cf. Binford 1962, "ideo-technic function"; McKie 1977, "Ritual and ideology"; Renfrew 1982, "cognitions"; etc.).

Note that theoretical presuppositions, metamodels, metaphysical blueprints, and the like, much discussed in the recent philosophy of science, also receive much attention in archaeology because they play a major role in theory, methods, and even classification. Because there is no agreement about them, the need is to make them explicit, a central theme in current archaeology. One example is the differences between the ecological/adaptive view of the "new archaeology" (Binford 1981, 1983a,b; Gibbon 1984; etc.) as opposed to the more ideational/symbolic/meaning approach that I emphasize (Deetz 1977; Hodder 1981, 1982a,b,c, 1986; Isaac in Sieveking *et al.,* 1976). The latter argues that even tools originated as cult objects rather than utilitarian objects, that latent rather than manifest functions were dominant. It is clearly important to state these presuppositions, derive the implications, and test them against the empirical evidence, making the inferential process explicit and being prepared to modify inferences. This often seems to be the minimal shared view of what "being scientific" means.

Binford (1981, 1983a,b) is the most insistent on the need to be "scientific," although in its various guises this is very common in the literature. He also uses this term in its strongest form; for example, insisting on the covering-law model of explanations and the hypothetico-deductive approach from Hempel. This is criticized by many (e.g., Hodder 1982a,b; Moore and Keene 1983; Renfrew 1984; Salmon 1982) as is the U.S. "new archaeology," of which Binford is one of the founders and that more generally tends to emphasize a "strong" version of science (e.g., Fritz and Plog 1970; Watson, LeBlanc, and Redman 1971, 1984).

Binford typically insists on the need for hypothesis testing and theory building and strongly emphasizes methodology. He is concerned with the philosophy of archaeology and "archaeological metaphysics," arguing that a careful consideration and analysis of these matters will lead to positive developments in the field, changing the domain (in a manner analogous to Chapter 1); influencing methodology, models, and concepts.

Much of the debate concerns the issue of deductive or inductive methods of inference and explanation. There is increasing awareness that this debate is often misconceived because both inductive *and* deductive reasoning are used in the continuous feedback between data and hypotheses (e.g., Connah 1981). Central to the debate is, of course, the role of models and theories and, increasingly, much of the debate concerns these (e.g., Binford 1981, 1983a,b; Gardin 1980; Gibbon 1984).

As already repeatedly emphasized, the specifics of the debate are less important than the effort explicitly to examine the philosophical underpinnings of the discipline and to criticize the metaphysics of traditional archaeology. The very concern with such issues and examination of alternatives leads to the growth of archaeology as a science without prejudging the specific approach. Even critics of the value of such philosophical concerns (e.g., Plog 1982) admit that it has had a major and beneficial impact on archaeology in terms of explicitness, self-conscious improvement, increased rigor, and so on. The point is made that it is essential for archaeology, as for any productive science, to operate with an explicit awareness of the ideas, assumptions, and presuppositions by which it proceeds (Binford 1983a, p. 31). I would argue that given such a position, the specific assumptions become much less significant and the debate more productive. The very argument and need to be explicit and self-conscious about such matters results in being more scientific and becoming more theoretical. The argument itself has been a major advance, has transformed the discipline, and has led to spectacular results.

Increasingly there is less concern with proposing specific theories and more with exploring what form such a theory might take, that is, with discussing how the past can be known and what forms of explanation are most promising (cf. Renfrew, Rowlands, and Segreaves, 1982, especially Part I). In this debate those who strongly disagree about the forms of explanation (e.g., Binford 1983a; Renfrew, Rowlands, and Segreaves 1982) agree about the need for archaeology to be theoretically and epistemologically aware and also to define the subject matter of its domain. Also emphasized is the need to analyze examples of what forms of explanation are actually used in archaeology (Kaplan's "logic in use"); that could help to decide what forms should be used ("reconstructed logic"). In that sense even that debate is becoming more realistic: What researchers actually do is relevant to what they should do.* One can see this as an empirical base, as it were, for the debate. It also shows a rise in sophistication by emphasizing the interplay of what is done and what should be done, a realization that traditional philosophy of science is concerned solely with reconstructed logic so that there is a major gap between working scientists and philosophers. In fact, science does not happen the way many philosophers view it. At the same time, however, reconstructed logic is important. This is because, in order to explain, it is necessary to have some conception of what might be an acceptable form of explanation (Renfrew 1984). The increased sophistication of debate is also evident, as in the case of deductive versus inductive methods of inference, in attempts to seek common ground among various approaches to explanation, suggesting that it might be too early to commit oneself to any specific logical form of explanation (e.g., Renfrew 1984).

If one looks at the progress of the debate over the past 20 years this increasing sophistication is quite clear. Not only is the debate less shrill and less categorical, more open to nuances, but there are also periodic attempts to summarize the debate to that point. Such attempts (e.g., Salmon 1982) often then propose their own specific proposals that are, in turn, criticized (e.g., Renfrew, Rowland and Segreaves 1982; Renfrew 1984). Yet attempts to discuss and summarize the position are most useful. Such

---

*Note that this is very much the case in the recent philosophy of science literature.

summaries are also useful in that they show what the arguments are about; for example, that much of the debate is between a *narrow* interpretation of science, which involves accepting a particular philosophical view about correct standards of confirmation or explanation, and a *broad* view that sees science as well-founded, testable "reliable knowledge" (Ziman 1978) as opposed to guessing or asserting.

The consensus seems to be that in spite of all the problems, the discussions between philosophers and archaeologists have been beneficial because they have led to an increase in the conceptual clarity that results from isolating and separating issues in complex problems (Salmon 1982, p. 179)—what I call "dismantling" (Rapoport 1976, 1977, 1980a, 1989a; cf. Plog 1982; Renfrew, Rowlands, and Segreaves 1982; Renfrew 1984).

Even when one disagrees with the specific approaches advocated, such as linguistic, "hermeneutic," Marxist, or structuralist (Hodder 1982a), one finds highly significant characteristics that show the effects of the debate I have been discussing. One finds that, even starting from these totally different philosophical presuppositions, positions are reached with which one can agree—and that one has reached from a totally different starting point. This suggests more agreement than might be apparent. More important, one also finds ways of debating basic philosophical, theoretical, and empirical issues that are both comprehensible and useful. Even attacks on "scientific archaeology" (Hodder 1986) show many of its characteristics.

Even when particular views are rejected, there is agreement that the attempts were useful because they led to sounder methods, more confidence in conclusions, and a more accurate knowledge of the past (Hodder 1982a, Chapter 4; cf. Renfrew 1984; Renfrew, Rowland, and Segreaves 1982; Salmon 1982). This is similar to recent, more balanced views in cognitive psychology that behaviorism, before it went too far, did much good that survives, mainly in rigor, methodology, and the like, which can be applied to topics such as imagery that behaviorists would reject (Kosslyn 1980, 1983).

Thus all the attempts to be explicitly scientific, even when they lead to specific models of inference and confirmation (e.g., the hypothetico-analog method [Smith 1977]) have certain common characteristics. These typify the impact of the debate that I have been discussing. They all refer to the philosophy of science; they make the argument domain-specific; they explicitly argue their case; they discuss the pros and cons of their position; and they consider and evaluate different models of inference. There is agreement that archaeology must become rigorous and theoretical, that its arguments must be explicit, that they must be testable empirically, that they must be justified philosophically, that epistemological issues must be debated. This consensus, and the effects on archaeology of the efforts so far, support my argument that this is an essential step in developing a better approach to the study of the history of the built environment.

REDEFINING THE DOMAIN

It is significant that many of the developments in archaeology with which this chapter deals occurred after its domain had been explicitly redefined in ways identical to my proposal in Chapter 1 (e.g., Dunnell 1971; Salmon 1982, p. 140). These redefinitions emphasize the connections between human behavior and social, political, and economic organization and the remains of material culture, the relations among the remains themselves, and between them and where they are found, that is, the context of the cultural landscape. These developments are related to a move away from a consideration only of elite and ceremonial structures to a consideration of the total record of the past; what Binford (1981) characterizes as a move from "a relic and monument phase" to an "artifact and assemblage phase." This was due at least partly to new questions being asked (again, analogously to Chapter 2); these were more anthropological, concerned with social and cultural behavior, cognition, meaning, and the like. In order to be able to answer such questions, archaeology had to look at those data that best showed how people lived, behaved, and thought; those data were the whole of material culture.

Temples and palaces, the principal concern of traditional archaeology and still, at least metaphorically, that of architectural history, are only the starting point for overall surveys that go further and further from them (e.g., Folan, Kintz, and Fletcher 1983; Freidel and Sabloff 1984; Millon 1973). This mapping may take years; excavation is by multidisciplinary teams and takes even longer, involving much material—vegetable, animal, and mineral—not previously retrieved. The analysis becomes extremely sophisticated; in it, contemporary data—ethnographic, experimental, and other—are used in controlled analogy—as discussed later. That also amounts to a form of redefinition of the domain.

Not only have art objects, palaces, and temples become relatively less important, all artifacts have become less significant in themselves; they are significant for what they can tell us about the people who produced and used them, their culture, social arrangements, their behavior, ideas, and meanings. The common theme is a concern with "contextual analysis" (Flannery 1976; Hodder 1982a,b,c, 1986), with the whole settlement pattern; the settlement and all its elements, its larger regional setting and beyond; the concern is with the *cultural landscape,* the sum total of the human imprint on the fact of the earth, which becomes the data. This has been called "landscape archaeology" (Folan, Kintz and Fletcher 1983; cf. Blanton 1978; Flannery 1976; Freidel and Sabloff 1984; Hodder 1982b; Lambrick 1988; Longworth and Cherry 1986; McKintosh and McKintosh 1980; Millon 1973; Nicholas *et al.* 1986; Parsons *et al.* 1982; Pieper 1980a; Taylor 1983, etc.). This emphasis on the cultural landscape and on tracing the relevant boundaries of the system is equivalent to my concern with the system of settings, the extent of which needs to be discovered rather than assumed (Rapoport 1977, 1980c, 1985b, 1986a, 1989a; Vayda 1983).

Note that this does not just happen, it is explicitly argued. One example is the argument (Deetz 1977) that archaeology, which he also relates to history (dealing with the archaeology/history interface) must study the total record of the past, the whole of material culture defined as broadly as possible. This includes high-style elements, folk culture, popular culture, buildings, artifacts, settlements, fields, and animal and human remains. Also emphasized is the need to study elements that rarely, if ever, survive in the archaeological record: human bodies, their proxemics and nonverbal communication, clothing, music, speech, dance, and so on. This, of course, necessitates the use of research on contemporary people and of contemporary and past data together, as I have already suggested, and has led to the development of ethnoarchaeology and other forms of controlled analogy to be discussed later. It also leads to an increasing emphasis on comparative data, also encouraged by the need to generalize and develop theory.

The point is also made that whereas greater theoretical and epistemological awareness and explicitness are necessary to archaeology, the primary issue is, in fact, substantive and concerns *what is the subject matter of archaeology?* (e.g., Renfrew, Rowlands, and Segreaves 1982, p. 3; Salmon 1982, e.g., pp. 37, 140). Equally essential to the need to define the field of inquiry is the need to formulate questions; together these correspond to what I called defining the domain. These apparently simple-seeming tasks are in some ways the most difficult and decisive in the whole of the enterprise of research (Renfrew, Rowlands, and Segreaves 1982, p. 9; Renfrew 1984; cf. Northrop 1947; Shapere 1977, 1984). This was, of course, my argument in Chapters 1 and 2.

These are important because epistemology partly depends on ontology and because the forms of explanation may differ depending on what one is explaining, for example, classes of events, processes, patterns, and so on. In much of the recent literature the emphasis is, in fact, on the importance of moving from specific events to classes of events. In order to generalize, it is necessary systematically to formulate and apply propositions to a wide variety of cases (e.g., Renfrew, Rowlands, and Segreaves 1982, p. 21). This requires, as I have argued in Chapters 1 through 3, the broadest possible body of evidence and comparative work (cf. Bullock 1984; Kent 1984; Mayr 1982). This use of generalizations and broad samples and comparisons also reveals patterns that even a contextual analysis of unique cases can never do.

POSING CLEAR AND EXPLICIT QUESTIONS

Almost the whole of the literature discussed in this chapter emphasizes the importance of formulating clear questions about the past. This has already been discussed; moreover, every reference in this chapter makes this point. I will thus summarize the argument and just list a few of the relevant references.

The point is made that most generally there must be precision and clarity in concepts used and in terminology; the meaning of statements must be clear (e.g., Dunnell 1971). The general problem areas being addressed in any given case and the resulting questions must be defined carefully and explicitly and must be specific enough to enable alternative hypotheses to be stated (e.g., Smith 1977, 1978; cf. Moore and Keene 1983, p. 137). Because formulating questions is one of the most decisive steps in the whole enterprise of research (Northrop 1947; Renfrew 1984, p. 9), the importance of asking interesting and significant questions cannot be overemphasized (Moore and Keene 1983, p. 7). Because interest, significance, and importance are highly subjective matters, this means, I would suggest, that the case for posing particular questions needs to be made explicitly, argued, and justified. This emphasis on question formulation is because one must define what one wants to know about the past, what one is trying to do. Questions must be defined before research can be done (Binford

1983b, p. 26)—a general point, of course, relevant to all research. One can, in fact, interpret the literature in archaeology over the past 20 years partly as an argument leading to a progressive refinement and periodic redefinition of the critical questions.

EXPLICITNESS, RIGOR, LOGIC, CLARITY, AND PRECISION

There is a continual emphasis, already briefly mentioned (e.g. Dunnell 1971), on the use of clear concepts and terms, on reasoned argument, on internal and external consistency, on a concern with logic. There are many programmatic calls for rigor with the implication that this leads to the development of a discipline. The recent, rapid development of archaeology supports that position. Above all, one can detect an emphasis on making all arguments and their steps as explicit as possible. In some ways, I think, this is the essential requirement, because if an argument and all its steps are made explicit, most of the other *desiderata* will follow immediately. Even if they do not, reasoned criticism and eventual improvement become not only possible but likely in the longer run.

Thus one can describe explicitly how one goes about making reconstructions and three-dimensional visualizations of past environments (e.g., Sorrell 1981). It is also possible to evaluate the accuracy and reasonableness of previous visual reconstructions by studying them over time, seeing how they have been done by different authors, how they have been done for various purposes, and so forth (e.g., Piggott 1978). There can, therefore, be considerable sophistication even in converting data to visual material.

This process can, of course, be used to carry out detailed analyses. Thus, based on large bodies of evidence, and explicitly citing pros and cons for each step, one can build up larger and larger patterns starting with individual elements; these then lead to inferences. The data include not only surveys and excavations but whatever is already known in the literature. One can also use a variety of methods, techniques, and knowledge as long as this use is made explicit. A good example of this process, in the case of a "traditional" study of a single place, is the study of Cozumel (Freidel and Sabloff 1984). Multiple lines of evidence are used, the reasoning is made explicit, and both pro-and-con arguments are provided. As they move to higher and higher levels of generalization, they "normalize," that is, accept *the most likely conclusion*—explicitly giving their reasons. These conclusions then become the basis for still higher level or order generalizations or reconstructions. In this way they move from elements to various groupings to the overall settlement patterns for the island and show that linkages beyond must be taken into account; the boundaries of the system are discovered, not assumed (Rapoport 1980c, 1986a, 1989a; Vayda 1983).

Throughout, at whatever level, whether formal plaza groups, residential arrangements, or the settlement pattern, the arguments are highly explicit and can be followed step by step. The various weaknesses of the study are also pointed out; for example, lack of chronological control so that contemporaneity of features is not yet established. In effect, as they argue, they have done the best possible using the *best available data* (cf. Lewis-Williams 1981, 1983). Their argument is that their interpretation of the patterns they claim to have found (which themselves are explicitly discussed) are both plausible and testable, principally because the chain of reasoning can be tested but also because their interpretation is based on a vast amount of other work in the area and is also supported by recent work elsewhere in southern Mesoamerica. This, of course, parallels the argument in Chapter 3 on multiple lines of evidence.

Comparable characteristics are found in other studies of specific places, for example Cobá (Folan, Kintz, and Fletcher 1983), in analyses of particular built environments for more theoretical purposes (e.g., Renfrew 1983) and in theoretical studies (e.g., Sabloff 1981). In fact, there is rather remarkable agreement on these points. Thus in dealing with simulation (to be discussed later), the emphasis is on precision, being systematic, clear analysis, the use of explicit models, and also explicitly demarcating those areas of knowledge where there is current agreement and those where there is still doubt and controversy (Sabloff 1981, p. 188). This last point is, of course, equivalent to the suggestion that the distinction is not between hard and soft science but between well-established knowledge and that which is not yet well established (Boulding 1980), and that it is useful to study the extent of closure (e.g., Cornell and Lightman 1982). In analyzing megalithic monuments, Renfrew (1983) reviews a large body of evidence, showing implicitly that in making inferences, broad evidence and comparative work are necessary. He then explicitly argues through a series of steps about the relation of megalithic monuments to population density (denied by Moore and Keene 1983), migration, and so on. He then derives a hypothesis about the use of such monuments for social control, social order, and stability. Explicit suggestions are made about how such a hypothesis might be tested and the relative difficulty of various tests.

Clearly, as one moves from reconstructing a particular place to making inferences about behavior in its broadest sense, the difficulty goes up, yet the same *desiderata* seem to apply. The use of a systematic argument that is made explicit is essential (e.g., DeBoer and Lathrop, 1979). Also critical becomes the use of comparative work with broad bodies of evidence so that patterns may be recognized and generalizations made. Recall that inferences from archaeological material to behavior imply the use of data on behavior. This involves the controlled use of analogy, in which explicitness is essential (e.g., Hodder 1982b) as argued in Chapter 3 and as we will see later. In fact, turning implicit into explicit assumptions has the effect of allowing testable hypotheses to be derived. That single step, therefore, changes the whole enterprise (e.g., Kent 1984). The clarity and precision of argument that follow has already changed the field (Renfrew, Rowlands, and Segreaves 1982).

I have already suggested that explicitness may well be the most important issue because even if the other characteristics do not follow immediately, they will do so eventually. Of course, it seems clear from these examples that the other consequences will follow quickly. This is at least partly because by explicitly structuring the steps of one's argument, its logic is improved. In fact, one could argue that to be made "more logical" is equivalent to making one's argument in the form of chains of explicitly defined operations exposing the reasoning that underlies it; greater explicitness leads to clearer thinking (e.g., Gardin 1980). This also applies to *constructs;* this has led to an emphasis on classification and terminological precision. When things are made explicit, they typically need to be explained; they then become testable. By being made explicit, a discipline also becomes a body of knowledge that can also be taught and learned explicitly (Dunnell 1971). Moreover, we saw in the section on history in Chapter 4 that there is always some theory of human nature used, and this needs to be made explicit. This is also the case in archaeology, and the case is made that, although one may disagree with any such given theory, as long as it is explicit and the argument can be followed step by step, it can be criticized, tested against empirical data, hypotheses based on it can also be tested, and so on (Hodder 1982a).

This means, of course, that the hypotheses themselves need to be made explicit. Thus, from carefully and explicitly delimited problem areas one can derive explicit, preferably multiple, hypotheses. From those, different sets of predictions can be deduced and explicitly stated as can the test implications. The links between a hypothesis and an observational prediction typically involve a "bridging argument," that is, a logical argument. This once again means explicitly stated premises and auxiliary hypotheses. Because observational predictions that are falsified may not invalidate the main hypothesis but these auxiliary hypotheses, it is essential that these later be made explicit rather than be left hidden, unstated, and unrecognized (Smith 1977, p. 612, 1978).

The point is that in any field, and archaeology is no exception, many implicit assumptions are made. These can be discussed only when they are made explicit; only then can other studies be brought to bear on them, other literature shown to be relevant, controlled analogies (such as ethnoarchaeology) used, and so on. All these developments follow from making such assumptions explicit. This is, of course, very similar in general terms to the argument discussed in Chapter 4 about history. Because all history assumes some model of human nature and behavior, the important step is to make such implicit models explicit. For example, Tilley (in Hodder 1982a) proposes a particular model of human nature based on "dialectical structuralism." The issue is not whether this is a reasonable model; what is important is that it is an example of a model that is made explicit and is then related to material culture. In fact, other chapters in this book (Hodder 1982a) also make explicit certain implicit theories of human nature and behavior, and these are then discussed. Note also that implicit models of human nature have been shown to be involved in planning (Wood *et al.* 1966) and in architecture (Ellis 1986–1987). Once made explicit, they can often be shown to be wrong. More important, once made explicit they can be evaluated and research can be brought to bear on them; EBS can be seen as an effort to do precisely that. The point is even made (Hodder 1982a) that the relationship between the past and the present is itself often taken as a given, and hence left unexamined; once made explicit, it can then be subjected to analysis and examination.

Note that models of human nature, once made explicit, can be evaluated, and the literature in anthropology, psychology, sociobiology, cultural geography, ethnoarchaeology, and other disciplines can be brought to bear on them. Any specific model can then be evaluated and its relevance shown for interpreting the patterning of archaeological material and making inferences. For example, Moore (1983) develops an explicit model based on the notion that archaeological cultures represent information fields. On that basis, he articulates an explicit model of human nature and behavior: certain fundamental human characteristics involved in dealing with communication and information. It is significant that this work is very close to EBS (e.g., Rapoport 1977, 1981a, 1982a, 1986c, among many others). It can also be related to a very large and sophisticated literature in human geography.

It was pointed out before that much of the discussion of the philosophical bases of archaeology was concerned with debating the precise logical form of explanation adopted. It was also suggested that the specific model was less important than the explicitness of the debate itself. This is also the point being made in this section. In the final analysis, what makes archaeology (or any field) scientific is not the precise logical form of the explanation adopted but rather the explicit statement of the models used, assumptions, questions, hypotheses, and so on and, hence, the possibility of testing them systematically, logically, openly, and in replicable ways (e.g., Renfrew 1984, p. 46).

OBJECTIVITY

This last point bears on the effort to be as objective as possible, which characterizes any science. This is because objectivity is intimately related to the careful, unequivocal, and explicit statement of presuppositions and assumptions, such as uniformitarianism. When accompanied by a careful definition of terms, isolation of useful conceptual issues, explicit statement of criteria, and the like, maximum objectivity follows.

Although the need for objectivity is commonly emphasized, Binford is particularly insistent on this. He suggests that this conscious attempt to be objective involves clarity and explicitness not only in the assumptions and arguments but also in identifying and describing data. The explicitness of this process allows one to reduce the ambiguity of observation (e.g., Binford 1983a, pp. 47–49; cf. Dunnell 1971). This, in effect, leads to consensibility which is a prerequisite for consensuality (Ziman 1978), which many now identify with scientific objectivity.* This attention to data is meant to ensure an empirical base for any inferences and an empirical content of any theories by ensuring that archaeology describes the real world (e.g., Renfrew, Rowlands, and Segreaves 1982; Renfrew 1984; Salmon 1982).

I have already discussed the need for the explicit statement of assumptions, premises, auxiliary hypotheses, and the like. This tends to ensure maximum objectivity. In general, it is suggested (Smith, 1977) that by formulating alternative multiple working hypotheses, one can remain more objective than if one only has one. For one thing, there is less likelihood of becoming excessively enamored of it. Objectivity is also enhanced by carefully considering and clarifying the concepts that are being used (Moore and Keene 1983, p. 136).

For example, if concepts such as culture, behavior, and material culture are carefully defined and dismantled (Rapoport 1976, 1977, 1980a, 1986c, 1989a), models can then be constructed of their interrelationships. Because more than one model can be identified, greater objectivity results (e.g., Kent 1984; Renfrew 1984).

In the former study, two models are identified—a traditional ethnographic model and an archaeological model (Kent 1984, p. 13, Figure 1). These are then decomposed into their components and the interrelationships among them discussed and carefully specified. From this, the model to be used in the study is derived explicitly. It is then tested against three groups; once again, reasons for the selection of the particular groups are very explicitly and carefully argued. The characteristics of these groups are

*I have already commented on the problems caused by the lack of descriptive languages for environments.

then described in great detail ethnographically; this is equivalent to being made explicit. The use of multiple ethnographic sources helps ensure objectivity; explicitness allows evaluation and criticism. As a result, clear finds emerge about the problem area defined—the cultural use of space. These are applied to three hypotheses that are derived from the archaeological literature where they are implicit; they are stated explicitly at the outset of the study. One hypothesis cannot be evaluated due to various confounding factors; the other two are shown to be incorrect for the specific cross-cultural comparison. It follows that, at the very least, they cannot be assumed to be universal. The conclusion, *that the use of space is culture specific* (Kent 1984) is, of course, one that I have argued for years (e.g., Rapoport 1969a, 1976, 1977, 1980a, 1983a, 1985b, 1986c, etc.). The implication for making inferences about the past is clear and important: One cannot use assumptions based on the hypotheses implicit in the archaeological literature for all groups at all times and in all places. Note that this important finding follows from making the hypotheses explicit and the use of an explicit argument. It also makes necessary explicit arguments about the specific groups being studied in any given case, the context, and so on. All these increase the objectivity of the analysis.

Note how consistent this study is with all the others discussed in this chapter; in approach, in being explicit, carefully defining terms, concepts, and entities, rigor, and so on. This is the case even when studies apparently disagree rather fundamentally (e.g., Binford 1983a; Hodder 1982a). This consistency, of course, extends to the other fields discussed in Chapter 4 and supports my argument in Chapter 3.

RIGOROUS MUTUAL CRITICISM

The emphasis on explicitness is, of course, meant to allow rigorous criticism, the use of previous work, and work from other related disciplines. This leads to greater cumulativeness and, hence, the rapid growth and development that has characterized archaeology over the past 20 years. This also, as we have just seen, leads to greater objectivity and hence more reliable knowledge through convergence to consensuality through consensibility (Ziman 1978).

I have already commented on the debate regarding epistemology and other philosophical issues (e.g., Binford 1981, 1983a,b; Fritz and Plog 1970; Hodder 1982a; Kelley and Hanen 1988; Renfrew 1984; Renfrew, Rowlands, and Segreaves 1982; Watson, LeBlanc, and Redman 1971, 1984). I have also pointed out how previous work and varied sources of evidence are used in reconstructions of particular places and in making inferences. This use is implicitly, and often explicitly, a form of criticism: Previous findings or studies are shown to be incomplete or inadequate, and progressively improved. In general, the whole of the literature in archaeology that I reviewed shows a high frequency of rigorous mutual criticism—almost any study in this chapter could serve as an example. Reading it in more or less chronological order also shows progressive development. Programmatic statements become implemented or given up. Reconstructions of past environments show development and improvement through constructive criticism. Inferences show ever greater sophistication (e.g., Longworth and Cherry 1986). Frequently, over time, very useful criticisms and proposals are in turn criticized.

In the case of epistemology, as we have already seen, one common criticism concerns the premature selection of a specific model of explanation or criterion for being "scientific." Equally common is the criticism of a single method in favor of a multiplicity of methods (cf. Rapoport 1970a, 1973a). The emphasis on particular methods has recently been likened to Kaplan's (1964) "law of the hammer" (also known as the "law of the instrument"). This states that a boy who has a hammer finds many things that need hammering; the same criticism is made about particular models, concepts, and theories and about direct and literal borrowing from other disciplines (Moore and Keene 1983). The debate itself, its explicitness, rigor, sophistication, and reasoned character, is more significant than its specifics. This is because it leads to progress and cumulativeness. Thus, the emphasis by some scholars on taxonomy (e.g., Dunnell 1971) has been criticized by others (e.g., Salmon 1982) on the grounds that one cannot initiate theory construction by constructing a system of definitions (which she accuses Dunnell of doing). In turn, Renfrew (Renfrew, Rowlands, and Segreaves 1982, Part I; Renfrew 1984) criticizes both Dunnell and Salmon, proposing a "better view." Moreover, works such as these provide periodic summaries of the state of the debate.

SELF-CORRECTION, CUMULATIVENESS, AND RAPID PROGRESS

The emphasis on explicitness and rigorous mutual criticism have led, as we have just seen, to the beginning of a major characteristic of science: a field that is self-correcting in the long run and, as a result of this, cumulative in nature. This is related to a sense of the continuity of the work of all members of the discipline. They use each others' work to do their own, know the literature, and keep up with the latest developments in their field and in fields related to their work. Combined with the critical process, which is different to arguing about basics, and all the other characteristics being discussed, this leads to the rapid progress of the field with which this chapter began. It is, of course, my argument that if the study of the history of the built environment were approached in the way I am advocating, progress could be equally rapid, measured in decades rather than glacial or, worse, nonexistent.

Although this whole chapter, in effect, describes this rapid progress, consider some examples. In the study of past man—environment interactions at the ecosystem level, one can observe not only how sophisticated recent studies are (e.g., Binford 1981; Connah 1981; Jochim 1976, 1983; among others). One can also find, rather quickly, the beginning of development of more general models and theory; a synthesis based partly on the application of findings, methods, and theory from a number of different disciplines (Butzer 1982; cf. Bilsky 1980). One can also see the rapid development of specific areas of study. For example, synthesis and review of work on prehistoric Europe, including settlements, dwellings, and the like, show striking progress in one decade (1968–1978): rapid rate of change, improvement in the level of synthesis achieved, and so on. There is a clear sense of a cumulative and progressive field (e.g., Phillips 1980). When in late December 1986, I visited an exhibition at the British Museum, "The Archaeology in Britain: New Views of the Past," the same thing was clear: In the 40 years reviewed, that particular subfield has shown extremely rapid progress (cf. Longworth and Cherry 1986).

Throughout this chapter I have emphasized not only this characteristic of advances and cumulativeness in theory, knowledge, and method but also the development of new topic areas, some of which are very close to EBS and this book. Almost any recent study even of a single site will also make this clear. Thus, the study of Cozumel (Freidel and Sabloff 1984) makes clear that the interpretations (i.e., inferences) are supported by a vast amount of other work on the area. It is also supported by recent work elsewhere in southern Mesoamerica. I would emphasize that every example of *that* work is also based on vast amounts of work. Also, the study of Cozumel is explicitly an interim report; the point is also made explicitly that when the final reports are published, these, in turn, will be compared with studies already published of related places (e.g., Mayapan) and also with other forthcoming studies of "Decadent" period settlements. This not only shows a knowledge of ongoing work in the field; it also shows an explicit view of this specific work as part of, and set in, a larger, ongoing, and cumulative whole.

This also becomes clear from a study of Cobá (Folan, Kintz, and Fletcher 1983). This is a paleosociological study of landscape archaeology. They adopt this concept from the previous work of Armillas but take it much further and greatly develop it. This study will, in turn, serve as a base for future work. Once again, there is an explicit sense of the continuity of the work and its cumulativeness. Moreover, because Cobá is a Classic Maya site and Cozumel a "Decadent" site, the two together (as well as other work in the Maya area) can be predicted to lead to progress in larger, more general theoretical questions; for example, the whole sequence of development of the Maya. Beyond that it will become an element of even more theoretical studies, that is, culture change, trajectories, and the like.

This point is also made by considering a recent discussion of simulation in archaeology (Renfrew 1984). This explicitly takes earlier work on this topic (e.g., Sabloff 1981) as its point of departure. By reading the two in sequence, the progress, development, increasing sophistication, and cumulativeness become clear. The same conclusion results from a comparison of two studies on a similar topic published 9 years apart (Fairservis 1975; Renfrew 1984, especially introduction, Chapter 4 in Part II, Parts IV and V, especially Chapters 11 through 13). The first study is, in itself, an advance on earlier work (e.g., Childe's). It is an attempt, first, explicitly to conceptualize the term *civilization* and then to use three sites of increasing complexity to show the development toward civilization. In each case, many variables and conditions are impressively reconstructed, including site, climate, economy, food, social organization and kinship, religion, permanence of settlement, and so on. A rigorous argument is then made, on the basis of explicit criteria and marshalling the evidence, that one of these sites reached the level of civilization. The chapters in the latter study only 9 years later (Renfrew, 1984) deal with the same topic of the "jump" into what can be called civilization. The argument is much more explicit and rigorous, being based on new quantitative models, systems modeling, and simulation. Many more data are used because they have become available; there has been rapid progress there also.

I began this chapter with the point that archaeology and prehistory are relatively recent—they began in the nineteenth century. The progress, particularly in prehistory since then, has been described as one of the great intellectual and scientific achievements of this period; it is attributed to the application of scientific techniques and theoretical constructs from many disciplines to the interpretation of the material remains of past (and hominid) groups (Fagan 1983, p. 3). I disagree with his argument that whereas archaeology can apply scientific methods, it cannot *be* scientific (p. 17); this disagreement most likely is an example of the disagreement about the meaning of "scientific" discussed in Chapter 3. This is likely given the fact that there seems to be agreement on the undeniable fact of the progress itself, and even about the reasons for it. Moreover, as I try to show in this chapter, the "great awakening" in archaeology in the 1960s (Renfrew, Rowlands, and Segreaves 1982, p. 7) is, in fact, due to a contrary view—and a series of actions based on that view, that archaeology can and must become a science.

NEED FOR AN EMPIRICAL BASE

One finds, in the archaeological literature, a commitment to an empirical content of any inferences, generalizations, hypotheses, and theories. There is, consequently, a commitment to and a concern with empirical data. One of the criteria distinguishing science and scholarship from other forms of human endeavor is the testing of ideas against empirical data. This is the case whatever the particular form of the ideas, models, or whatever.

Thus, it is argued, even in the case of approaches that explicitly reject "positivism" that the archaeological record is empirically constraining; that the empirical data prevent the adoption of just *any* theoretical positions (Hodder 1982a; cf. Bunge 1983; Jacob 1982). Hence, models are not completely speculative, and there is a continuous interplay of theory and data. All the papers or books that develop models or theory always explicitly and, in detail, apply them to the relevant bodies of evidence. Thus, after Hodder (Hodder 1982; cf. Hodder, Isaac, and Hammond 1981) arrives at his particular model of symbolic and contextual archaeology, *he immediately applies it to a body of empirical evidence*—in that case the Orkney Islands.

Because the task of archaeology is now seen as that of "decoding" the information revealed by the patterns present in the record, it follows that discovering the data, describing them and classifying them are important, indeed essential steps. One must have an adequate empirical base. The problem of not having the data one needs will become quite clear in my case study. It is also quite clear in my experience of my own research and teaching. The converse is also true—often the availability of a large and varied body of data suggests major developments, models, and theories—as discussed in Chapter 4. It seems clear that the recent flowering of archaeology is partly due to the availability of a major database. The development of new approaches, in turn, leads to the collection of more, and more valid and relevant, data. This is, of course, why the definition of the domain (Chapters 1 and 2) is so important—it suggests what data are needed. Once these become even partly available, a field can progress through the constant interaction of ideas, concepts, hypotheses and theories, and data.

It follows that such laws, generalizations, and theories in archaeology must have empirical content; they must refer to the real world. This then also raises basic questions concerning the kind of empirical data that constitutes evidence for or against a given hypothesis (Salmon 1982, pp. 31–32)—a matter much discussed in the literature

(Binford 1981, 1983a,b; Renfrew 1984; Renfrew, Rowlands and Segreaves 1982; Salmon 1982; Smith 1977, 1978; etc.).

This literature presupposes that the link with empirical data is essential; its emphasis on evidence is an aspect of this. So are the constant calls for large bodies of data and their clear and unambiguous description and classification. We have already seen that this is even the case with work in the structuralist, hermeneutic, and similar traditions (e.g., Hodder 1982a, 1986). The theories are always tested against empirical evidence even, as is essential, they go beyond it. This, of course, is also the case with work in traditions more sympathetic to science. Thus, after developing a model of inference, Smith (1977) tests it, at much greater length than the model described in the paper, in a book (1978). In the paper (1977), after carefully discussing the logical links between the model and hypotheses on the one hand and predictions regarding observational data, five criteria are given for evaluating alternative hypotheses, that is, relating them to the data discovered. In that way the testing is quite explicit and unambiguous; in fact, it was published before the publication of the tests. At the same time, although two of the criteria are very general, involving the *significance* of predictions and the *simplicity* of hypotheses, the other three criteria for evaluating hypotheses bear on my arguments about the desirable characteristics of evidence. Positive evaluation, that is, confirmation, is related to:

1. The *empirical falsity* of empirical predictions. Those hypotheses for which test implications are false are less preferable. It is useful to note that the recent discussions in philosophy of science on falsification suggests that this is rather more complicated than it seems.
2. The *number* of observational predictions supported. Other things being equal, the hypothesis with the greatest number of supporting observational predictions is preferred.
3. The *variety* and *independence* of observational predictions shown to be empirically true becomes the criterion, other things being equal (Smith, 1977, p. 613).

Although "independence" is a difficult notion and all the criteria present problems that Smith (1977, 1978) and the literature generally discuss, it would appear that the three criteria taken together strongly support my discussion earlier about the need for the largest and broadest bodies of evidence, the greatest number of relevant disciplines, agreement with data and theories accepted in other fields, and so on.

Correcting misconceptions is always related to data. Even when reanalysis of already existing data is involved in the first instance, it usually demands new data in order to support the corrected view. Often, the original misconception is due to a "superficial reading of limited historical sources" (Morris and Thompson 1985, p. 7). In this case, the study of the Inca empire involves several shifts. First, from the capital (Cuzco)—where most work has been done—to a provincial town (Huanuco Pampa). This is explicitly similar to my domain definition, being concerned with a wider range of settlements and evidence about the daily life of ordinary people (see also history, in Chapter 4). Second, it asks new questions about the functioning of complex preindustrial societies in economic and political terms. Third, it uses a wide range of methods applied to a wide variety of evidence, including written sources (ethnohistory), ethnographic analogy, and archaeological data, emphasizing daily life. The strengths and weaknesses of each method and body of evidence are carefully discussed, and a great number of very interesting findings and interpretations follow, many having to do with city planning, compounds, dwellings, elite areas, and the like. Activity systems are reconstructed as being central to the understanding of the architecture, city, villages, and the whole region (cf. Rapoport 1977, 1985b, 1986a, 1989a).

THE HANDLING OF DATA*

The emphasis on empirical content has a number of major implications. First is the need for careful attention to the nature and availability of data; for example, many kinds of data other than artifacts are now retrieved. There is also a concern with their completeness and the effects of time and differential preservation (e.g., Binford 1981, 1983a,b; Clarke 1978; DeBoer 1983; Smith 1977), sampling, the collection and retrieval of data, their identification, description, measurement, taxonomy, and classification (e.g., Dalton 1981; Dunnell 1971; Smith 1977; cf. Salmon 1982). Second, there must be careful attention to how such data are used, a reliance on multiple bodies of evidence both from archaeology and from without. The latter includes the use of non-archaeological evidence such as oral traditions, ethnography, ethnohistory, and so forth; as well as the development of new approaches such as experimental archaeology, ethnoarchaeology, and the like.

We have seen that in the more traditional field of reconstructions of particular places the scale and extent of concern have changed, as has the level of generalization (e.g., Flannery 1976). This often involves the use of other kinds of data. For example, in studying prehistoric settlement patterns in the Valley of Mexico, archaeological method and theory are very successfully combined with ethnohistoric and historic data to build up a multidimensional picture of a past cultural landscape (Parsons 1982; cf. the Oaxaca Project: e.g., Blanton 1978; Nicholas et al., 1986; cf. Flannery and Marcus 1983).

The need to sample much larger areas is shown in the study of an early site in Britain (*The Times* 1982a). In this case the whole site had to be studied, sieved, and carefully sampled in order to be able to make *negative evidence* become usable and important; it was the *absence* of something expected—waste. The expectation itself, of course, depended on a knowledge of much evidence and an acceptance of uniformitarianism: The pattern expected was derived from a large sample of sites, that is, comparative work across sites that showed that waste and garbage are typically present (cf. Hadingham 1987, p. 157). Its absence then became an anomaly to be explained; the inference was that this was a sacred site. Thus what characterizes "sacred" depends on careful sampling (cf. Folan, Kintz, and Fletcher 1983; Coogan in Shanks 1988).

In a more general study of early Britain, multiple methods and different bodies of evidence are used to make major inferences. Hypotheses are proposed to explain the data and integrate explanations. By studying a great variety of evidence about the past, very complex and sophisticated reconstructions become possible about Britain 4,000 years ago. The society is described: its metrology, geometry, astronomy, settlements,

*This section combines Sections 9 and 10 in the list in Chapter 3.

and buildings, and so on (Mackie 1977). Note that using data and concepts from a great variety of fields not only increases the amount of data available, it also increases its variety and independence, and provides new and different concepts, methods, and approaches; although, as we have already discussed, these latter must not be used literally (Moore and Keene 1983; cf. Rapoport 1982b, 1983a).

Another instance of using a variety of data from diverse fields is a diachronic study of the Lake Chad region in Africa (Connah 1981). The evidence, specifically from archaeology, history, and ethnography, is used to study the interaction of people and their *ecological* environment—different from, but analogous with, EBS. Although archaeology and history both study the past, they do so differently, using different data, sources, and methods. Bringing them together is, therefore, useful. In this study, a very large number of methods are applied to a specific place, and a history of man–environment interaction in this place over 3,000 years is traced. The study transcends the specific locale, however, and is more general: It is an instance of history at its broadest, in the sense of the study of the past. It is significant that this study assumes continuity (i.e., uniformitarianism) in human history and that included are the ecological and geographic settings in the past and changes in them over time in relation to humans. The archaeological landscape is established and interpreted by relying partly on ethnographic material on settlements, buildings, and the like. The earliest evidence is based on a very systematic and quantitative excavation. A series of sites of a particular soil type on which settlements tended to be concentrated is studied, as is the change from the earliest responses to urbanization and state development.

Two things are clear in addition to the points already made. First, the study of the past can be very sophisticated indeed when the types of approaches I am advocating are used. Second, in such studies models, theories, methods, and taxonomies, derived from archaeology as well as other fields, play a central role: It is the interplay of data and theory that is important. The more varied *both* are, the more reliable both reconstruction and inference become. Thus reconstruction, understanding, and explanation of the ways of life of ancient people, whether these be specific events, activities, institutions, thought, beliefs, meanings, or whatever, become possible when a wide variety of data and methods from many fields are used. It is also greatly facilitated by going beyond single cases and even narrow classes and using the total evidence, or at least appropriate samples both of archaeological and other (analog) data (e.g., Gardin 1980; Renfrew 1984, e.g., Part III; Sieveking *et al.* 1976, p. 40; Smith 1977, p. 607, 1978; etc.).

I have already commented on the obviously important issue of the differential survival of material. This becomes of importance as soon as the domain of the history of the built environment is extended to include preliterate, vernacular, and other comparable environments. It is clearly also central in archaeology (e.g., Clarke 1978) who discusses predepositional, depositional, postdepositional, and retrieval theory, all of which precede analytical and interpretive theories. In this connection also models from other fields have been found useful. For example, paleontology also deals with material likely to have survived differentially—and recovered randomly. Models have been developed there to help make inferences from the recovered residues to living systems. The inferential chain goes from a sample assemblage (involving sampling issues—cf. Kent 1987) through a burial (archaeological) assemblage and discard assemblage, to a behavioral assemblage. This can be illustrated as follows (DeBoer 1983, pp. 20–21; see Figure 5.1).

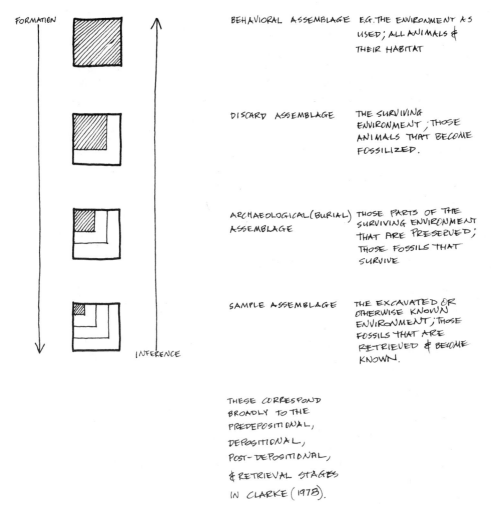

FORMATION

BEHAVIORAL ASSEMBLAGE E.G. THE ENVIRONMENT AS USED; ALL ANIMALS & THEIR HABITAT

DISCARD ASSEMBLAGE — THE SURVIVING ENVIRONMENT; THOSE ANIMALS THAT BECOME FOSSILIZED.

ARCHAEOLOGICAL (BURIAL) ASSEMBLAGE — THOSE PARTS OF THE SURVIVING ENVIRONMENT THAT ARE PRESERVED; THOSE FOSSILS THAT SURVIVE

SAMPLE ASSEMBLAGE — THE EXCAVATED OR OTHERWISE KNOWN ENVIRONMENT; THOSE FOSSILS THAT ARE RETRIEVED & BECOME KNOWN.

INFERENCE

THESE CORRESPOND BROADLY TO THE PREDEPOSITIONAL, DEPOSITIONAL, POST-DEPOSITIONAL, & RETRIEVAL STAGES IN CLARKE (1978).

Figure 5.1. The formation of, and inference from, known data from the past (based on DeBoer 1982, Fig.2.1, pp. 20–21).

Although data are clearly important for culture studies, typically the higher the level of generalization, the more theoretical any particular work is, the more important do the characteristics of data being discussed and the range of methods become. Thus, for example, the inference made from megalithic monuments—that they are associated with the rise of centralized political control—depends on the availability and use of a large body of evidence from a wide variety of places so that this hypothesized association can be demonstrated (e.g., Renfrew 1983). At a more detailed scale, the inferences also depend on knowing the scale of projects, for example, that Stonehenge represents an investment of 30 million man-hours (cf. Hadingham 1975; Hammond 1972; Longworth and Cherry 1986). The argument further depends on data on population density (a relationship denied in Moore and Keene, 1983). It also depends on the ability to relate the presence of monuments to the absence of migration and vice versa, that is, again a large body of evidence in many places, cross-culturally, and over time.

The extent of such data can be major. Thus in recent studies of Cobá in Meso-america, data over an area of 8,000 square kilometers are considered, multiple kinds of evidence and a variety of studies from different fields; moreover, comparative work plays a major role (Folan, Kintz, and Fletcher 1983). Comparative work was also very important for the major study of Teotihuacán, where a comparison between that city and Tikal made it clear that two towns of the same scale in the same area had very different densities (Millon 1973).

The point is often made that one cannot use single sources (e.g., Binford 1981, 1983a,b; Morris and Thompson 1985; Renfrew, Rowlands, and Segreaves 1982). One example is in estimating past populations: Many and diverse sources, data, and meth-ods must be used (e.g., Sumner 1979, pp. 164–165). They must, however, explicitly be shown to be relevant to the problem area in question, or the argument. Many different forms of relevance may be used; for example, data sources and methods may need to be within the same cultural frame, however that is (explicitly) defined, and justified. Thus, in testing a model developed of the roles of symbols, Hodder (1982c) applies it to the Orkney Islands. Then, however, he emphasizes the comparison of information from various domains within the same cultural frame. In this case the domains include settlements, dwellings, ritual sites, burial sites, and so on. This enables the identifica-tion of a common theme, that is, pattern (cf. Isbell 1978). That, in turn, leads to sophisticated and convincing hypotheses not only about settlements, dwellings, and tombs but also about attitudes about life, death, and sacred, and so on.

Renfrew (1984 p. 362) makes an important specific point that is highly relevant to my general argument. He criticizes the "new archaeology" in the United States on the basis of what he perceives as a major shortcoming, that it began in the southwestern United States and never really moved out of it. I am not concerned with the validity of this criticism but with the implicit point: that the need is for comparative studies working with, and testing any methods, inferences, and generalizations against a vari-ety of different areas and places. Recall the point that *sample variety* may possibly be more important than *sample size* in the confirmation of disconfirmation of hypotheses (e.g., Salmon 1982, pp. 31–32). It follows that comparative work based on the broadest possible range of situations and approaches is a major way to study anything—and certainly the past.

Given differential survival and "random" recovery, it follows that archaeology tries to infer past behavior, social organization, and cognition from patterns in bad samples of material culture and, often, indirect traces of it. This raises questions about how these inevitable lacunae and shortcomings are to be handled. One response is to be as explicit as possible about the problems. Another, central to this book, is the need to use all the data available. This raises another problem, which we will encounter in my case study—inadequate data. It has been argued explicitly that one can use early excavation material known to be "bad" together with newer data. All available data can be used; what is important is how they are used, the reasoning and the argument, its rigor and explicitness (Chapman 1981). Although new, "good" data and research are very impor-tant, the "bad samples" derived from previous work (even in the nineteenth century) must also be used. This is because in order to obtain the broadest sample, *all* the material must be used: One cannot afford to lose any. From the various kinds of data one can derive models, hypotheses, and theories. As I have repeatedly argued, one must use the *best available* data, not *perfect* data. Their quality needs to be reexamined; they can be reanalyzed in the light of new methods and techniques, and one often comes across the reanalysis of material from museums, previous publications, and elsewhere. This is clearly also the case with ethnographic, travel, and other data. Above all, one needs to make explicit how data are being used, why, the pros and cons discussed, and so on.

METHODOLOGICAL SOPHISTICATION, MULTIPLE METHODS, AND TAXONOMY

There is a major emphasis in the literature on the use of multiple methods, on methodological sophistication including explicit discussions about their choice, their pros and cons, their relation to theory, and to philosophical presuppositions. There is also an emphasis on, and debate about, the importance of taxonomy and classification (e.g., Binford 1983b; Clarke 1978; Connah 1981; Dunnell 1971; Gardin 1980; Gould 1985; Mackie 1977; cf. Salmon 1982).

The point is made that in order to develop more rigorous, logical, and explicit approaches and methods, classification and taxonomy become important. Different classifications are possible; which one is chosen depends on the goals (Gardin 1980). Note, however, that there is a recent body of work in taxonomy (which I will not review here) that argues that classifications are not quite as arbitrary as had been thought, that there are natural joints, as it were, in the world that fit some classifications better than others (e.g., Johnson-Laird and Wason 1977; Rosch 1978; Rosch and Lloyd 1978). This work, however, deals with classification in a somewhat different sense (for a review see Rapoport 1986c, in press).

Gardin (1980) suggests that the cataloguing of material precedes classification. This may be questioned because cataloguing already involves classification. Whether a thing is an axe, a house, or a something depends on classification; even traits depend on some taxonomy; grouping even more so. Taxonomies and classifications depend, at least partly, on goals and models. Those, in turn, depend on theoretical presuppositions and metamodels; for example, "simple explanations are best," "the ecological environment is most important," "ideas influence material culture," and so on (e.g., Deetz 1977, p. 23; Gibbon 1984). It is useful to recall, however, that the real world constrains taxonomies as it does hypotheses, models, inferences, and theories.

The form of any artifact, and certainly the built environments with which I am concerned, is the result of a variety of causes: tradition, materials, instrumental requirements, latent functions such as aesthetics, meaning, and so on (Isaac in Hodder, Isaac, and Hammond 1981, p. 40; Rapoport 1969a, 1980a, 1982a, 1985b). A multidimensional model such as this leads to different classifications and taxonomies so that, for example, "typology" becomes less acceptable, whereas style (Rapoport 1977, 1980d, 1982a; Wobst 1977) and morphology become more salient (Isaac in Hodder, Isaac, and Hammond 1981). If typologies are used, one needs morphological, technological, and both manifest and latent functional evidence (cf. Freidel and Sabloff's 1984 "form/function correlation"). It also seems to follow from such arguments that monothetic classifications and ideal types become less useful and that the polythetic approach advocated by Clarke (1978) seems more useful (Rapoport 1982b, 1988b, 1989a, in press). It is of interest to note that although there has been much discussion pro and con regarding the use of polythetic classifications, in general it has been better received in Britain than in the United States (Clarke 1978, p. 41).

The strongest and most extreme position on the importance of taxonomy in archaeology is that of Dunnell (1971), which is strongly criticized by Salmon (1982, pp. 140–154), who argues that one cannot initiate theory construction by constructing a system of definitions. It is, however, the case that many sciences begin with a "natural history" stage that involves categorization and classification; hence taxonomy and systematics (e.g., Kaplan 1964; Mayr 1982; Northrop 1947; among others). One could even argue that such a natural history stage is essential (Rapoport 1986c). Note also that classification plays an important role in defining the subject matter of studies, that is, defining the domain. It is, therefore, the case that, in spite of the argument about the relation of taxonomies to theory, Dunnell's emphasis may not be altogether misplaced.

His argument about the importance of systematics (although he uses it in a mono-thetic rather than polythetic sense) is due to the need to create units within a discipline. It follows that systematics must be used to define a field, in this case prehistory, rather than using content or subject matter. He argues that to be able to make inferences about people from artifacts, one must first find patterns in material culture that result from recurring human actions. That, in turn, requires the identification of recurring at-tributes that, in turn, requires grouping and that depends on taxonomy and classifica-tion. He also argues for the importance of clarity and precision in units and termi-nology in order to develop a field into a science. For that, systematics are essential (Dunnell 1971). Dalton (1981) implicitly agrees with Dunnell and hence implicitly disagrees with Salmon, that the careful definition of terms and isolation of conceptual terms is an important first task in the development of the power of a discipline (cf. Sanders 1984). This is because that brings order to a confused situation. Often, the careful reclassification of existing data, thus increasing their *quality,* may be more useful than increasing their *quantity.* I would argue, however, that both are essential and linked.

As briefly mentioned, Salmon (1982, pp. 140–154) criticizes Dunnell's position. She argues that an emphasis on terminological clarity may make things worse when it is seen, as in this case (and much social science) as a prerequisite to conceptual clarity and theory construction *rather than as part and parcel of the latter* (cf. Kaplan, 1964, on the danger of too early closure). She argues that theory construction cannot be initiated by constructing a system of definitions and classification. Yet that this is precisely how biology began (Mayr, 1982), and this proved very useful indeed (cf. Northrop, 1947). The difference may be in the fact that biology, for example, began with observation. The further point is made (Salmon 1982, p. 154) that concept formation, which includes definitions; the construction of classifications and theory formation need to proceed in a highly interactive way. *Because classification is in itself a theoretic activity it cannot simply PRECEDE theory construction.* In my view the debate is really one of emphasis and does not diminish the importance of classification *per se.* There is also a basic paradox involved (Kaplan, 1964) which is that one must begin somewhere and often classification is the starting point and theory follows; as in the case of biology, astrono-my, and probably other fields. Yet one can argue that there cannot be classification without theory. In any case, as theory construction develops, it informs classification that changes and, in turn, leads to further theory development. In that sense the two are clearly interactive.

INTERDISCIPLINARY AND MULTIDISCIPLINARY APPROACHES

Recent archaeology emphasizes the need for a maximum relationship with other disciplines that share a concern with similar methods, their data, methods, concepts, and theories. The result is an emphasis on interdisciplinary and multidisciplinary approaches and on the need for archaeology to be compatible with other disciplines. Among disciplines explicitly linked to archaeology are linguistics, ethnology, toponymy (place names), physical anthropology, paleoanthropology (in itself a mix of many disciplines such as archaeology, paleontology, paleoecology, geology, animal behavior, evolutionary science, systematic zoology, etc.), economics, geography, history, sociology, psychology as well as a range of natural sciences ranging from botany and climatology through geology and mineralogy to chemistry and physics. Note that EBS (and many sciences) are also a result of such a mix and that, in general, the most interesting work has been at overlaps among disciplines as I have already suggested—hybrid disciplines have shown hybrid vigor.

This is not to say that all these disciplines are used or referred to in any one study, although some recent teams have been both very large and extremely diverse in terms of disciplines. More common is the application of concepts, ideas, and data from one or two relevant fields although, of course, because studies build on each other, there is an implicit, but nonetheless real, growth in the impact of many more disciplines than the explicit applications would suggest.

A specific recent example of the explicit linkage between archaeology and another discipline at a conceptual, theoretical level argues for the development of archaeological location theory. This involves the application of concepts and theory from location theory in geography (in itself a hybrid field) to archaeology. The intention is then through historical and cross-cultural analysis to be able to make better and more general inferences about social structure on the basis of spatial form (Portugali 1984).* Note that this makes explicit what has implicitly been done in much recent archaeology (e.g., Freidel and Sabloff 1984, p. 97; Hodder 1978a; etc.).

There are also examples of the application of data, techniques, methods, and theories from a number of disciplines to studies of large areas such as Europe (e.g., Phillips 1980) or specific places, for example, the region of Jenne in Mali (McIntosh and McIntosh 1980). In this latter, for example, the findings of authorities from various disciplines regarding a given topic (e.g., cultigens originating in Africa) are plotted and related. Early travel reports are also used, and their quality is critically evaluated. Then follow various European reports and oral traditions. (This is very similar to my programmatic statement in Rapoport 1970a, 1973a; cf. 1982a.) Finally there is systematic excavation of various sites. The very diverse body of data is then cross-checked and

---

*Johnson (1980) tests the rank-order rule for settlements using archaeological and cross-cultural evidence, also explicitly linking geography and archaeology.

synthesized and related to multiple methods. This study and others (Butzer 1982; Hassan 1981; Parsons *et al.,* 1982; Phillips 1980; Platt 1978; Schiffer 1982; etc.) show that when theories, techniques, and methods are borrowed from other disciplines and applied to the study of the past, the rapidity of development and achievement that follow have been quite remarkable. Yet such approaches have never been used in architectural or even urban history, nor have the variety of data available.

Note that such borrowing of concepts and methods must be done carefully and is rarely literal or direct. When reference is made to other comparable disciplines, there is an explicit review of the pros and cons of using concepts, methods, or models, and attempts to establish the dangers and pitfalls, to identify those that work and those that do not. The point is made that such theory, concepts, and models need to be *adapted* rather than adopted (e.g., Moore and Keene 1983). In that book, for example, DeBoer in Chapter 2 explicitly applies models from paleontology; in Chapter 6 Keene criticizes the way optimization theory and foraging theory have been applied to archaeology; and Jochim in Chapter 7 also criticizes the use of optimization models. In Chapter 8 Moore criticizes the use of certain decision-making models in archaeology and recommends the use of others from geography, for example, the diffusion of information models. We thus see not only the interaction of different disciplines but also the rigorous mutual criticism discussed earlier.

There are also many programmatic statements about the need for interdisciplinary fertilization. For example, Binford (1983b) argues for the need to study contemporary people, experimental situations, historical documents of all sorts, as well as archaeological data that include faunal analysis, lithic analysis, and many other specialized disciplinary contributions. The need is to use many lines of evidence in order to achieve "robust methods" of inference—a point accepted even by Binford's critics (e.g., Hodder 1986). The use of such multiple lines of evidence also implies the use of multiple methods because each body of evidence requires and uses different methods. Moreover, because it is now almost a truism that methods are related to concepts, models, and theory, this also implies that a variety of the latter are used.

SEARCH FOR PATTERN

Note that the last five or six sections can all be seen as having one major goal: to aid in the identification of regularities, that is, in the *search for pattern.* This is, of course, a major topic in recent discussions about science and, as the debate about making archaeology into a science has developed, so has an emphasis on the importance of pattern recognition. Recall my argument that a major consequence of such a search and a precondition for it is the use of the broadest and most diverse bodies of evidence and comparative analysis.

Thus in dealing with the issue of faunal remains, the point is made about the importance of recognizing distinctive ("diagnostic") patterns. Only from those can one recognize or identify the agents responsible (Binford 1981). Although the specific emphasis in the latter is on bones, the argument about the importance of pattern recognition is, as pointed out above, very general. For "bones," one could substitute other "things"; the specific indicators, measures, methods, and inferential arguments would be different, but the importance of pattern recognition is general would remain because from patterns one can derive the *expected* composition (in this case, bone assemblages) and compare these with the *actual* ones. As discussed in Chapter 3, anomalies can only be identified if there are expectations. Having identified patterns and having concluded that they are interesting, one can then ask what they mean and make inferences about what happened in the past to result in such patterns. Ideally, of course, alternative explanations should be sought.

This theme of looking for patterns, one of the most basic activities of all scientists (and scholars), is a common theme in the recent literature in archaeology. This may be called repeated patterns of coocurrence (Smith 1982), or regularities (Salmon 1982). They may bear different labels (Binford 1981, 1983a,b; Renfrew 1982, 1984; Renfrew, Rowlands, and Segreaves 1982; Smith 1977), but the essence is the same: the importance of identifying patterns in the material record, in the relationships among attributes, in behavior and in environment–behavior interaction for the purpose of using such regularities as analogs for inference. Clearly patterns are also central for uniformitarian assumptions and were the key to my discussion of studies of rock paintings in Chapter 3. Note that in that discussion, as well as my discussion generally, and in archaeology, the point is made that pattern recognition involves expectations. This not only presupposes the characteristics of the evidence necessary already discussed; it also implies at the very least generalization and, more commonly, a move toward models and theory (e.g., Chapman 1981; Clarke 1978; Renfrew 1984; Smith 1977, 1978; etc.).

GENERALIZATION, EXPLANATION, AND THEORY

I began this book by suggesting that its approach derives essentially from a concern with using a broad body of evidence for purposes of generalization and explanatory theory. Those latter, I argued, are essential to EBS as they are to any scientific discipline. A similar concern with the emphasis on generalization, explanation, and theory construction characterizes recent archaeology. Although this should be clear from the emphasis given to discussing epistemological and comparable issues, it deserves more detailed discussion.

When large and varied bodies of evidence are used, as are comparative studies, multiple methods, cross-checks with other disciplines, controlled analogies, explicit arguments and inferences, empirical generalizations become possible. Although these are not sufficient (e.g., Binford 1983a,b; Willer and Willer 1973), they are necessary, both to suggest patterns and regularities as well as in order to be able to move on to inference, explanation, model building and, ultimately, theory construction. Although there is disagreement about the specifics of those, and even whether that is possible, there is, I think, much less disagreement about its desirability. It is fair to say that the totality of the literature reviewed in this chapter would agree with that; sometimes implicitly but often explicitly. Recall also the arguments discussed in Chapters 3 and 4 that generalizations are always used in the study of the past (or anything else), but that they need to be made explicit.

Thus Hodder (1982c; cf. Hodder, Isaac, and Hammond 1981) explicitly makes two generalizations that, if true, have major implications for archaeology and the study of the past. The first is that material culture plays an active rather than a passive role in relations among various groups and, I would add, in culture generally (Hockings 1984; Rapoport 1986c). The second is that reasons for the choice of particular elements of material culture in various social strategies need to be seen in the cultural context, that is, in systematic terms (cf. Leach 1976). This of course emphasizes a cognitive approach, the role of schemata (Rapoport 1972, 1980a, 1982a). In Hodder's view, these draw attention to important general principles that need to be considered; they also tend to prevent simplistic and hence incorrect inference directly from material patterns to social patterns of behavior, ignoring context, meaning, and the like. Such generalizations not only follow from analyses of broad, varied data; they also allow data from very diverse places to be used. Thus Hodder (1982c) is able to apply data based on studies done in contemporary Africa to a specific archaeological site in prehistoric Scotland. This clearly marks a major change in archaeology as a discipline and makes possible even broader generalizations.

Similarly, starting with work on the !Kung (Lee 1979; Lee and DeVore 1976; Yellen 1977), one can generalize for other hunter–gatherer societies such as Eskimo (e.g., Binford 1981). This use of analogy, to be discussed in the next section, depends on the uniformitarian assumption. This can be tested either in a large sample of cases or in a specific place over a long time period, for example, in Iran where it was shown that nomadic patterns today are very much like those in 8,000-year-old sites; there is a long, constant history (Hole, 1979).

These are, of course, examples of ethnoarchaeology. Although development is for purposes of controlled, explicit, and rigorous use of analogy, it goes beyond that. Ethnoarchaeology becomes *a generalizing science of human culture* that looks at material culture (which clearly includes built environments and cultural landscapes, including semifixed elements) in contemporary (developed) and traditional (nonindustrialized) societies in order to help with a more accurate reconstruction of the past (Hodder 1982b). This particular definition is not universally shared as we will see in the next section (e.g., Kent 1987). Also, although this supports my argument in Chapter 3, the latter goes beyond it: My emphasis is somewhat different; using more accurate data of the past, which is only the first step, as well as contemporary research to arrive at a more accurate, valid, and convincing general theory of EBR relevant to the present and the future.

Goals comparable to this more general goal are also to be found in archaeology. Thus a group of contributors to a book (Hodder, 1982a) try to derive some general propositions that can be used, for example, to state a theory of human behavior. In effect, they are generalizing by asking of past societies the same questions that one might ask of living ones. As in my argument, present and past data are used together.

A major emphasis on generalization is characteristic of Binford (1981, 1983a,b) as one would expect from his reliance on the covering-law model of explanation. It is not necessary to accept the latter in order to agree with his emphasis on the need to move from an idiographic to a nomothetic approach. Others also emphasize the need to generalize, no matter how difficult; one needs to discuss why this is necessary, how it can be achieved, and how legitimate generalization might be (Renfrew, Rowlands, and Segreaves, 1982); moreover, this is not enough as Binford repeatedly emphasizes (cf. Willer and Willer 1973).

Renfrew (1984) agrees that the issue of idiographic versus nomothetic must be considered because it is central both to history and social science. The question is to what extent history (and human nature and behavior) can be generalized and to what extent it must be restricted to the consideration of specific, and unique, cases. Recall, however, that even the study of specific cases benefits greatly by employing a science metaphor. Renfrew further argues that this must be constantly reconsidered, because the issues, and certain possible misconceptions, become clearer as more work is done and as there is more mutual criticism. He suggests that most historical explanations have both a specific (and particular) component related to the case in question, and a much more general assertion that must be relevant to the actual case it is desired to explain. This is similar to the argument about history in Chapter 4 that some generalization is always present in history implicitly and that it needs to be made explicit. In thus relating the general and the specific components of historical explanation, Renfrew tries to find a middle ground (as do many philosophers of science in their discussions; e.g., Hanson 1971). Renfrew cites the similar point made by Collingwood (1939, p. 140) who, although a traditionalist, argues that all historical explanation involves more general principles. As I argued in Chapter 3, at the very least one must ask what one expected. This is related to our discussion about pattern recognition and expectations without which one cannot even recognize uniquenesses or anomalies. The same point is made by Drake (1974) when he speaks of "implicit quantitative terms" of which "normal" is a perfect example (see Chapter 4).

The common ground that Renfrew is seeking is, therefore, the one already discussed: that the general assertion about which there seems to be agreement among both advocates of scientific explanation and, it would seem, even those of more traditional explanation, must be a necessary component. The need is, as usual, to make it explicit, and of this, recent archaeology was aware from the start. Note also that there is significant agreement that generalizations do not have to be universal (i.e., for all societies, times, and places). They may be for particular subsets or groups (for example of societies). Whether they are applicable to others is then a matter for empirical investigation as well as theory building. The latter usually involves the specification of limits, of conditions under which any laws apply or do not apply.

If statements are really only applicable to a single, specific case, then no comparative work would be possible, no social science, no theory. One is left with pure description, that is, the strongest form of empiricism (of which, ironically, current phenomenology is the only surviving example) (cf. Freese 1980; Lopreato 1984; Nicholson 1983, Wallace 1983; etc.).

Salmon (1982, pp. 21–22) argues that explanation requires reference to laws, although these do not have to be universal but can be restricted to particular classes and subclasses that makes generalization both easier and more relevant to social science, where the inability to derive universal generalizations has generated much not very useful argument (p. 18). Note also that even the arguments of those who question the value and validity of the science metaphor in, for example, anthropology, arguing that it is a discipline based on interpretation, that is, in search of meaning, rather than an empirical science in search of laws (e.g., Geertz, 1975, 1983) still imply laws at least at the level of statistical generalization; one must still know what "normal" is. Laws are essential in order to see the regularities that, in turn, are needed to see uniqueness or anomaly.

It needs to be admitted, however, that so far this has been more programmatic that actual, more a matter of intention and hope than of achievement.

CONTROLLED USE OF ANALOGIES

In Chapter 3 I discussed the implications of the almost self-evident fact that some form of analogy is inevitably involved in making inferences about the past. I discussed some of the criticisms of the use of analogy and the responses to criticism that can broadly be summarized as involving the careful, explicit, and controlled use of analogy.

In archaeology this issue has been much discussed. A number of people have criticized the use of analogies (e.g., Fletcher 1977; Gould,, 1980; Orme, 1981; among others). One criticism is that the past cannot be reconstructed according to the present if one wishes to find out *whether* the past was like the present. There is also the problem of much individual variation in behavior even in the most traditional societies, although this occurs within undoubted patterns (Rapoport 1980a, 1986c, 1989a). But note that much of this knowledge, as well as that about group variations (e.g., Hill and Gunn, 1977; Kent, 1984) has been discovered at least partly by the use of analogies.

These and other criticisms suggest that one can be wrong in one's behavioral inferences. This is because archaeologists deal with the results of behavior, not behavior itself. They have typically used implicit models of behavior that need to be made explicit (see Chapters 3 and 4; cf. Gordon 1978). One needs to discuss how behavior relates to artifacts and the degree of variability that needs to be tested (e.g., Cross 1983). Ultimately, however, the only way to relate artifacts to behavior and cognitive structure is to use analogy (e.g., Hadingham 1987). The use of analogy is also involved in the development and testing of models that, as we have seen, are frequently seen as analogies. Because analogies must be used, the question is how best to use them.

Essentially the development of ethnoarchaeology and also of experimental archaeology is the current answer to the problems raised by ethnographic analogy. Thus, whether the criticisms (e.g., Fletcher 1977; Gould 1980; Orme 1981) have been valid or not, they have been most useful because they have forced the use of analogy and the implicit connection between archaeology and ethnography to be made self-conscious and explicit. This is equivalent to a move to the *controlled use of analogy*. The development of archaeological ethnography and ethnoarchaeology (Kent 1987) has been a particularly important attempt within archaeology in that direction, of particular interest because it is frequently indistinguishable from EBS (e.g., Binford 1981; Gould and Schiffer, 1981; Kent, 1984; Kramer 1979, 1982; etc.).

Ethnoarchaeology is usefully defined as the formulation and testing of archaeologically oriented methods, hypotheses, models, and theories with ethnographic data (Kent 1987). Kent's point is precisely to distinguish this from the use of ethnographic analogy in general and specifically from the development of *archaeological ethnography* and *anthropological archaeology* that are the reverse, involving the testing of hypotheses, models, theories, and so on derived from ethnography with archaeological data. She criticizes the use of "ethnoarchaeology" for all three. This usage is common, however, and I will continue it here for simplicity. I will also use it to emphasize what I regard as *the essential point: the explicit use of data from both the present and the past.*

This development not only illustrates the closest current link of archaeology with EBS, but it also shows how the criticism of haphazard ethnographic analogy has led to new approaches self-consciously and explicitly intended to overcome the shortcomings revealed by critics. In that respect, archaeology has become self-correcting like some other sciences.

Anthropological, ethnographic, ethnohistoric, and other data have been used in archaeology as the only way to derive inferences about behavior from material culture. There is a virtual consensus on the need to use comparative archaeological, ethnographic, and other contemporary research (e.g., Binford 1983b). This is because, like in science generally, archaeological inference inevitably takes the form of an argument by analogy (Smith 1977, p. 607). In archaeology this always comprises, on the one hand, the archaeological data and on the other, ethnographic and other relevant social science information regarding human behavior. Some of the criticisms of this relate to the uniformitarian assumption. This assumption also underlies ethnoarchaeology as it must because it is essential for any understanding of the past (Binford 1981, 1983a,b). And, as we saw in Chapter 3, its use has been explicitly reviewed and justified in archaeology (Dalton 1981).

The purpose of ethnoarchaeology and other forms of controlled analogy is to test models about the relation of behavior to the variability of material culture not only against data from the past (a different if not impossible task) but also in existing situations, where behavior, material culture, and their relationships can be observed, tested, and studied (Longacre 1981).

In effect, one can think of ethnoarchaeological data as being applied *cross-temporally* as it were; it can also be applied cross-culturally. By its very nature, therefore, the use of such data involves (or implies) comparative work; after all, analogy *is* comparison. As an example, we find the controlled use of ethnographic work from contemporary New Guinea being used to illuminate or even explain the archaeological record from prehistoric Europe (Steensberg 1979). Although in that study the emphasis is on technology, conclusions are drawn about land clearing, villages, houses, and fences—about the cultural landscape—that are highly relevant to the history of the built environment.

The use of controlled analogies based on ethnoarchaeology now has a sizable literature. I will consider primarily the work of five people and their associates (Binford 1981, 1983b; Gould 1978, 1980; Hodder 1981, 1982b,c; Kent 1984, 1987; Kramer 1979, 1982). These show increasingly close conceptual linkages with EBS and thus provide a particularly strong supporting argument. Ranging, as they do, from 1978 to 1987, they also show the rapid growth in sophistication of this relatively new subfield.

The goal of ethnoarchaeology is to link material remains from the past to the human behavior from which they resulted. The material traces surviving in the present have, as we have seen, been greatly affected by human and natural forces (Gould 1978, p. vii; cf. Clarke 1978; DeBoer 1983). Thus, there are both gaps in the material record and no past behavior surviving or observable. The gaps are filled by controlled analogy with contemporary ethnoarchaeological data. Case studies typically use a wide range of methods to study past societies, for example Tasmanian aborigines. Major projects also investigate contemporary garbage behavior, as in the well-known ongoing Tucson study, because garbage middens are a major source of archaeological data. The range of case studies is extensive, and in all cases hypotheses are made and studied over time using multiple methods (cf. Gould 1978).

Two years later, Gould (1980) is proposing a unified theory of ethnoarchaeology based on a case study of Australian Aborigines. This study, as we have already seen, strongly opposed the use of analogy and proposes the use of *argument by anomaly*. This consists in trying to identify anomalies and trying to explain them by seeking alternative explanations. Recall, however, that to notice anomalies, and for the concept to have any meaning, one must have expectations that are based on patterns and regularities. In this sense, the study begs the question about *how can regularities be established without analogies*. The answer is that they cannot; it follows that ethnoarchaeology is the use of controlled analogy (cf.Gould and Schiffer 1981).

This is because the observation of contemporary behavior can facilitate the development and refinement of inferences about past behavior from material culture traces particularly if there are strong regularities (or similarities) *especially over long periods of time* (Kramer 1979). As suggested earlier, the reverse is also the case: Data of the past, if like those of the present, give one much more confidence in generalizations. Note also that similarities or regularities depend on the use of some form of classification and hence imply taxonomies.

In this view, ethnoarchaeology investigates aspects of contemporary social, cultural, and psychological behavior from an archaeological perspective: It is the study of contemporary societies using the time and space perspective central to archaeological research. (This makes it closer to "archaeological ethnography" [Kent 1987].) This becomes particularly useful if the study of contemporary behavior is through the testing of hypotheses. I would add that these must bear on material culture that, I have long argued (Rapoport 1976, 1986c), ethnographers have ignored for many decades and neither documented nor considered (cf. Renfrew 1984). It therefore follows that the development of ethnoarchaeology is relevant not only for archaeology but to anthropology. This is also the case with EBS *vis-à-vis* other social sciences and with the type of history I advocate in this book. This is useful to both the history of the built environment and EBS.

Many empirical and theoretical case studies in this book (Kramer 1979) study various aspects of behavior including how one can make inferences about the cognitive basis of archaeological data through controlled analogy with ethnoarchaeology. Similar studies published elsewhere reinforce this point. Thus ethnoarchaeology can study "native" (emic) cognitive classifications, for example, in an aesthetic framework (e.g., Hardin 1983). This not only shows that aesthetics can be studied scientifically, as we saw in Chapter 3 (cf. Crozier and Chapman 1984) but also how such knowledge can be applied to inferences about the cognitive bases of past artifacts (cf. Hanson 1983).

That controlled analogy is central to ethnoarchaeology is also shown by two other chapters in Kramer (1979). Thus Sumner shows how controlled analogy can be used to estimate past populations (cf. Hassan 1981). Hole argues that the use of such analogy ideally requires continuity with the past. This may be difficult to find (or to establish). The solution suggested is to look for situations analogous in important respects to the archaeological situation in question (see Figure 3.11). As the discussion in Chapter 3 suggests, these need to be *specified, made explicit, and reasons given*. That requires a rigorous disciplined approach, with clear questions and hypotheses from the outset.

Hole (1979) is able to show in Luristan (Iran) that the present-day nomadic patterns are very much like those found in 8,000-year-old sites; that is, they have a long history. Having established that continuity, the argument based on the analogy can then proceed.

My point about the characteristically rapid development of fields that adopt the approach I am advocating can also be seen in the work of individuals. For example, Kramer's study of a contemporary village in Iran that forms one of the case studies in the 1979 book she edited is further developed in a book 3 years later (Kramer, 1982). This book is a good example of what can be done because it has effectively applied EBS to the study of the past. The book addresses the question of *architectural variability* (Rapoport 1969a, 1985b,c,d, 1986c) and how it may or may not relate to variables such as family size and wealth. These relationships are shown to be complex. The point is also made that different areas of the dwelling need to be clearly distinguished (cf. Rapoport 1980c, 1989a) because some features of dwellings relate to population size and others to wealth. The point is clearly part of EBS: that different motives and choices operate leading to variability *within* environments as well as *across* them. Kramer (1982) then uses these contemporary data together with archaeological data from the past. She also uses paleoecological data about the region and site that clarifies climate, vegetation, land use, food, economy, settlement patterns at the regional level, villages, and their areas (sites). These, combined with the data on dwellings, make inferences possible, particularly because the climate, settlement patterns, villages, houses, and so on 7,000 years ago are shown to be very similar to those today; that is, continuity is established. Given this great continuity, it becomes possible to do a variety of detailed analyses and comparisons using different methods of villages and their internal organization, the percentages of roofed and unroofed areas, dwelling sizes, and their areas per person and so on (Kramer 1982, pp. 248–258). The conclusion of this controlled analogy is that one must look for variability because there is no typical anything even in such traditional, "simple" settlements. This means that *one cannot assume a normative or typical view* (a conclusion also supported in other situations by other studies; see Kent 1984, later discussion). This pattern means that there is no single architectural measure for "wealth," although a *group* of architectural variables, artifacts (semi-fixed elements), food, and so on can identify wealth. This is, of course, in accord with Clarke's (1978) view that polythetic rather than monothetic approaches should be used (cf. Rapoport 1988b, in press). In spite of this variability, the value of studying vernacular environments and the great utility of ethnoarchaeology for archaeology are emphasized and further studies suggested. More generally, in terms of this book, the conclusion is, first, how close this approach is to EBS and to the type of study of the past I am advocating. Second, how useful is this combination of a great variety of research on contemporary and past data when this is done through the use of controlled analogy.

Hodder (1981, 1982c) has done a number of studies on a regional scale in Africa based on a series of hypotheses that need to be tested in a contemporary ethnographic context, the topic being the role of material artifacts as symbols. The artifacts included pots, stools, and the position of hearths in dwellings, among others. A variety of sophisticated methods were used to identify those items used as "symbolic markers." Also studied was the use of different items concurrently (cf. Wobst 1977) to increase what I call redundancy. One method used was to plot items in space, that is, plotting plans of houses and their decorations, plans of compounds and elements in them, village plans, and the distribution of villages at the regional level. This is identical to the domain as I have defined it, and it also reinforces my argument about the need to identify systems of settings rather than assuming them (Rapoport 1969a, 1977, 1980c, 1986a, 1989a).

The specific findings of these studies do not concern us here. The complexity of the relationships, however, and the fact that these studies enabled the various social strategies of different groups to be understood prevents the assumption (without explicit limitations) that material culture variability *directly* reflects social patterning and behavior. More complex models must be used that incorporate beliefs, ideas, and meanings (e.g., Rapoport 1976, 1980a, 1982a, 1986c). In terms of this section, it also follows that the relevance of these models to studies of the past is through their controlled use as analogies. Models are not derived only on the basis of the fieldwork in Africa; they are clarified and reinforced by comparing that work with many other studies in different areas. This comparative work suggests further questions and also leads to some generalizations. The result is not just that symbolic archaeology is possible but the inescapable need for it and for contextual archaeology. The models developed are then applied to a study of the archaeology of the Orkney Islands. Note not only how close this approach is both to EBS and my own work over the years, and to the thrust of this book, but also how different this (and the other studies) are to simple and direct, that is, uncontrolled, analogy.

Specific suggestions can be found about the kinds of analogies that are needed (Moore, 1982). This explicitness, of course, means that one can then disagree with any specifics while agreeing with the necessity for the controlled use of analogies. It is significant that ethnoarchaeology is one of the fastest growing areas in archaeology (Moore, 1982). This is because it addresses one of the principal problems faced not only by archaeology but by any historical inference or explanation—the valid use of analogy. As already mentioned, the study of "refuse," "rubbish," or garbage—or its *absence* (Hadingham 1987, p. 157; *The Times,* 1982a) is a major source of data in archaeology. It can also be used in contemporary situations, as in the well-known Tucson Project, or more informally. For example, in periodic visits to India one geographer relied on quick counts of litter in standard-sized areas in cities, villages, or by the roadside, as a quick and informal measure of economic growth: the less litter, the higher the standard of living (David Sopher, personal communication). There are also examples of the use of litter to study the quality and status of urban areas, whether they are declining and so on (Rapoport 1982a).

In this connection, Moore (1982) does a study among the Marakwet of Kenya. He tries to link refuse patterns to other types of data and to a cultural context. Such insights can then be applied to the archaeological record. I would add that by comparing this to other groups, cultures, situations, and so on regularities and patterns could be traced and generalizations possibly made. This would make this material even more useful.

As ethnoarchaeology in all its guises has grown and developed attempts have been made, and continue to be made, to pull such work together, to review it, and to begin to generalize and develop theory (e.g., Kent 1987). Thus Hodder (1982b) pulls together his own work on ethnoarchaeology as well as reviewing the use of ethnographic data and anthropological concepts more generally. He makes the point here being emphasized, that *the past can only be studied through analogies with the present-day world; it is inevitably based on analogy.* This point has, of course, been made by others in archaeology (e.g., Mackie 1977) and, as I have already mentioned, about science in general (Boyd 1984; Hesse 1966; Leatherdale 1974; McMullin 1984; Ortony 1979; Schön 1969; Swanson 1983; etc.). Hodder shows (1982b, Chapter 1) that *any* inferences made from material culture to behavior are essentially analogical—*even identifying an artifact as a pot, axe, or house.* He both argues and shows how *careful, explicit, rigorous, and analytical* the use of analogy can be (1982b, Chapter 3). (Recall my argument in Chapter 3 that explicitness is the key to all the other characteristics.) Thus by explicitly and carefully specifying the attributes of agreement, this resemblance can suggest agreement about other attributes (e.g., Smith, 1977; see Figure 3.11). Note that Smith explicitly develops a model of inference based on analogy—the hypothetico-analog (h-a) model.

In the study by Kent (1984), the overlap, indeed identity, between ethnoarchaeology and EBS is almost complete. In this, three hypotheses about the use of space that are implicit in archaeology are made explicit. These are then tested by comparing such use of space by Navajo, Hispanic-Americans, and Euro-Americans in modern, typical "suburban" dwellings. This is done in such a way as to be almost publishable as an EBS study as it stands.* The conclusion, *that culture is the main influence on the use of space,* is, of course, central to my work since 1969 (in *House Form and Culture*). Some specific findings can be questioned and would themselves not stand up to a wider cross-cultural analysis. The conclusion, though, is most relevant to the study of the past. This is (confirming Kramer 1982) that the hypotheses about the use of space used in archaeology cannot be applied universally: They vary with cultures, that is, from group to group.

This is implicitly clear if one compares our familiar use of space with, for example, the !Kung San (e.g., Draper 1973; Lee 1979; Lee and DeVore 1976; Yellen 1977). Ethnographic work regarding the !Kung not only shows this variability but also has ethnoarchaeological uses. It leads to the insight that by observing how and where the !Kung San set up their camps it is possible to discover where to look for ancient campsites (an example of *predicting* the past, see later discussion) and also how artifacts found are to be interpreted. It also becomes possible to begin to generalize for other hunter–gatherer societies. This is also shown by Binford (1981) as he combines ethnoarchaeology (without calling it that), observational, and sometimes experimental, testing of hypotheses and hence uses analogy in controlled ways. He also discusses the use of analogy through an analogy! (Binford 1983a, p. 420). This is that in order to identify animals from their footprints, it is first necessary to study the footprints of *identified* animals, so that the relationship between the animal and its spoor is controlled (i.e., known). Ethnoarchaeology, experimental archaeology (e.g., Coles 1973; Ingersoll, Yellen, and MacDonald 1977; Yellen 1977), and the like try to do exactly that. The use of this knowledge is, of course, equivalent to the use of a controlled analogy.

*It is both interesting and unfortunate that disciplinary boundaries are still strong enough so that no EBS research is used or cited in this book.

When applied to the past there is, of course, an additional criterion—that the same relationship between animals and their tracks obtained in the past exists now (Binford 1983a, p. 420). This is, of course, the uniformitarian assumption. It follows that a crucial problem facing anyone studying the past, whether historian, archaeologist, geologist, paleontologist, or whoever, is how to justify the uniformitarian assumption (see Chapter 3).

PREDICTION

It has often been argued that the emphasis on prediction in much discussion of science is irrelevant in the study of the past. After all, the past cannot be predicted. I argued in Chapter 3 that it *is* possible (Rapoport 1986c, 1988a, 1990), and in fact, in recent archaeological research, predictions are frequently made along with indications of how these are to be tested. I have already mentioned a very elementary example—predicting where ancient hunter–gatherer camps are to be found. The test here is equally elementary: dig in the predicted locations. But the making of predictions goes far beyond that.

One way in which such predictions are made is when hypotheses are generated, often by using the convergence of multiple lines of evidence, on the basis of observation, surface surveys, ethnographic and other such materials, analogy, reasoning based on work from other disciplines, and so on. These hypotheses can then be tested by controlled excavation (e.g., Jochim, 1976). They can also be tested by using other kinds of data from the past, for example, historical accounts, paleoecological data, and other. In many cases, clear and explicit predictions are accompanied by indications of how they can be tested (e.g. Folan, Kintz, and Fletcher 1983; Smith 1977, 1978).

In a major study of the late Maya site of Cozumel, hypotheses are generated on the basis of two major possible lines of independent information: visible surface remains and ethnohistorical data (including travel description). This is supplemented by background knowledge based on previous work, analogy with ethnographic data, logical reasoning, and so forth. These hypotheses then become predictions to be tested by a third major possible line of evidence—controlled excavation (Freidel and Sabloff, 1984). For example, one type of prediction concerns certain structures visible on the surface; these are deemed to be sacred structures. This is based on a set of characteristics (what I called a *repertoire* [Rapoport 1982a]). These include location, restricted visual access, use of substructures, specific plan forms, presence of altars, use of certain materials, such as masonry, as opposed to others (e.g., perishable). They also list possible problems and complications with their reasoning (an example of an explicit, logical argument). These predictions are to be tested by future excavations.*

An overall prediction, in the form of a hypothesis, is also made about Cozumel as a whole: that it was primarily a long-distance trading center. To test this, Freidel and Sabloff (1984) argue that it will be necessary to use all the evidence, the full range of

---

*In a different context, Michael Coogan establishes four criteria for identifying a cultic site: isolation, exotic materials, continuity over time (itself requiring diachronic data!), and parallels with unquestionably cultic sites (i.e., the use of analogy). (In *Palestine Exploration Quarterly,* June/July 1987, in Shanks 1988).

structures (not only masonry ones), the whole island, and beyond (cf. Rapoport 1980c, 1986a, 1989a; cf. Vayda 1983). There are thus suggestions about what evidence is to be sought and where. Even when particular predictions, for example, about residential arrangements, are not supported by excavation, several other lines of evidence are adduced to argue that the hypothesis is still tenable because the absence of confirmation is not disproof (Freidel and Sabloff 1984, p. 117). This is, of course, a familiar point; it also illustrates the importance of explicit argument.

So far I have been discussing the testing of predictions by controlled excavation and also the use of other lines of evidence. Predictions can also be linked to simulation; that is, *one can simulate the past:* One can assume certain models and ranges of variables and run simulations (usually, although not necessarily, on computers). These simulations can then be compared with data: spatial, archaeological, ethnographic, and so on (e.g., Widgren 1978). They can also be compared with predictions. At the same time, such simulations can be taken as predictions to be tested by further work in the field (e.g., Hodder 1978b). It is interesting to note the close resemblance of this to the discussion of simulation in cosmology in Chapter 4 (Cornell and Lightman 1982). Simulation has also been used in other historical sciences, for example, biological evolution (Dawkins 1986; Sabloff 1981, p. 121).

Simulation is useful because it encourages, indeed forces, explicitness, clarity, precision, objectivity of thought, and the like. If one can make something work in simulation, one has understood it because, effectively, one has predicted it (Sabloff 1981). This point has also been made in recent work on mental imagery: Being able to simulate mental activity on the computer gives one confidence that one has understood how the brain works (Kosslyn 1980, 1983). Simulation is also useful in finding important, patterned regularities even in the case of processes that involve chance. They encourage process statement, that is, dynamic models that can link theory with empirical data and also help move from static archaeological data in the present to dynamic behavior in the past. It follows that simulation helps in the exploration of problems too complex for more usual forms of analysis (Sabloff 1981, pp. 121, 128).

Through simulation, which involves modeling a system, one effectively makes predictions and then checks those retrodictively (Renfrew 1984, p. 288). Simulation can also be seen as a form of experiment, either for testing hypotheses or of an exploratory ("what if") form (Renfrew 1984, p. 344). In all the studies discussed in the books just cited (Hodder 1978b; Renfrew 1984, Chapters 10–12; Sabloff 1981), many detailed examples and case studies can be found of the many and ingenious forms of simulation possible.

I have already referred to the different ways of checking predictions: excavation, ethnoarchaeology, the use of multiple diverse and independent lines of evidence. There is one other way that needs to be briefly considered. This is what has been called *experimental archaeology.* I have already given an example of that (e.g., Steensberg 1979) when discussing ethnoarchaeology.

The principal idea behind this approach, on which there is a sizable literature (e.g., Coles 1973; Ingersoll, Yellen, and MacDonald 1977; Yellen 1977) is that artifacts (cloth, tools, buildings, and so on) can be made or duplicated by actually using the methods or processes believed to have been used. The relevance for construction methods, how feasible they were, how many man-hours were involved in given cases, is clear. This could be applied to anything from dwellings and tombs to major earthworks and cathedrals, for example, to test recent suggestions about the construction of Chartres and other French cathedrals (James 1979, 1981, 1982) or precolumbian structures in Peru (Hadingham 1987). One can also test processes involved in felling trees (e.g., with stone axes), agriculture (with digging sticks), or animal husbandry to see how feasible they were, how long they would take, and so on. Habitability and likely behavior can also be tested, for example, by constructing farms and having people live in them, using contemporaneous clothing, food, and so forth as has been done at Lejre (Denmark) and Petersfield (the Butser Project) in England, for example. It is also possible to test issues of differential survival and the relation of traces to artifacts by first making and then destroying those in various ways. The traces created in known ways can then be compared with excavated data to help with interpretation. This is remarkably close to the example discussed of how one learns to identify animals from their spoors (Binford 1983a, p. 420).

There are many overlaps between experimental archaeology and ethnoarchaeology. Many studies published in the two bodies of literature are quite similar (e.g., some of the studies in Kramer 1979). This point is also made by Gould (1978). Not only do the kinds of research overlap, the two can be combined to strengthen controlled analogies. For example, experimental archaeology and ethnoarchaeology have been combined by doing limited controlled experiments in traditional contexts (e.g., Forge 1973b; Washburn 1983; cf. discussion on art in Chapter 3) and in contemporary contexts (e.g., certain aspects of the Tucson garbage project; Hodder 1982b, Chapter 9). In fact, all three methods discussed—ethnoarchaeology, experimental archaeology, and simulation can be related, as they have been in terms of "control" versus "realism" (Hodder 1982b, p. 30; see Figure 5.2).

THE DEVELOPMENT OF THEORY

I have already referred several times to the central role of theory in science. It is the need to develop theory in EBS that led to the argument in this book. It is, therefore, significant as part of this supporting argument that recent archaeology is characterized by conscious and explicit attempts to develop concepts and theory.

I will not be discussing the general role and importance of explanatory theory in science here. In relation to archaeology, the point has been made that the only way to link material remains of the past with behavior is through building *bridging theory* (Sabloff 1981, p. 7; cf. Binford 1983a,b). Thus, Ebert (1979) argues that one cannot simply and directly translate form into assemblages and types and then into inferences based on them. One must use bridging assumptions in order to relate artifacts to behavior. Once such assumptions are stated explicitly, they can be tested and they also begin to suggest methods (cf. Smith 1977). It also follows that methods are needed to provide data relevant both to the creation and testing of middle-range theory and ultimately the confirmation or rejection of general archaeological theories (Sabloff 1981, p. 7).

Although the emphasis on theory development is a result mainly of the concern with epistemological issues discussed earlier, it is also partly a result of the stage of

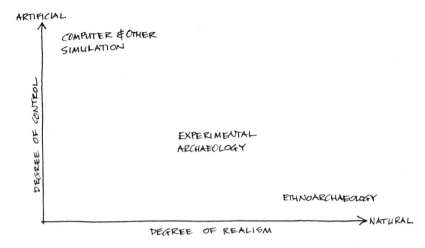

Figure 5.2. "Control" versus "realism" in archaeological research (based on Hodder 1982b, p. 30).

development of the field. In the past, archaeology was concerned with describing data; now there already are a great deal of data. One example is the astounding amount of data in a case study of ancient Rome (Sabloff 1981, p. 149). These data form a basis for hypotheses and help in the development of theory. This is because "theory flourishes best when significant collections of facts present themselves for explanation and interpretation" (Sabloff 1981, p. 188). This point has been made regarding biology (Mayr 1982) and neuroscience (Bullock 1984), and I have argued that the lack of such data has been inhibiting in EBS (Rapoport 1986c).

Adequate data are thus important to provide an empirical base; one must have it as it is a necessary condition for theory. It is not, however, sufficient. Once it is available, one needs to go beyond it—to move from the *quantity of data* to the *quality of conceptual apparatus* (Dalton 1981). Theory never emerges from data, nor is even generalization sufficient (Willer and Willer 1973)—concepts and theory must be developed consciously and actively. This is not only because theory is the best form of generalization; from theory one can derive problem areas, questions, and so on, and, more important, specific hypotheses that can be tested. Because one inevitably needs to go beyond the evidence at the very least in making inferences (let alone explaining data), theory becomes central: Only theory enables one explicitly and logically to go beyond the evidence by specifying connections and linkages, models and mechanisms. Because archaeology always goes beyond the material traces found, the emphasis on theory in the literature is not surprising. It can also be understood as an attempt to move from the more traditional studies of single places or groups (however well done) to more general theory.

That emphasis is primarily on what has been called middle-range theory* as opposed to grand theory, which Binford (1981, 1983a,b) has called general cultural,

---

*Although that is used in a somewhat different sense from the better-known use of that term by Robert Merton.

ideological, or paradigmatic frameworks. Note, first, that he uses the latter term very differently from its more common usage based on Kuhn's work. Second, he identifies these frameworks by one major characteristic—that they lead to "endless polemical, emotional debate which also typifies alternative cultural beliefs and values." These have also been called "cosmologies" (Douglas 1982, p. 35; Nash 1963). The centrality of such middle-range theory is due to its role as a guide to explaining the world. This theories do by trying to answer "why" questions (e.g., Binford 1981, p. 25). Theory and theoretical laws are also distinguished from empirical generalizations. The former say how the world is by asking why things are as they are, whereas empirical generalization merely summarizes experience based on observation (Binford 1983a; cf. Willer and Willer 1973).

Although Binford may be the most insistent, the need for, and importance of, theory is largely consensual (e.g., Gardin 1980; Hodder 1982a; Hodder, Isaac, and Hammond 1981; Moore and Keene 1983; Renfrew 1984; Renfrew, Rowlands, and Segreaves 1982; especially Part I; Salmon 1982; Smith 1977, 1978, etc.). All seem to agree that, for example, scientific archaeology must go beyond description or even empirical generalization to concepts and then to theory as systems of concepts, constructs, and the like linked to empirical reality. Recall that the beginnings of "scientific archaeology" began with criticisms of the lack of theory in archaeology (e.g., Clarke 1978 [1968]; Watson, Leblanc, and Redman 1971) that one still encounters today (e.g., Binford 1981, 1983a,b; Salmon 1982, p. 140). A theory is then defined quite typically as "a set of interrelated rather high-level principles or laws that can provide an explanatory framework to accommodate a broad range of phenomena" (Salmon 1982, p. 140; cf. Margenau in LeShan and Margenau, 1982; Papineau 1979; Zaltman, Lemasters, and Heffring 1982). It is also pointed out that without theory no method or measurement has any meaning (e.g., Moore and Keene 1983; cf. Kaplan 1964). Although there is fairly general agreement that such theory does not yet exist, although it is badly needed, there is much effort to develop at least beginnings of it. The clear, explicit statements of propositions regarding human behavior with the consequences traced and tested are one example (e.g., Hodder 1982a; Smith 1977, 1978; etc.).

The need and, indeed, centrality of theory even at the level of measurement and method, of course, also agrees with much recent social science (e.g., Freese 1980; Lopreato 1984; Nicholson 1983; Papineau 1979; Wallace 1983; Willer and Willer 1973; among many others). It is also a common theme in the philosophy of science where, whatever other disagreements one finds, there is consensus not only on the importance of theory but on theory as the distinguishing characteristic of science. In archaeology the point is made that the problem is not with inadequate data or methods but with the lack of theory. Only with theory can one have meaningful expectations with which to approach the data; without expectations, there can be no problems or even puzzles. Recall, however, the equally emphasized position that theory must be anchored to empirical data.

MODELS AND MODEL TESTING

I have already mentioned the recent views that identify models as what links theory to empirical evidence, which provide the dynamics. In archaeology, one also finds a major emphasis on the role of models, the explicit building of models to link theory with empirical evidence. It can be argued that the development of careful operational linkages between concepts and data is also clearly part of this. Also, models are not merely generated or built—they are tested in a variety of ways, some already discussed earlier. Moreover, there are attempts to generate multiple and alternative models. Finally, models try to deal with matters such as human behavior, cognition, meaning, cultural evolution, and the like. There is agreement that models are essential and central in the development of a more scientific archaeology (e.g., Gibbon 1984; Moore and Keene, 1983). It is also clear that the explicit nature of model building makes it possible to criticize them and encourages such criticism that, once again, is a theme in the literature.

For example, in a study of availability of salmon in a prehistoric setting in Europe, Jochim (1983) develops three different models, including ones relating to how such fish could be caught. The purpose is to trace the implications (via their food value) for group size, the largest possible settlement size, settlement patterns, and the like. The problems with the published archaeological data and of the record are carefully discussed. Given that it is all that is available (what I called the *best available* information), this information regarding group size and settlements is then used to test the implications of the models—what one could call predictions, postdictions, or retrodictions. One of the three models is then chosen as providing the "best fit." Although this is not done in this particular study, it is clear that simulation (as discussed earlier) could be used.

Frequently models are derived or adapted from other fields. Although I have already discussed the dangers of this (cf. Moore and Keene 1983), one finds models from biology, evolutionary science, and other disciplines being applied with interesting and stimulating results (cf. Renfrew, Rowlands, and Segreaves 1982, Parts II and III). In any case, many different models tend to be tested (e.g., Dalton 1981; Renfrew 1984). This is because one needs to *invent* (not discover) conceptual models because they help to extract what is important and nonobvious from the vast mass of diverse data. They also provide a rationale for what is extracted, reasons for why it is important. Because one does not know what models will work best, one needs to invent, borrow, or adapt many alternate ones (e.g., Clarke 1978). For example, Binford (1983b, p. 120) lists 10 models of trade with their spatial implications (i.e., predictions); these are then tested, quantitatively and in detail (in his Chapter 5). A multiplicity of models, as long as they are constrained by empirical evidence (e.g., Bunge 1983; Jacob 1982) is not a disadvantage. This is because, although models are empirically underdetermined, they are not purely speculative. There are constraints not only of empirical data but also of background knowledge, plausibility (e.g., Salmon 1982; Smith 1977) from theories in other fields and so on. The point is also often made that there are no true or false models, only revealing or misleading ones (Gibbon 1984; Nicholson 1983). Note two things: First, that models are most frequently identified with analogies (modeling = analogy); it follows that my discussion of the controlled use of analogy is relevant. Second, taxonomy is also an aspect of this, because grouping and classification can also be seen as the equivalent of modeling for specific purposes.

I have already mentioned several times that even work that questions so-called "positivistic" science still exhibits many of the characteristics I am describing. This also applies to models. For example, one finds discussion of explicit criteria for models in a structuralist framework (e.g., Hodder 1982a, especially Chapter 4; cf. Wylie 1982). It follows that the use and construction of models is not necessarily related to any particular analytical framework (e.g., "positivist") which that book in fact explicitly rejects.

Models are not necessarily about human behavior or about the relationships between such behavior and material culture. One also sees more abstract models being developed and tested. Thus an explicit model of inference and confirmation is developed and then tested against empirical data in a major case study (Smith 1977, 1978). Note, once again, that the whole process of inference, which is central to any study of the past, is based on controlled analogy. Because models are increasingly seen as forms of analogy, the centrality of models and the testing of models of inference is hardly surprising.

INFERENCE

I have already discussed the centrality of inference in any study of the past and in science generally. From the early days, the "new" archaeology, in Britain, the United States, and elsewhere, has seen its task as "decoding" the "congealed information" in the archaeological record (e.g., Clarke 1978) or of "decoding" the patterns observable in the archaeological record (Binford 1983b).

It is a basic assumption of archaeology(as it is of EBS, and must be of the history of the built environment) that because some behavioral elements of the sociocultural system have material correlates, it follows that *inferences can be made about the behavior with which these material elements are associated.* The issue is *how;* and we have seen that whatever the specifics, such inference is essentially analogical. The development of ethnoarchaeology is merely one specific response to the more general requirement that analogy as a form of inference involves using information from one phenomenon to understand another, on the basis of some relation of comparability between them. Thus, on the face of it, an analogy between prehistoric Britain and the Classic Maya seems unlikely; yet even used in a not rigorously controlled way, it leads to some sophisticated inferences (Mackie 1977). One can strengthen such analogies and differentiate among competing analogies by using controlled analogy, as discussed. In this way, it becomes possible to form propositions about the past, to deduce consequences, and to test the implications against independent archaeological evidence. The careful use of relational analogy is, in fact, the nearest archaeology can get to a rigorous method (Hodder 1982b, especially p. 27). That requires both a use of models and links with various other disciplines, including the social sciences.

The results can be convincing and interesting. Many examples have already been discussed. But consider how many characteristics of a Mycenean kingdom, including the number of sheep, can be reconstructed (Chadwick 1983) and how inferences can be made about human populations of the past (e.g., Hassan 1981) or about privacy from a study of apartments in Teotihuacán (Millon 1973). In that case, inferences are also made about special foreigners' neighborhoods, the presence of what can only be called urban renewal, and so on. Inferences about privacy are also made about dwellings in the Bronze Age Aegean (Sanders 1985). This latter is also one of the very few examples of relating EBS to archaeology (cf. Rapoport 1988a, 1990).

Inference in archaeology, and in the case of many of the historical data of EBR, is difficult for reasons already discussed: Inferences must be made about human behavior from physical artifacts alone; the data are always in the present, and inferences must be made about the past; such inference is always of past dynamics from static contemporary phenomena. This makes such data *more similar to the data of natural science than of history* (based on writing) (Binford 1983b, pp. 21–22) or of social science where behavior can be observed and people questioned.

Note that the possibility of making such inferences has been explicitly denied (e.g., Douglas 1972). The pitfalls of doing so have been amusingly illustrated both regarding the domestic bathroom in the United States (Miner 1956) and the motel (Macaulay 1979). The fact that it is constantly being done is because such inferences, however difficult or even questionable, must be done. This is because often even in the recent past (e.g., vernacular environments) and always in prehistory, only archaeological material is available. As just one example, there are no written or other sources on Harappan culture, yet inferences must and can be made (e.g., Jansen 1980).

The quality of such inferences, how well-founded they are, and the possibility of improvement depends on making implicit assumptions explicit, testing the inferences against both contemporary and archaeological evidence, and using the broadest range possible of evidence and disciplines (e.g., Binford 1981, 1983a,b; Hodder 1982b; Kent 1984; Kramer 1979, 1982). In fact, over the past 20 years archaeology has moved from "pessimism, nervousness and timidity" about this goal to optimism (Renfrew 1984, p. 6). This is a result of all the characteristics that I am discussing in this chapter, which have led to careful, sophisticated, and hence convincing inferences. They become tested rather than untested "leaps of faith" and never mere assertions. This supports my argument that the approach I am advocating will also lead to much more reliable and convincing inferences about cognitive and symbolic behaviors. The point has now been reached where making inferences about human *cognitive* behavior is deemed possible (Renfrew 1982; cf. several studies in Kramer 1979). The point is made that inferences about social structure have been made for some time (e.g., Redman 1978c); the possibility of doing that had also been questioned, as had its desirability (Renfrew 1982, pp. 10–11). More specifically, it had been said that one cannot reconstruct social organization earlier than 1700 C.E. Yet many examples exist, some of which we have discussed, showing that this can be done even for prehistory (cf. Renfrew 1984, where examples are given of such inferences for Malta [from temples], Europe [from cities], Mesopotamia, Mycenean Greece, the Maya lowlands, Etruria, Northern Europe in 2000 B.C.E., etc.; cf. Chadwick 1983; Hodder, Isaac, and Hammond 1981). Once that was shown to be possible, there was a shift to attempting to make inferences about human cognition, what Renfrew (1982) calls "archaeology of the mind" (cf. "symbolic archaeology" [Hodder 1982a,c]). What this means is that there has been a double shift, first to social archaeology, then to behavioral, cognitive, and symbolic archaeology. It is now deemed possible to make inferences about the intelligent and purposeful behavior of the makers and users of material culture in the remote past, to understand and explain their minds and their thinking.

Renfrew (1982) uses a nonenvironmental example showing the chain of reasoning that allows many inferences to be made from some colored stone cubes at Mohenjo-Daro to the minds of those who made them. Among these, that they had concepts of weight like ours; used units and hence modular measures; had a system of numeration the ratios of which can be derived; that they used these for practical purposes; that there was a notion of equivalence among different materials on the basis of weight and hence possibly a ratio of value among such materials, which implies some constant rate of exchange among commonalities (cf. Schmandt-Besserat 1983 on writing; Marshack 1971 on astronomy, counting, and symbolic notation). One could add that social inferences are also possible from the same analysis: for example, that there was regulation of trade and control, which is planning. This is, of course, confirmed by the obvious planning, layout, and design of cities like Mohenjo-Daro and Harappa (e.g., Pieper 1980a). One can also derive planning ideas from the study of prehistoric settlements (e.g., examples in Rapoport 1979b); religious ideology from the study of sanctuaries and depictions of them. Recall that this is precisely what was done with African rock painting (Lewis-Williams 1981, 1983; Vinnicombe 1976; cf. Leroi-Gourham 1982). As another example among the many possible, inferences have also been made about the spatial and temporal categories of the Zapotec and Mixtec (Flannery and Marcus 1983) and of the Maya cosmos from an analysis of their settlement patterns (Freidel and Sabloff 1984, p. 158; cf. Marcus 1973, 1976).* It even proves possible to deduce changes in models of the cosmos from the Classic to the Decadent periods (Freidel and Sabloff 1984, pp. 184–185). From these changes, implications can be deduced, such as the replacement of a single center by movement among many centers. One result is the need to study the whole settlement pattern so that these ritual movements can be traced. This is, of course, the point made earlier about the domain definition and the need to study the whole system of settings (Rapoport 1980c, 1986a, 1989a; cf. Vayda 1983).

More detailed inferences about such ceremonial or ritual movements—which *in themselves* left no traces—have also been made (cf. Vogt, 1968, 1976). These ritual movements in real space and time also reflect the spatial and temporal structure of the Maya cosmos and thus reinforce the finding from the analysis of settlements. Inferences about movement, of course, required the use of ethnographic data thus showing links with anthropology (and ethnohistory). In addition to links to anthropology, inferences of this type also involve links with other social sciences, with the more traditional humanities, including history, as well as with psychology, ethology, sociobiology, evolutionary biology, and others.

Although ethnoarchaeology and the like developed specifically to generate new data in the inference process, making the analogies more controlled, other data can be found and used. It is often the case that such data do exist even if very few (e.g., Bibby 1970) and can be used for analogy. For example, a major synthesis of life in medieval England relates settlements, buildings, and semifixed elements, links the study of social conditions, written history, social science, and archaeology, and uses both high-style and vernacular design (Platt 1978). It is almost an exemplar of my argument but applied to social history. In other cases, even when the data is potentially available, it is ignored. Thus an attempt to make inferences about the use of public spaces in Harappa (Jansen 1980) ignores what can still be observed in India today, as I personally observed and described recently (Rapoport 1987a; cf. Jain 1980a). This also implies that

---

*Broda *et al.* (1987) do this for the Aztec based, in the first instance, on an analysis of the Great Temple of Tenochtitlán, but using a great variety of other data and sources, and extending it to the city and the world.

there may be continuity; establishing whether that is the case, where and when, to what extent and so on, is, in fact, the purpose of historical study. My case study will deal with precisely this specific issue. Recall, that to establish continuity, that is, identify patterns, one needs comparative studies using large and varied bodies of evidence. For example, in India continuities can be shown to exist in pilgrim movement, its role, and its effect on the cultural landscape (Bhardwaj 1973; Sopher 1969). More important, for the present argument, these can be related to the studies of the Maya discussed earlier (Freidel and Sabloff 1984; Vogt 1968, 1976); they can also be related to others, for example, of the Mayo Indians (Crumrine 1964, 1977) and Australian Aborigines (Rapoport 1975a). Taken together, and with others, they reveal a particular way of linking people and settings that is still relevant today in understanding cultural landscapes and has design implications (Rapoport, 1981a, 1986a, 1989a). It also has implications for the very origins of humans and built environments (Isaac 1972, 1983; Rapoport 1979a). This discussion can also serve as an example of my point about linkages with other bodies of evidence and other disciplines.

It seems clear that inferences are best made by the convergence of multiple lines of evidence (e.g., Jochim 1976) as well as of multiple methods, disciplinary frameworks, and so on. In doing that, one can correlate different variables. For example, the making of inferences from past built environments to behavior can be understood in terms of correlating formal features such as size, height, materials, space organization, and the like, with use or function (including latent functions), that is, correlating "formal types" and "functional types" (Freidel and Sabloff 1984). This is equivalent to making inferences from the former to the latter. Each formal type has variability, that is, there is a range of possible things that can be done, and the specific conditions and constraints determine the choices made (what I have called the "choice model of design"— Rapoport 1977, 1983b, in press; cf. Rapoport 1983a, 1985b). In other words, as already discussed, there is a repertoire in any culture from which one can choose (cf. Rapoport, 1982a). This idea of a repertoire, although it is not called that, is used to infer that certain buildings at Cozumel are for ritual (Freidel and Sabloff, 1984) as discussed earlier. The same notion of a repertoire of basic elements, how they are arranged into a spatial configuration and further elaborated by a system of spatial mediation, is used to study ritual at Suchindram (Pieper 1980b).

Making inferences of this kind becomes easier and more reliable the larger and more diverse the body of evidence and the lines of evidence. In the case of the individual elements used to infer different functions and uses at Cozumel, multiple lines of evidence are used, and the pros and cons of each are specified in some detail. Before moving to higher levels of generalization, the inferences are "normalized," that is, *the most likely conclusion is accepted.* These are then used in the higher order reconstructions such as the various groupings, the overall settlement and settlement patterns for the whole island and beyond. In effect, generalized inferences from particular structures are used to make inferences about groups of structures. Generalizations from those, in turn, are applied to particular sections of the settlement, the whole settlement, and the overall settlement pattern (Freidel and Sabloff 1984, pp. 114–115). This can be visualized as follows (see Figure 5.3).

The purpose of all this is to *recognize basic patterns* (on p. 179 the project is described as a "pattern recognition study") and then to ask a series of questions that

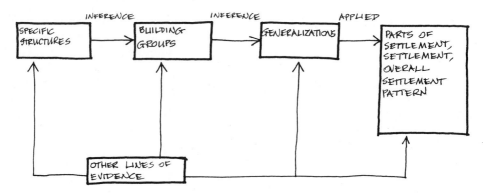

Figure 5.3. Inference and generalization in archaeology: from buildings to settlement patterns.

help in trying to make correlations, and hence draw inferences, about social, behavioral, and even cognitive patterns. A similar process of making inferences about cognition from dwellings and various artifacts is also found elsewhere (e.g., Deetz 1977, pp. 42–43, 59–60).

Note how closely this relates to concepts such as the concrete object—symbolic object sequences and the notion that activities can be analyzed into four levels from instrumental (manifest), to meaning (latent) discussed in Chapter 3 (Rapoport 1977, 1982a, 1988a, 1990). This, in turn, is related to the idea in archaeology of the three levels of function also discussed in Chapter 3 (technomic, socio-technic, and ideotechnic) (Binford 1962 and later work). Although far from universally accepted, this similarity is interesting and the analysis useful and suggestive.

This is also the case with those artifacts called built environments where, I have recently argued, there are not only four levels of function but also three levels of meaning (Rapoport 1988a). In all these and other ways inferences are being made about past human cognitions and cross-checked by using contemporary data (e.g., Miller and Tilley, 1984). As already pointed out, in connection with this and other works, even when one disagrees with specific inferences, the process is useful and cumulative. Also, as discussed in Chapter 3, inference becomes more difficult as one moves from instrumental (manifest) functions to latent (or from technomic to ideo-technic).

The recurring emphasis on pattern is also important in order that absence of something can become as important as the presence of something and can be used in decoding material culture. This is, of course, only possible if expectations exist, and they, in turn, depend on recognizing patterns and regularities. Those in turn depend on some constancy, revealed through comparative work using large and diverse bodies of evidence. These involve uniformitarian assumptions that must be made explicit; without those no inference is possible. In one study that I have already mentioned, a very large and complex ceremonial structure produced no finds after extensive and careful sieving. This "cleanliness" is unexpected because the site was used for several centuries and the norm (pattern or expectation) is that such sites contain lots of detritus and waste. Thus this anomaly, the *absence* of artifacts, becomes highly significant and leads to inferences about the ritual nature of the site and its importance (*The Times,* 1982a). This inference from negative evidence only became possible as a result of the scale, extent, and care of the sampling and because there were generalized expectations based on a large and diverse body of previous work.

Note that inferences become stronger and more convincing, and more feasible for social and cognitive variables, when all the other characteristics discussed in this chapter (in support of my argument in Chapter 3) apply. This is when the work is comparative and interdisciplinary; when large and diverse bodies of evidence are used and there is adequate description and classification; when premises are explicitly stated; when multiple methods are applied; when the argument is explicit, logical and, if possible, formalized so that it can be followed and criticized; when there are attempts to generalize and construct explanatory theory, and description and explanation are clearly distinguished, and so on. All these make inferences stronger and more reliable. This is because they are tested rather than untested "leaps of faith." This is the reason why models of inference themselves are explicitly stated, evaluated, discussed, and tested, as already discussed (e.g., Renfrew, Rowlands, and Segreaves 1982; Salmon 1982; Smith 1977, 1978; etc.). As a result, there is an awareness of the need for multiple working hypotheses (an example of "*strong inference*" [Platt 1964]) and of how one can limit potentially unlimited alternative hypotheses. There are sophisticated discussions of the kind of evidence needed (broad and diverse) and the problems of inadequate and biased data, problems caused by the fact that not only is one making inferences from material culture to behavior, one is doing so from very limited portions of material culture that happen to have survived and then to have been recovered (see my discussion earlier in this chapter and Figure 5.1; cf. Clarke 1978; DeBoer 1983, Figure 2.1).

Done in this way, inference also becomes more believable. This is because *inferential arguments* can be seen as *warranting arguments,* that is arguments that tend to warrant to others one's beliefs about the world (Binford 1983b, 1983a, pp. 414–419). Given the characteristics being discussed, they can become plausible and progressively more plausible.

In this process, as I argued before, one moves from domains that are better known to those that are less known and less well established; from those domains where inferences are easier to those where they are more difficult (Hawkes 1954 cited in Binford 1983b, p. 16). More generally one also moves from areas of knowledge and disciplines dealing with comparable topics, which are better established to those that are less well established (Boulding 1980). This is, of course, the essence of the use of models and of all arguments by analogy.

SOME EXAMPLES

Many examples have already been given in the course of the discussion in this chapter. Examples were given in each of sections that show new approaches in archaeology. The discussion was intended to support my argument by showing, first, how closely these approaches fit those suggested for the study of the history of the built environment. Second, I argued that by adopting these approaches archaeology developed rapidly; it follows that the study of the history of built environments is also likely to do so. In ending this review, it therefore seems useful to consider a few selected studies, treated solely as examples, to show what can be inferred from the evidence available. There is no suggestion that these examples are better than others: In fact, I have already referred to all. Their choice is thus arbitrary in that most of the studies could have been used.

Many studies, even of specific places, generalize in several senses. They are potentially generalizable in terms of epistemology, being tests of models of inference. They are also generalizable in substantive terms, being case studies of broad questions; for example, prehistoric patterns of human behavior. One such study (Smith, 1978) provides a useful example because it illustrates both these types of generalizations. First, this book-length study is explicitly used as a test case for a model of inference published earlier as a paper (Smith 1977). Second, although apparently a study of a specific area, the Mississippi Valley, its intention is to begin to generalize about patterns of human behavior. It begins with small sites, so that one can learn from them and then tackle larger and more complex ones. This not only shows an explicit commitment to growth and cumulativeness; it also nicely illustrates my argument over the years of starting with clear-cut, and possibly easier, situations and then moving on to more subtle and ambiguous, and hence possibly more difficult ones (e.g., Rapoport 1969, 1980d, 1982a, 1983a, in press). It is worth recalling that Smith asks five major questions, generates three alternative hypotheses for each, and makes observational predictions before excavation. This enables relevant data to be collected. After the description of the data found, the final chapter returns to the hypotheses and the questions. The predictions and data uncovered are clearly distinguished. This makes it possible to test the former in terms of the criteria for evaluation and degrees of plausibility described in the model of inference being tested. In this way, this study can almost serve as an exemplar of a rigorous and explicit approach to research about the past.

In other studies the generalization is of a different sort. For example, in studying the Mesoamerican village (Flannery, 1976), a developmental sequence is developed for a large area. Moreover, an explicit research strategy is used, organized by scale— starting with the household,* through dwellings and dwelling clusters, to communities and regions. Thus the whole system of settings and the full range of patterns can be discovered. As a result, inferences can be made about a whole range of behavioral issues, including social, cognitive, and ritual. In this process many lines of evidence are used from a vast amount of previous work, and hypotheses from the present work are tested. Much of this previous work is also critically reviewed and, finally, an overall synthesis is made (cf. Parsons et al., 1982, for another example, in the southern Valley of Mexico; cf. Blanton 1978; Flannery and Marcus 1983 on the Mixtec and Zapotec; Nicholas et al. 1986; and others).

Also generalizing is a study of Neolithic France over a 4,000-year period (Scarre, 1983). Here the inferences are about past cultural landscapes, that is, it is a form of prehistoric cultural geography. This is, of course, the domain of the history of the built environment as I have defined it. The object is to study the diachronic development of settlement patterns in France between 6000 and 2000 B.C.E., including agricultural strategies, trade, and social organization. Although the evidence available is clearly partial, it is the best available. Thus the inferences are clearly interim and require

---

*Note recent explicit arguments about the use of this unit in comparative and historical studies (e.g., Netting, Wilk, and Arnould 1984; Wilk and Ashmore 1988).

testing. The results of this testing, in turn, lead to new hypotheses and new data. The conclusions, which reveal high complexity, are less relevant than the fact that it is possible to study cultural landscapes 8000 years B.P. while at the same time not only not simplifying but taking into account greater complexity. Also of interest is the way in which the study is done. Among other characteristics, it shows cumulativeness and progress; uses multiple lines of evidence and a large body of previous work; explicitly uses sophisticated taxonomy and classification and multiple methods; introduces sophisticated new techniques; engages in explicit argument; emphasizes the need to achieve intersubjective agreement (i.e., consensuality that leads to consensibility [Ziman 1978]). In other words, this study shows most of the characteristics under discussion and, at the same time, the results this makes possible.

In discussing epistemological debates in archaeology, I suggested that these are often concerned with the meaning of "scientific" meant, rather than whether one needed to be scientific (e.g., Kelley and Hanen 1988; Salmon 1976, 1982). In other words, the debate was about the adoption of this or that specific model, or whether one should adopt a narrow or broad view of science. I also referred to the ongoing debate between some British and U.S. archaeologists and, specifically, the criticisms of Binford's rather strong position that one finds in the British work (e.g., Hodder 1982a, 1986; Renfrew, Rowlands, and Segreaves 1982; Renfrew 1984).

Yet, when one reviews specific examples of Binford's work one finds that they show many, if not most, of the characteristics that I have identified as characteristics of almost all recent work in archaeology, whether in Britain, France, or the United States. A good example is provided by Binford's (1983b, especially Chapters 4–7) review of the controversy about the Neanderthals. Essentially he reviews a body of previous work, especially that of Borde, and identifies the essential problem that is posed; he then tries to solve, or answer, that problem. The emphasis throughout is as much on showing *how* one might go about answering or resolving the problem as on the answer itself. The importance of classification, taxonomy, and systematics becomes clear: Without them one cannot deal with the central problem, which is *the meaning of the variability of the record of the past*. This requires identifying patterns. To recognize patterns, one must define and refine in an ordered manner those relevant properties that can be observed systematically (what one might call indicators). This requires many observations and analyses of large bodies of data in order to be able to recognize the patterns in them. When this was done, the patterns discovered shocked people because they could not be integrated into the traditional view. The anomaly was in conflict with views about cultural variability and forced people to revise their views. More data, facts, and patterns produced by doing correlations strengthened the anomaly; they, in fact, made the problem worse.

At this point in what is a highly explicit argument, the critical question became whether the past was really so different from conditions documented in the modern world, or were there characteristics in the way archaeologists were looking at the past that created the apparent differences between the modern world and inferences about the past.

Binford tackled the problem by doing ethnoarchaeology among the Eskimo, that choice being explicitly argued. The way in which that work was done is very close to EBS, for example, my work on systems of activities in systems of settings (Rapoport

1969, 1977, 1980a,c, 1982d, 1986a, 1989a). In fact, my model is exactly what he discovered to apply to the Eskimo. A great deal of detailed data were collected at various scales from parts of single sites to the overall, and very extensive, home range. In other words, he discovers the boundaries of the relevant system. He then used comparative material on systems of settings among the !Kung, the Seri, and other Indian tribes (cf. Kent 1987; Lee 1979; Lee and DeVore 1976; Yellen 1977). By using these multiple lines of ethnoarchaeological data, he gradually built up a most comprehensive analysis of Eskimo behavior and use of space from parts of sites (e.g., around a hearth) through dwellings, sites, regions, and all the way to the home range. He then develops a "theory of site structure"* that he applies to the archaeological record of the Neanderthals, asking the question "What does it mean?" In doing so the full range of evidence is examined—artifacts, site structure, faunal remains, and so on—the whole Neanderthal cultural (or archaeological) landscape. Then one can proceed to the question: Why did it happen?

My final example (among many sophisticated studies I could have used) is a *paleosociological study* and *landscape archaeology* already mentioned several times. Again, the important point is that such studies become possible—and how. This is a study of Cobá, a Classic Maya metropolis (Folan, Kintz, and Fletcher 1983). At the very outset a point is made that is very relevant to the current argument in architecture about "holistic description" [*sic*]. Contrary to the use of this term in architecture, Folan *et al.* identify *holism* with use of the maximum evidence possible, emphasizing that this is quite different from the use of that term in phenomenology, humanistic psychology, and so on (Rapoport, 1985a).† In their study "holism" means studying the maximum range of evidence possible and to study landscape archaeology from a paleosociological perspective.

In this case it involved detailed surveys (95% complete) of an 8,000 km² area with 20,000 structures. This allowed Folan and coworkers to delimit the area; a detailed architectural inventory then determined the number of structures. Inferences are then made about the number of inhabitants in Cobá, using explicit criteria and a review of the literature on the urban status of Maya settlements, their constituent elements, their morphology, and so on. This is equivalent to a contextualization of Cobá both in terms of existing knowledge and spatially (cf. Rapoport 1986a, 1990; Vayda 1983). This approach also allows them to relate Cobá to the region and study the distribution of settlements, areas, and cities. This study is very different from architectural history that studies single buildings out of context, or the work of one designer, but it is very close indeed to my approach.

The Cobá study also looked at geology, soils, hydrography, fauna, climate, and the like, and changes in some of those over time (i.e., history), as well as the effects of those on the Maya generally and on Cobá specifically. On the basis of this large body of diverse evidence it becomes possible to make many and detailed inferences about the city, residence patterns, social organizations, and the like. For example, the investiga-

---

*I am not *evaluating* this theory with which I have problems as being too materialistic. What seems important is how it is derived and used and the fact that it is explicit and can be debated.

†I will postpone a critical examination of the term *holism* to the book on theory, but cf. Efron 1984; Phillips 1976.

tors were able to derive both general Maya and specific Cobá characteristics that allowed them to correlate social status with vegetation. This is particularly interesting because this is also the case in contemporary U.S., Australian, and Canadian cities where status and vegetation (landscaping) are very clearly correlated, sometimes more clearly than status and buildings (see the many examples in Rapoport 1977, 1981a, 1982a; cf. Duncan, Lindsey, and Buchan 1985).

An extraordinary number of highly convincing (or plausible) inferences became possible due to the use of much and varied evidence, sophisticated methods, explicit argument, and the other characteristics of current archaeology. It became possible to derive detailed taxonomies of social classes, the household composition of compounds, and demographic, economic, and political characteristics. It became possible to study the family, the community, and the regional/interregional levels in both social and plan terms. The extent of the initial survey and the identification of classes and of important features and the quantitative analysis of remains made it possible to study the whole cultural landscape, including its emergence and its relation to new and more complex forms of organization. These inferences were cross-checked by using ethnographic, ethnohistoric, and ethnoarchaeological data. Also, early hypotheses in the literature (for example, from 1954) are tested against the new, improved data. Due to better and more data and new methods, it became possible to propose more complex models.

Cobá is also compared with Mayapan in terms of the mapping of surface features in both. Note that the study of Cobá is based *only* on surface mapping; *no excavation has yet been done*. This means that the conclusions of this study are effectively predictive hypotheses that can be tested through excavation. In fact, suggestions are made about how excavations should be carried out in order to test the predicted model at different scales (cf. Freidel and Sabloff 1984).

Much more could be said about how the analysis of the data, ethnohistoric evidence, and ethnographic evidence converge and come together. It even becomes possible critically to evaluate early travel descriptions. One could also describe the comparisons that are made with other studies elsewhere in the Maya realm. In all cases where inferences are made, the problems are pointed out, the pros and cons discussed, lower and upper limits and ranges set, if appropriate, and so on. However, enough has been said to make the principal point. The significance of this example is that this kind of sophisticated and convincing analysis of settlement patterns, a city, neighborhoods, a variety of buildings and groupings, architectural elements and features, plants, and so on becomes possible. Moreover, on the basis of these analyses of material culture and the use of many other lines of evidence, numerous inferences can be made about the inhabitants of this place, their behavior, rituals, social organization, economics, politics, and so forth. The principal point of this study, the others discussed, and the many not even mentioned, and of this chapter, is that the way that I have proposed of studying past built environments has been applied to a domain similar to ours. The method seems to work well and to lead to rapid and cumulative growth of knowledge. As a result, archaeology provides a particularly convincing supporting argument.

Part III

Case Study
Pedestrian Streets

Chapter 6

Pedestrians and Settings

INTRODUCTION

This case study deals with a single, specific topic; as such it does not fully illustrate the approach discussed in this book. It is an early and relatively simple example, a first step in a lengthy process of developing a new approach to the study of the history of the built environment. For one thing, like other scientific fields, it will inevitably be a group endeavor. It does, however, illustrate some of the most important issues discussed in Chapters 1–3; neither is it at odds with the principal issues raised in the supporting arguments.

This study is about *the use of historical data to derive precedents useful to design generally and to urban design specifically*. In urban design, such data have been of some interest to analysts of urban landscapes and spaces (e.g., Bacon 1967; Cullen 1961; Rudofsky 1969; Sharp 1968; Sitté 1965 [1889]; Specter 1974; among others). More recently, there has been increased interest in this subject by designers and writers on design (e.g., Anderson 1978; de Arce 1978; Krier 1979; Rowe and Koetter [n.d.], among others). The present study, however, adopts an EBS perspective (Rapoport 1969b, 1977, 1979a,b, 1987a). This, of course, was central to the argument of the earlier sections of this book. It follows that because my purpose is to illustrate a very different approach both to design and to history, I use a larger and different set of data. More important, the approach and methods are quite different. Thus although much of the argument in this case study can be seen as first testing and then confirming some intuitive conclusions of certain design writers on pedestrian streets (e.g., Cullen 1961; De Wolfe 1966, 1971; Nairn 1955, 1956, 1965; Sharp 1968; Sitté 1965), there are some major differences. Thus:

1. The commonalities among those writers are traced to some concepts derived from EBS research, primarily those related to the notion of *complexity*. The conclusions are thus based on empirical research in a variety of fields and on a more theoretical base; they enable one to derive testable hypotheses. They also relate contemporary interdisciplinary research to the study of past environments.

2. The material used in testing these hypotheses is explicitly cross-cultural; it also covers a much longer time span and a much larger range of environments. Thus the hypotheses are tested against a much broader body of evidence that, as we have seen, has major advantages.
3. There is also a particular orientation toward learning from past environments. It is assumed that such learning, that is, precedents, cannot be done directly by copying forms but must proceed by analysis using concepts from which generalized lessons can be derived (e.g., Rapoport 1982b, 1983a, 1989b). Only in this way can one relate the past to contemporary (and future) work.

Recall that it is an essential part of my argument that in order to generalize validly about environment–behavior interactions one needs concepts, theories, and models based on the broadest possible evidence. This is necessary in order to trace regularities in apparent chaos and also understand those differences that are significant. In this way one is more likely to see patterns, which is the most significant thing for which to look. Being able to establish the presence of such patterns may help one deal with the problem of constancy and change and to establish certain baselines that might guide environmental design. If humans as a species have certain characteristics and if they have done certain things in certain ways for a very long time, then there may be very good reasons for these things, particularly if even animals show these patterns. At the very least, this is the *prima facie* hypothesis to be falsified. Moreover, if apparent change and variability are an expression of invariant processes, this is extremely important because the reasons for doing apparently different things remain the same (Rapoport 1977, 1979a,b, 1982a). This also enables comparative evaluations of different solutions in terms of how they meet the same objectives.

This, of course, has been my argument for the need to broaden the evidence as far as possible: to use all of history, not just the recent past, and thus to use archaeological data as well as research in existing environments; to go beyond the Western tradition and include the full range of other cultures; to use all that humans have built, including nonliterate cultures and vernacular environments.

It has been my contention that broadening the evidence in this way leads to more confidence in any generalizations made. *If any patterns and regularities occur in otherwise very different cultures and persist over long periods of time, it is likely that they are basic or, at least, important and significant.*

The topic of the present case study is pedestrian spaces. Thus, if such spaces exhibit certain characteristics across a set of environments as described, these characteristics should be important. This becomes more important if one can also relate it to concepts and findings based on contemporary work. It also becomes more significant when we specially concentrate on preindustrial societies. First, such societies tend to show *more extreme variations* along many dimensions than do postindustrial societies (e.g., Udy 1964, p. 162).* Using those is then equivalent to varying settings more, making any regularities more telling. Second, and at the same time, as one concentrates on preindustrial societies and goes further into the past, the more purely pedestrian do

---

*This is analogous to my argument more generally for beginning with extreme situations and then moving on to more subtle ones (Rapoport 1983a, 1986c).

the settlements become. They can thus be expected to reflect much more directly those *physical* characteristics related to walking than do later settlements. The cultural variability of actual activities and settings combined with the fact that they are all largely pedestrian provides the best combination; this is further reinforced by the range of cultures, periods, climates, and so on.

There is another more personal reason for the choice of the topic. I have been interested in it since my student days. My undergraduate thesis dealt with pedestrian behavior; my graduate thesis, with the nature of urban design for pedestrians (Rapoport 1957). Although long before I knew of any work on complexity, implicitly and intuitively, I was dealing with this very subject. At the same time in a purely personal and emotional way, I have always found vernacular urban environments (as well as other types of vernacular environments) of preindustrial societies much more satisfying, attractive, interesting, and walkable than those of any designers and wondered why. *Because subjective preference is no basis for evaluation, analysis, or design, a more objective study seemed essential.*

One of the themes that I have been emphasizing is the need to make one's argument fully explicit and to relate it both to empirical data and research and to concepts and theory. These come from EBS. Thus in addressing the first step of the argument, *the relation between pedestrians and settings,* I begin with a brief summary of the nature of EBS because it provides the framework within which the topic is analyzed.

ENVIRONMENT–BEHAVIOR STUDIES

This approach begins with the general proposition that design must be based on knowledge and that this knowledge concerns environment–behavior interaction. It follows that design consists of making human and environmental characteristics congruent with each other. As already discussed, the field can be understood in terms of three general questions.

Because any specific question can be seen in terms of these, it is useful to elaborate them to some extent (cf. Rapoport 1977, pp. 1–4). Human characteristics, in this case, comprise two major aspects: *cultural* (desire, habit, and propensity to walk) and *perceptual* (those characteristics that need to be satisfied for a setting to be supportive of walking [Rapoport 1981b, 1987a]). Another useful formulation is the following: There are some *invariant, general, species-specific (i.e., panhuman)* perceptual characteristics, although with idiosyncratic variations based on individual history, personality, and the like. There are also *group-specific, cultural* characteristics that are more variable and also include idiosyncratic variations (no group is fully homogeneous [e.g., Rapoport 1980a, p. 25; 1989a]). Because one cannot deal with individual, idiosyncratic variations, although one is interested in the range of variability, one needs to generalize. One is thus left with perceptual and cultural and subcultural variables.

Regarding panhuman characteristics, some recent ideas on possible evolutionary baselines for design already discussed, are of interest (cf. Kaplan 1987; Rapoport 1975b, 1979b, 1986c). This makes walking a particularly interesting activity to consider in the first case study, because *it is essentially unchanged since the origins of the species*—or even before. Unlike other travel modes, pedestrian behavior is millions of years old—

bipedalism is now widely accepted as the first step in human evolution, as having occurred long before the emergence of *Homo sapiens* (e.g., Ruse 1986). This suggests the possibility of a clear evolutionary baseline; the uniformitarian problem is taken care of. It follows that it is useful to identify the characteristics of settings in which it evolved, although I will not do that (Boyden 1979; Boyden *et al.* 1981; Dubos 1966, 1972; Fox 1970; Geist 1978; Hamburg 1975; Tiger and Fox 1971; among many others reviewed earlier). More important, any characteristics of settings discovered are directly applicable to present-day pedestrian environments. Also, because the evolutionary *perceptual* baseline is still generally valid, different characteristics derived for other travel modes are equally relevant, because they are based on the same perceptual needs and mechanisms. Thus it is likely that even if people are "unaware" of their needs, they will respond to characteristics of settings appropriate to walking, select such settings for walking, and walk more in settings possessing these. This congruence between walking and supportive characteristics should be most in evidence in settings specifically created for walking, when that was the only, or principal travel mode. Because this activity has been invariant over time, it becomes a particularly suitable subject for historical and cross-cultural analysis. Conversely, the much lesser incidence of walking today is essentially "cultural" and may possibly be reversible by changes in culture and also by designing settings possessing the appropriate characteristics.

This, however, brings up the second of the three questions of EBS—the effect of environments on people. That hinges on the validity of the notion of environmental determinism. The evidence makes it quite clear that the design of the physical environment alone cannot lead people to engage in any activity. Moreover, it is easier to be negatively determining, that is, to block given behaviors by making them impossible or very difficult, than it is to be positively determining—to generate activities or behaviors. In other words, motivation may overcome unsuitable environments, at a cost, but environments cannot generate motivation (Rapoport 1968a, 1969c, 1977, 1983c). Given the motivation or predisposition to walk, for example, the question becomes, What physical characteristics of the environment are most congruent with, and supportive of, such behavior?

There are two things wrong with the way the second question is generally posed. First, it is wrongly formulated. People are not, in general, put into environments that then affect them. Rather they *select* environments, leaving those they find undesirable and seeking out desirable ones; there is a choice of settings based on preference (Rapoport 1977, 1980b, 1983c). This notion of *habitat selection* is very basic and, although derived from ecology, seems to apply very well to humans (e.g., Rapoport 1985b). One can study how different individuals and groups make different choices using different priorities and consequently are distributed differentially across different settings; not all settings attract people equally. This choice process involves an interplay of inborn characteristics and experience (in humans—culture); all species live in characteristic habitats, selecting those physical and social attributes that suit them. It is necessary to understand how choice operates, which characteristics are chosen and why, the specific environmental cues perceived, the schemata against which they are matched, and hence the resulting choices (e.g., Rapoport 1977, p. 48). Of course, although wants generally reflect needs, there may be cases where preferences have unintended, undesirable consequences (Rappaport 1979); it then becomes necessary to distinguish between wants and needs, but this needs to be done very cautiously and on the basis of reliable knowledge (Rapoport, 1980b, 1985b).

The question being considered is which kinds of habitats, that is, settings, will be (or would be) selected for pedestrian activities. Note, however, that pedestrian activities consist of two major kinds: dynamic, for example, walking and strolling; and static, for example, sitting or resting (e.g., DiVette 1977; Rapoport 1977, pp. 246–247). Both are pedestrian in the sense that people are not riding or driving; in that sense both are likely to differ from settings for traffic. Given that there are many possible types of space (Rapoport 1970b, 1977), both the preceding examples are pedestrian in contradistinction to traffic/motorist spaces, that is, human space rather than machine space (Horvath 1974).

Dynamic and static spaces are likely to have, or require, different characteristics. Movement spaces tend to be linear, narrow, and to have high complexity levels so that they entice with hidden views, encouraging walking, strolling, and sauntering. Rest spaces tend to be more static and wider, frequently contain greenery, require sitting facilities, and so on. Such spaces, whether plazas or avenues, encourage visual exploration from one spot—mainly of other people; they need to act as stages for social behavior, for people who become objects of interest and provide the requisite complexity levels. Thus streets that should appear excessively wide in terms of this study, such as the Champs Elysées in Paris, Dizengoff Street in Tel Aviv, or Paseo Colon in Barcelona, provide human interest in their outdoor cafes. Also, by narrowing the pavement through the use of semifixed feature elements, appropriate movement spaces are created that become even more complex due to the presence of people (i.e., of nonfixed feature elements).* Moreover, pedestrians and seated people see and are seen. In many cities one finds the contrast, as in the Plaka in Athens, between narrow shopping and pedestrians streets and wider rest spaces, often with trees, where tables of cafes and taverns, as well as markets, are found—although these spaces can still be remarkably small.† There is also evidence that different perceptual processes operate in linear spaces (streets) and nonlinear spaces (squares, plazas, and the like) (Vigier 1965). This study will be concerned with dynamic pedestrian spaces, leaving static pedestrian spaces for analysis elsewhere (cf. Rapoport 1986a).

In the present case, as usual, the choice of settings depends on the group (culture), the nature of the activity (walking or sitting), and the nature of the setting, for example, its perceptual characteristics. These can be studied for static pedestrian spaces (e.g., Ciolek 1977; Crowhurst-Lennard and Lennard 1984; DiVette 1977; Gilfoil and Bowen 1980; Grey et al. 1970; Rapoport 1986a; Whyte 1980), for vehicular spaces, and for dynamic pedestrian spaces (Rapoport 1977, 1981b, 1986a, 1987a).

The second problem with the way the general question about the effect of environment on behavior is posed is that it is too broad. Instead of asking whether there are effects, and what these might be, one should rather ask which environmental charac-

---

*Note that this is the way in which large spaces are turned into markets and otherwise excessively wide spaces turned into pedestrian movement spaces—as in many U.S. pedestrian malls, the Ramblas in Barcelona, and so on.

†In Athens one also finds vast spaces, like Constitution Square, which are the equivalents of the Champs Elysées or the Grands Boulevards. But this is another topic that, however, still fits into the conceptual framework based on complexity.

teristics have which effects on which groups; under which sets of conditions and in what ways. In this connection, it is also more useful to consider environments or settings as being inhibiting or facilitating for given activities or behavior, or as being catalysts (e.g., Wells 1965); these can be interpreted as releasing previously inhibited behaviors, that is, are still not *generating* behavior (Rapoport 1983c).

In this study, inhibiting settings are not considered in any detail, that is, I do not analyze settings lacking essential and key attributes. The emphasis is on identifying these attributes in supportive environments (Rapoport 1979c, 1983a)—those settings that are supportive of pedestrian movement. Thus given a predisposition to walk in a given culture and given a set of cultural rules, settings to be supportive need certain characteristics. What are these? It will be seen in Chapter 7 that a set of specific attributes, to be tested against the evidence, can be derived from the key concept of complexity. That, in turn, is related to the third question: The mechanisms involved are primarily perceptual.

PEDESTRIAN BEHAVIOR

From an EBS perspective, the relation of environments and people is the result of complex interactions among cultural, perceptual, and environmental (physical) variables. This, then, also applies to the specific activity "walking." In terms of the three questions of EBS, in the case of walking, human characteristics include the speed of walking, the nature of perception, information processing, the need for interest and hence appropriate levels of complexity, certain evolutionary characteristics, cultural rules, and so on. The settings are selected according to their appropriateness: cultural, perceptual, and physical. They can be supportive or act as catalysts, making walking more interesting and more pleasant, or encouraging it. In these terms, a study such as this begins with a question that is relatively simple in principle: *Under what conditions would one expect the maximum pedestrian use of streets?*

A number of factors are clearly involved. For example:

1. *Technology.* In cases where wheels, or even riding animals, are unknown, or known but little used, walking was extremely prevalent, because the only other alternative—being carried—was only available to a small minority.
2. *Safety,* whether from traffic or crime, although the major impact is, of course, of *perceived* rather than actual safety (Rapoport 1977; Stoks 1983). Another example is the effect of town size on children's mobility affected by parents' perceptions of safety (Gump and James [n.d.]).
3. *Environmental variables* such as noise, fumes, congestion, quality of paving (if any), and so on.
4. *Climate and weather,* for example, season, rain, snow, heat, and so on, although these are clearly modified by culture; these would also include shade or sunlight where needed, and so on.
5. *Topography,* such as hills and slopes, which may, however, affect people differentially, for example, the elderly or handicapped. We are thus dealing with "*perceived* topography."

6. *Distance* to a given goal or, more correctly cognized, subjectively defined distance.

7. *Presence and availability of services* such as shops, cafes, food, kiosks, toilets, seats, and so on.

8. *Culture* that defines the acceptability of walking, rules for appropriate behaviors, settings, and so on.

9. *Certain physical, perceptual characteristics,* for example, adequate complexity levels, adequate interest, and so forth, supportive of walking.

In general, though, walking like most activities is mainly a function of the last two, particularly because many of the others can be related to those or subsumed under them. In other words, it is a function of two major sets of factors: cultural and physical.

Any activity such as walking is primarily culturally based in that it is the result of unwritten rules and customs, traditions and habits, the prevailing life-style, and definition of activities appropriate to given settings (e.g., Rapoport 1969a, 1977, 1979b, 1987c).* This has to do not only with accepted levels of physical exertion but with attitudes to sociability or reserve. For example, if reserve and anonymity are the accepted norm, then settings encouraging sociability will be seen as inhibiting; if the contrary—as supportive. They will thus be evaluated differently. This is partly a matter of unwritten rules and partly of design. Thus designing the best Italian street or Greek island plaza will have no effect if people regard using streets or plazas as unacceptable. But if there is a desire to use streets for walking or promenading, or plazas for sitting and socializing, then certain physical configurations are much more likely to achieve this than others, whereas those that are completely antithetical may be so inhibiting as to stop such behavior (Rapoport, 1969c).

I hypothesized as early as 1969 (Rapoport, 1969a) that there were two major kinds of space use, two major "styles" of urban space use. This seems to be supported by empirical work (Becker 1973; Thakudersai 1972; cf. Rapoport, 1977, Chapter 1, Chapter 2, pp. 91–96, Chapter 5; etc.). Thus members of certain groups seem predisposed to use streets and plazas for many more activities than others, seeking either greater public involvement or more privatization of such spaces (Rapoport 1987a). For example, in comparing a small town in Britain ("Yoredale") with one in the United States (the "Midwest"), it was found that in the former there is much more reliance on walking than in the latter; walking is more vigorous, and a whole set of behavioral and design implications follow (Barker and Barker 1961, p. 458). "Streets and sidewalks" are used a total of 77,544 hours/year in the Midwest as against 300,000 in Yoredale—a 400% greater use in Britain. In terms of the popularity of public settings, Yoredale streets come first, with 21.33% of behavior in public, whereas in the Midwest they are third, with 7.76% of behavior (Barker and Schoggen 1973, pp. 415, 425). Because these totals include cars (the settings being "traffic ways"), the difference is likely to be even greater. It is, in fact, stated that there are many pedestrians in Yoredale and few in Midwest (Barker and Schoggen 1973, p. 425).† Note at the same time that the Midwest is warmer

---

*Note that, strictly speaking, technology ([1] above), that is, the availability of animal, mechanical, or other means of transport, is part of culture.

†In general there seems to be more walking in Britain than in the United States. It is also interesting to note how much walking was still being done in Britain in the late nineteenth and early twentieth centuries—at least judging by novels.

(p. 29). In general, perceptual variables are ignored in these studies, but some suggestive indications in terms of this case study can be found. Thus, in Britain, there are many more winding paths and roads (Barker and Barker 1961); more settings in Yoredale are "more beautiful"; the environment there is older and more historic; the street pattern is "unsystematic" rather than being a grid; streets are lined with abutting rather than spatially separated buildings; some types of construction are more common (with textural implications), and more colors are used (Barker and Schoggen 1973, pp. 415–425). All of these are, of course, physical variables that, as will be seen in Chapter 7, fit the hypothesis to be developed extremely well. At this point the important finding is the interplay of cultural and physical variables: Pedestrians have certain cultural predispositions, the settings provide cues about expected use and possess perceptual characteristics supportive of this use.

Note that compared to other cultures, streets in Britain are not much used. Thus

> when an Englishman first visits Spain, he is surprised to find the public spaces of the towns and villages thronged with people in the early evening, all wandering up and down in an apparently random way. The immediate reaction is not that this is their cultural equivalent of the more familiar cocktail parties, but rather that it is some sort of strange local custom. (Morris 1969, p. 258)

This can be compared to Greece (Thakudersai 1972), Italy (Allen 1969b), Australia versus Montreal (Rapoport 1977), Middle East and Asian cities, such as in India (Rapoport 1987a). It has even been pointed out that compared to Brazil, streets in France are not much used (Lévi-Strauss 1955). In Cali, Colombia, most social interaction traditionally takes place in the streets (Rubbo 1978).

Of course the point is not that these other groups are more gregarious than Englishmen (although they may be) or that Englishmen are not gregarious. It is rather that the unwritten rules of the culture are *different* and hence that gregariousness occurs in different settings within the system of settings (Rapoport 1969a, 1977, 1980c, 1982d, 1986a, 1989a). In the case of Britain, not only at the "cocktail parties" as mentioned, itself a highly culture-specific notion, but in pubs and elsewhere. In this connection, Gordon Cullen's "dream of the gregarious Englishman" is significant regarding the use of streets: It is not done in streets: "What will people think?" (Carter 1969; Curl 1969).

I would argue that in England there is a great deal of eating outdoors, partly at stalls (possibly unwritten rules allow eating *standing up*). Moreover, there is an amazing number of outdoor settings for eating and drinking, pubs, restaurants, and so on, *but they tend not to be visible;* they are back rather than front regions and, again, may respond to specific unwritten rules (e.g., Goffman 1963, 1971). Currently changes in life-style (and hence culture) have led to more visible outdoor behavior (for example, outdoor dining in streets). This is still far more limited than, for example, even Continental Europe. Although this all applies to static behavior, the argument is comparable regarding walking.

Although I am not discussing cultural variables in detail, they are primary for any activity including walking, but they have been ignored. Yet cultural variables generate behavior and help to explain the use or nonuse of settings, including streets and other urban spaces.

Given a culturally based predisposition toward walking and the associated unwritten rules, perceptual variables are very important. The particular perceptual qualities of spaces, broadly having to do with complexity, then characterize settings as supportive for walking and become facilitating rather than inhibiting. Within a given cultural context, they should then encourage walking.

WALKING AND SETTINGS

The preceding section may be summarized in terms of two major propositions:

1. The pedestrian use of streets is primarily culturally based.
2. Given a particular cultural context, certain perceptual characteristics are needed to provide that environmental quality appropriate for pedestrian streets.

This can be shown thus; see Figure 6.1.

A major problem with establishing the nature of those characteristics supportive of walking is the need to disentangle the cultural and the perceptual variables. This is the rationale for this case study. By restricting the evidence largely to those environments where walking was the only, or the principal, mode of travel, that is, preindustrial settlements, one has places where walking was not only culturally acceptable but expected—the norm. Walking then became a major determinant of design. Within these settlements the emphasis is on their vernacular parts, where pedestrian movement was more prevalent than on the more monumental portions, where horses and other animals, carriages, or sedan chairs were more likely to be used. Moreover, in these monumental parts other purposes and objectives become more significant: status, power, impressiveness, parades, explicit symbolism, and the like; this is another reason for concentrating on vernacular areas because they should show greater congruence with the pedestrian activities occurring in them. Finally, as discussed in Chapter 1, such settings comprise by far the greater part of what has been built; they provide the bulk of the evidence. There are thus quantitative and qualitative reasons for my emphasis on vernacular, preindustrial settings.

By adopting this approach one is, in effect, "eliminating" the cultural component, and one can concentrate on discovering those attributes of settings that, almost by definition, are supportive of walking and other pedestrian activities. There is one

Figure 6.1. The relationship among culture, walking, and settings.

complicating factor that, although noted, is neglected. This is that although streets are defined morphologically in Chapter 8 as the spaces to be analyzed, there have been preindustrial societies that, although having settlements, large villages, and even urban settlements, did not have "streets." They had compounds, plazas, and other spaces linked by paths (as in parts of sub-Saharan Africa, Asia, or Latin America) or by causeways (as in Maya cities). This anomaly will eventually need to be considered. However, the widespread distribution of streets, their persistence, and their importance from a contemporary design point of view makes concentrating on them a valid exercise.

In effect, by dealing only with settings designed for walking, one is likely to discover expressed environmental preferences. This is particularly the case if it is accepted that vernacular design, because it occurs over long periods of time and through group processes, becomes highly congruent with activities and preferences: It has been selected for those. Such environments created in a "selectionist" way then provide the precedents necessary, now that design has become more "instructionist" (Rapoport 1986c). Any such conclusions derived from such settings become even more credible when the evidence is broadened and becomes varied geographically and cross-culturally and over long time spans. By using such a *broad* sample (not necessarily a *large* sample, although both are highly desirable), any regularities discovered have much greater potential validity and can be accepted with much more confidence as providing precedents for supportive environments.

Settings can be supportive either culturally (Rapoport 1978b, 1979c, 1980a, 1983a) or perceptually. In either case, one can ask the following questions about such settings:

• What is being supported?
• How is it being supported?
• By what is it being supported?

Then, in the present instance, the hypothesized answers are, briefly:

• Walking is being supported.
• This is done, other things being equal, by maintaining high levels of perceptual interest.
• This is achieved through high levels of perceptual complexity.*

There is another aspect to this. The perceptual characteristics producing complexity are hypothesized as operating by increasing the *pleasure* of walking through environments, stimulating exploratory activity. One is thus not dealing only with manifest or instrumental functions of safety, convenience, and the like but with latent ones, such as pleasure, delight, interest, exploration, ludic behavior, and so on. (On this use of "manifest" and "latent," see Rapoport 1977, 1982a, 1989a). It could be argued that for nonrecreational, nontourist behavior, the ludic element is being overemphasized. It is, however, taken to be an important component of environments that encourage walking. I have already suggested that environments are not only supportive but can also be seen as catalysts (Wells 1965). In that sense they release behavior previously inhibited. Note

---

*Note that we are dealing with *perceived* complexity, which is a subjective quality derived from designed complexity; an important linking variable is speed of movement (Rapoport 1977; 1987a; see Chapter 7).

that this can sometimes be achieved by mere physical separation, particularly where, as in many traditional cities (for example, in Europe), the perceptual characteristics of pedestrian areas are often very much like those being investigated in this study (unlike those of pedestrianized malls in the United States). Those latter, however, can also be expected to tend to be closer to the characteristics being discussed than other streets in U.S. cities.

Note another distinction. I am concerned with *perceptual* variables as opposed not only to *cultural* but also as opposed to *cognitive* (having to do with orientation, imageability, and the like) (e.g., Lynch, 1960; Rapoport 1977); and also opposed to *associational*. This means that very little explicit consideration is given to the *meaning* of such settings. In practice, of course, these two aspects cannot easily be disentangled. It is likely that settings with perceptual characteristics supportive of pedestrian movement are also settings that signify or communicate their appropriateness for such behavior and activities (e.g., Rapoport 1979d, 1982a; see also Chapter 8 for an elaboration of this point).

Yet meaning is important because, as already indicated, one is dealing not only with manifest and instrumental functions but also with latent aspects, among which meaning plays the most important role. In choosing such settings, certain perceived characteristics are matched against certain expectations, norms, images, and so on, which makes this process conceptually similar to other forms of habitat selection based on perceived environmental characteristics (Rapoport 1977, especially Chapter 2; 1980b). It can, however, also be argued that meaning is more important in the case of pedestrian static settings (such as plazas) than of settings for pedestrian dynamic behavior, such as streets (Rapoport 1986a).

Given that preference and the evaluation of environmental quality are culturally extremely variable (Rapoport 1977, 1979c, 1980a,b, etc.) and assuming further that they can be seen as attributes that are liked or disliked (the pulls and pushes respectively of habitat selection), one can ask if there are any regularities that appear—in different periods, locations, and cultures. This is, of course, the reason for using a broad sample. Any such regularities will then be highly significant. The question to be answered is: What are the invariant human characteristics associated with walking and which attributes of settings are congruent with these?

There is another type of constancy that helps make this type of analysis particularly useful. It seems likely, as already discussed in Chapter 3, that there is more environmental constancy at the microscale than at the macroscale (Rapoport 1977). In other words, megalopolises, metropolitan areas, and cities are more variable than neighborhoods; these in turn may be more variable than pedestrian streets. That also applies to human behavior, which has changed a great deal more at the macroscale than at smaller scales where it exhibits more constancy. Because the analysis is of walking and at the very small scale—the street or immediate pattern of streets—the various examples should, in fact, be comparable.

LITERATURE REVIEW

Because I have argued in Part I of this book for the need to use the best available knowledge and to strive to make work cumulative, I begin with a literature review.

In the literature on pedestrians, some behavioral characteristics of both dynamic and static activities are described (e.g., Berkowitz 1969; Ciolek 1977, 1978a,b; Garbrecht 1971a,b, 1973, 1977, 1978; Hill 1974, 1978, 1984; among others).

In examining a selection of the literature on cities generally one finds:

1. Many tend to be at the macroscale and at high levels of abstraction (e.g., the geographical literature, history of urban form).
2. If they are at the microscale, they tend to be based only on personal, intuitive, and aesthetic criteria (e.g., the planning and design literature [Ling 1967; Seymour 1969; Sharp 1968; among others]).
3. Such analyses rarely include vernacular spaces. Even when these are used for illustrative purposes, the discussion is never developed nor related to theory (e.g., Ritter 1964; especially pp. 38–55).
4. Even when there is reference to history, it is not considered in any detail, nor *used* (e.g., Anderson 1978). Neither is the historical material used to examine EBS hypotheses (e.g., Breines and Dean 1974).
5. Such studies are never cross-cultural, nor do they generally include non-Western examples: The role of culture is also neglected. Most examples of pedestrian malls are European and, occasionally, from the United States (Antoniou 1971a; Brambilla 1973; Brambilla and Longo 1977a,b,c; Brambilla, Longo, and Dzurinko 1977; OECD, 1974; Rubenstein 1979; Wiedenhoeft 1975).
6. Those studies based on social science approaches (sociology, anthropology, history) tend to ignore the physical environment.
7. Studies tend to be more concerned with traffic than with pedestrian movement; if the latter is considered, it is in terms of safety. Hence the need for the physical separation of pedestrians and traffic is most frequently discussed and comprises the bulk of the literature. It seems to be implicitly assumed that physical separation of pedestrians and motorists, or the control of crossings is sufficient by itself (Churchman 1977; Doxiadis 1979, pp. 479–489; Sivengard 1968). The major hypothesis on which this study is based is that the problem is more closely related to the need for totally different *kinds* of spaces (Rapoport 1981b, 1987a).
8. Very rarely do any studies deal with the necessary or desirable perceptual characteristics of pedestrian spaces from an EBS perspective—or indeed with perceptual characteristics at all (e.g., Anderson 1978; Antoniou 1968, 1971a; Appleyard 1981; Benepe 1965; Brambilla 1973; Brambilla and Longo 1977a,b,c; Brambilla, Longo, and Dzurinko 1977; Breines and Dean 1974; Buchanan 1963; Churchman and Tzamir 1977; Doxiadis 1975; Doxiadis *et al.* 1974; Fruin 1971; Hoel 1968; Morris and Zisman 1962; Ogden 1970; Pushkarev and Zapan 1975; Regional Plan Association, 1969; Robertson 1973; Stone 1971; Thompson and Hart 1968; *Traffic in Towns* 1963).

In analyzing perception and information in the city, including pedestrian behavior, Carr (1973) neglects the kinds of variables considered in this study. Some studies of the aesthetics and perceptual character of freeways can be suggestive (e.g., Appleyard *et al.* 1964; Carr and Schissler 1969; Tunnard and Pushkarev 1963). For example, because particular forms of highways slow down or speed up traffic, this can be interpreted in terms of information processing and appropriate levels of complexity. This fits my hypothesis discussed later; in fact, the relation between speed of movement and complexity will prove most important to this argument. In some ways the present study is a pedestrian equivalent of the previously mentioned studies, even though the approach, data, and method are totally different.

McCoy (1975) describes, in some detail, the planning of a new part of Canberra, Australia. Pedestrian spaces or perception are ignored; in contrast to the detailed analysis of traffic spaces, no attention is given to the perceptual needs of pedestrians beyond vague, general statements that "liberal provisions are made for pedestrian traffic" (p. 148). Allison *et al.* (1970) generally do not consider human characteristics and, although dealing with improving the environmental quality of pedestrian settings, neglect perceptual variables.

Aravatinos (1963), in dealing with shopping in cities, also does not specifically deal with such variables; all one finds are passing references to liveliness, intimacy, and so on and mention of the "lessons of the past" (p. 68). Moreover, included are schemes by Neutra and Hilbersheimer that conflict with the foregoing—and are at odds with my whole argument. Marchand (1974), in discussing the planning of pedestrian spaces, takes a *cognitive* approach, dealing with orientation, subjective distance, and the like rather than with perception—although he calls it that! (cf. Rapoport 1977, Chapter 1). Fruin (1971) discusses pedestrians and considers, among other factors, those human characteristics related to pedestrian planning and design and even mentions history. However, there is very little on perceptual needs or on historical evidence, and the whole approach is quite different. In fact, among the many recent studies of pedestrian spaces reviewed, there is a complete, and striking, lack of any analysis of the nature of such spaces, their character, and perceptual qualities. If history is mentioned at all, it is merely in passing. Moreover, such studies tend either to be of the "I like it/I don't like it" variety or typically exhibit two other characteristics. First, a lack of a theoretical background and hence of an understanding of why and how any settings work (or whether, indeed, they do). Second, as a result of a lack of explicit criteria that could be used in analysis, they tend to offer analogies or propose copying examples rather than learning by analyzing examples by means of concepts, models, or theories; *there is a very different view of the nature of valid precedents.*

Although some of these studies consider historical urban spaces, they are generally confined to Western Europe and are generally in the high-style design tradition. Illustrations tend to emphasize their *aesthetic* qualities as being self-evident (e.g., Seymour 1969; Sharp 1968; Specter 1974) or to concentrate on street furniture (e.g., Design Council 1979). Also, open spaces for static behavior rather than streets tend to be analyzed (Churchman and Tzamir 1977; Crowhurst-Lennard and Lennard 1984; Hecksher 1977; Whyte 1980). Neither those nor Appleyard's (1976, pp. 181ff.) evaluation of pedestrian schemes in European cities* deals with perceptual qualities, nor are streets in other parts of the world discussed. The emphasis is largely on traffic volumes (cf. Appleyard and Lintell 1972). This is also the case with a special issue on "People-Oriented Urban Streets" (*Ekistics* 1978).

Hitzer (1971) deals with streets and roads both historically and cross-culturally.

---

*Fifty-nine of 300 cities in OECD countries had such schemes at the time of that study.

The emphasis, however, is on roads (highways) rather than streets. Regarding these latter, there is nothing on the variables I am discussing; also, the evidence and the method are quite different.

In reviewing the various documents of the Pedestrian Research Network that, for a while, summarized ongoing EBS research on pedestrians and current research interests of members; in examining the workshops on pedestrians at EDRA conferences through mid-1980 and a newsletter (PEDNET, edited by Michael R. Hill), one finds nothing on perception, perceptual qualities, or complexity: These are not even found in the key word list in a computerized bibliography on pedestrian movement (Pednet/4, July, 1978). Similarly, the program of the 1980 midyear meeting of the Pedestrian Committee of the U.S. Transportation Research Board ignored these topics. Topics covered included: what are pedestrians; where are they found; what problems do they face; changing pedestrian behavior by education and training. Environmental issues covered include urban planning for pedestrians; pedestrian malls; engineering approaches to accommodating pedestrians; regulation and enforcement; walking as a viable transportation mode; new research techniques; and integrated approaches in pedestrian programs.

A 26-item bibliography on pedestrians in the city (Martin 1969) contains nothing on the topics I am considering. A review of a study by Monheim of 300 pedestrian precincts in West Germany (*Architecture Australia* 1976, p. 12) points out that although pedestrian volumes go up with closure to traffic (for example, in Kauffinger Strasse, Munich from 70,000 to 116,000 pedestrians a day) and business improves, there is no mention of the attributes of such spaces other than the absence of fumes, noise, and vibration (and hence the resulting "peace" and "attractiveness"). Similarly, the *Urban Design Newsletter* (1979, p. 5) lists five suggestions for making people more inclined to walk:

1. Improve sidewalk space planning by allowing extra width where there is street furniture, at bus stops and at corners where people are crowded.
2. Use street-level glass in buildings to allow around-the-corner visibility.
3. Consider better solutions to snow and ice removal than piling these on sidewalks.
4. Increase pedestrian-oriented signage.
5. Provide more toilets, seating and entertainment in public spaces such as shopping malls and plazas.

These do form part of the list of attributes discussed that play a role in pedestrian use of settings. However, they do not seem to be derived from any theoretical approach, nor do they consider the perceptual attributes of pedestrian spaces or the related human characteristics.

Because this study tries to relate environments to human characteristics and does so by adopting an EBS approach, it becomes necessary to review some studies of pedestrian behavior. Much of this deals with pedestrian flows, trying to establish characteristics of how pedestrians tend to move through space; for example, taking the apparently shortest, easiest most convenient route; moving from feature to feature; moving along borders and other flow tendencies (e.g., Ciolek 1977, 1978a; Garbrecht 1979a,b; 1973, 1977, 1978; Hill 1974, 1984; Hoel 1968; Khisty 1985; Porter 1964;

Weiss and Boutourline 1962; among others). Also discussed are the need for a comfortable underfoot surface, avoiding major changes in the elevation of the walking surface, and so on. There are brief and peripheral references to the need for perceptual interest, stimulation, and entertainment: for interesting paths, areas with shop windows, stalls, and other people. These are selected, whereas empty, barren spaces are avoided. Although this is in agreement with the present study, it is neither developed nor related to any theory. These remain isolated intuitions or, at best, isolated empirical observations. Also, many studies are of the *static* behavior of small groups (e.g., Ciolek 1977; Stilitz 1969) that also neglect perceptual needs.

In a major cross-national study of pedestrian behavior (Berkowitz 1969), 155 sampling points in cities in Turkey, Iran, Afghanistan, England, West Germany, Sweden, Italy, and the United States were used. Considered were grouping, interaction, and physical contact patterns of 21,316 pedestrians, classified by gender and five other categories inferred from their appearance. Perceptual needs and the characteristics of settings were ignored. Similarly Wolff (1973) looked at pedestrian behavior in terms of how it is coordinated to avoid collisions or clashes; subcultural variables, gender, and the like are considered but settings ignored.

Garbrecht (1971a,b; 1973, 1977, 1978) deals with some pedestrian flow tendencies and emphasizes the importance of pedestrian traffic even in modern cities. In general, however, characteristics of settings are ignored. Their potential importance can, however, be inferred from the fact that in some cities the percentage of pedestrians has increased following environmental changes. The nature of such changes are not explicitly described nor considered in terms of perceptual needs. The role of such needs can sometimes be inferred from studies dealing with different issues. Thus Baum, Davis, and Aiello (1978) comment that streets may be more important than green open spaces for people and that more people are to be found in streets with small markets or pharmacies. This can be interpreted, at least partly, in terms of greater interest of settings and the likely presence of other people (e.g., Ciolek 1978a; cf. Rapoport 1986a); however, safety (in terms of crime) probably plays a role.* It is also found that streets with shops lead to more walking, that is, less static behavior, which will be seen to be significant later in terms of complexity and "interest" as opposed to "liking."

A study of walking in downtown Seattle (Grey *et al.* 1970) shows that in addition to neatness, tidiness, and lack of visual chaos, pedestrians prefer large amounts of building area coverage, low amounts of sky area, vertical building configuration, considerable amounts of foliage, and few, if any, signs. Low buildings that leave large sky areas visible and have a horizontal feeling are liked less. This overall apparent preference for crowded, densely developed urban settings is not derived from any theory, nor is it explained. These findings are, however, predictable from the notion of complexity (Rapoport 1967, 1971a, 1977, 1981b, 1987a; Rapoport and Kantor 1967; Rapoport and Hawkes 1970; see also Chapter 7 and the hypotheses developed in Chapter 8). The Seattle study also found that the age or style of buildings is less important, that is, that relationships are more important than the nature of the elements (e.g., Rapoport 1969b,

---

*Note that currently the fear of crimes such as mugging and rape may override all the other considerations and inhibit walking, particularly because some of the variables we will find to be supportive of walking have negative implications for safety (e.g., Stoks 1983).

1977). Elements may, however, be important in terms of their effect on the perceptual character of the setting. For example, they may be important in terms of the texture or richness of the buildings, that is, the enclosing elements. This can be inferred from preferences for three-dimensional facades (Krampen 1979; Westerman 1976).

There are some studies congruent with the present one. Khisty (n.d. but received in mid-1979) emphasizes the need to study pedestrian behavior by tracking people and observing those settings that pedestrians tend to frequent. This is equivalent to studying habitat selection and is a useful approach partly congruent with mine, although clearly only applicable to contemporary work. The point is also made that all settings have both social and physical attributes that include, first, the patterns of activity; second, the spatial form of the setting, such as its dimensions, shape, bounding surfaces, connections to other spaces, and so on (these are closest to the variables with which I am concerned); third, the ambience of the setting, qualities relating to microclimate, light, sounds, smells, and so on, which we will see are also related to my argument; finally, communication, for example, signs and signals that provide the required information about behaviors, meanings, facilities, and so on. It is further suggested that environments supportive of pedestrians can be understood at four levels: survival level, efficiency level, comfort level, and pleasure and enjoyment level.*

Much of this is congruent with my argument. The spatial form variables are not, however, related to interest, pleasure, and enjoyment, nor are these notions developed further; for example, there is no detailed consideration of the attributes of such spaces. Moreover, these insights are never tested against a broad body of evidence.

There are even studies that explicitly deny the significance of perceptual needs (e.g., Gray 1965). This denies the need for "narrowness, coziness and crookedness" (p. 48) and, in reviewing pedestrian streets in Europe, Gray argues that none of them are "excessively wide." However, it is never quite clear what is "narrow," "cozy," or "wide," nor are these assertions tested against a broader body of data.

It is noteworthy that, as already pointed out, static spaces have been rather more studied than dynamic spaces in terms of their attributes as they relate to human preference and behavior. Although this will be elaborated in Chapter 7 in terms of theory, it can be suggested that static spaces are more related to *liking*, whereas dynamic spaces are related to *interest* (Rapoport 1986a). In studies of static spaces, design characteristics are sometimes included (e.g., Ciolek 1977, 1978a,b, 1979; Crowhurst-Lennard and Lennard 1984; Im 1984; Joardar and Neill 1978; Mills 1978; Rhodes 1973; Seymour 1969; Stilitz 1969; Whyte 1980; etc.). Generally, however, the physical attributes are not analyzed in detail, nor are they the ones I consider. The spaces are, of course, not streets, nor is the approach historical, cross-cultural, or related to vernacular design. The variables mainly studied tend to be weather, for example, sun and wind (e.g., Cohen, Moss, and Zube 1979; Mills 1978). De Jonge (1967–1968) includes protection, views, ability to territorialize, and the proximity or distance of other groups as characteristics of settings for sitting in a park (cf. Rapoport 1977, for a review). DiVette (1977) examines a series of static spaces in Amsterdam (called Pedestrian Rest Areas) and identifies some of their physical characteristics: location, availability of seating and food, greenery, relation to movement, ability to see activity, children's play

---

*These are clearly inspired by the well-known hierarchy of needs proposed by Abraham Maslow.

areas. Gilfoil and Bowen (1980) suggest that time spent in plazas is largely *independent* of specific environmental design features. Generally, however, many of the attributes identified by DiVette (1977) are found to be important. For example, Whyte (1980) emphasizes provisions for sitting, availability of food, having activities visible, presence of shops, copresence of other people, sun, trees, and water, shelter from wind and, I would add, rain. Share (1978), in comparing two static spaces in San Francisco, focuses on natural elements (trees and grass), relation to streets and surrounding activity, exposure to sunshine, and the placement and design of seating. Purcell and Thorne (1976) compare static pedestrian spaces used during lunch breaks by office and shop employees. They describe their ideal space as spacious, relaxing, peaceful, colorful, with "soft-paving" (i.e., grass), many trees, little pedestrian traffic, and removed from vehicular traffic, with neither many nor few activities to watch or do, and neither very public nor very private. These characteristics (in Sydney, Australia) are rather different from other studies of static spaces, possibly because of cultural differences and the emphasis on lunchtime use. They are, of course, very different indeed from the lively, bustling spaces already emerging as needed for pedestrian movement and investigated in the present study. This difference is, of course, also related to the difference between settings for dynamic and static pedestrian behavior with the latter not only more related to liking but, as a result, likely to be more variable culturally (Rapoport 1986a).

For dynamic spaces, that is, pedestrian streets, there are scattered suggestions without data and without explicit analysis of those data used. For example, Parr (1967, 1969a,b) was among the first to suggest that the conflict between pedestrians and motorists was *not* to be understood in terms of people versus cars but *rather in terms of slow versus fast movement, that is, that the conflict is in terms of perception*. His analysis, however, was mainly introspective and anecdotal. At a similar, intuitive level there is a form of consensus among certain design writers who imply, or suggest, that certain kinds of spaces seem to be suited to walking and hence preferred by pedestrians (e.g., Cullen 1961; De Wolfe 1966, 1971; Nairn 1955, 1956, 1965; Rudofsky 1964; Sitté 1965; Sharp 1968). These, however, are never studied from an EBS perspective, nor are explicit hypotheses made. Neither are these suggestions by sensitive observers tested against a historical and cross-cultural body of data. This body of writing is, however, highly suggestive. Points similar to mine are sometimes made *implicitly*. For example, it is suggested that it is not enough merely to close a street to motor traffic in order to pedestrianize it—it also needs to be changed (Lerup, Brambilla, and Longo 1974). The changes described are not derived from any theory but can be interpreted in terms of what is called complexity (cf. Sims and Jammal 1979).

One recent review of pedestrian behavior does refer to hypotheses on complexity (Hill 1984). There is also one attempt to explain the visual attractiveness of urban areas in terms of the structure of the human brain (Laverick n.d., received 1980). Much of the discussion is concerned with what I call associational aspects and liking, but interest and arousal are considered. The conclusions agree with my emphasis on complexity and, above all, on enclosure. This is partly in terms of its contrast with exposure, that is, in terms of what I call *noticeable differences* and thus in terms of complexity. Laverick gives an historical overview of the importance of curved streets, that is, those providing enclosure, from Aristotle and Vitruvius, through Alberti to Sitté, Cullen, and De Wolfe. Some of his examples are quite similar to the ones I have used (e.g., Rapoport 1977,

Figure 4.8, p. 216; Figure 4.10, p. 218). Although the approach is different, *there is some overlap at a conceptual level.* One study by a student of mine closely relates to many of the general characteristics of pedestrian settings that I discuss. These include safety, noise and pollution, climate, public facilities, and so on, as well as aesthetics (Llewellyn-Smith 1972). This last category is closest to what I call complexity; included are scale, change of direction, surprising vistas, changes of texture and of spatial elements. These will be seen to be part of my hypothesis also.

Another exception to the lack of consideration of perceptual variables, which is conceptually related to the present study (Lozano 1974), explicitly uses the concept of complexity (as developed by Rapoport and Kantor 1967; Rapoport and Hawkes 1970; and Rapoport 1971a, among others). The visual qualities of good urban spaces are defined in terms of complexity, variety within an order, diversity, surprise, and the like. These are very much like the variables this case study addresses, but the analysis is not developed, it is not applied explicitly to pedestrian streets, and the examples are very different. Neither does the analysis in any way resemble that in the present study because its mutual relevance to historical study is not part of it.

This literature review has shown three major things. First, that there are some suggestive hints about the qualities that pedestrian streets should have; second, that there has been very little formal study of such streets; and, third, that none of the studies adopt the present approach nor use the evidence used in this study. I, therefore, turn to the development of a specific general hypothesis derived from the EBS notion of environmental complexity. From this, specific hypotheses will be derived that will then be tested against the historical evidence.

Chapter 7

The Perceptual Characteristics of Pedestrian Streets
The General and Specific Hypotheses

INTRODUCTION

The argument underlying this study can be simply summarized. For any given behavior patterns or activities, there are settings that are supportive. In general, when they have the opportunity, humans, like other organisms, select habitats having appropriate characteristics even if the matching processes occur beyond awareness. In this process, preference tends to be based to a large extent on associational aspects (Rapoport 1982a). Cognitive aspects are also important; in the case of walking, good imageability and clear orientation at the large and middle scales, because an organism can hardly perceive if it is not oriented (Bruner 1951; Lynch 1960, Appendix A; Sandström 1972).

Settings also possess appropriate perceptual characteristics. Walking, with its clear relationship to evolution-based characteristics, and the subsequent invariance, provides good reasons for taking an approach based on perceptual characteristics, partly because perception also is relatively invariant as compared to cognition and especially to meaning and evaluation (Rapoport 1977, 1986a). Thus a study based on perceptual characteristics of settings for an activity like walking will be easier to approach cross-culturally and over long periods of time. This makes it useful for a first attempt.

The task is, therefore, to determine which specific perceptual attributes characterize settings suitable for walking. Recall that the invariance due to scale suggests settings at fairly small scales, reinforcing the same consequence of considering perceptual variables. It should be possible to predict these attributes and, if the hypothesis is correct, they should then be found in a sample of the kinds of pedestrian spaces already described.

Because my purpose is to analyze a wide range of pedestrian streets, primarily in preindustrial settlements, the characteristics that they have in common could be derived inductively. In fact, my interest in the topic began in this way. *It appears that many environments in different areas, eras, and cultures that tend to be liked and preferred by pedestrians have one thing in common: They all seem to be perceptually interesting, complex, and rich.* It is, however, useful to try and derive these characteristics from the concept of complexity based on EBS research. It provides the framework for the analysis and, in line with the argument in this book, it needs to be developed explicitly. This I will do as briefly as possible and then derive the expected characteristics as a set of hypotheses to be tested against the broad body of evidence. From a design perspective, the question can be made normative: Which characteristics should spaces used for walking exhibit? *It is this set of characteristics that constitutes precedent, NOT THE STREETS THEMSELVES.*

COMPLEXITY

The starting point, from which the general and specific hypotheses are to be derived, is that pedestrian settings should exhibit certain appropriate levels of *complexity.* This concept has undergone a variety of changes and developments since it was introduced into EBS (Rapoport and Kantor 1967). I have discussed these in considerable detail with many references dealing with a large range of issues and with many examples (Rapoport 1977, especially pp. 207–247). Here I will merely summarize some of the more important points to the extent necessary for the case study.

The underlying notion, based on an extensive research literature from EBS and other fields, is that human beings, like most organisms, process information. In so doing they seek certain levels of information input, certain perceptual rates, which depend on the individual and his or her culture, adaptation level and the like, the nature of the activity, and the context or situation. These preferred levels constitute complexity. They can be contrasted with the extremes of sensory deprivation, and hence boredom and chaos and, therefore, sensory overload. That level of complexity is equivalent to "the pacer," that level between chaos and monotony that challenges without exceeding capacity (Rapoport and Kantor 1967), a level of arousal leading to diversive exploration (e.g., Wohlwill 1976) and providing moderate motivation and an absence of unnecessary frustration (Nahemow and Lawton 1973; Tolman 1948). These levels of complexity seem to be necessary for human well-being; people, including infants, need changing and complex environments. Animals also need such environments (e.g., Willems and Rausch 1969), and even organisms as primitive as planarians seem to prefer complexity (Best 1963). It is, therefore, quite likely that this need has an evolutionary basis (e.g., Geist 1978). Part of the reason for this need may be the well-known fact that perception itself is dynamic, that there is spontaneous activity throughout the central nervous system so that constant change of stimulation is essential for perception to occur. With exposure to an unchanging environment, "stimulus satiation" and eventually stimulus aversion occur, so that the environment comes to be disliked, avoided, and other settings sought. This is habitat selection and, in fact, complexity is a particular aspect of environmental quality leading to environmental preference. People seek *some* novelty and uncertainty, that is, information, both in the physical and social environments.

At the outset three major points can thus be made:

1. Recent ethological and psychological research shows that animals and humans (including infants) prefer complex patterns in their visual field.
2. There is a preferred range of perceptual input, related to a maximum rate of usable information, with both too simple and too chaotic visual fields disliked.
3. There are two ways of achieving complexity: through ambiguity (in the sense of multiplicity of meanings rather than uncertainty) and hence using allusive and open-ended design; or through the use of varied and rich environments and those that are not visible from a single view but unfold and reveal themselves and thus have an element of surprisingness, unexpectedness, mystery, and so forth. Although some researchers regard these as distinct characteristics, I regard them as aspects of complexity.

There proved to be problems with the use of the concept of *ambiguity*. Although it is one way of achieving "complexity in the mind," it is difficult to handle because meanings are in the associational realm and hence extremely variable.

Ambiguity itself has two interpretations. One is uncertainty, which is a perceptual quality and can still be handled. Thus a space or form that cannot be fully seen or grasped all at once, and is thus uncertain, is ambiguous and more complex than one that can (e.g., Venturi 1966). This, however, can then be understood as part of complexity: uncertainty, serial vision, unfolding, or mystery. The other sense of ambiguity—multiplicity of meanings—is a literary and hence associational quality. The same environmental elements can then have very different meanings; moreover, although associations and meanings were shared and predictable in the past, particularly in traditional societies, they are highly idiosyncratic today, unpredictable, and consequently difficult for designers to manipulate. In effect, past environments communicated; they do not do so today (Rapoport 1975a, 1976, 1979b, 1982a). As a result, designers, at least currently, can far more easily manipulate the perceptual elements (complexity) than the associational ones (ambiguity). In any given situation, it may be possible to discover the most widely shared associations (e.g., Duncan, Lindsey, and Buchan 1985; Rapoport 1982a); there may even be archetypal associations, that is, common responses to certain stimuli (e.g., Jung 1964; McCully 1971). Also, by the consistent use of associations such as form/activities, form/location, hierarchy/location or form and so on, they might be learned so that, in the long run, they may become stable and usable. However, both for analysis and design, ambiguity seems less useful than those perceptual elements in the urban environment that designers can manipulate. In summary, complexity is perceptual, multisensory, and is related to the number and organization of elements; ambiguity is associational, may be nonsensory, and is related to the meanings attached to elements and their relationships.

The distinction between associational and perceptual qualities becomes immediately useful in explaining certain apparent anomalies in the preference for complexity that arise when one compares natural and man-made environments. It explains why in the United States, nature scenes are greatly preferred to urban scenes regardless of complexity. Complexity predicts preference *within* each category but not between them (e.g., Kaplan and Wendt 1972). The preference for natural over man-made is then due to a set of associational values related to nature, so that perceptual characteristics such as complexity are not used to discriminate between the two sets; associational criteria are used. The nature of the elements and thus their meaning or associational qualities play a role (e.g., Rapoport 1982a; Walker 1970, p. 638; cf. Wohlwill 1968, 1976, 1983). The preference for natural settings is partly associational because complexity is just one component of environmental quality or preference, which are multidimensional concepts. Although I will be concentrating on perceptual aspects, the role of associational variables needs to be borne in mind. It is quite likely that high levels of enclosure, being clearly in a nonmotor space, and seeing only other pedestrians are associational aspects.

For example, arcades are effective as pedestrian settings because of both perceptual and associational variables. Not only are the levels of complexity inside and outside the arcade very different (and appropriate for pedestrians and traffic respectively), the highly enclosed, sheltered, almost nurturant arcaded space; the shops, restaurants, and the like; the enclosure and separation from traffic while in view of it; the freedom of movement and slow tempos contrasting with the adjoining speed and linear flow of traffic all provide a clear contrast in meaning. One is clearly in a human rather than nonhuman space (Rapoport 1970b), a pedestrian rather than a machine space (Horvath 1974; Rapoport 1977, p. 221; 1987a).

Similarly, the raised sidewalks of Paris (Rapoport 1977, Figure 4.32, p. 246; 1987a) provide different and appropriate perceptual characteristics for motorists and pedestrians. By putting the latter *above* rather than underground, and in what is clearly an important setting (rather than an overpass that is clearly a second-class setting), meanings are communicated about relative importance, status, and the like. It sets a context and defines a situation (Rapoport 1979d, 1982a) and thus also works in the associational realm.

One can also suggest that many of the characteristics that provide perceptual complexity also tend to communicate appropriate meaning, most simply: *This is a desirable pedestrian space.* This is also implied by the suggestion (Prak 1977, pp. 47–57) that certain spatial arrangements help in the formation of "conceptual space," although this is seen more usefully as the definition of a setting with the appropriate meaning, that is, a setting for culturally acceptable and appropriate behavior (Rapoport 1982a, 1987a).

Defining complexity in terms of a maximum rate of usable information implies an active rather than passive receiver, and allows, among others, for the effects of culture, learning, and adaptation. It also follows that the effects of overload and deprivation become subjectively similar (Rapoport 1977, Figure 4.5, p. 210). Organisms cease to respond to both repeated and chaotic stimuli. The former is response saturation, the latter a defense against overload; one result is tunnel vision that impairs peripheral matching (Mackworth 1968) and makes the environment perceptually poorer. The result is equivalent to monotony. Complexity results from such interactions between people and environments. The latter typically involve variations within an order; learning and experience increase the ability to grasp the order, leading to more emphasis on the relationships among elements than on the elements themselves. People organize and code stimuli into bits, chunks, and larger structures in order to increase or reduce information to the maximum usable rate sought. Thus there is a reduction of information to complexity, not a reduction of complexity, and even apparently simple environments can sometimes be made more complex. At the same time, some environments make achieving complexity much easier; it is far from an entirely subjective matter.

Usable information is related to stimuli that are detectable variations within a system of expectations; departures from such expectations constitute variety. In this sense, complexity is related to variations within an order, which is equivalent to *noticeable differences*. These are important in perception generally: It is changes in stimuli (in all sensory modalities) rather than the stimuli themselves that are significant (e.g., Gibson 1968; Rapoport 1971d). Noticeable differences will be discussed later but clearly in addition to ambiguity, an environment can have low usable information for three reasons: There is little variety among elements; the elements, though varied, can be predicted in advance; there is thus no surprise or novelty and hence no information; the elements are so numerous, varied, and unrelated that there is no order; the perceptual system is overloaded, and no usable information results.

It can be shown that complexity is related differentially to *interest* (time of exploration) and *liking* (pleasingness) (Rapoport 1977). The former increases monotonically with complexity, although it is likely that there is an upper limit when overload is approached or reached. Liking, on the other hand, tends to be an inverted U-curve (Smets 1971; Wohlwill 1971). Thus neither very complex nor very simple environments are liked (e.g., Acking and Küller 1973; Rapoport 1977, Figure 4.7, p. 212).

Because the inverted U-curve seems to apply to preference and the straight line to exploration, it is likely that they describe the two different kinds of pedestrian spaces: static and dynamic, respectively. Liking is then the principal criterion for static spaces (e.g., plazas) whereas interest is the principal criterion for dynamic spaces (e.g., streets). It follows that these spaces may have different desirable characteristics, particularly because liking is more influenced by meaning: It is more a matter of associational qualities than is interest (although not exclusively). Interest, on the other hand, is more a matter of perceptual qualities (although not exclusively either) (Rapoport 1986a).

In that connection the distinction between *arousal* and *affiliation* (Mehrabian and Russell 1973) is also relevant. I would suggest that the former is related to streets, the latter to static spaces (e.g., Thakudersai 1972). It is also noteworthy that perceptual strategies and processes differ in streets and plazas (Vigier 1965).

Another possible difference has to do with a greater emphasis on other people in static spaces. These are, as it were, a set for "theater" (Sennett 1977). In order to emphasize people as both scene and audience, *lower* (but not low) complexity levels are needed than for dynamic spaces; for example, it has been noted that the Paris Opera is so complex that people are barely seen (Sennett 1977, p. 207). Static spaces need to become a background for people—whom they also need to attract. Such spaces also need different enclosure levels than do streets; there need to be different kinds of spaces in which people watching is the main activity. Thus unlike the long, narrow, twisted dynamic spaces, static spaces tend to need to be wider. With many people, and the outdoor cafe tables, vendors, and the like frequently associated with them, these spaces are narrowed, and great complexity is produced in terms of *people*, seeing and being seen (Rapoport 1986a).

Of course, seeing and being seen is not a completely static activity—promenading is important as, for example, in eighteenth century parks, the Champs Elysées, and so on. Particularly if the promenade is grand, the spaces also become grand, very different from the equivalent vernacular spaces, as in the Grands Boulevards of Paris; although on these, cafe tables spread onto the sidewalks (Sennett 1977, p. 216). These are then not spaces primarily for movement through the environment but for walking (or riding) as a social activity (of elites).* In that sense they are more like static than dynamic spaces.

The different levels of enclosure and complexity, then, have to do with the difference between interest and liking; with the emphasis on information (and, hence, complexity) from people and people watching rather than from the setting itself—with the greater emphasis on associational aspects, both people and natural elements, in attracting people who in turn attract other people to watch them and yet others to watch the watchers. The result is that dynamic and static pedestrian spaces need different perceptual characteristics because the activity is different (Rapoport 1977, Figure 4.33, p. 247; 1986a).

Moreover, because pedestrian spaces for walking may be related primarily to interest, whereas settings for static activities may be more related to liking, it follows that for walking, perceptual qualities play a larger role than for sitting where associational qualities become more prominent. As already suggested, this should lead to greater preference for natural environments and it is, in fact, found that for static spaces greenery is highly important (e.g., Mills 1978; Purcell and Thorne 1976). Given the relative strength of associational and perceptual elements, one may prefer simple natural settings to complex urban ones (see Rapoport 1977, pp. 317–318).

We have seen in Chapter 6 that individuals and groups vary in their preferred interaction rates in urban open spaces. They may also vary in their preferences for complexity or arousal levels (Berlyne 1960, 1974). These latter differences, however, should be less than the former because they are less related to culture. They may be partly due to the effects of learning and experience (i.e., adaptation levels); there may also be sensation-seeking as opposed to non-sensation-seeking people who have different requirements for complexity (Markman 1970; Mehrabian and Russell 1973, 1974). Some people, who give structural descriptions of environments, get bored with them more rapidly than those who give experiential descriptions (Nahemow 1971); this may also relate to the distinction between a scientific and aesthetic experience of cities (Gittins, 1969) with the latter retaining their interest longer and perceiving higher levels of complexity. In all these cases, however, one would expect similar variability in most groups (unlike in the case of social interaction). Moreover, in all cases, *objectively more complex environments would still be perceived as more complex by all*. Differences among individuals tend to be far smaller than differences among environments; and rich, vivid, and complex environments tend to be preferred to uniform and monotonous or chaotic ones (Lowenthal 1972).

The activity and the context also play a role in that there are also likely to be associational effects due to being in an "exotic" city, a medieval city, or a city with specific meaning to a person. These will have effects on curiosity and exploratory behavior over and above those due to perceptual qualities (cf. Wohlwill 1976). There

---

*There are, of course, vernacular equivalents, such as the *passagiata* in Italy (Allen 1969b). The vernacular spaces associated with them are also somewhat different from "normal" streets, being wider and more like a stage; also, typically plazas are used for the *passagiata* more than streets.

are also situational effects; for example, being a tourist or going walking for pleasure are different than job-related behavior. Other things being equal, however, complexity is hypothesized to be of major importance generally, and particularly in the case of pedestrian streets as opposed to spaces such as plazas, where associational qualities seem to be more important (Rapoport 1986a). However, activity and context also play a role in preferred levels of complexity, and a series of important analytical and design implications follow, for example, in distinguishing between residential and recreational, central and peripheral areas (see Rapoport 1977, pp. 207–247 and references; 1986a, 1987a).

In the present study this is less important because it is suggested that walking is a basic activity co-terminous with humans, essentially unchanged, and hence with a very strong evolutionary base. If that is indeed the case, one would expect the variability both of the activity and the settings appropriate to it to be low, so that complexity levels should be similar and hold cross-culturally. There will still be cultural variations, for example in the importance of walking and readiness to engage in it: Streets to be used must be seen as appropriate settings (e.g., Rapoport 1982a, 1986a, 1987a; Schak 1972). The variability, however, in the complexity of settings intended for walking should be much reduced (cf. also work on complexity related to art, e.g., Crozier and Chapman 1984).

Play is also a relevant activity to consider. Not only does it tend to be at the pacer level, being neither boring with too few requirements for ability or desires nor anxiety provoking with too many (Csikszentmihalyi and Bennett 1971). It also often requires settings that can be explored in which one can get "lost," and the like (e.g., Hart 1979; The Sun 1971). The fact that children frequently do not play in settings provided is partly a function of such lack of complexity (Cooper 1970; D.O.E. 1972; Friedberg 1970; Rapoport 1969c; Whyte 1968). The kinds of settings that are picked for play in urban areas always tend to be complex (e.g., Brolin and Zeisel 1968); those play spaces that are designed with complexity and are open-ended enough to allow for complexity over time do tend to be used (Ellis 1972; Moore 1966). In such cases, complexity is related to multiple uses, choice, and diversity of activities at one time and over time; many physical elements; varied materials, surfaces, shapes, textures, heights, colors, light and shade, smells, sounds, and so on. Such environments offer interesting analogs, not only because they are for pedestrians, rely on interest, and have a major element of exploratory activity but because walking also involves ludic behavior. The analogy is also strengthened because, like walking, play has a strong evolutionary base.

The richness and complexity of settings may be partly due to subliminal perception that receives all input, whereas conscious perception only deals with selected stimuli. The effects of subliminal perception are also probably strongest in the non-visual, autocentric senses (Ehrenzweig 1970). The normal state is one of homeostasis between cognitive schemata and perception, and stimuli only become consciously noticed when they change that state, that is, when they are novel or unique. Complexity thus has beyond awareness components (Dixon 1971; Smith 1972). Also involved, among other things, are the number of elements, their intensity against the background, their novelty, surprisingness, incongruity, mystery, temporal variations (and, hence, effects of speed), and so on; always present are also associational aspects. In effect, variety depends on the number and character of the elements (which is partly subjective) and the number of interpretations possible, that is, ambiguity, which is neglected here.

Changes in perceived complexity with experience and adaptation are an important problem. The result of exposure is satiation and although there is some recovery, there seems to be a continuous long-term progression toward seeking more and more complex stimuli.* The result is that although at first relatively simple stimuli are preferred, more complex ones are preferred later and stimuli that at first appear chaotic become acceptable with time; this is clearly a serious problem for designers.† On the other hand, learning also leads to the discovery of higher levels of complexity, affecting noticeable differences: With experience, ever finer differences, cues, and details are discovered in the environment. Thus what is boring to an outsider may be rich to the native—the desert to an Australian Aborigine, the Arctic to an Eskimo, the prairie to its resident. A forest is more complex to a botanist than to a layman—a city to a designer. However, as already emphasized, *objectively more complex environments more easily provide greater amounts of potential information.*

Complexity is also the result of movement through environments. People do not spend their time in environments closest to "optimal complexity" (Walker 1972). Although they may select such optimal levels for particular purposes, they move through the whole range of environments, from the simplest to the most chaotic—this maximizes transitions and overall complexity (Rapoport 1977, pp. 207–247; cf. Rapoport 1986a, 1989a). In this context, design can help prevent satiation and adaptation. There is also a difference between novelty and complexity: Novelty is a short-term phenomenon, whereas complexity endures through time and is related to activities, uses, and so on. This relates to the notion of "operational complexity" (Appleyard 1970)—that resulting from the varied uses to which an environment can be put. In this connection, also, complexity is not a result solely of differences, changes, and variety of physical characteristics at any one time *but also over time.* Thus in terms of areas traversed and routes taken, sets of events are never quite identical: Routes change, the environment itself changes over time, with seasons, weather, rebuilding, changes in uses and activities, and so forth (and this can be encouraged through open-ended design); the context and the state of the observer also change. Thus the psychologically expected results of habituation and adaptation seem much less severe in real environments than in the laboratory. The urban environment is also large, so that one does not see it all at once; it is also seen in different sequences. Neither are large portions of it seen regularly, so that one has time to forget its full richness and resensitize to it.

At the same time the memory of an environment is always much poorer than reality, especially in nonvisual senses. One remembers relatively little even after repeated exposures so that every time an environment is experienced it will be complex in comparison and will always provide new information. One remembers generalized schemata and the order and and form of things, not the detail (Bartlett, 1967, p. 195). Memory is greatly simplified and affected by filters, expectations, and mental set. There is always some loss of memory and hence always some surprise—and complexity is

---

*This may go down at higher ages when greater redundancy is required.
†Personal communication from E. L. Walker, Psychology Laboratory, University of Michigan.

due to surprises, to departures from an expected order. New information is always present in the experienced environment. Moreover, both subliminal perception and nonvisual senses add to the richness of the experienced scene.

The amount of information in the urban environment always exceeds channel capacity so that different sets of cues are selected at different times, and hence different parts of the potential environment are used at different times. Although familiar scenes tend to be seen as simpler than unfamiliar and novel ones and complexity is reduced somewhat over time (Bartlett 1967; Walker 1972), enough information remains for further notice. Even when we know what to expect even after just one trip, and certainly after many, the perceptual experience is still there and has an impact—even if reduced (Rapoport 1977, Figure 4.8, p. 216, Figure 4.9, p. 217). That impact, however, still depends on the nature of the environment, so that objectively complex environments retain a larger impact for longer.

In the case of streets, characteristics related to the number of perceived attributes are important. These are equivalent to noticeable differences and hence lead to complexity by means of movement through distinct settings. The gradual unfolding and emergence of such differences is, of course, equivalent to notions such as surprisingness, mystery, and the like, which form part of the concept of complexity.

Complexity results from the availability of a variety of possible movement paths, the juxtaposition of varied elements and areas, the location, mix, and changes of activities, the presence of open-ended design that allows changes over time, and so on. These seem sufficient to overcome habituation and adaptation effects. For example, vernacular environments generally tend to be more complex than high-style design, and to remain richer, because there is a greater mixture of activities in space, more activities in streets and hence more multisensory experience of these, more changes of activities over time and also changes in, and additions to spaces and building because of open endedness. The result is a much greater richness both in space and time. This is enhanced by the very strong order that is easily "read" and understood so that very minor variations from it become noticeable and contribute to complexity (Rapoport 1977, 1988b, in press). Uniform environments with little contrast tend not to be complex (e.g., Rapoport 1977, pp. 220ff.).

The availability of possible movement paths in an urban area already mentioned may encourage exploration independently of the nature of the spaces themselves. This is not only because they lead to some uncertainty about where one will finish and how one would get there—an example of lostness at the small scale within a clear orientation context. It is also because different ones can be chosen over time by any one individual, leading to greatly increased complexity by allowing memory to fade before any one is seen again and by allowing many possible combinations and permutations; see Figure 7.1.

It is interesting to compare a grid to an apparently more complex path system. A grid actually allows large numbers of alternative paths, but these have high redundancy and hence less information; they thus tend to be simpler. Grids are made more complex by slopes (for example, compare San Francisco with Midwestern cities), by variety in eye-level elements (Vernez-Moudon 1986), that is, the texture or grain of the urban fabric (Rapoport 1977, Figure 4.4, p. 205), vegetation, and the like. It is significant that in general, the grid has tended to disappear when strong central authority and power

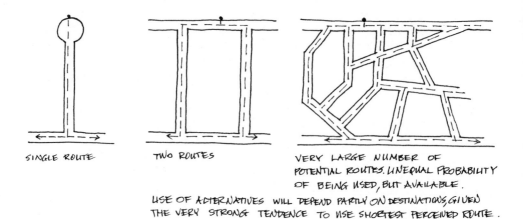

SINGLE ROUTE   TWO ROUTES   VERY LARGE NUMBER OF
                                          POTENTIAL ROUTES. UNEQUAL PROBABILITY
                                          OF BEING USED, BUT AVAILABLE.
                      USE OF ALTERNATIVES WILL DEPEND PARTLY ON DESTINATIONS, GIVEN
                      THE VERY STRONG TENDENCE TO USE SHORTEST PERCEIVED ROUTE.

Figure 7.1. Complexity at the level of areas (from Rapoport 1977, Fig. 4.10, p. 218).

have weakened (Stanislawski 1961). In this way the desirable levels of complexity are restored as grids give way to more complex patterns. For example, in Nubia the narrow streets of the grid "began rapidly to break down . . . into an 'organic' layout" (Kemp 1972a, p. 65) when central power waned. This also happened in Mohenjo-Daro where the wide straight avenues were replaced by an "untidy muddle" (Ward 1977, p. 106), and in China (Tuan 1968); it is typically a recurring "problem." For example, in Paris as early as 1607 attempts were being made to prevent projections into streets (Tricaut, n.d., p. 14). Nearly 1,500 years earlier a similar process took place in Damascus after the decline of Roman rule (Eliséeff 1970), even though the Roman street with its arcades had reasonably high complexity levels; see Figure 7.2.

This tendency for grids to revert to a particular pattern needs to be explained: Why the pre-twentieth century "organic" pattern that is close knit with narrow twisted streets, market places, courts, and so forth is so prevalent—a fact that only becomes known when a large and varied body of evidence is examined. One needs to ask why planning controls must "force" people to the grid and why the nongrid "higgledy-piggledy" layout seems to be preferred (Morris 1979, pp. 8–17). Thus grid streets are modified when used by pedestrians, and "natural" growth replaces the grid when central power breaks down; the grid rarely reappears (Stanislawski 1961). The grid generally seems ephemeral, but it did not disappear in the United States. Its survival there may be due to the fact that most U.S. cities are "postpedestrian," and pedestrian use tends to be underemphasized. The more typical changes can then be hypothesized as being related to pedestrian behavior because pedestrians always prefer more complex environments (Rapoport 1977; Wheeler 1972). This suggestion is reinforced by the fact that some early U.S. urban areas, such as downtown Manhattan and Boston, exhibited the expected traditional patterns (e.g., Clay 1978).

Another case where grids have tended to persist is Spanish Latin America where almost all settlements were planned under the Laws of the Indies.* Their persistence is

*For example, in Mexico there are only three exceptions (Guanujuato, Taxco, and Zacatecas [Hartung 1968]).

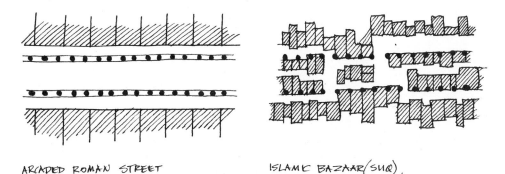

ARCADED ROMAN STREET   ISLAMIC BAZAAR (SUQ).

Figure 7.2. Transformation of a Roman street into an Islamic bazaar, Damascus (from Rapoport 1977, Fig. 4.11, p. 219, after Eliséeff 1970).

an anomaly less easy to explain and poses an interesting problem, although, in fact, there were constant efforts at encroachment against which streets and the grid itself had to be defended (Hardoy and Schaedel, n.d., p. 249; cf. Kinzer 1978). Moreover, the contrast with Brazil is instructive: There the Laws of the Indies did not apply, and the expected pattern is found. This difference becomes even more significant due to the fact that towns in Spanish Latin America were radically different from the home country, whereas in Brazil the Portuguese used the medieval and traditional forms also used in Portugal (Hanke 1973, pp. 265ff.). Because they used no planning and no code, these towns grew in "picturesque confusion": The model was Lisbon, a city of steep, narrow twisting streets with no vehicular traffic—only pedestrians or sedan chairs (Smith 1973). This view is not universally accepted, however, and it is pointed out that many grid-pattern towns were laid out in Brazil (e.g., Delson 1979).

The notion of complexity and the gradually emerging hypothesis lead to an explanation of these regularities. They are due to perceptual qualities of the pattern that, in turn, is related to pedestrian activities. In cases where streets are used primarily for walking one can expect changes, such as those to the grid, which increase perceptual complexity.

The effects of changes like those in Damascus, that is, the introduction of greater enclosure, turns, changes of direction, blocked views, and the like has an impact on perceived complexity partly due to the fact that each such change increases uncertainty and hence information content. This increases even more if decision points are introduced. In such complex, rich urban environments one can walk for long periods without becoming tired (Parr 1969b). The same distance through an open parking lot would seem endless because of inadequate rates of information (Rapoport 1977, pp. 241–242). Varied environments provide much information, with some always left over, and they become interesting. The effects on walking may be due to changes in subjective distance. I have suggested (Rapoport 1977, Chapter 3) that the evidence of the effects of turns, bends, and route segmentation on subjective distance is equivocal; there is evidence both for an increase and decrease in subjective distance. In terms of complexity, both views can be deduced. One would be that the more information per unit of route the longer the route should seem, given that route length is estimated in terms of information per unit time. The other view is that the more information, the more interest and the shorter does elapsed time appear; because distance is estimated in terms of time, the shorter subjective distance. Experientially high information environments seem shorter to traverse than low information ones, but this is reversed in memory: Complex routes are experienced as short and remembered as long and vice-versa (Cohen 1967; Steinberg 1969). This effect and a model of the processes responsible for it based on information storage has, in fact, been developed (Sadalla and Magel 1980; Sadalla and Staplin 1980a,b).

Although my concern here is at the microscale, urban complexity is also related to areal differences—in uses, activities, social homogeneity or heterogeneity, changes over time, and levels of complexity of the areas themselves. Distinctive areas, whether regional or within cities, contribute to complexity. Within cities, this is a result of a variety of areas with diverse character in terms of uses, people, and physical character in all sense modalities. Transitions among them become noticeable. The examples at all scales have one thing in common: Complexity is the result of becoming aware of differences; it is a matter of *noticeable differences*.

NOTICEABLE DIFFERENCES

Overload, deprivation, and complexity depend on the number of distinct elements that can be perceived; it is these elements that constitute variations from a perceived order. This helps explain the peculiar richness of much vernacular design: Because the order and the rules are so strong and consistent, small variations are noticed and become important. In addition to limiting the vocabulary in this way, and being highly consistent in any given area, vernacular design varies among areas producing further noticeable differences at larger scales—the transitions among distinct areas. In high-style design the rules are much more idiosyncratic and variations therefore less apparent. Moreover, variations become more difficult to achieve because such design is not as open ended as vernacular (Rapoport 1988b, in press). In the case of a roadside strip, very large variations are needed—if, indeed, any variations will be noticed (Rapoport 1977, pp. 222–240).

Note that we are dealing with *potential* noticeable differences and my discussion is confined to those. Which of these are actually noticed is an empirical question: It needs to be discovered whether they become noticeable to users. If they do not, they are not noticeable differences. If one accepts signal detection theory (e.g., Daniel *et al.*, n.d.; Murch 1973), the idea of thresholds becomes suspect because of the extremely important role of the readiness or willingness of the perceiver to make judgments on the basis of minimal or uncertain cues (the perceiver's "criterion state"). But such personal and group variables, adaptation, and learning are beyond designers' control. All they can do is to try to control thresholds and the strength of signals, to provide requisite levels of potential noticeable differences for various groups using a variety of elements and sense modalities; the characteristics of such potential noticeable differences can be specified. It then becomes more likely that in such environments a sufficient number of elements become noticeable and result in appropriate levels of *perceived complexity* (to be discussed later).

In essence, noticeable differences are related to the background: They can be understood in terms of figure–ground relations; they can also be understood in terms of changes in the state of borders of the stimulus (Gibson 1968). This reinforces the idea that relationships among elements may be more important than the elements themselves (Rapoport 1969e): It is the juxtaposition of elements that leads to noticeable differences. Thus in a brightly lit area, the most noticeable element may be a dark shop; in a dark area it may be bright lights, and the level of brightness will depend on the background (Rapoport 1977, pp. 225–228). Cues also become more noticeable as they become more salient and meaningful to the perceiver, who tends to develop high sensitivity to such elements (e.g., Duncan 1973; Rapoport 1981a; Royse 1969; Schnapper 1971, p. 91). This is, however, culturally extremely variable and mainly a function of associational qualities, although it has perceptual implications. For example, in a culture where colors, sounds, and smells are important they will tend to be more noticed than in cases where they are unimportant (Rapoport 1975a; Stea and Wood 1971; Wood 1969).

Noticeable differences are stronger when perceptual qualities in various sensory modalities and associational ones are congruent and reinforce one another. For example, small parks in a residential area would be noticed and would not merge into the larger predominant use pattern because they provide breaks in the urban fabric and transitions at the edge. This effect would be stronger in a densely built-up area without street trees than in a low density area of single family dwellings with many street trees. Changes in the thermal, olfactory, and acoustic cues would reinforce this and would also be stronger in the former case (Rapoport 1977, Chapter 3). This is then reinforced by the salience and meaning of such elements that often play a major role in the judgment of the quality, status, and social identity of areas (Rapoport 1977, Chapter 2, 1982a). In cases where such variations are not noticed, they do not produce complexity. The clearer, stronger, and more salient the contrast, the greater the likelihood that they will be noticed.

It is possible to list cues among which people choose, the effectiveness of which depends on several cues operating simultaneously and reinforcing each other: a repertoire, as it were, of elements that are potential noticeable differences and that can apply at various scales (Rapoport 1977, pp. 124–125, 228–240; 1983d).

Physical differences

Vision

Objects—shape, size, height, color, materials, texture, details.
Space quality—size, shape; barriers and links—merging, transitions, etc.
Light and shade, light levels and light quality, temporal changes in light.
Greenery, man-made versus natural, type of planting.
Visual aspects of perceived density.
New versus old.
Order versus variety.
Well maintained versus badly maintained or neglected.
Scale and urban grain.
Road pattern.
Topography—natural or man made.
Location—prominence, at decision points, on hills, etc.

Kinesthetics
Changes of level, curves, speed of movement, etc.
Sound
Noisy versus quiet.
Man-made sounds (industry, traffic, music, talk and laughter) versus natural—
wind, trees, birds, water, etc.
Dead versus reverberant.
Temporal changes in sound.
Smells
Man-made versus natural; plants, flowers, sea, etc; foods, etc.
Air movement
Temperature
Tactile
Mainly texture underfoot.

Social differences

People—languages spoken, behavior, dress, physical types.
Activities—type and intensity; clubs, restaurants, churches, fairs, markets, etc.
Uses—shopping, residential, industrial, etc.; uniform versus mixed.
Cars versus pedestrians, other means of travel; movement versus quietude.
Objects—signs, advertisements, foods, objects used, fences, plants, and gardens, decorations, etc.
How the city is used—street use versus nonuse; front/back distinctions; private/public distinctions, etc.; introverted vs. extroverted. These are all related to cultural barriers and rules for behavior (the cues that have to be understood and obeyed as well as noticed).
Hierarchy and symbolism, meaning, signs of social identity and status.

Temporal differences

Long term: Changes over time from State A to State B: changes in people, in maintenance, in uses, etc. This is a whole set of cues indicating change versus continuity and stability, and this change may be seen as positive or negative. Many of the potential noticeable differences listed can be read and interpreted as social indicators of good or bad, deteriorating or upgrading areas. They are culture specific.
Short term: Type of uses day and night, days of week, weekends, intensity of use over time. Tempos and rhythms of activities.

In the development of the hypotheses to be tested in this study, I will only be concerned with *visual cues mainly in the fixed-feature realm.* This does not mean, however, that the others are not significant. The contrary is the case and, eventually, they will need to be considered (cf. Rapoport 1977, pp. 184–195).* In the case of historical data, however, visual fixed-feature elements are the only ones generally available. If they alone, in fact, encourage pedestrian activity, this should be even more the

---

*The process of perception of complexity is additive. This applies to different characteristics, such as space and form (e.g., Pyron 1971, 1972), and this process is strengthened when all sense modalities are involved.

case when both semifixed and nonfixed elements and the attendant nonvisual cues reinforce the supportiveness of such settings.

As already suggested, it is likely that potential cues are noticed more, and noticeable differences become clearer and stronger as more cues in more sensory modalities become congruent (Pyron 1971, 1972; cf. Rapoport 1977, Figure 4.6, p. 211, Figure 4.19, p. 228; Southworth 1969; Steinitz 1968). The strength of the figure–ground relation, the cultural or personal significance and salience, and the congruence of various cues and sensory modalities all play a role in potential noticeable differences becoming actual ones; in this, movement and living things are very important (Bartlett 1967; Gulick 1963; Maurer and Baxter 1972; Sieverts 1967, 1969; Steinitz 1968; Weiss and Boutourline 1962; Whyte 1980; cf. Rapoport 1977, pp. 207–247).

Nineteenth-century New York, in spite of its uniform grid of streets, leveling of all hills, and uniform housefronts, was very diverse because of the presence of many specialized areas—some just a block in extent, others the size of a small town each with different religious life, language, newspapers, restaurants, holidays, and street life: To visit all was almost like a trip to Europe (Jackson 1972, pp. 205–206). Moreover, many streets showed the use of semifixed elements to create higher levels of enclosure, complexity, and so on, as in the pushcarts and outdoor markets of the Lower East Side or the more designed elements visible in old photographs of Park Avenue. These, in turn, attracted many people further adding to complexity.

Again, these elements can work at various scales from the largest to the smallest. Examples at large scales can be given of Edinburgh, London, San Cristobal de las Casas (Mexico), Tokyo, Zanzibar, and others where these types of elements can be found and frequently work in concert (Rapoport 1977, pp. 207–247). Copenhagen provides an example closer to our present concern. Starting at Kongens Nytorv there is a particular scale, complexity, and age of the old part. A clear transition is then provided by the pedestrian area of the Østregade that shows many of the characteristics of pedestrian streets—it is narrow (33 feet) and highly enclosed but periodically opens up into plazas; has different uses, shops, paving materials, music groups, dining and drinking, flowers, and different tempos of movement. There is also a complex network of other pedestrian routes. It is significant that an area that possesses the characteristics emphasized in this case study *was* selected to become the pedestrian area, as is the case in most cities in Europe and elsewhere. There is then another clear transition to the Town Hall square and the area beyond with its wide streets, tall buildings, neons, heavy traffic and bustle—a modern city like all others (see Rapoport 1977, Figure 4.24, p. 234).

Noticeable differences can also play a role in driving, although that does not concern us here. At the scale of streets, a larger range of potential cues from the repertoire can be used, given the characteristics of pedestrian perception already discussed. This can be done even when pedestrians and traffic share streets. The various systems of signs, lights, information, and so on can be distinguished and analyzed (Carr 1973); so can the changes in paving used in the Dutch *Woonerfs*. All can be interpreted in terms of noticeable differences (cf. Lynch and Rivkin 1970). Architectural elements that bound streets play a role also so that shape, age, color, materials, texture, height, and the like can be used, as can walls, gates, fences, and the like. These can reinforce other sensory cues already mentioned; if congruent with "general character," planting, uses, maintenance level, and so forth, noticeable differences become clearer.

There is thus a hierarchy of potential noticeable differences at many scales and of all kinds and using all sensory modalities. If noticed, they lead to complexity; if not noticed, they are effectively absent. However, they may be noticed subliminally and, in any case, they are ultimately *always related to elements in the environment.*

A variety of research suggests which of the elements of the repertoire are the most likely to be effective. Strong rather than subtle space variations, the texture, color, and geometrical alignment of enclosing elements; height of enclosing elements, and personalization all seem important. Studies that identify failures can be as useful as those that illustrate successes (Cooper 1965, pp. 100–106; Pyron 1971, 1972; Wilmott 1963, p. 4). Given current knowledge such cues cannot be assumed *a priori* but need to be discovered. The present case study is an attempt to identify a small subset of such cues and test them against a wide range of past environments. Further empirical research would broaden this and move into nonvisual senses, semifixed elements, and so on. Research on nonvisual noticeable differences is particularly lacking. Here the use of cross-cultural data and travel descriptions as well as empirical work with subjects are the only ways of broadening the sample (see discussion in Rapoport 1977; cf. Rapoport 1964–1965). Together even stronger patterns and regularities could be uncovered and broader generalizations become possible. In this sense, this case study illustrates the mutual relation of contemporary research and the analysis of historical data.

What already seems clear is that the concept of noticeable differences is useful in a wide variety of situations. Thus it is significant that Australian Aborigines, who define places largely symbolically, do so much more clearly and distinctly where there are noticeable differences in the landscape—water, rocks, major trees, and the like. Moreover, ritually important places (which tend to be the most important) are those that are most distinct from their surroundings (Rapoport 1975a). Sacred places generally often seem to be noticeably different; moreover, such differences tend to be reinforced through design and construction (e.g., Scully 1962), and also through behavior, dress, sounds, smells, and others (see examples in Rapoport 1977). In all these cases the physical settings, including semifixed and nonfixed elements, provide potential stimuli that can become noticeable differences. Because no one set of elements is used by all, the more possible elements are provided, the more chance that at least some will be so used; their congruence is also important. In fact, this becomes a hypothesis best studied by using the largest possible sample of sacred places from varied cultures over long periods of time. One can also study any commonalities among elements that tend to be widely used. This becomes *another potential historical study, related to those discussed in Chapter 2, which well illustrates the argument of this book.*

Noticeable differences thus seem to be a very useful, general, and unifying concept. Much of the urban design literature can be interpreted in terms of the manipulation of potential noticeable differences (although, except in EBS, these never receive empirical study). In general, *the more potential noticeably different elements and settings exist, the greater the chances of people perceiving them and experiencing complexity.* I have argued that appropriate complexity is the key characteristic of pedestrian settings. It thus seems that the concept of noticeable differences provides the most useful way to derive hypotheses regarding the perceptual characteristics of pedestrian spaces; this can best be addressed by contrasting them with those of spaces for motorists.

EFFECTS OF SPEED OF MOVEMENT

One major and obvious difference between pedestrians and motorists is their speed of movement. Together with notions such as information processing, channel capacity, boredom and overload, complexity and noticeable differences, this suggests that *rate of information* is important, that we are dealing with the *number of noticeable differences per unit time*. It follows that speed plays an important role in the perception of noticeable differences and hence of complexity. For example, pedestrians and motorists should differ greatly in their perception of the city because the city is experienced over time. This makes such perception dynamic and sequential; it is made up of short scans and involves the integration of successive partial views. These, of course, are only significant if there are noticeable differences among successive views and some uncertainty as to the next view (Rapoport 1977, Chapter 3 and references; cf. Johnson 1965; Pyron 1971; Thiel 1970). This integration of partial views, that is, the rate of noticeable differences per unit time, is affected by speed.

Speed is in itself judged in terms of the rate of information flow: Low information trips seem slower than high information ones. Speed also influences how often noticeable differences occur, how long they are in view, and hence whether they are noticed. Subtle cues need slow pace; driving is not only fast but demands concentration, leaving no time or channel capacity to appreciate the environment. Thus pedestrians have a much better awareness of places and clearer ideas of the significance, meaning, and activities in the city than either drivers or users of public transport. Because of the lower speed and lower criticality of their movement, pedestrians can perceive many more differences in form, activity, and so on. Pedestrians are also less insulated from multisensory information, and the active nature of walking increases the dimensionality of information. Thus the relative effects of different transport modes on the knowledge of the city (Rapoport 1977) can be partly interpreted in terms of the perception of noticeable differences.

This implies that settings for different speeds require different cues and different levels of complexity. I have already mentioned streets where two different levels of complexity can be provided for motorists and pedestrians. Long pedestrian underpasses with white, shiny tiled walls are monotonous at pedestrian speeds; the apparent rate of progress is reduced by the lack of noticeable differences and adequate levels of information. The roadside strip, at driving speeds, becomes chaotic. There is a relation between levels of complexity and speed, so that the same roadside strip that is full of parking lots and large elements is extremely open spatially and provides inadequate information to pedestrians (Rapoport 1977, pp. 241–242). Because of the distance to goals and lack of visible changes in cues, at slow speeds there is a low rate of meaningful information, few noticeable differences, and the environment is boring.

There is clearly a continuum of means of travel, such as walking, bicycling, slow driving, fast driving, as well as various public transport modes—bus, train and underground. Each has an associated desirable complexity level, that is, difference perceptual requirements that can be specified. Moreover, driving on a narrow road lined by trees is different from driving on a similar road through an open plain, or through mountains; an urban freeway is different from a rural one; driving on a surface or elevated road is different from driving through a tunnel—always in terms of appropriate perceptual characteristics. The point can be made most clearly, however, by concentrating on the two extremes: walking, strolling, and sauntering on the one hand, and fast urban motor traffic on the other, and trying to derive the perceptual characteristics desirable in these contrasting settings. Then it will become easy to elaborate those perceptual characteristics associated with pedestrian settings. In discussing pedestrian streets and urban highways, I will mainly consider visual variables although, as already emphasized, environmental perception is multisensory (Rapoport 1977, pp. 184–195). Moreover, the nonvisual senses become particularly important in pedestrian settings because pedestrians are more exposed to them and also have more opportunity to experience and appreciate them. They thus add greatly to the potential noticeable differences and hence desirable complexity of such settings. They are, of course, highly characteristic in pedestrian settings of surviving preindustrial or traditional settlements. Finally, although complexity in all settings, and particularly pedestrian settings, results from fixed-feature elements, semifixed feature elements, and nonfixed feature elements (i.e., people), only fixed-feature elements will be discussed. Clearly, however, the others fit easily into the conceptual framework.

Perceived complexity is related to the number of noticeable differences per unit time and hence to speed. As speed increases, the task becomes more demanding, and concentration increases. Several other things also happen. Thus speed influences how discrete stimuli are organized into groups. At high speed, elements are grouped into simple chunks, whereas at slow speeds more discrete elements are perceived. Foreground detail begins to fade, due to the rapid movement of close objects. The earliest point of clear view recedes from 30 ft at 40 mph to 110 ft at 60 mph. At the same time detail beyond 1,400 ft cannot be seen as it is too small, so that the range is between 110 and 1400 ft—and that is traversed in 15 seconds (Tunnard and Pushkarev 1963, pp. 172–174). Elaborate detail is thus both useless and undesirable. Thus high speed makes an objectively complex environment chaotic; a simple environment, interesting at high speed, becomes monotonous at slow speeds. Complexity in a traffic tunnel and simplicity in a prison yard are both perceptually undesirable (Chang 1956, p. 20).

At high speed, the point of concentration (or focus) recedes from 600 ft at 25 mph to 2,000 ft at 65 mph. As a result elements must become larger. Also, although objects perpendicular to the road become prominent, those parallel to it lose prominence.

Peripheral vision diminishes so that although at 25 mph the horizontal angle is 100°, it reduces to less than 40° at 60 mph. One result is "tunnel vision," which may induce hypnosis and sleep. Side elements need to be simple and subdued and perceived semiconsciously in the blurred field of peripheral vision, with the main features on the axis of vision and the point of concentration periodically moved laterally to maintain attention (Tunnard and Pushkarev 1963, pp. 172–174).

Central vision is essential for fine detail and small differences in contrast, texture, color, and the like, whereas peripheral vision detects movement. Hence the presence of elements close to a rapidly moving observer, who must focus ahead, can be most distressing by greatly exaggerating apparent speed, because seen through peripheral vision; this is particularly the case if these elements are complex. In effect, space perception becomes impaired so that near objects are seen, get close, and disappear very quickly. They thus tend to "loom," which is extremely stressful (Coss 1973), and elements too close to the edge or overhead, and sudden curves, should be avoided.

Movement itself, in terms of sequential perception, can be understood in terms of noticeable differences. Movement through an environment can be described in terms of transitions, "emergence from behind," sequences, and transformations (Gibson, 1968, pp. 206–208). This can only happen when there are noticeable differences; on a featureless plain or in a completely featureless tunnel, and in a vehicle with no kinesthetic cues about speed or movement, the apparent rate of movement will be much lower than through an environment rich in noticeable differences and transitions. It follows that complexity depends on the number of noticeable differences per unit time, changes of any attribute: rate, direction, slope, curvature, color, texture, enclosure, smell, sound, light quality, or whatever. Analyses of urban environments by artists in terms of transitions and sequences (e.g., Kepes 1961) can not only easily be interpreted in terms of noticeable differences, but must also include a consideration of how fast they occur. Given a certain number of noticeable differences per unit length of an environment, it is clear that at lower speeds the environment would tend to be simplified and tend toward monotony and sensory deprivation, whereas at higher speeds it would tend toward chaos and overload; see Figure 7.3.

Note that due to the active role of observers, some compensation is possible through "chunking" for the latter and "decomposing" for the former. In the case of pedestrian speeds, the perceiver is free to explore the environment, to use all sensory modalities, and to seek detail, all of which leads to some increase in complexity. However, achieving requisite levels is much easier if the environment provides potential noticeable differences. In any case, settings for high and low speed, for motorists and pedestrians should be quite different perceptually. It would be impossible to appreciate the subtleties of Katsura Imperial Villa, Fatehpur Sikri, or any other traditional high-style or vernacular architectural or urban design sequences at high speed; it is also impossible to appreciate a freeway at walking speed.

At car speeds the time available to notice and "read" information is greatly reduced whether this information be spatial, eikonic, verbal, human, and so on—or a combination of those. This demands large, infrequent, broad and smooth rhythms; car-oriented environments require large-scale elements. Urban light and sign systems are greatly affected by speed, as anyone who has traveled along a highway strip at night at high

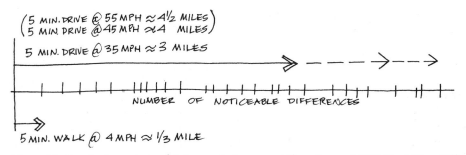

Figure 7.3. Perceived complexity and the speed of movement (from Rapoport 1977, Fig. 4.29, p. 242).

speed knows. Sequences of views in the design of highways have been discussed (Appleyard, Lynch, and Meyer 1964; Carr and Schissler 1969), but these ideas, based on the effects of speed on perception, were not based on noticeable differences. Neither have they been generalized nor applied to pedestrian settings, as pointed out in Chapter 6. Unlike motorists, pedestrians can appreciate fine grain, can vary their rate of movement, can look around, and stop to observe detail; they are potentially aware of the environment around them and in all sense modalities. The important conclusion is that, quite apart from problems of safety and pollution, and hence *physical* separation, pedestrian and high-speed environments are *perceptually* incompatible; that is, the conflict is not merely between cars and pedestrians but between fast and slow speeds and also between types of movement—smooth or jerky, straight or irregular (Rapoport 1981b, 1987a).

To repeat: An environment comfortably stimulating from a car becomes monotonously boring on foot, whereas what is interesting on foot becomes chaotic in a car. The Shambles at York are a good pedestrian environment, whereas the Pyramids are ideal at car speeds (Parr 1969a). More generally, traditional settlements suit pedestrians; Ville Radieuse and its progeny are meant for motorists. The two environments need to be quite different in terms of noticeable differences and perceptual organization: At high speed one needs large scale, simplicity, and relatively distant views, whereas at slow speed one needs enclosure, small scale, intricacy, and complexity.

The result are perceptual needs for high-speed settings that are the opposite of those for pedestrians. Motorists' perception is also affected by the length each element is in view and also by the criticality of the task. Pedestrians, on the other hand, have each element in view as long as they wish and can satisfy their interest in it because of the low criticality of the task—walking. When pedestrians are harassed by traffic, their task becomes more critical, and they cannot perceive the environment in ways appropriate to their speed: This is a common problem in contemporary cities.

The nature of the environment structures motorists' perceptions within the limits of their task, speed, and car characteristics; thus environments can be more or less congruent with needs. Given the changes described in the visual field as speed increases, it follows that for motorists, building setbacks should be greater than for slow speeds; although less varied than for pedestrians, setbacks should not be uniform. Uniform and consistent surroundings confound orientation, confuse destination location, and reduce interest because they fail to provide noticeable differences. The shapes of the visual fields on either side of the road should be similar, that is, roughly symmetrical, although one side is always dominant—the visual field never expands equally on both sides. Distances between peripherally or foveally viewed elements should be reduced; buildings should be regular in height, and rhythms should be large scale and simple (Pollock 1972).

These are clearly all factors derived from complexity/speed relations. Elements along roads should provide information at rates intermediate between a freeway and a pedestrian speed, with gradual transitions. Sudden contrasts between high and low information environments should be avoided; although areas of differing complexity are still needed, transitions among them should be gradual. There should be a smooth, continuous succession of such areas, with their complexity and intensity decreasing as speed increases. Generally, then, as speed increases, the number of noticeable differences in the environment should decrease and setbacks should increase; also as traffic intensity increases the perceptual complexity of the environment should be reduced (Pollock 1972; cf. Rapoport 1957; Rapoport and Kantor 1967, Rapoport and Hawkes 1970; Rapoport 1977). This happens "naturally": When highway design increases complexity, speeds tend to go down; perceptually simple highways tend to lead to higher speeds (Tunnard and Pushkarev 1963).

In many traditional cities, buildings tend to be rather narrow. Because each building is treated individually, the result is a rather fine grain, high texture and complex, rich character of the elements; see Figure 7.4.

The change from that to large, simple elements happens very commonly with "modernization"; this works well for traffic but badly for pedestrians. An example is the change in Brussels from old areas (which are like the left side of Figure 7.4) to the new NATO and OECD buildings that are like its right side. Comparable changes are found in most cities; one example is the City in London. A recent attempt to prevent such changes from occurring in San Francisco (Vernez-Moudon 1986, especially pp. 191–242) identifies and tries to preserve the components of the traditional system even in new development: a narrow module (25–35 ft), individual facade treatment, irregular setbacks, rich treatment of each individual building (bay windows, "gingerbread," etc.) and the like.

Pedestrians can use, and desire, much more acute and abrupt transitions—spatial, kinesthetic, textural, in color, in light levels, sound, and all other sense modalities. Only they can notice, react to, and respond to the variety of stimuli that rich, opulent environments can provide. They also *need* them in order for the settings to be supportive by encouraging exploration through interest. This follows from all that has been said: the different rates of speed and the different ways of perceiving—free and flexible for pedestrians, constrained and "tunnel" for motorists. The characteristics of pedestri-

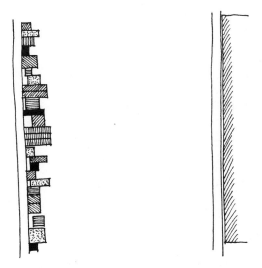

Figure 7.4. Grain, texture, and complexity (from Rapoport, 1977, Fig. 4.4, p. 205).

an settings and high-speed settings can be illustrated and contrasted as follows; see Figure 7.5.

The characteristics of pedestrian spaces shown here are the basis for the more specific hypotheses of this study. Without ignoring the need for physical separation for purposes of safety and amenity, or the other variables (listed in Chapter 6) likely to be related to pedestrian behavior, these characteristics describe the *perceptual* differences between high-speed and pedestrian settings. The latter should be rich, full of detail, complex, with clear transitions and small-scale, irregular elements. A highway should be simple, with large-scale, more widely spaced elements. Generally, the higher the speed, the less information per unit length is needed in the environment. There are clearly also differences between a scenic road and a freeway because context will modify even high-speed movement. Thus ludic and exploratory as opposed to grimly purposeful behavior will need very different levels of complexity *at given speeds*. Walking for pleasure or to a job are very different; this is also the case for the two forms of driving. People may thus select different settings for apparently similar activities depending on context, *but the basic difference between slow and high-speed environments will persist.*

In general, pedestrians rarely look above eye level in enclosed urban spaces; thus the perception of detail becomes inevitable, and this becomes what such settings require. Given the needs of drivers as described, their movement channels should be simple, and it is free-standing elements and tops against the sky, and clusters of tall buildings that become important (Heath 1971; Worskett 1969, p. 98; cf. Rapoport 1977, Figure 4.32, p. 246). When all these characteristics, together with the previously discussed distinction between dynamic and static pedestrian spaces are combined, it becomes possible to propose a hierarchy and a continuum of movement and rest spaces, related to noticeable differences (Rapoport 1977, Figure 4.33, p. 247, 1987a).

THE GENERAL HYPOTHESIS

On the basis of the argument thus far based on EBS research, the general hypothesis to be tested against a varied body of evidence from traditional settlements can be derived. Given that complexity is a function of the number of noticeable differences per unit time, one can deduce an even more specific set of characteristics—the specific hypotheses—to be expected in settings supportive of pedestrian movement. Before those are derived, however, a few additional considerations need to be introduced.

One is the elaboration of a point already made briefly that as a function of speed the *perceived* complexity of environments is more constant than their *designed* (or potential) complexity. Whereas high-speed environments need to be designed with much lower complexity than those for slow speed, at their respective speeds their perceived complexity may be more similar. The perception of complexity is thus subjective, but certain environments produce the specific experience much more easily, reliably, and unequivocally than others *at specific (and appropriate) speeds.*

This is a very likely interpretation of a study on the enjoyment of interpretive trails through natural settings (Gustke and Hodgson 1980). Involved were the rate of travel and the contrast between what had been experienced and what was being experienced;

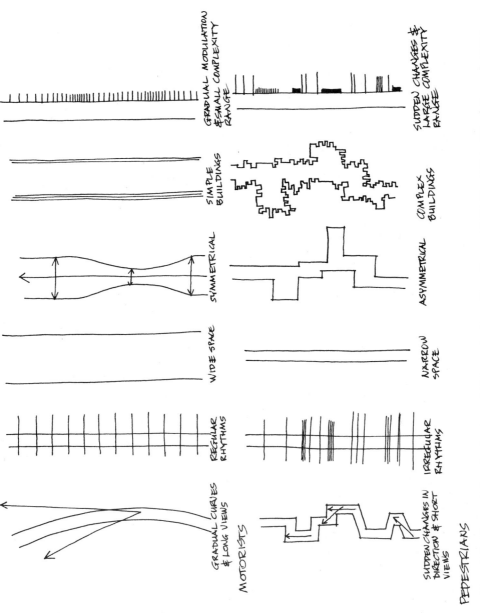

Figure 7.5. Characteristics of pedestrian and high-speed settings (from Rapoport 1977, Fig. 4.30, p. 244; cf. Rapoport 1987a).

this was as important as the actual composition of the scene. The trail thus becomes a sequence of landscapes—woods, meadows, lakeshore, woods again, and so forth; a sequence of vistas open and closed. Smells, textures, sounds, temperatures, breezes, amount of light, and the like are constant in any one setting, but change as new settings are entered. Each of these, *or rather the transition between them,* becomes a noticeable difference (in my terms). Moreover, as for highways (Tunnard and Pushkarev 1963), richer environments (i.e., those with more such transitions and hence more information) tended to slow down the rate of movement along the trail. This constancy at the level of process follows from my argument about evolutionary constancy. Also, the description of the characteristics supportive of pedestrian movement in the case of a "natural" environment (in which humans evolved!) has clear implications for man-made settings.

This is an example of how the ideas already discussed can help explain a particular study that is not based on them; in fact, to confirm a hypothesis that could be derived from the previous discussion. It illustrates the role of complexity that results from hidden views, transitions, changes, and the like, the richness that can be understood in terms of a relation between noticeable differences and speed. This idea of a sequence of places also suggests the need to break up, or articulate, pedestrian spaces into segments, avoiding long vistas or excessively large, empty spaces. Not only will such articulated spaces be more complex than ones seen at a glance; movement through them results in many more new views, as hidden views are revealed, more transitions and the like; hence many more noticeable differences and hence higher complexity; thus high enclosure levels should be characteristic of pedestrian settings. This is reinforced by the observation that rich texture and elaboration of detail is important for pedestrians. There is, however, also another line of argument that greatly strengthens the likelihood that pedestrian spaces should be characterized by high levels of enclosure.

A starting point is the observation that walking in appropriate pedestrian settings, and particularly in preindustrial cultures, is much more social than other forms of movement (Rapoport 1987a). This still tends to be the case today, so that pedestrian behavior tends to be a social event, involving trading, eating, promenading, socializing, or exchanging information. Even when no active interaction takes place, the co-presence of the other people is still an important component of walking—as is the case in successful contemporary pedestrian malls, for example. Thus settings need to be supportive of this co-presence as well as, in the case of many cultures, the active interaction that accompanies pedestrian behavior. This needed to be even stronger in preindustrial cultures. Most such co-presence or social interaction tends to occur in small, well-bounded settings (e.g., Goffman 1963, 1971) that could be described as "architectural" in scale. Those tend to have high enclosure, relatively small dimensions, and a limited length of view. It follows that pedestrian spaces should have similar characteristics, that is, be "roomlike" and highly bounded. It is interesting to note that many of the design writers cited (e.g., Cullen 1961; De Wolfe 1966; Sitté 1965; among others) often speak of "outdoor rooms." The commonly expressed idea that pedestrian spaces, particularly static ones, can be conceived as "stages" or theaters is also relevant (e.g., Crowhurst-Lennard and Lennard, 1984; Rapoport 1982a, 1986a; Sennett 1977). This suggests that the scale (i.e., dimensions) appropriate to *intimate theater* are relevant; it also suggests that dynamic spaces can also be conceptualized as a sequence of settings each of which is effectively a stage (cf. Rapoport 1982a, Chapter 7, e.g., p. 182; 1977, e.g., Figure 6.7, p. 35, and elsewhere).

Most outdoor spaces tend to be different from architectural ones in size, boundedness, and "ambience"—the latter being a difficult characteristic to define but an important one in environmental perception (cf. Ittelson 1973; Rapoport 1989d). For one thing, the lack of an overhead boundary makes the character of enclosure different. It is, however, likely that vernacular spaces that were developed exclusively for pedestrians in preindustrial contexts should show characteristics as close as possible to the architectural scale.

A further line of argument that supports this is the long-standing attempt to describe some limiting dimensions for the perception of other people, and hence for comfortable pedestrian urban spaces. For example, in 1884, Maertens suggested an upper limit of 3,990 feet for perceiving other people. Hughes (1974), in discussing military situations, suggests that at 5,100 feet troops cannot be seen without binoculars. Fletcher (1953) suggests that an upper limit for hearing the human voice is 6,600 ft. All these limits, of course, are extreme *upper* ones. Gilinski (1951) suggests that stationary people in outdoor spaces act as though their horizons (or "infinite distances") are relatively close. He suggests that the limit of one's visual field is of the order of 48–300 ft depending on the people and the situation. Hall (1966) sets the far phase of the public distance as starting at 25 ft. Other suggestions that these distances are quite small come from McBride (1972); Smith *et al.* (1975); Whyte (1980). Similarly, attempts have been made to define the physical dimensions of squares, lengths of streets, and even large important buildings (such as civic or religious buildings) (e.g., Blumenfeld 1953; Doxiadis 1975; Lynch 1971; Spreiregen 1965). All of these are suggestive.

One can also begin with perceptual data. The human field of view normally has a 180° angle of peripheral vision horizontally and 150° vertically, with a clear field of vision 27° high and 45° wide, although, as we have already seen, these angles decrease as speed increases (Lynch 1971; Pollock 1972; Tunnard and Pushkarev 1963). Figures can be derived for what can be seen and recognized at various distances, although these also change with speed. For pedestrians, with 20/20 vision, under normal lighting conditions an angle of at least one minute must be subtended. Thus, one can see 3.5 in. at 1,000 ft, whereas at 465 ft an object ½ in. can be seen, that is, facial features. Thus, at approximately 4,000 ft a human figure can be detected, at 400–500 ft one can tell whether it is a man or woman and discern gestures, at 75–80 ft a person can be recognized, his or her face becomes clear at 45 ft, and one can feel in direct social contact within the proxemic range (3–10 ft). Thus static pedestrian outdoor spaces are excessively close at 3–10 ft, intimate at 40 ft, and still at "human scale" at 80 ft. Most successful urban squares of the past rarely exceed 450 ft (Hoskin 1968; Lynch 1971). These dimensions also influence whether tall buildings are seen as landmarks, or small parts and details of them are perceived. This depends on the relationship between height and distance or degree of enclosure: An object the major dimension of which is the same as its distance from the eye cannot really be seen as a whole and details dominate; when it is twice as far it can be seen clearly as a whole; when three times as far, it is still dominant but seen in relation to other objects, whereas at distances exceeding that it becomes part of the general scene (Lynch 1971). Data like these can be used to test rules of thumb such that enclosure is most comfortable when the enclosing elements are ½–⅓ as high as the width of the space, that below ¼ there is not enclosure, whereas over ½ the space becomes a trench or pit (Sitté 1965). All these rules, however, ignore the nature of the elements, the role of kinesthetics and other sense modalities, the dynamic nature of perception, and observer characteristics. They have also never been tested cross-culturally or using other forms of diverse evidence. Another complicating factor is that both height and distance are subjectively estimated and depend partly on the size of the space itself. The lighting conditions are also related to the various ways in which depth and spatial extension are estimated—relative size, overlapping of objects, parallax of stereoscopic vision, motion parallax, height above horizon, textural gradient, shaping, linear perspective (which involves estimates of absolute size), and atmospherics (Gibson 1950).

Thiel (1970, p. 601) suggests that 40 ft is an upper limit for seeing facial expressions, 80 ft for facial recognition, 400 ft for seeing people, and 450 ft for discerning action. This corresponds with the data in Hall (1966, p. 119) and elsewhere that general vision subtends an angle of 60°, clear vision one of 12°, and detailed vision 1°. Assuming that a tall, well-built male with shoes and clothes on is 6 ft tall and 2 ft wide (Fruin 1971), 300 ft becomes critical for the 1° cone of detailed vision. It should be emphasized that in most preindustrial cultures, people tended to be significantly shorter than this 6-foot figure; moreover, not all people are adult males. Also, in many, if not most, preindustrial cultures various indicators of group identity such as hats, clothing, hairstyles, and facial markings were very important (e.g., Lofland 1973; Moore 1983; Wobst 1977; among others). It follows that even shorter distances become appropriate.

Ciolek (1977, 1978a,b,c) points out that interaction (like perception) occurs through all senses. The radius for hearing is 100 ft (which is also the limit for good vision within the 12° cone), for smell it is 30 ft, for touch via "tools" is 10 ft, and for direct touch 3 ft (cf. Hall 1966). There are thus five interlocking zones decreasing by a factor of 3, and it is suggested that this agrees well with the neurophysiology of the human sensory apparatus, as well as his analysis of the actual spatial interaction of pedestrians. Thus, this space within which people interact is limited, subdivided, and structured. These figures for static behavior tend to reinforce those for dynamic behavior and prescriptions for urban spaces.* The convergence of these various lines of argument tends to give one more confidence. Also, in all cases these figures are totally different from those appropriate for motorists.

There is another set of possible reasons why smaller dimensions should characterize pedestrian spaces in addition to the fact that actual walking distances are shorter and that such spaces tend to be perceptually more complex at walking speeds than large, open spaces. These have to do with the experience of time. On the one hand, there are suggestions that interest influences subjective time experience, so that in complex, interesting settings, time seems to pass more quickly and walking is easier (Rapoport 1977, pp. 44, 169, 173–174). We have also briefly discussed the opposite suggestion. A more recent suggestion is that subjective time experience is related to scale (e.g., DeLong 1981, 1983a,b, 1985; DeLong, Greenberg, and Keaney, 1986). There seems to be evidence that small spaces (in this case in model form) lead to subjective

---

*Note that, as pointed out earlier, there are many more data, prescriptions or rules of thumb on *static*, rest spaces than for pedestrian streets, which have largely been neglected.

time lapses proportional to their scale. Although this has not been studied in full-scale small spaces as far as I know, it is at least possible that pedestrian streets as sequences of small spaces may shorten one's subjective experience of elapsed time, making for easier walking, particularly when this phenomenon interacts with the effects of complexity.

THE SPECIFIC HYPOTHESES

Combining the results of the discussion so far in this chapter and partly in Chapter 6, it follows that, ideally, a particular set of characteristics should characterize vernacular pedestrian spaces in preindustrial cultures; these, then, provide the precedents for designing appropriate pedestrian settings. These are the variables that should be studied if the information were available. Note again that whereas noticeable differences occur in all sensory modalities (smells, sounds, kinesthetics, temperature, air movement, touch, etc.), this study only deals with *vision*. This is partly because of the reliance on indirect data, rather than empirical fieldwork, and because it is difficult to measure, record, and discuss other sensory modalities. It is mainly due to the fact that for archaeological or historical examples these are not available. They can, however, be derived through controlled analogy; for example, research in relevant contemporary settings (e.g., Rapoport 1987a), early travel descriptions, and so on. They can also possibly be inferred from the archaeological record (e.g., Sanders 1985). Note also that whereas complexity is due to a considerable extent to semifixed and nonfixed features (awnings, displays, etc., and people and their activities) as well as fixed-feature elements, *this study deals almost exclusively with the latter*. Moreover, temporal aspects are neglected, that is, complexity over time. For example, in various places, streets may be closed to traffic at night (or other times), semifixed elements such as display stands, dining stands, and the like introduced that narrow the space and produce great complexity. The overall complexity becomes even greater because, in effect, two different settings are experienced in the same street at two different times. This is, of course, the typical situation with outdoor markets, which transform urban (or rural) spaces in this way. Also underemphasized is the important role of complexity at the scale of *areas*. This involves characteristics such as a large number of possible paths, large numbers of choice points, many hidden views hinting at further spaces, a large variety of contrasting spaces beyond the street itself, leading to many possible varied sequences, and so on. Although it is this complexity at the level of areas that is important for the pedestrian use of urban areas more generally, this first study deals largely with only one of its components—the street itself. It is meant to illustrate the argument about history. The approach, the method, and the data lend themselves, however, to further studies at larger scales. Hypotheses can be evaluated at that scale also (e.g., Rapoport 1977) and can provide precedents, based on large bodies of evidence, of lessons for planning and design. Note that at the area level, that is, in connection with the overall urban fabric and even at the level of the street, there is an alternative hypothesis based on a different set of concepts, body of knowledge, and method. This has to do with the social effects (and purposes) of such spaces (Hillier and Hanson 1984). Note, however, that even if correct, that hypothesis is not at odds with the present hypothesis but complementary. Moreover, the alternative view would also benefit from being tested against a broader and more varied body of evidence: It is also a candidate for the type of approach advocated in this book.

The specific characteristics that follow fit into a hypothesis of the form: "For activity x (walking), settings with characteristics a,b,c,d. . . .n, will be supportive." These are the key and essential attributes of dynamic pedestrian settings that one would expect to find widely shared by examples drawn from many cultures and over a long time span, principally from vernacular environments in preindustrial cultures. In terms of our previous discussion they form a hypothetical repertoire of visual fixed-feature elements for achieving complexity in pedestrian streets.

A) Likely to have high
   levels of enclosure

    (i) Enclosing
       elements likely to
       be tall

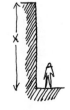

    (ii) Vertical
       width/height
       ratio

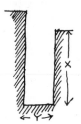

    (iii) Low percentage
       of sky visible

B) Likely to be narrow   (i) Relatively low
                 width

C) Likely to have
   complex spaces, i.e.,
   many potential
   noticeable

differences (sudden
changes, irregular
rhythms, transitions
of various sorts)   (i) Variation in
                             width, hence
                             variation among
                             minimum,
                             maximum, and
                             average width

                       (ii) Many turns and
                            twists per unit
                            length within a
                            given space

                      (iii) Articulation of
                            space—hence
                            space made up of
                            a sequence of
                            subspaces

                       (iv) High contrast
                            among these
                            spaces and in
                            those sequences

(v) Presence of major projecting elements (buildings, trees, doorways, etc.)

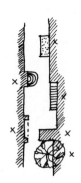

(vi) Large number of projecting elements per unit length

D) Likely to have short or blocked views

(i) Short subspaces

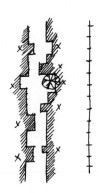

(ii) Limited length of views, hence division into segments, defined by horizontal blocking or by use of angles or overlapping planes

(iii) Use of level
    changes to block
    views vertically

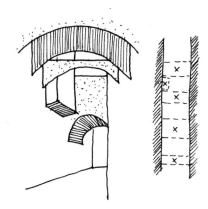

(iv) Use of overhead
     elements

(v) Use of bends,
    curves, and
    angles

(vi) Use of cross-
      streets

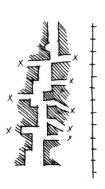

E) Likely to have
highly articulated
surfaces of enclosing
elements
(1) Side planes   (i) Large number of
elements or units
per unit length,
hence fine grain
of enclosing
surfaces (small
module,
variegated
treatment,
irregular
setbacks, etc.)

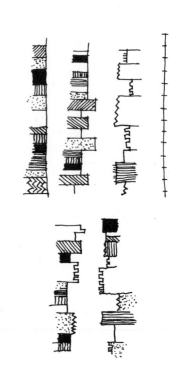

(ii) High overall
visual texture of
enclosing
surfaces

(iii) Rich treatment of
each individual
unit, hence rich
details, cornices,
steps, porches,
doorways,
balconies,
windows, and
other projecting
or three-
dimensional
elements

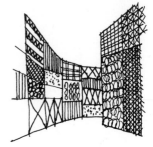

(iv) Use of highly
textured
materials

(v) Use of a variety
of materials

(vi) Use of different
colors

(vii) Use of irregular
rhythms

(viii) Use of sudden
and/or abrupt
changes

(2) Underfoot plane

(i) Use of highly
textured materials
compatible with
walking (or to
indicate
nonwalking
areas)

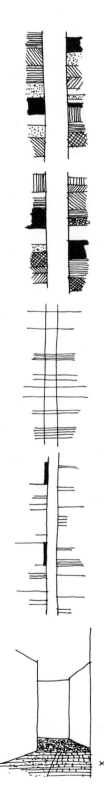

(ii) Use of a variety
of textures and
materials

(iii) Changes in
level—ramps,
steps, slopes,
vertical curves

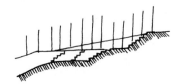

(iv) Changes in light
and shade

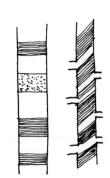

(3) Overhead plane

(i) Presence of
projecting
elements
overhead—roof
overhangs,
awnings, arches
and bridging
passages over
street; balconies,
etc.

(ii) Large number of
overhead
elements per unit
length

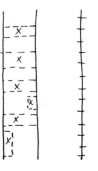

(iii) Complex and
intricate roof
lines, chimneys,
etc.

F) High complexity at
the *area* level

(i) Large number of
possible paths

(ii) Large number of
choice points

(iii) Indirect views
hinting at further
spaces (streets,
courts, plazas,
etc.)

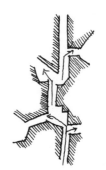

(iv) Sequences of
different spaces at
the area level

(v) High contrast
among spaces at
area level

Should these or some of them, in fact, be found in the examples to be considered, they would then become the precedents that might guide the design of pedestrian streets and other pedestrian movement systems.

We would not, of course, expect to find *all* these attributes in each place: Cultural variables will still play a role; so will climate and various constraints that may distort the "ideal type." For example, in some cases we will find grids (as in Latin America). The question then becomes whether the grid itself tends to show as many as possible of the attributes as it *can* show. Also, it is more than likely that these characteristics constitute a polythetic set (Clarke 1978; Rapoport 1982b, 1988b, in press; Sokal 1977, p. 190). This not only means that one looks for multiple characteristics rather than single ones, which are more commonly used in traditional architectural history. The distinction between monothetic and polythetic sets is that the latter assumes that one gets overlapping classes with fuzzy boundaries. In monothetic classifications, the classes established differ by at least one property that is uniform among all members of each class. Polythetic taxa are groups of things that share a large proportion of their property but do not necessarily agree on any one property. No single feature or even subset is peculiar to the class—it is the *particular conjunction of features* that defines it. One thus moves away from ideal type definitions based on a single characteristic, or a small set of characteristics deemed to be both necessary and sufficient to define the type. Rather the definition (or, in this case, identification) is based on a significantly larger number of characteristics. Also, *it is not necessary that every member of the group possess all qualifying attributes*. Rather, one finds a range of variations within defined limits* such that each member of the type (settings for walking in this case) possesses a large number of the attributes or characteristics and each attribute is shared by large numbers of members of the type. No single attribute (or small set of attributes) is both sufficient and necessary for group membership. Even when multiple attributes are used, there is only one form of monothetic group, but many varieties of polythetic groupings according to various criteria (Clarke 1978, pp. 36–37).

Finally, given the opportunistic sample used and the lack of concern by other researchers with the variables that I am emphasizing, it is also highly unlikely that we will be able to obtain all, or even most, of the information at this stage.†

---

*In this case I am not defining these limits; I am using the concept generically and loosely, although it can be made more rigorous.

†This was written before the examples were analyzed. This, in fact, proved to be the case: Only very few of the hypothesized variables or attributes could be studied. The photographs and drawings included, however, suggest that they are present.

Chapter 8

The Evidence
The Sample and the Method

INTRODUCTION

The case study approaches its subject matter from two sides, an approach of great importance in research generally. On the one hand, it begins with pedestrian behavior and its relation to complexity, based on current research from EBS. This has been the subject matter of Chapters 6 and 7. On the other hand, it begins with the observation that most past vernacular environments seem to have a particular form. In that sense, it begins with what seems to be a pattern. It also seems, on the basis of personal intuitive feelings, anecdotal evidence, and the writing of certain design writers that this form of vernacular streets is highly supportive for pedestrians.

The purpose of the case study in starting from both ends is to link these three sets of data and thus eventually to show that they are related in the ways and via the mechanisms discussed. The case study thus takes the position emphasized in this book, that present-day data and data from the past must be used concurrently because they are mutually enlightening. In this specific case, the hypotheses were derived from present-day EBS research. They are tested against evidence consisting of a wide range of past pedestrian settings (streets). It will be seen that this study exhibits many of the characteristics described in Chapter 3 (and also subsequently found in Chapters 4 and 5), although as yet in a rather primitive form. It will also be seen that the lack of data to which I have referred means that only very few of the attributes suggested as important could be tested.*

---

*Not only are the data unavailable when using existing material (although they could be obtained through fieldwork in existing environments and simulations of archaeological remains), the exigencies of publishing make it impossible to include the 3,335 slides I have of relevant streets, make color reproduction impossible, and so on. There is, however, a form of informal testing possible: Any one of my readers could test the argument by walking through appropriate pedestrian streets or looking at the many thousands of photographs, in books (e.g., the eight volumes of Gutkind (various) magazines (e.g., the 100 years of *National Geographic*), and many other sources.

Two related questions are being studied. First, are the generalizations about the need for complexity generally, and in walking particularly, valid when tested against a broad sample of streets designed for such activity? Second, do the perceptual qualities derived from these generalizations account for the character of such streets? If the answers to both are "yes," then the attributes shown to be present can provide precedents for design, precedents of *principle* rather than by *copying,* which I take to be the only valid way of using precedent from historical (and cross-cultural) evidence. Because the point has often been made that almost anything can be proved by historical examples if those are chosen tendentiously, the selection of evidence needs to be made explicit (cf. Rapoport 1986c).

THE EVIDENCE

I have already argued that the particular approach to EBS makes a broad sample (in location, cross-culturally, and through time) *necessary.* At the same time, the essentially unchanged nature of walking makes it *easily* possible. The use of such broader evidence has proved useful in other cases; for example, in considering the nature of dwellings, the origins of built environments, the meaning of privacy, the equivalence of apparently diverse urban forms, the energy efficiency of settlements, how meaning is communicated by settings, and so on (Rapoport, 1969a, 1977, 1979a,b, 1980d, 1986a,b). However, its use becomes particularly straightforward in this case, both because the activity is what it is and also, as already pointed out, because we are considering settings at the microscale.

The decision to concentrate on physical perceptual variables, keeping cultural ones as constant as possible, and hence to concentrate on settings created when walking was the norm, defines the universe from which the evidence is selected. The material consists of streets in *preindustrial settlements,* that is, those where no major traffic other than walking was to be found and also where any such traffic was not greatly at odds with walking, either in terms of spatial requirements or speed. It is also obviously confined to settlements that had *those morphological units called "streets."** Both these concepts or terms require discussion.

Preindustrial is not an absolute term. Many portions of large contemporary metropolises are spatially creations of preindustrial societies if they survived nineteenth- and twentieth-century modifications. This is the case with Boston (Clay 1978), London (Adburgham 1979), and many European and other cities. Moreover, many settlements in developing countries were until recently, and in many cases still are, fully preindustrial in terms of the concerns of the study (see Figure 8.1).

It is also far from clear what is a street. That depends on how it is defined; for example, morphologically or as a setting for particular activities (Rapoport 1973a,

---

*For example, there are settlements, such as those of sub-Saharan Africa, the Maya realm, and elsewhere that have no streets, only paths; in Africa linking compounds or causeways linking dwelling clusters in the Maya area. One would, in general, expect exceptions to be found in such areas, lacking streets in the morphological sense. One would also expect exceptions, or weaker support, in planned areas such as Spanish Latin America and in postindustrial cities, such as those in the United States, Canada, Australia and so on.

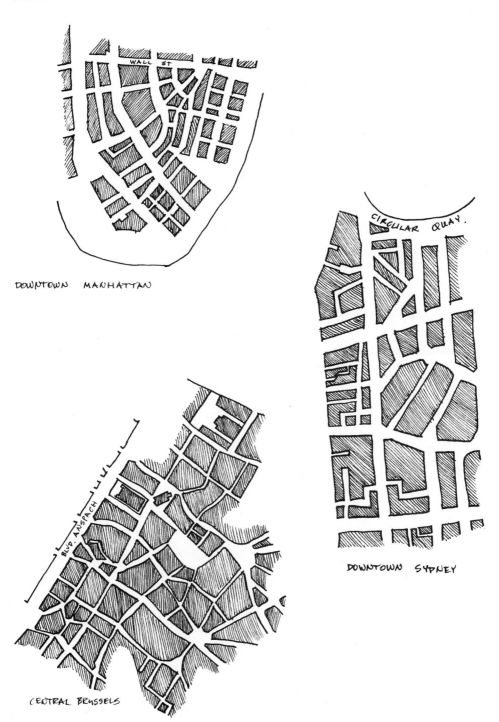

Figure 8.1. Traditional street patterns in contemporary cities. (Different scales; arcades etc. not shown. Many other examples—e.g., Geneva, Basel, Copenhagen, Stockholm, and Cambridge—are not drawn.)

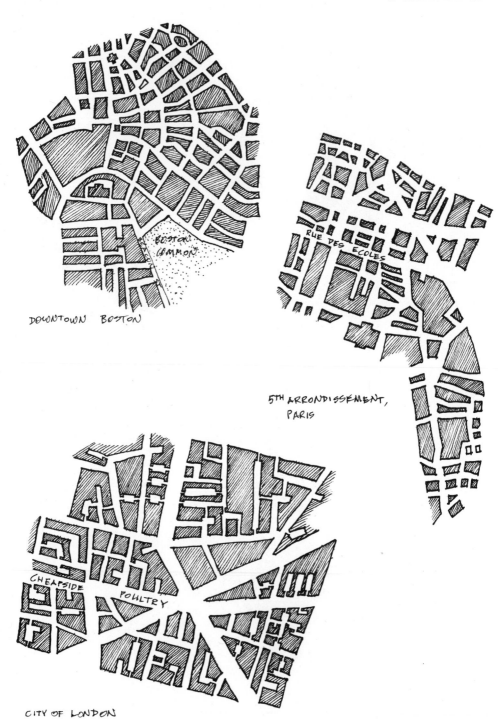

Figure 8.1. (*Continued*)

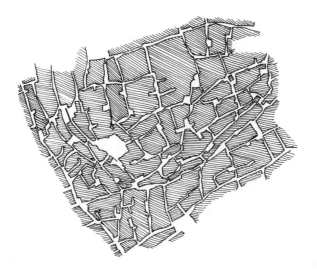

PART OF GRANADA (SPAIN) (DISTRICTS OF LOS AXARES & DE LA CAURACHA)
BASED ON GUTKIND (VARIOUS) VOL 3, FIG. 465, p 460.

ACROPOLIS

THEATRE

PLAKA, ATHENS

Figure 8.1. (Continued)

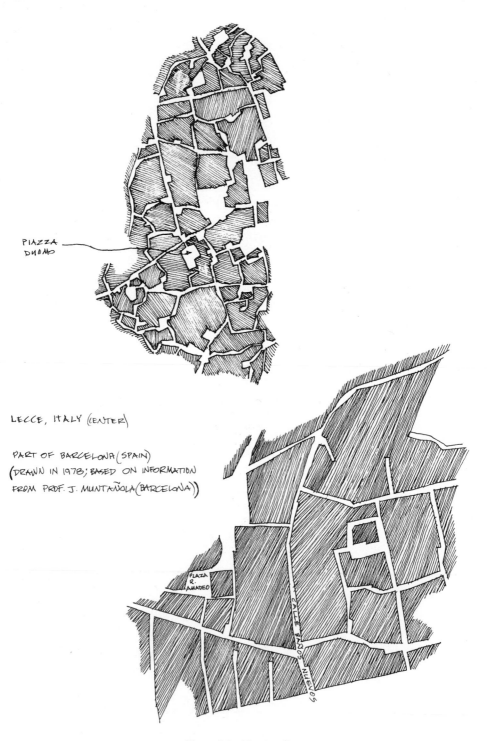

PIAZZA
DUOMO

LECCE, ITALY (CENTER)

PART OF BARCELONA (SPAIN)
(DRAWN IN 1978; BASED ON INFORMATION
FROM PROF. J. MUNTAÑOLA (BARCELONA))

PLAZA
R.
LLAMADED

CALLE BAÑOS NUEVOS

Figure 8.1. (*Continued*)

1977, 1980c). This would determine whether *street* is an adequate descriptive category or whether it should be defined in terms of a public–private continuum or in terms of use. In other words, is a morphological definition in terms of "a linear space between buildings" a useful category or would a definition in terms of a behavior setting be more useful, in which case a compound may become an analog. Other analogous functions may take place in shopping centers, restaurants, pubs, coffee shops, or even dwellings (Rapoport 1986a), and this would affect the discussion of the street as a behavior setting. Any such discussion would then have to include the cultural definition of rule systems for public behavior and also different "space splits," corresponding to what have been called cognitive domains in cognitive anthropology. It is this problem of what one could term functional nonequivalence and failure to relate environmental form to cultural norms and norms of the environment that is at the root of the weakness of otherwise insightful studies of the street (e.g., Rudofsky 1969).

In the present case, I assume that even though streets are difficult to define in ways that are valid cross-culturally, as are cities (Rapoport 1977), suburbs (Rapoport 1980b), or dwellings (Rapoport 1980c), there are such units that are comparable across cultures. I will use an *imposed etic* (e.g., Rapoport 1980a), even though this is less desirable than using a *derived etic*. The definition adopted is based on the observation that "street" is a widely observed, well-known, and accepted setting (Barker and Wright 1955; Barker and Barker 1961; Barker and Schoggen 1973 etc.). It is also the case that in contemporary urban design, that is, the situation to which any precedents will be applied, streets will be universally present. This definition is *morphological*. Thus, for the purposes of this study, *streets are the more or less narrow, linear spaces lined by buildings, fences, walls, and the like, found in settlements and used for circulation and frequently other activities* (Rapoport, 1989c). The concern is with *pedestrian streets,* the major question being *whether the characteristics hypothesized to be supportive of walking will be commonly found in such settings.*

The analysis is also restricted to the vernacular portions of settlements. This is for two reasons, already mentioned. First, in preindustrial settlements, such areas were much more settings for purely pedestrian movement than were elite areas. Second, spaces in elite areas more often reflect not their purpose as settings for everyday activities such as walking but purposes such as status, grandeur, symbolism, processions and parades, rituals, crowd control, and so on. They are thus less suitable than vernacular portions for the purpose of establishing attributes of pedestrian spaces.* Vernacular urban design is also the result of many decisions by many individuals and thus reflects shared schemata. The result is that such design expresses, in physical form, what people need and want. High-style spaces, on the other hand, often designed by individuals for individual patrons, may be highly idiosyncratic. They are also greatly influenced by design "theories" most of which have little or no relationship to human behavior. They also tend to change much more frequently and abruptly than do vernacular designs (Rapoport, in press). Moreover, the choice process in terms of which

---

*To give just one example: When the vernacular and planned parts in, say, Chan Chan (Peru) are compared, there is a clear distinction between the planned avenues among *ciudadelas* and streets within them as opposed to the circulation spaces within the small irregular groups of rooms (cf. Moseley and Mackey 1974). I take only the latter to be part of the sample.

design can be understood (Rapoport 1977) operates over long periods of time to arrive at forms. This means that they have been *selected* over long periods of time (Rapoport 1986c). Such forms, therefore, once again tend to reflect consensual needs and wants of large aggregates of people; they have become congruent with activities and are highly supportive of them (Rapoport, in press). If many such examples from many places and cultures, and over long periods of time, show the same, or similar, characteristics, one can be confident that these patterns are highly significant.

It is clear that just as today most streets are designed *for traffic,* in the past they were designed for pedestrians. Moreover, this was more likely in the case of smaller, residential streets than more major ones. This is seen clearly in the case of Iranian cities (Bonine 1979, 1980), in other Middle East cities, the boulevards of Paris versus the smaller side streets, Indian bazaar streets versus the streets in a residential cluster, and in many other cases.

In the case of major streets, other travel modes were used *in addition* to pedestrians—animals or carts. Generally animals are less in conflict with pedestrians than carts; hence areas without wheeled traffic, such as Nepal, Venice, and the like, are useful. The best documented are Middle East cities where donkeys, and particularly camels, were determinants for major streets (Bulliet 1975; Hakim 1986, Figure 2[a], p. 21). There the evidence is even more compelling because the camel actually *replaced* wheeled carts; with this shift, streets changed from *broad and right angled* to *narrow and twisty* (Bulliet 1975). This agrees with my point that the width and layout of streets in most settlements were influenced by the needs of wheeled transport (Bulliet 1975, pp. 224–228). These made for a minimum width of one or, better, two axles, plus clearance; corners could not be too sharp, dead-ends had to be avoided, and encroachments on the public way had to be prevented.* In the Middle East and North Africa, after the camel replaced the wheel, one finds narrow streets,† tight corners, a labyrinthine urban fabric, blind corners, and encroachments on the public way. Thus when the constraints of wheeled traffic were missing, streets evolved to meet more closely human needs because they only had to satisfy the needs of pedestrians and pack animals. This not only was better for pedestrians; it also suited cultural norms of privacy and seclusion of women (Wheatley 1976); helped make possible greater clustering by homogeneity and kinship; and also worked better climatically, in terms of shade, exposure, heat, wind, and the like (Rapoport 1986b). All these characteristics in turn further helped pedestrian movement and should thus most closely reflect the expected order, next to prehistoric settlements, Venice, Nepal, and so on. The perusal of the voluminous literature on Middle East and North African settlements supports this view (e.g., Gonen 1981, p. 18).

We have already seen the figures quoted by Hakim (1986). In the old city of Jerusalem, the "typical" street today is comparable, 15 ft wide. The monumental Cardo (Byzantine, but possibly coinciding with the Roman) is 75 ft wide. On the other hand, among preindustrial settlements, those showing least correspondence with the expected characteristics, should be planned settlements. This is the case in the medieval Bastides as compared to other medieval towns and cities. Another instructive example

---

*Note, however, that one would expect these constraints to be weakest in residential areas.
†These range from approximately 3.5 meters for through streets (allowing two fully loaded camels to pass) to just under 2 meters for residential culs-de-sac (Hakim 1986, pp. 20–21).

is Spanish Latin America, where most cities were designed following the Laws of the Indies of July 3, 1573.* Under these laws, streets were to be 14 *varas* wide—approximately 40 ft (39.75 ft) or 12 m (11.9 m) and in a grid pattern. Due to inexact surveying, the widths vary slightly. Also, streets were to be wider in cold than in hot areas, that is, they varied with climate. In general, greater width was encouraged, and transportation was a consideration. At the same time, the building line was strictly regulated to prevent encroachment (e.g., Nelson 1963, pp. 78–79). Note that in those settlements not developed according to the Laws, the expected patterns do seem to be present, for example three mining towns in Mexico—Zacatecas, Taxco, and Guanujuato (Hartung 1968). Even in grid-pattern Spanish Latin America, one would still expect residential, "vernacular" and older preindustrial areas to show the expected characteristics to the maximum extent possible and more than more high-style (e.g., monumental) areas or, specially, newer areas postdating the widespread use of carriages and, then, motor transport—cars, buses, and trucks.

The material is thus gathered from any and all places and periods that meet these relatively simple criteria—pedestrian movement being the only, or principal, travel mode.

THE SAMPLE

A major problem encountered, given the decision to use "indirect" methods, that is, to work with secondary sources, is the difficulty of finding material for places where they would be expected to be common. It is indeed striking in reviewing all possibly relevant sources on Venice to note *the absence of material on streets* (e.g., Briganti 1970, shows overall views or plazas; Lasig *et al.* 1967, monumental streets and plazas; Perocco and Salvadori 1975, plazas and canals).† In general, studies of streets in cities have been neglected; more specifically, there has been an almost complete neglect of vernacular streets. In those cases where such streets are shown, it is usually in the form of photographs; in the rare cases where plans are given, scales are often absent (e.g., Morris 1979), which makes it difficult to test the hypotheses. In fact, it will be seen that *most of the hypotheses could not be tested* due to lack of data, although just looking at the visual material included, or the innumerable other examples that could have been included,‡ for example, the hundreds of published photographs and hundreds, if not

---

*These laws are published in full in the *Hispanic American Historical Review,* Vol. 5, 1972, pp. 249–254. Note that the view that most cities were designed following these laws, while extremely common, is not universally accepted (see discussion in Nelson 1963, p. 79). These streets still show many of the expected attributes. For example, in Bogotà (Colombia), main streets are 35 ft wide, secondary ones 25 ft; building heights are between 25 and 54 ft; textures are rich and blocks short (Salcedo Salcedo, personal communication). In Guadalajara (Mexico) (excluding avenues), streets are 40 ft and 33 ft, buildings between 25 and 40 ft high, and block lengths vary between 185 and 295 ft (Hartung, personal communication). In Jaramijo (Ecuador) widths are 39, 33, and 26 ft, building heights between 10 and 20 ft, and blocks between 112 and 231 ft (Schavelzon 1981, p. 87–89).

†Even personal efforts through individuals proved unavailing: I would have had actually to go to Venice to obtain material.

‡My research assistant in redrawing the plans, A. Gupte, and I only used a very small fraction of the raw data on streets. I have also not used many data on arcades and, especially, on *markets*, which also show the same attributes.

thousands, of slides and sketches that I have made, would suggest that most of the characteristics do seem to be widely present. The comment on the difficulty of finding material and the excessively large number of examples seem contradictory. They are not. For certain places, there are vast numbers of examples, so that one can find hundreds in single books (e.g., Coppa 1968; Lavedan and Hugueney 1974; Nijst *et al.* 1973; the Gutkind series on city development). More importantly, *most of the examples that can be found do not contain the type of information required.*

Ideally one would want a sample of all areas, all periods, and all cultures, possibly even randomly selected. There is, in fact, a large technical literature on sampling for cross-cultural analysis (e.g., Brislin *et al.* 1973; Chaney and Revilla 1969; Köben 1969; McNett and Kirk 1968; Murdock and White 1969; Naroll 1970; Naroll and Michik 1980; Sudman 1976; among many others). However, given the uneven nature of the data, their lack of standardization and lack of full comparability across all attributes, rigorous sampling at this stage seems unnecessary and less than useful. One would also want detailed plans, sections, and perspectives/photographs of each street and urban area. Such data are generally not available and for certain areas (Latin America and particularly sub-Saharan Africa), they proved almost impossible to find in spite of much effort. Hence, the sample used is neither random, nor systematic or fully representative. Rather it is what one could call a "haphazard sample" or, more commonly, an opportunistic sample. In other words, it consists of accessible or published information, most of which became available in general reading for other purposes over a period of 15 years. This was supplemented by searches in books, journals, or magazines owned or in those libraries that were accessible either personally during extensive travels in many countries, or, for sources discovered or suspected, through interlibrary loans. A major effort involved contacting a number of people all over the world. Some did not respond, others responded by sending information; others referred me to other individuals; others yet to institutional sources or identified potential sources and references. (A list of these individuals is found in the acknowledgments.) All these leads were followed. Finally, I relied on my own slides as well as notes and sketches in travel notebooks. These, once again, were opportunistic in nature.

The constraints of book publishing and available help meant also that a relatively small number of the examples collected could be used. I estimate that only 10 to 15% of material collected appears in this study. For some cities or countries, two or three examples were used where dozens, or even hundreds, were available. Most plans of cities could produce thousands of examples; in the case of slides and travel sketches, my own and those obtained from other individuals or institutions, the percentage is much smaller still—most were not used at all.

In most cases, the full range of information was not available: I only had a plan, only a photograph, a plan without a scale, and so on. As a result, in many cases information for a single example had to be combined from several sources: a plan from one, a photo from another, paintings or drawings from a third, a verbal description from yet another. Verbal descriptions frequently are not accompanied by graphic material, plans, or the like. For example (in Qsar Es-Seghir in the twelfth through fifteenth centuries C.E.), "the streets were narrow, 2.5 m or less." They crossed at right angles but "made many short turns, so that the street plan was nowhere completely linear . . .

narrow, twisting alleyways" (Redman 1978b, pp. 15–16).* Sometimes dimensions had to be obtained from a known datum—a building, a human figure, a bicycle, an animal, and the like.

In view of all these considerations, the apparent "weakness" of the lack of systematic or rigorous sampling does not seem critical. The importance of being able to use an extremely broad set of historical and cross-cultural examples, however inadequately, outweighs this lack, as does the analysis of these examples in terms of a specific set of attributes derived from EBS theory. This is particularly the case given that this is a first and illustrative example: *It does not illustrate the type of historical study for which I argue in anything like its full development.*

THE METHOD

Again, a set of "apologies" for inadequacy follow.

The ideal would clearly have been to analyze all the examples in terms of all the variables listed at the end of Chapter 7. Given a more rigorous sample, any regularities could then have even been tested quantitatively. It should be emphasized that this study is largely qualitative and is not based on quantitative, statistical analysis; in any case, the quality of the data precludes this.

At the same time the use of some quantitative data and the approach taken makes possible a more statistical analysis should that be the goal. That would require fieldwork because, as shown, few of the published sources provide the information required. Two points, however, should be borne in mind. First, the use of such existing sources will remain essential for those examples from the past that are no longer in existence. Yet the use of such data is essential. In fact, the method used is a result of that decision, and one of the objectives of the study was to test the use of "indirect methods" using secondary sources. Second, even if funding for fieldwork were available, the amount of travel and time to collect fully adequate data for a sufficiently large and broad sample would be unrealistically high. Hence, once again, the potential utility of the present approach, particularly if it is meant to be generalizable to all kinds of material bearing on the use of historical data in EBS.

Given the nature of the published information, many of the attributes could either not be obtained, or only estimated. For example, the length of views requires dimensional plans. I am told that it is *theoretically* possible to estimate it from photographs or drawings using perspective methods, although its accuracy is likely to be low. Moreover, one then needs to know the characteristics of the lens used, the height of viewpoint, the type of perspective projection in a drawing, and so on. This is also the case with my own slides and sketches. This information is usually not available.

The number of turns, route structure, and similar characteristics can only be obtained from appropriate plans. Although grain can be estimated from appropriate

---

*I have collected many such verbal descriptions of vernacular streets in preindustrial settlements. I had hoped to content-analyze it as support for the analysis of visual material. Because the book is already too long, this will not be done here; I hope to publish this separately.

plans, or from photographs or drawings, it is difficult to measure this. One could count the number of distinguishable units, that is, vertical divisions, in a given view. However, this attribute like many others, has not been operationalized and, until this is done, only qualitative judgments are possible.

This is even more the case for visual texture. This can only be obtained from photographs and, occasionally, drawings and is almost entirely a matter of subjective, qualitative judgment, for example, "high," "medium," or "low." These are not terribly useful. Moreover, in this as in other judgments and estimates made, there is a major methodological weakness that should be borne in mind. All of the judgments and estimates have been made solely by me. As a next step, once some measures for the various attributes are operationalized, the use of a panel of judges is clearly desirable. Variables such as percentage of sky seen could not be used because the framing or cropping of views was not standardized, nor was the type of lens known.

The most easily obtainable attribute, which can be derived from plans, drawings, and photographs, is the *width* of streets. It is, in fact, the one used in every case. The next easiest, used in many cases, is the *height of enclosing elements*. Combined these give the *enclosure*.* Width, in the case of plans, is often based on estimates when the scale is very small (e.g., for the hundreds of plans in the Gutkind series on urban development). For larger scale plans, very exact widths and even *variations in width* could be obtained. When no scales are given, width can be obtained from known dimensions: a historical building or a plaza on a different plan, a verbal description elsewhere, or judgments based on photographs or drawings. In that latter case, judgments of widths and of height are made on the basis of known dimensions: a doorway, bicycle, cat (dogs are too variable in size, although in most preindustrial societies and developing countries there is apparently a fairly "standardized" dog).† Most commonly, however, such judgments were made on the basis of human figures taking into consideration period, culture, gender, apparent age, and/or clothing indicating status, and so on (see Spier 1967; cf. Fruin 1971). For example, in that place, at that time a male, female, or child of that apparent group was likely to have been "x" feet tall. Although my estimates were only "validated" in a very few cases by the largely independent estimates by a research assistant, they proved accurate when compared with plans, that is, cross-checked internally when both types of data were available.

Note that photographs have been used commonly and successfully in ethological, nonverbal communication and other forms of research dealing with nonfixed feature elements. Their use in this case, dealing with fixed-feature elements, is likely to be more reliable: They provide more objective data. This minimizes the problem raised by Worth and Adair (1972)—that cognitive preconceptions may be built into the very photographs. That problem is more likely to exist in the case of drawings (e.g., Gombrich 1961), but few of these have been used; however, they are potentially most useful, in fact essential, as a source in early travel accounts, by topographic painters, and so on.

It can be suggested that much research on pedestrian behavior *per se* has been in

---

*An assistant, A. Gupte, began to develop some statistics for the width and height/width ratio—means and standard deviations. However, this also was given up for the moment.
†*Nova* program on dogs on public television in 1986.

the quantitative tradition. On the other hand, the study of pedestrian spaces, when done at all, has been of the "I like it—I do not like it" variety. This study tries to avoid both extremes, although I am far more sympathetic to the former. The use of rather simple measurements is significant, not only because they make the insights and hypotheses more objective; they also help to demonstrate that patterns and regularities in fact exist, in the relationship between certain attributes of settings and their use for pedestrian behavior. The fact that these measurements *can* be used makes possible later statistical analysis, comparisons with other kinds of spaces (high-style, for traffic), the study of periodicities (e.g., Richardson and Kroeber 1940). The use of such data also makes *replication* possible; this is even possible in this qualitative study because the attributes being used are clearly articulated and the methods used in the study, and the assumptions, and weaknesses are both *explicit* and related to other work (e.g., Gilfoil and Bowen 1980; Shafer 1969a,b; Wagner 1979, specially pp. 154–158 among others). This last study uses simple counts that establish the *presence* of elements, an estimate of their *amount*, and the co-presence of central elements (cf. Redman, 1978a). In a more qualitative way, this is largely my approach.

The essential step consists of identifying the relevant attributes (Redman 1978a) of pedestrian streets, that is, the basic observable components of these streets. Because there are a potentially limitless number of attributes, *attribute recognition and selection is the crucial step* (Redman 1978a, pp. 162–163). These were derived in Chapter 7 on theoretical grounds as well as informal observation—mine and of writers on design. The issue then becomes which are *inessential* attributes (these are not considered), *essential,* or *key.* One is then interested first in the *presence* of elements that constitute the key or essential attributes in the sample considered. One is also interested in the quantity or amount of such elements, their co-presence, the absence of such elements in different spaces (Wagner 1979). This latter, for example, an analysis of settings for traffic, was done on theoretical grounds in Chapter 7; it will not be done in this study.

Thus, although the key attribute according to the hypothesis is *complexity,* this is made up of many elements and variables listed in the specific hypotheses. The purpose of the evidence that follows is to see whether these *essential attributes* are, in fact, present in the settings examined.

The analysis provides highly suggestive evidence that these attributes are present in the settings examined. With all the caveats expressed in this chapter, it seems that the hypotheses are strongly supported. If made more quantitative, the evidence could be plotted on graphs to show patterns of distribution. Other data available to readers— in books, journals, slides, in libraries, in the appropriate environments that are available to them, should further support my analysis. Their *absence* in other kinds of environments should provide further evidence. Thus the hypotheses can be tested further against new bodies of data and compared with the present analysis.

Note an interesting issue already briefly addressed that will *not* be further explored. This is: To what extent might one expect all the attributes to be present in all cases, and to what extent—or only some (and how many)? This is the issue of a monothetic versus a polythetic approach (Clarke 1978; Rapoport 1982b, 1988b, in press; Sokal and Sneath 1963; Sokal 1977). Also, specific historical contexts and conditions, and culturally specific variables still play a role; an example is the absence of streets in large parts of sub-Saharan Africa or in Maya settlements and other parts of precolumbian America, all of which were pedestrian. Even there, however, there were exceptions. One example is Cuzco, where the streets with Inca walls can still be seen and show many of the expected characteristics (see photographs). Another is Monte Negro, in the Zapotec/Mixtec area of Mexico, where there was an "apparent street," 100 m long and 4 to 6 m wide, running east and west, and lined on both sides by buildings (Flannery and Marcus 1983, p. 100, Figure 4.9). Others can be found in the literature, and in the evidence of the case study that follows. The role of constraints of various sorts are always present and may distort the ideal. Knowing the ideal is, however, most important in understanding existing settings and even identifying the constraints (see Figure 8.2).

The examples and analysis in this study can also be compared to others, of different periods and meant for other activities, for example traffic streets, roadside strips, high-style spaces; settings for static pedestrian behavior—plazas, markets, and the like. They could also be compared to those meant for pedestrian movement today, such as shopping centers and enclosed malls. In general, this is not done. Similarly, suburban areas, housing projects, or new urban areas will not be shown or analyzed, although the same methods could be applied and the various sets of examples compared and contrasted. It can be predicted that the characteristics found in preindustrial vernacular pedestrian streets will not be present in contemporary nonpedestrian spaces and only to a lesser extent in modern pedestrian settings; in fact, *the analysis is intended to provide precedents for improving the design of such modern spaces.* It can further be predicted that these characteristics will be present in pedestrian malls in older cities (Copenhagen, Munich, Rouen, Norwich, and so on), will be present to a lesser extent in those designed *de novo* (e.g., Rotterdam), and least present in pedestrian malls converted from traffic streets, such as in the United States. In these latter, one would still expect a tendency to choose those streets closest to the characteristics of the preindustrial vernacular examples in this study, for example, narrow rather than wide, with high

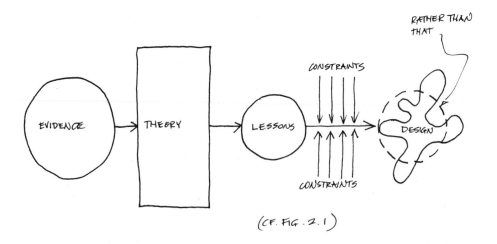

Figure 8.2. The role of constraints in design (from Rapoport 1983a).

enclosure. One could, for example, compare the streets so converted with all the streets in U.S. cities. Moreover, it can be predicted that the alterations made—in pavement, in semifixed elements such as kiosks, planters, displays, seats, and the like—will all tend in the direction of increasing complexity in terms of increasing the presence of the characteristics described in the hypotheses. Although none of these predictions will be tested in this study, the fact that they follow naturally from the approach used and can be tested using the same methodology is most significant: The ability to make testable predictions and suggestions for further work, and linking many kinds of settings with major generalizations from EBS are significant features of this approach to history and of this study, which illustrates it in a preliminary way.

Thus the material that follows consists of a sample of those examples that could be located and obtained. These are analyzed in terms of as many as possible of the attributes listed at the end of Chapter 7. These are in all cases width and, to a somewhat lesser extent, height (and hence enclosure). Also used, although less frequently, are grain, texture, and length of view. Other attributes are used less frequently still, as and when possible, or not used at all. Many of those are not explicitly analyzed but are shown to be present in the examples reproduced.

A selection of examples is reproduced—some single books have hundreds. Most have not been used. In each case the location and period (if known) are given. Note, finally, that only those sources where relevant and useful information was found are cited in the bibliography. The many hundreds of books, monographs, journals, and magazines that were checked but did not provide usable information are, in general, *not* cited.

The major general purpose of this study is to illustrate the argument of this book about the mutual relationship of EBS and history and about how history should be approached. It does that very inadequately and only to a limited extent; even so it is so different from the common architectural history that the point is made. The more specific but long-range objective is clearly to be able to derive useful precedents for design, that is, to specify the necessary spatial, perceptual, and other characteristics of pedestrian streets. The present study does not fully do even that—it has a more limited goal. This is to see whether a diverse sample of streets in largely pedestrian settlements from many periods, regions, and cultures will, in fact, possess those attributes to be expected on the basis of hypotheses derived from EBS.

I think the material that follows does show them to be present. The hypotheses thus receive preliminary support. As a result the set of characteristics becomes a first example of a *precedent for design*—a least generically: *the only type of precedent that can be validly sought and used.*

The Data

The evidence (or data) that follows consists of 137 pages of drawings—area plans, street plans, and street sections—and 92 photographs (selected from my collection of 3,335 slides of pedestrian streets). They represent a wide variety of places, totaling 192 (disregarding subareas and multiple examples from a single place) (see list of places), and they are from many cultures (see diagram showing geographical distribution). The earliest example, Çatal Hüyük, dates from the seventh millennium B.C.E.; the most recent—Louvain la Neuve—dates from ca. 1970 C.E. (see diagram showing temporal distribution).

Most of the drawings are considerably simplified: for example, plans in terms of grain, texture, details, and so on, and sections in terms of blocked views and so on. Also, very few, if any, of the examples are of "famous" streets.

LIST OF PLACES REPRESENTED
(In many cases there are multiple examples)

Acco (Israel)
Afyon (Turkey)
Aïn Leuh (Morocco)
Alalakh (Turkey)
Ampurias (Spain)
Amsterdam (The Netherlands)
Ankara (Turkey)
Antakya (Antioch) (Turkey)
Antalya (Turkey)
Antigua (Guatemala)
Apollonia (Greece)
Arberobello (Italy)
Arezzo (Italy)
Assur (Mesopotamia)

Athens (Greece)
Babylon (Babylonia)
Baghdad (Iraq)
Bangkok (Thailand)
Barcelona (Spain)
Basalú (Spain)
Basel (Switzerland)
Bergen (Norway)
Bhaktarpur (Nepal)
Bialystok (Poland)
Bogotá (Colombia)
Bukhara (Soviet Central Asia)
Bungmati (Nepal)
Burano (Italy)
Bursa (Turkey)

Büyükada (Turkey)
Cachoeira (Bahía, Brazil)
Cairo (Egypt)
Čakmakli-Depe (Southern Turkmenia, Soviet Central Asia)
Cambridge (England)
Canton (China)
Çatal Hüyük (Turkey)
Chan Chan (Peru)
Chester (England)
Chios (Greece)
Cisterino (Italy)
Copenhagen (Denmark)
Cordes (France)
Cuzco (Peru)

Dašlidži-Depe (Southern Turkmenia, Soviet Central Asia)
Daura (Nigeria)
Deir-El-Medina (Arabia)
Delft (The Netherlands)
Delhi (India)
Delos (Greece)
Diest (Belgium)
Dragør (Denmark)
Durham (England)
Džeitun (Southern Turkmenia, Soviet Central Asia)
Elsinore (Denmark)
Et Tûr (Israel)
Florián-Bogotá (Colombia)
Gedi (East Africa)
Geneva (Switzerland)
Geoksyur (Southern Turkmenia, Soviet Central Asia)
Gorée (Senegal)
Granada (Spain)
Guanujuato (Mexico)
Gunung Kawi (Indonesia)
Gurnia (Crete)
Hamā (Syria)
Hattusas (Bogazköy) (Turkey)
Herat (Afghanistan)
Hong Kong (China)
Hydra (Greece)
Ikhmindi (Nubia)
Isphahan (Iran)
Istanbul (Turkey)
Jaisalmer (India)
Jerusalem (Israel)
Kabul (Afghanistan)
Kairouan (Tunisia)
Kalibangan (Indus Valley)
Kam Tin (Hong Kong, New Territories)

Kano (Nigeria)
Katmandu (Nepal)
Katozakros (Crete)
Kharagpur (India)
Kommos (Crete)
Konya (Turkey)
Kota Gade (Indonesia)
Ksar Hallal (Tunisia)
Kula (Turkey)
Kuwait (Persian Gulf)
Lamu (Kenya)
Laon (France)
La Roquette (France)
La Spezia (Italy)
Lausanne (Switzerland)
Leuven (Belgium)
Lindos (Greece)
London (England)
Lothal (Indus Valley)
Louvain la Neuve (Belgium)
Lukang (Taiwan)
Luoyang (China)
Madura (Indonesia) (North Coast village)
Mampsis (Israel)
Marrakesh (Morocco)
Meinarti (Nubia)
Midnapur (India)
Mindanao (The Philippines) (village)
Misratah (Libya)
Mohenjo Daro (Pakistan)
Mugla (Turkey)
Murotsu (Japan)
Mykonos (Greece)
Nazareth (Israel)
New Herat (Afghanistan)
Nice (France)
Noci (Italy)
Nördlingen (Germany)
Nowa (Japan)
Olynthos (Greece)
Ouropreto (Brazil)
Oxford (England)
Pallokastro (Crete)

Palma de Mallorca (Spain)
Paratí (Brazil)
Phylakopi (Melos, Greece)
Poliochni (Lemnos, Asia Minor)
Priene (Greece)
Puno (Peru)
Quito (Equador)
Rabat (Morocco)
Ras Šamra (Ugarit)
Rio de Janeiro (Brazil)
Roquebrune (France)
Rosario (Argentina)
Safed (Israel)
Safranbolu (Turkey)
Saiwacho (Japan)
Salamanca (Spain)
Salisbury (England)
San Francisco el Alto (Guatemala)
San Juan (Puerto Rico)
Sankhu (Nepal)
São Paulo (Brazil)
Sekai (Ghana)
Selinas (Sicily)
Serra East (Egypt)
Seville (Spain)
Shaduppum (Tell Abu-Harmal, Iraq)
Shanghai (China)
Shustar (Iran)
Siena (Italy)
Singapore
Sirāf (Persian Gulf)
Solo (Indonesia)
Stockholm (Sweden)
Stone Town (Zanzibar, Tanzania)
Sumenep (Indonesia)
Surkotada (Indus Valley)
Taichung (Taiwan)
Takayama (Japan)
Tanimlama (Turkey)
Taxco (Mexico)

Taxila (India)
Teotihuacán (Mexico)
Tepe Gawra (Iraq)
Thebes (Western)(Egypt)
Thera (Santorini,
    Greece)
Tinerhir (Morocco)
Tlemcen (Algeria)
Tokyo (Japan)
Trondheim (Norway)
Tournus (France)
Troy (Turkey)
Tsinan (China)

Udaipur (India)
Unidentified town
    (Kutch, India)
Unidentified village
    (Punjab, India)
Unidentified town
    (Rajasthan, India)
Unidentified village
    (Nepal)
Ur (Mesopotamia)
Vence (France)
Villefranche sur Mer
    (France)

Viseu (Portugal)
Volterra (Italy)
Wad Lemeid (Sudan)
Wells (England)
Winchester (England)
Xian (China)
Yaffo (Israel)
Yeşil Ada (Turkey)
Yogyakarta (Indonesia)
Zacatecas (Mexico)
Zürich (Switzerland)

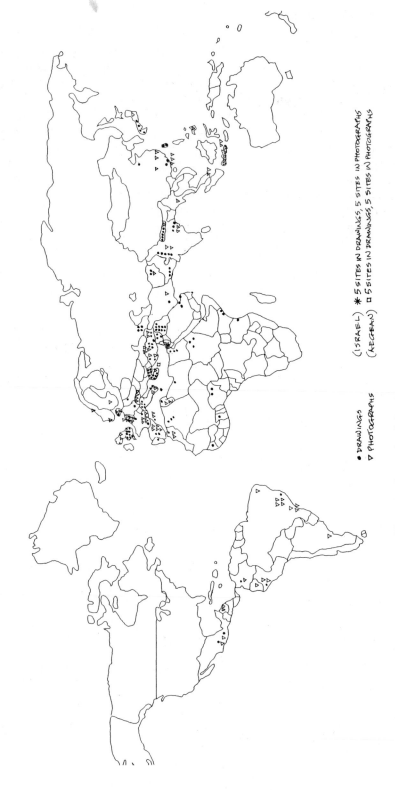

● DRAWINGS   (ISRAEL)   ✱ 5 SITES IN DRAWINGS, 5 SITES IN PHOTOGRAPHS
▽ PHOTOGRAPHS   (AEGEAN)   □ 5 SITES IN DRAWINGS, 5 SITES IN PHOTOGRAPHS

Figure A. Geographical distribution of evidence (sites).

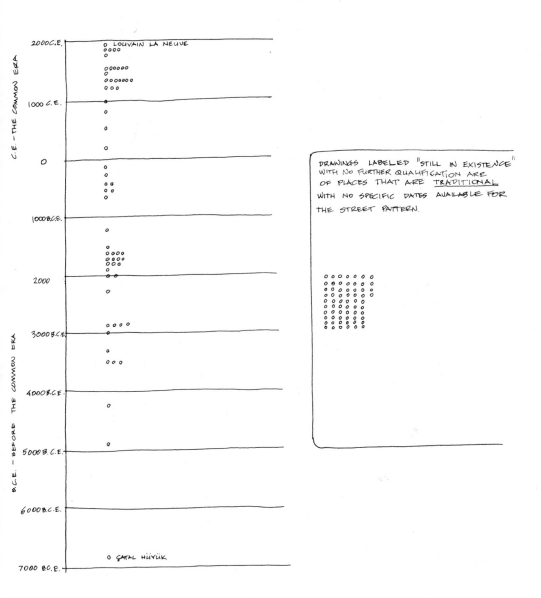

Figure B. Temporal distribution of evidence (sites) (*drawings only*).

Figure 1. Çatal Hüyük (Turkey) (Levels III, II, and I, 7th–6th millennium B.C.E.). Based on Coppa (1968), Fig. 73; Todd (1976); Mellaart (1967).

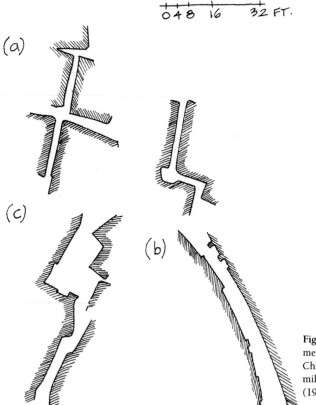

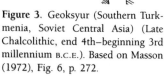

Figure 3. Geoksyur (Southern Turkmenia, Soviet Central Asia) (Late Chalcolithic, end 4th–beginning 3rd millennium B.C.E.). Based on Masson (1972), Fig. 6, p. 272.

Figure 2. (a) Džeitun (Southern Turkmenia, Soviet Central Asia) (Early Neolithic 5th millennium B.C.E.). Based on Masson (1972), Fig. 1, p. 266. (b) Čakmakli-Depe (Southern Turkmenia, Soviet Central Asia) (Anau IA period, 5th–4th millennium B.C.E.). Based on Masson (1972), Fig. 3, p. 269. (c) Dašlidži-Depe (Southern Turkmenia, Soviet Central Asia) (Chalcolithic, Namazgai culture, early 4th millennium B.C.E.). Based on Masson (1972), Fig. 4, p. 270.

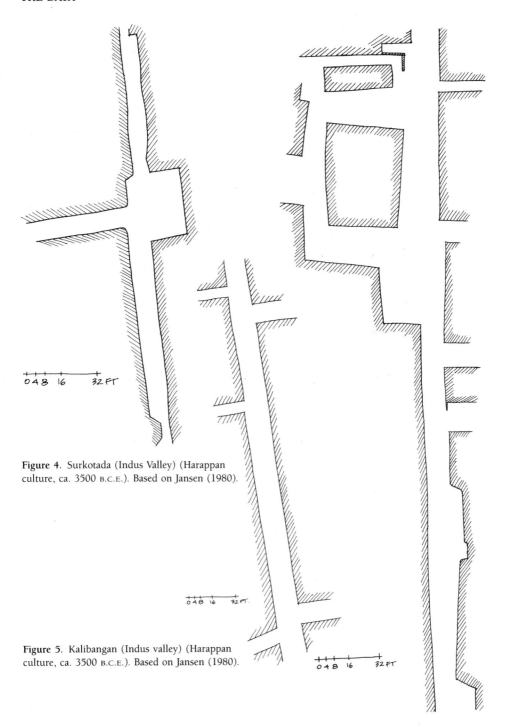

Figure 4. Surkotada (Indus Valley) (Harappan culture, ca. 3500 B.C.E.). Based on Jansen (1980).

Figure 5. Kalibangan (Indus valley) (Harappan culture, ca. 3500 B.C.E.). Based on Jansen (1980).

Figure 6. Lothal (Indus valley) (Harappan culture, ca. 3500 B.C.E.). Based on Jansen (1980), p. 15.

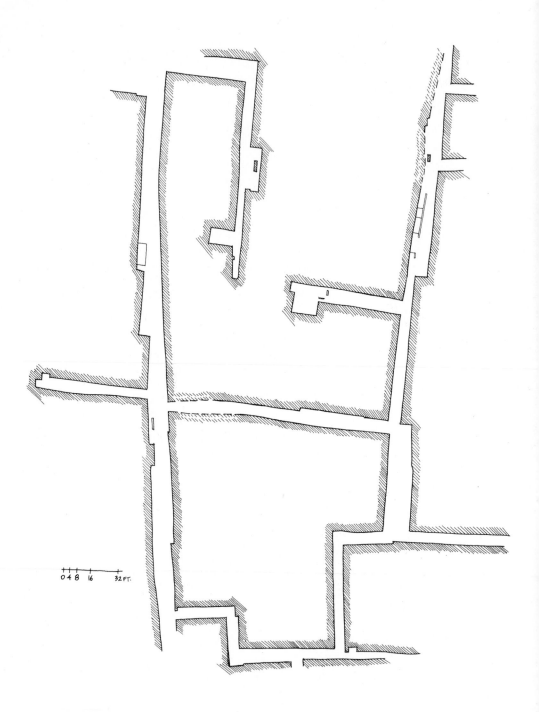

Figure 7. Mohenjo-Daro (Pakistan) (3500–3000 B.C.E.). Based on Piggott (1961), Fig. 21, p. 166.

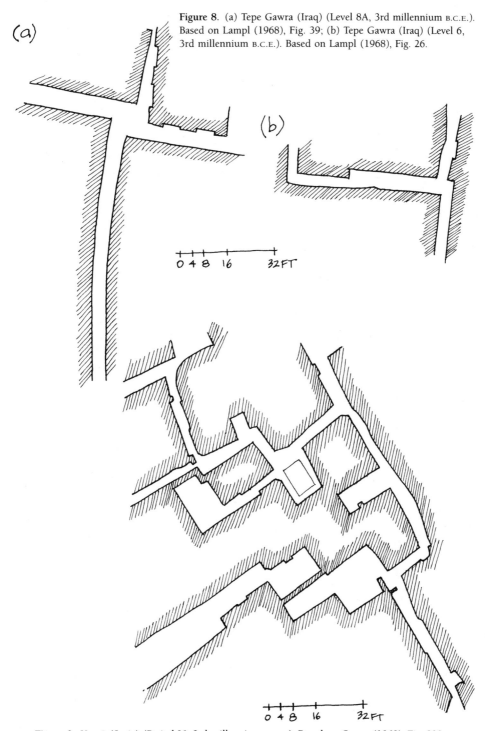

Figure 8. (a) Tepe Gawra (Iraq) (Level 8A, 3rd millennium B.C.E.). Based on Lampl (1968), Fig. 39; (b) Tepe Gawra (Iraq) (Level 6, 3rd millennium B.C.E.). Based on Lampl (1968), Fig. 26.

Figure 9. Hamā (Syria) (Period J6, 3rd millennium B.C.E.). Based on Coppa (1968), Fig. 238.

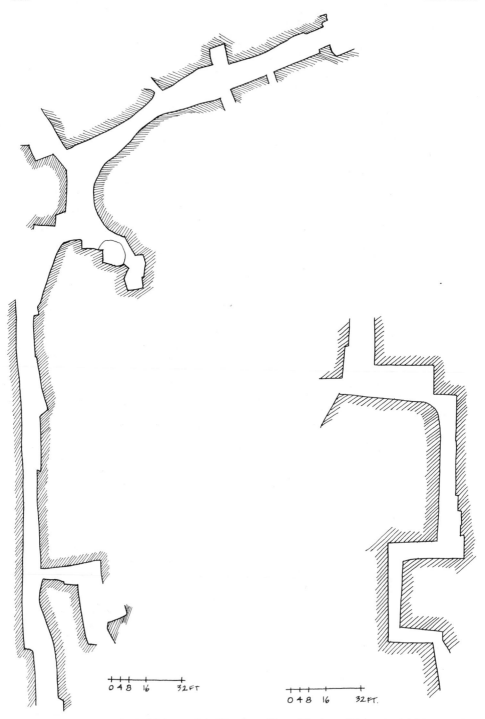

Figure 10. Poliochni (Island of Lemnos, Asia Minor)
(3rd millennium B.C.E.). Based on Coppa (1968),
Fig. 353, p. 391.

Figure 11. Assur (Mesopotamia) (ca.
3rd millennium B.C.E.). Based on Lampl (1968),
Fig. 57.

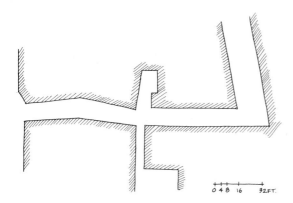

Figure 12. Troy (Asia Minor) (Level IIg, 2300–2200 B.C.E.). Based on Coppa (1968), Fig. 273, p. 319.

Figure 13. Serra East (Egypt) (ca. 2000 B.C.E.). Based on information provided by P. L. Shinnie (Calgary).

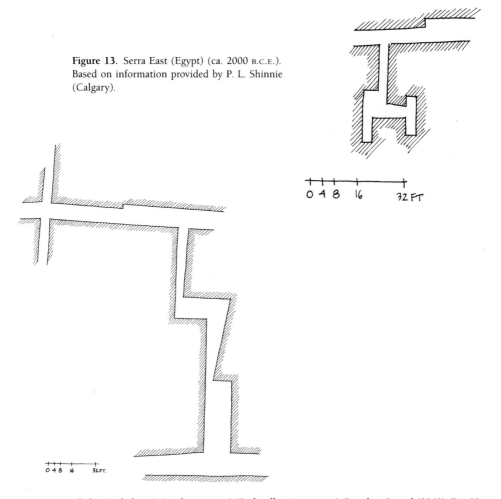

Figure 14. Babylon (Babylonia) (market quarter) (2nd millennium B.C.E.). Based on Lampl (1968), Fig. 58.

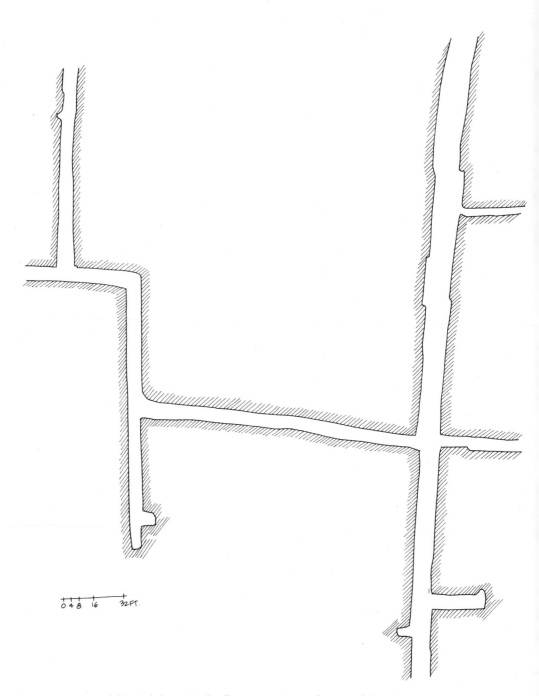

Figure 15. Babylon (Babylonia) (2nd millenium B.C.E.). Based on Lampl (1968), Fig. 58.

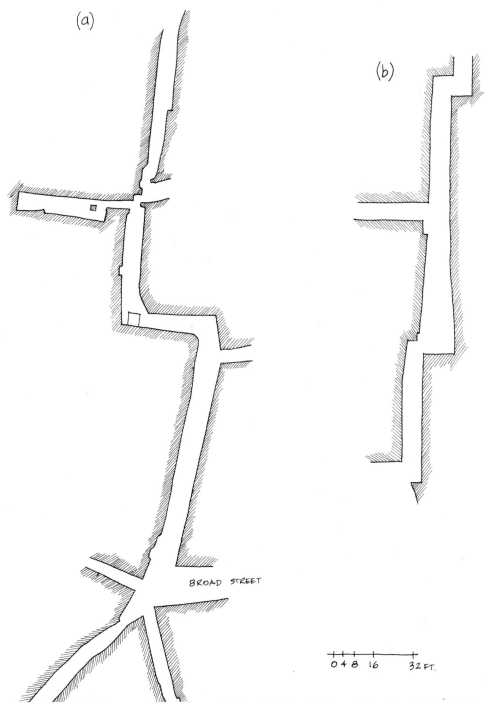

Figure 16. (a) Ur (Mesopotamia (eastern quarter) (Larsa Period, ca. 2000 B.C.E.). Based on Wooley (1955), Fig. 12, p. 176. (Note that a 14-ft-wide street was named *Broad* by archaeologists.); (b) Ur (Mesopotamia) (E. M. site) (Larsa Period, ca. 2000 B.C.E.). Based on Wooley (1955) and Lampl (1968).

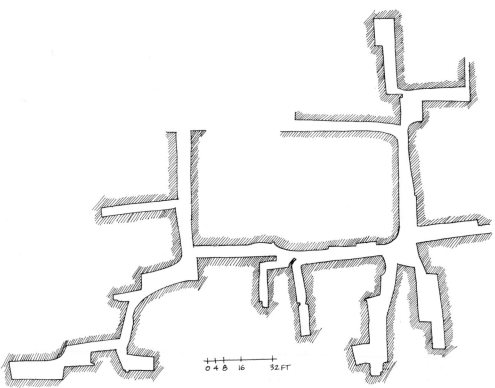

Figure 17. Western Thebes (Medīnet Habu) (Egypt) (second stage town, 12th Dynasty, Middle Kingdom, ca. 20th–18th cen. B.C.E.). Based on Kemp (1972b), Fig. 3, p. 665.

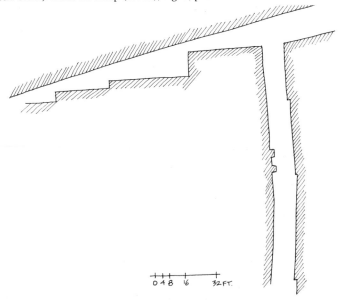

Figure 18. Alalakh (Turkey) (east of tell) (Levels 2 and 3, 2000–1600 B.C.E.). Based on Lampl (1968), Fig. 112.

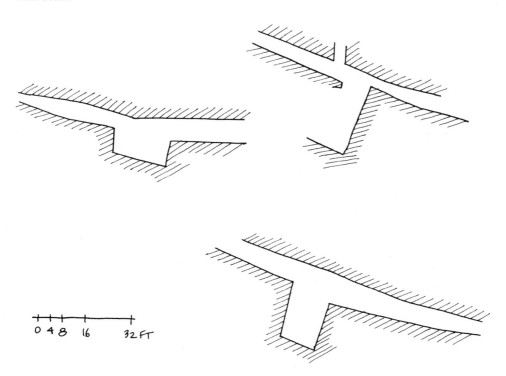

Figure 19. Phylakopi (Melos, Greece) (Levels II and III, ca. 2000–1800 B.C.E.). Based on Coppa (1968), Figs. 347 and 348, p. 387.

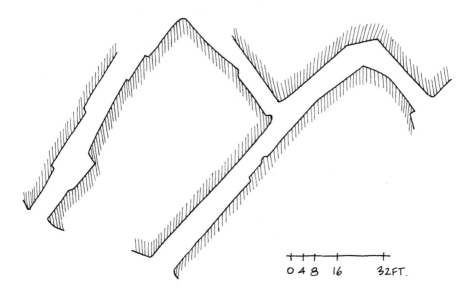

Figure 20. Hattusas (Bogazköy) (Turkey) (residential quarter) (Hittite; Level IV, ca. 1900–1700 B.C.E.). Based on Lampl (1968), Fig. 133.

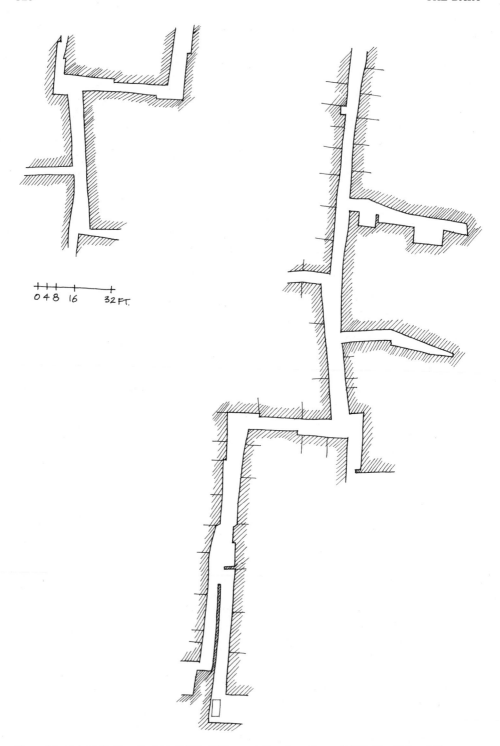

Figure 21. Deir-El Medina (Arabia) (2000–1785 B.C.E.). Based on Coppa (1968), Fig. 325, p. 361.

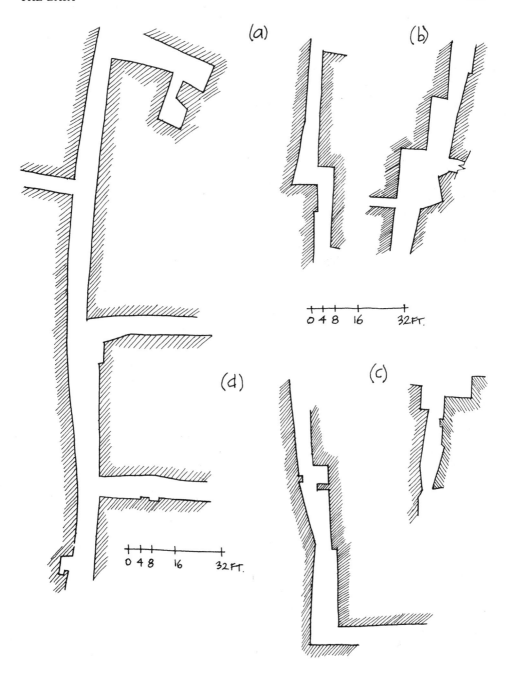

Figure 22. (*left*) Ras-Šamra (Ugarit) (18th cen. B.C.E.). Based on Coppa (1968), Fig. 264, p. 311.

Figure 23. (*right*) (a) Thera (Santorini) (Greece) (Telchines Street) (Minoan, 18th cen. B.C.E.). Based on Doumas (1983); Marinatos (1973). (b) Thera (Santorini) (Greece) (Minoan, ca. 18th–16th cen. B.C.E.). Based on Doumas (1983); (c) Kommos (Crete) (Minoan, ca. 18th–16th cen. B.C.E.). Based on Shaw *et al.* (1978), p. 116; (d) Katozakros (Crete) (Minoan, ca. 18th–16th cen. B.C.E.). Based on Coppa (1968), Fig. 396, p. 429.

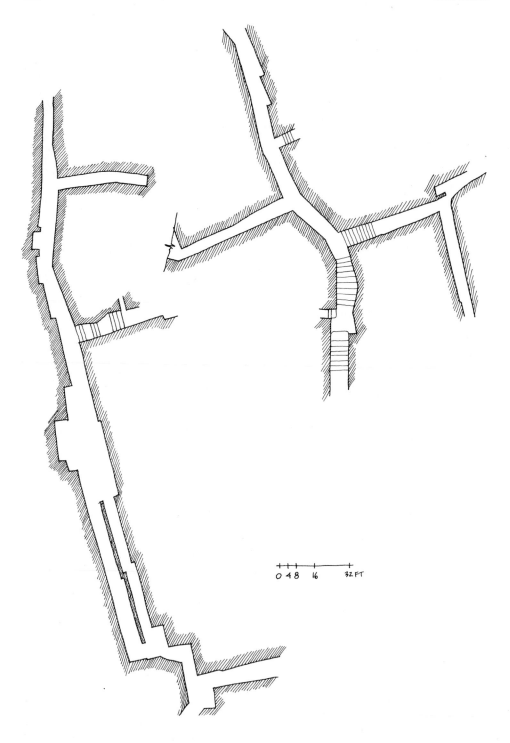

Figure 24. Gurnia (Crete) (Minoan, ca. 18th–16th cen. B.C.E.). Based on Coppa (1968), Fig. 401, p. 432.

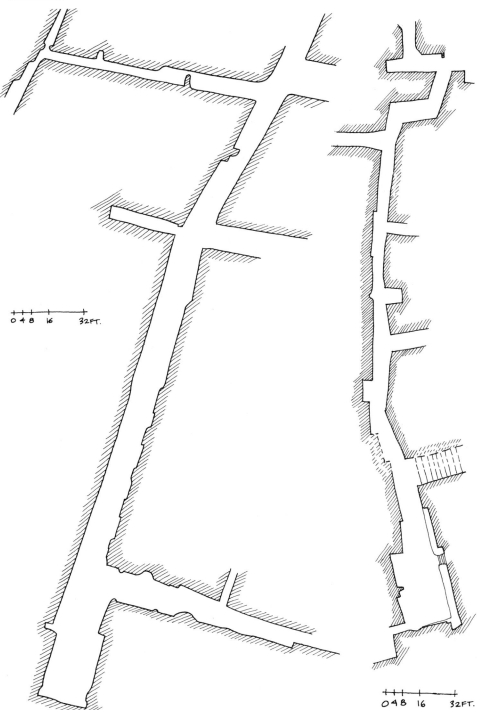

Figure 25. (*left*) Gurnia (Crete) (Minoan, 17th–16th cen. B.C.E.). Based on Hawkes (1974), p. 130.
Figure 26. (*right*) Shaduppum (Tell Abu-Harmal) (Iraq) (ca. 1800 B.C.E.). Based on Lampl (1968), Fig. 60.

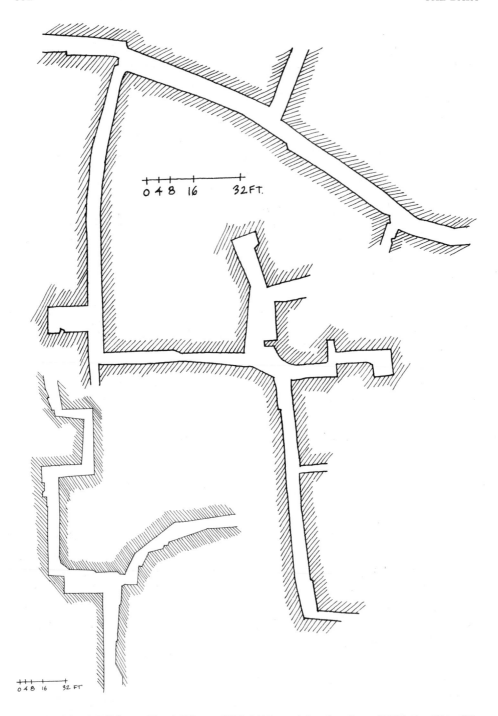

Figure 27. (*above*) Pallokastro (Crete) (Minoan, 1700–1400 B.C.E.). Based on Coppa (1968), Fig. 400, p. 431.
Figure 28. (*below*) Assur (Mesopotamia) (13th cen. B.C.E.). Based on Coppa (1968), Fig. 224, p. 278; Lampl (1968), Fig. 57.

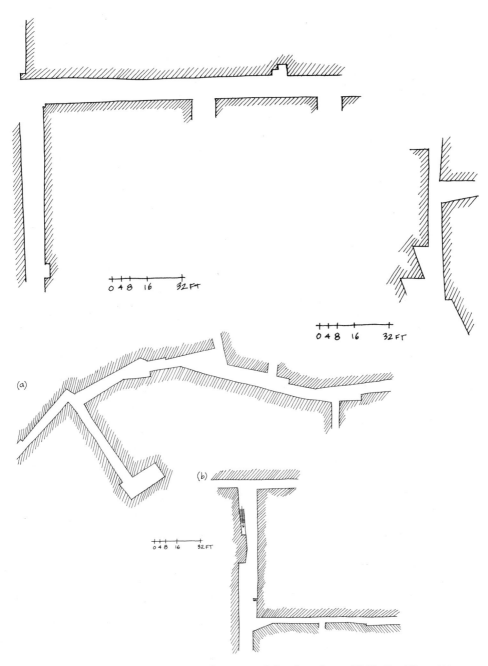

Figure 29. (*top*) Olynthos (Greece) (pre-5th cen. B.C.E.). Based on Coppa (1968), Fig. 953, p. 1098.

Figure 30. (*center*) Selinus (Sicily) (7th–6th cen. B.C.E.). Based on Lavas, in Doumanis and Oliver (1974), Fig. 4, p. 19.

Figure 31. (*bottom*) (a) Delos (Greece) (theater neighborhood) (7th–5th cen. B.C.E.). Based on Martin (1974), Fig. 49, p. 241. (b) Delos (Greece) (7th–5th cen. B.C.E.). Based on Doumanis and Oliver (1974), Fig. 5, p. 75.

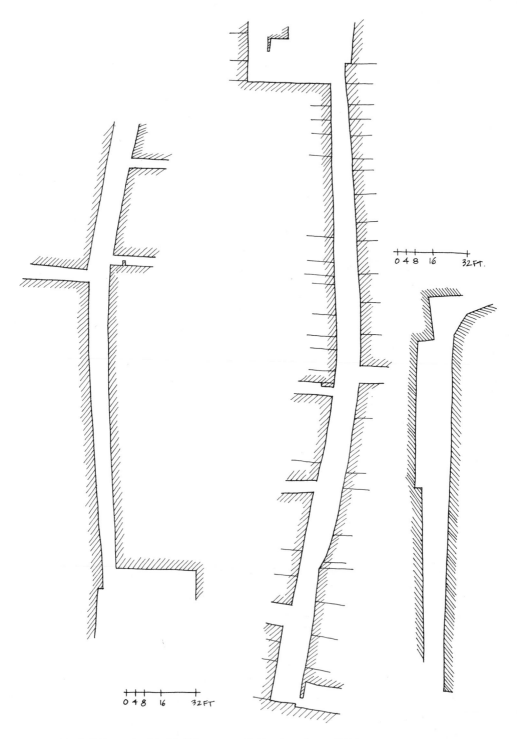

Figure 32. (*left*) Ampurias (Spain) (5th cen. B.C.E.). Based on Coppa (1968), Fig. 898, p. 1033.
Figure 33. (*right*) Athens (Greece) (classical, ca. 5th cen. B.C.E.). Based on Boyd (1981), Fig. 2, p. 144.

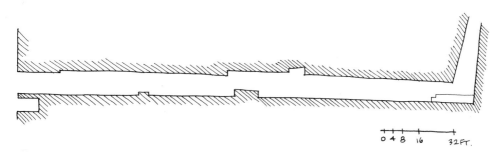

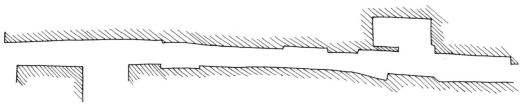

Figure 34. Taxila (India) (3rd cen. B.C.E.). Based on Coppa (1968), Fig. 969, p. 1131.

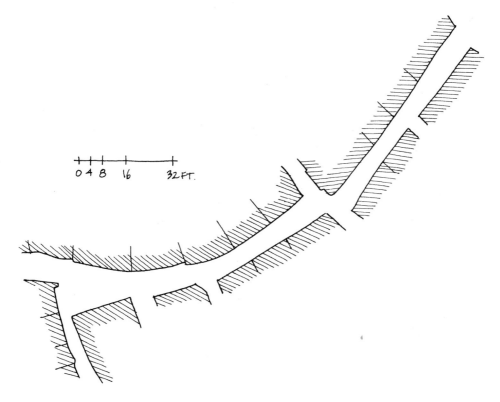

Figure 35. Delos (Greece) (theater quarter) (2nd cen. B.C.E.). Based on Coppa (1968), Fig. 1006, p. 1147.

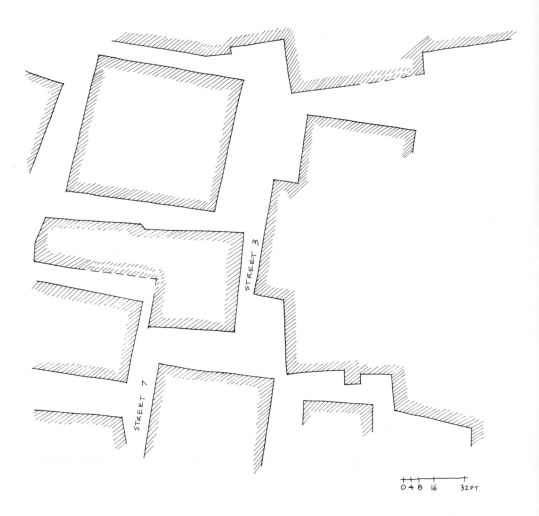

Figure 36. Mampsis (Israel) (S.W. quarter) (Middle and Late Nabatean, 1st–3rd cen. C.E.). Based on Negev (1988), Plan 9, p. 45.

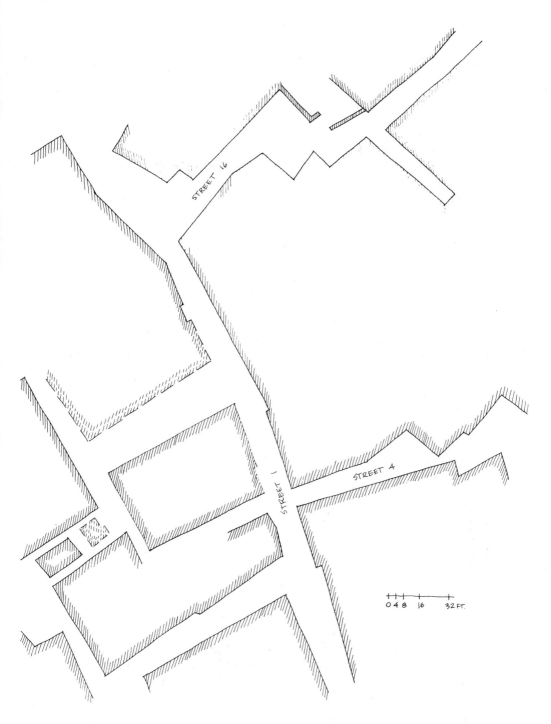

Figure 37. Mampsis (Israel) (S.W. quarter) (Middle and Late Nabatean, 1st–3rd cen. C.E.). Based on Negev (1988), Plan 1, pp. 28–29.

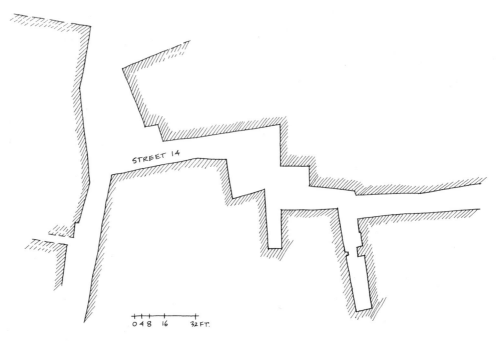

Figure 38. Mampsis (Israel) (N.E. quarter) (Middle and Late Nabatean, 1st–3rd cen. C.E.). Based on Negev (1988), Plan, p. 33.

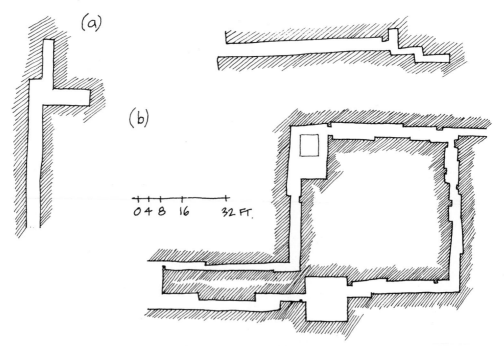

Figure 39. (a) Teotihuacán (Mexico) (Tetitla compound) (2nd–8th cen. C.E.). Based on Millon (1973); *Ekistics,* No. 271 (July/Aug. 1978), Fig. 10, p. 329. (b) Teotihuacán (Mexico) (Tlamilolpa compound) (2nd–8th cen. C.E.). Based on Millon (1973), Appendix 2; Coe (1977), Fig. 26, p. 64.

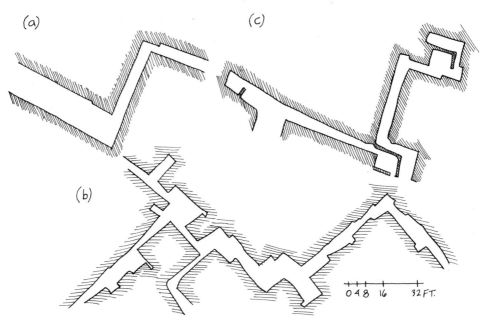

Figure 40. (a) Meinarti (Nubia) (Level 18, 1st–3rd cen. C.E.). Based on information provided by P. L. Shinnie (Calgary). (b) Meinarti (Nubia) (Level 15b, 5th–6th century C.E.). Based on information provided by P. L. Shinnie (Calgary). (c) Meinarti (Nubia) (Level 13, 7th–9th cen., C.E.). Based on information provided by P. L. Shinnie (Calgary).

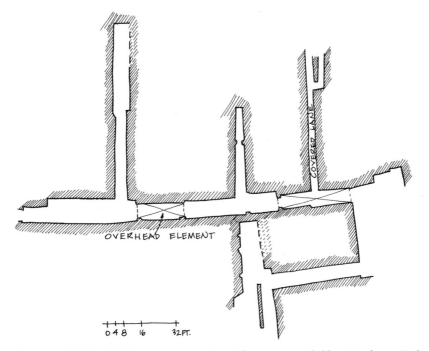

Figure 41. Ikhmindi (Nubia) (? 9th cen. C.E.). Based on information provided by P. L. Shinnie (Calgary).

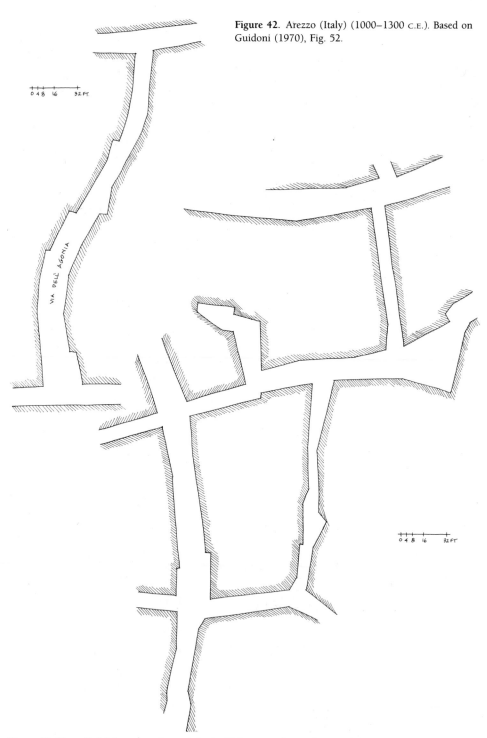

Figure 42. Arezzo (Italy) (1000–1300 C.E.). Based on Guidoni (1970), Fig. 52.

Figure 43. Siena (Italy) (area near Campo) (10th–13th cen. C.E.). Based on Guidoni (1970), Fig. 127, p. 256.

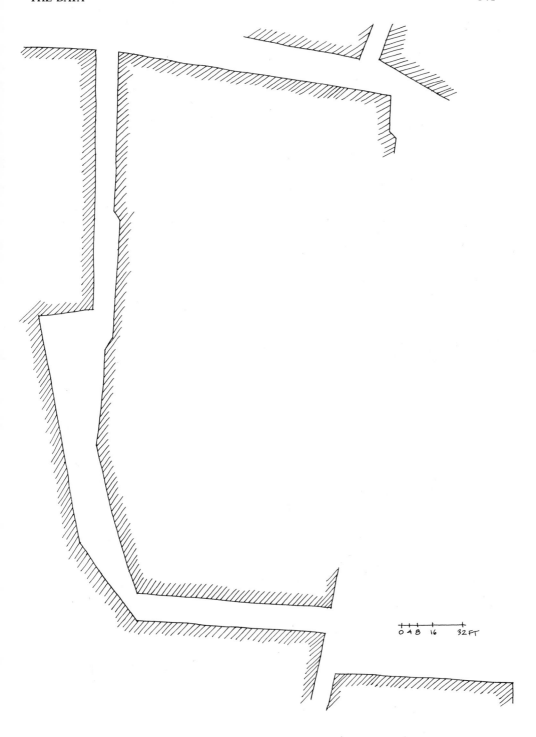

Figure 44. Volterra (Italy) (10th–13th cen. c.e.). Based on Guidoni (1970), Fig. 74, p. 114.

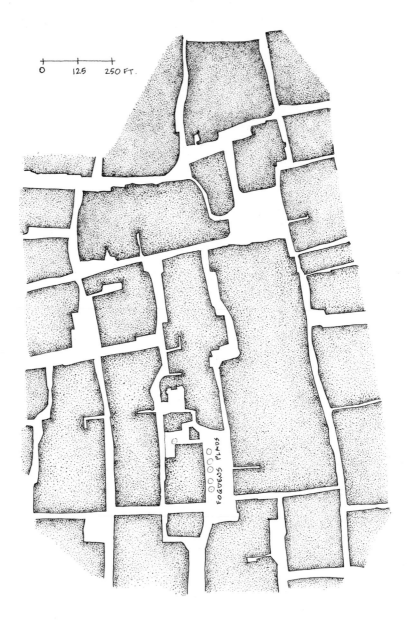

Figure 45. Dragør (Denmark) (13th cen. until present; still in existence). Based on Municipal Historic Building Council 1978 map.

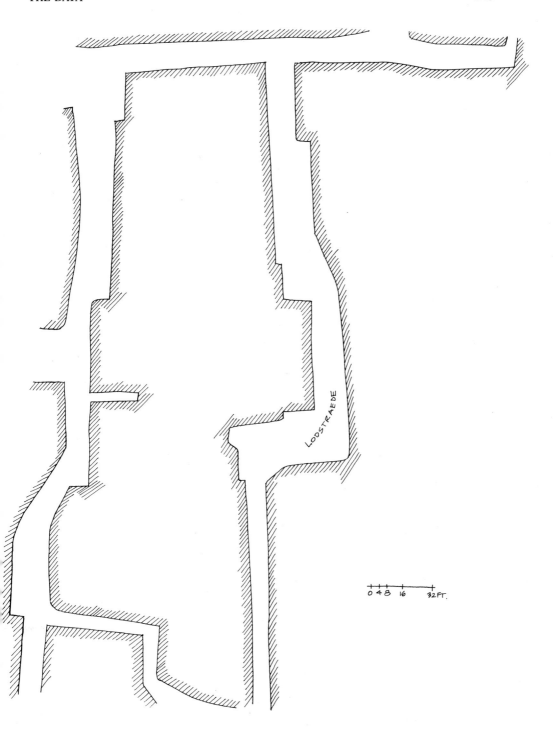

Figure 46. Dragør (Denmark) (13th cen. until the present; still in existence). Based on Municipal Historic Building Council 1978 map.

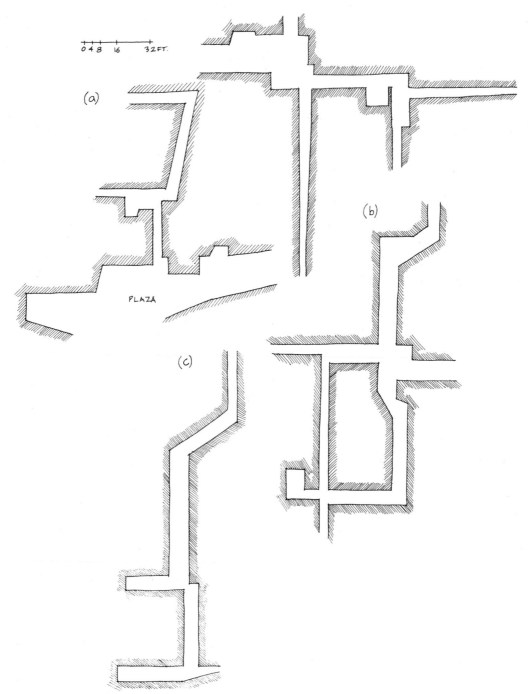

Figure 47. (a) Chan Chan (Peru) ("Tello" Ciudadella) (12th–15th cen. C.E.). Based on Moseley and Mackey (1974). (b) Chan Chan (Peru) ("Nuachaque Grande" Ciudadella) (12th–15th cen. C.E.). Based on Moseley and Mackey (1974); (c) Chan Chan (Peru) ("Laberinto" Ciudadella) (12th–15th cen. C.E.). Based on Moseley and Mackey (1974).

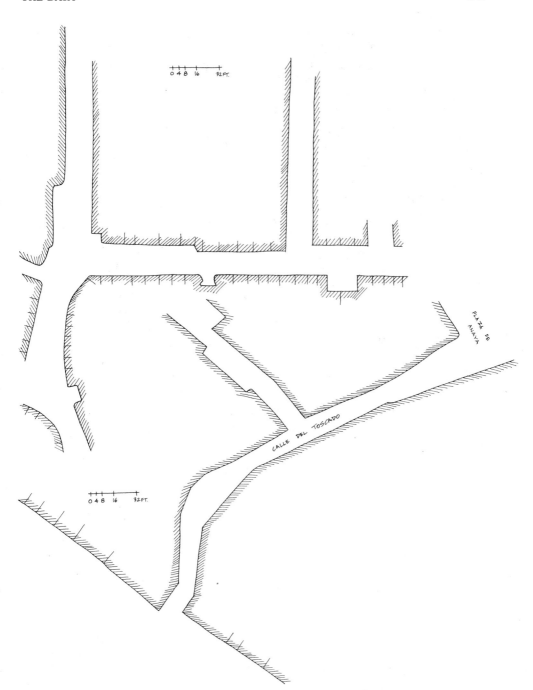

Figure 48. (*above*) Bukhara (Soviet Central Asia) (part of principal bazaar, hence probably at least partially professionally designed) (12th–16th cen. C.E.). Based on Gutkind (various), Vol. 8, Fig. 239, p. 401.

Figure 49. (*below*) Salamanca (Spain) (medieval—still in existence). Based on Auzelle and Jancovic (1954), Vol. 2, Fig. 414–416.

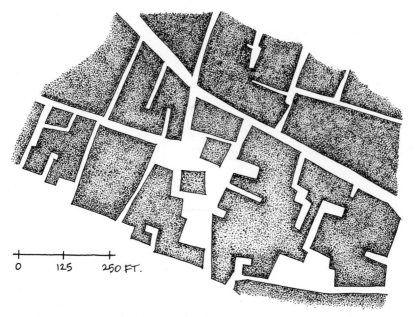

Figure 50. Misratah (Libya) (medieval—still in existence). Based on Blake (1968), Fig. 2.

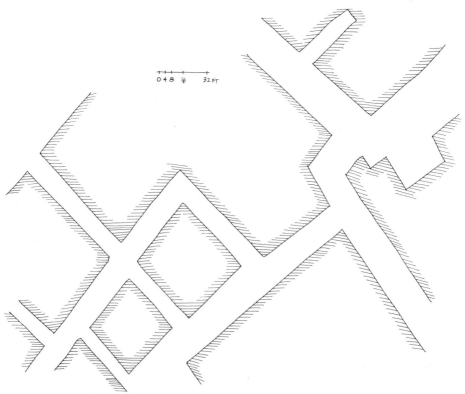

Figure 51. Misratah (Libya) (medieval—still in existence). Based on Blake (1968), Fig. 9, p. 33.

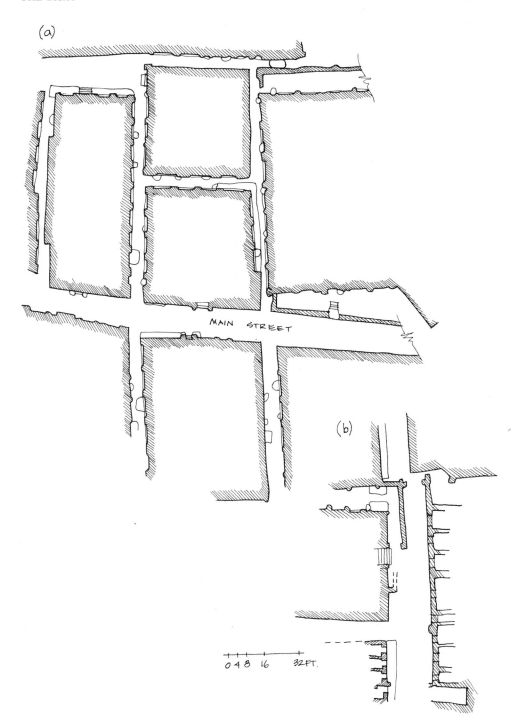

Figure 52. (a) Sirāf (Persian Gulf) (residential quarter, Site F) (medieval). Based on Whitehouse (1970), Fig. 7, p. 151; (b) Sirāf (Persian Gulf) (bazaar, Site C) (medieval). Based on Whitehouse (1970), Fig. 8, p. 152.

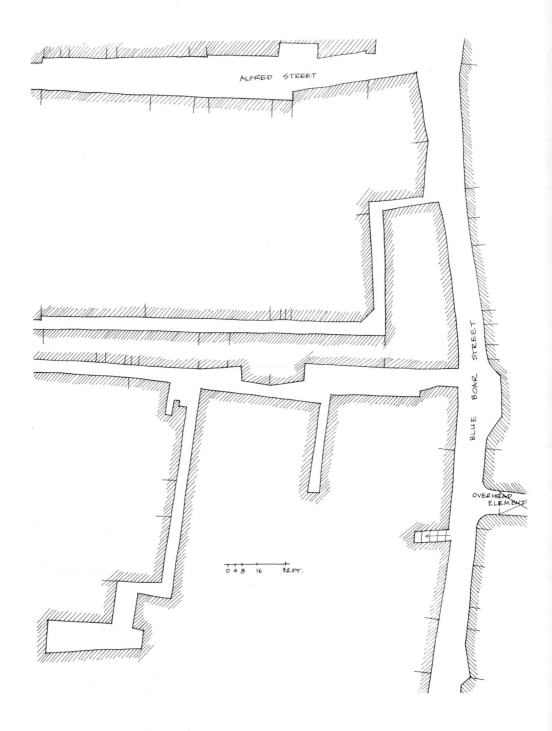

Figure 53. Oxford (England) (medieval—still in existence). Based on Ordnance Survey map.

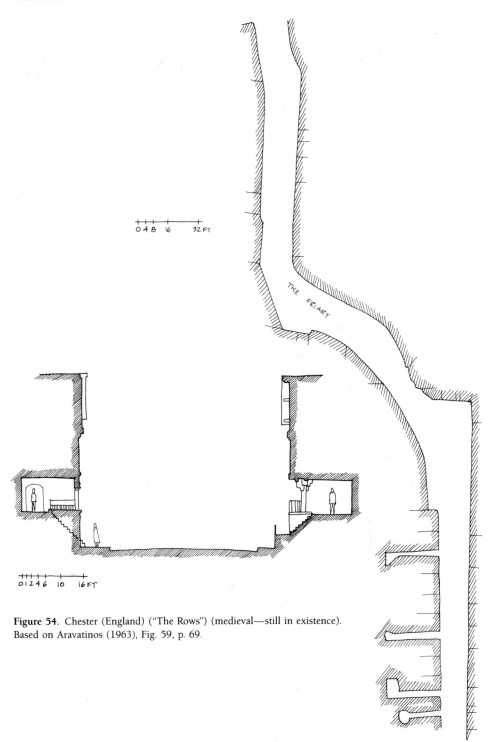

Figure 54. Chester (England) ("The Rows") (medieval—still in existence). Based on Aravatinos (1963), Fig. 59, p. 69.

Figure 55. Salisbury (England) (medieval—still in existence). Based on Ordnance Survey map.

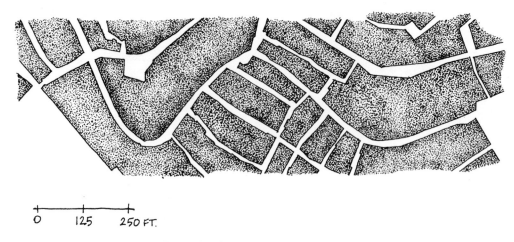

Figure 56. La Spezia (Italy) (medieval—still in existence). Based on Guidoni (1970), Fig. 61.

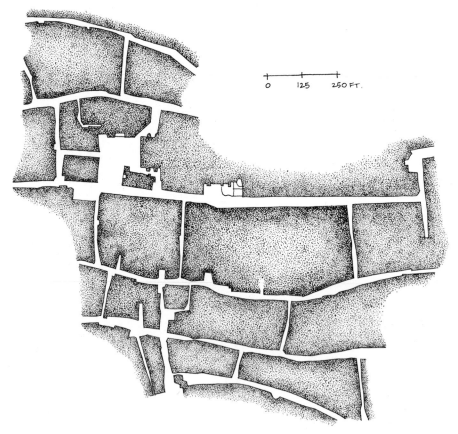

Figure 57. Laon (France) (medieval—still in existence). Based on Auzelle and Jancovic (1954), Vol. 2, Figs. 408–409.

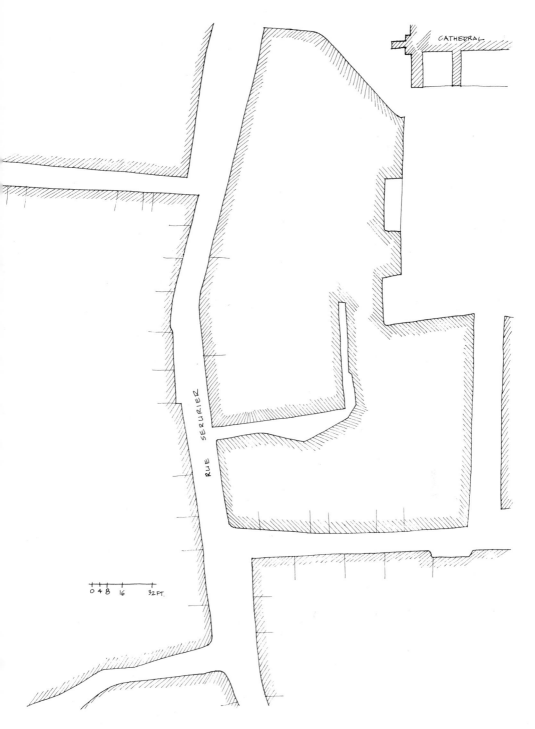

Figure 58. Laon (France) (medieval—still in existence). Based on Auzelle and Jancovic (1954), Vol. 2, Figs. 408–409.

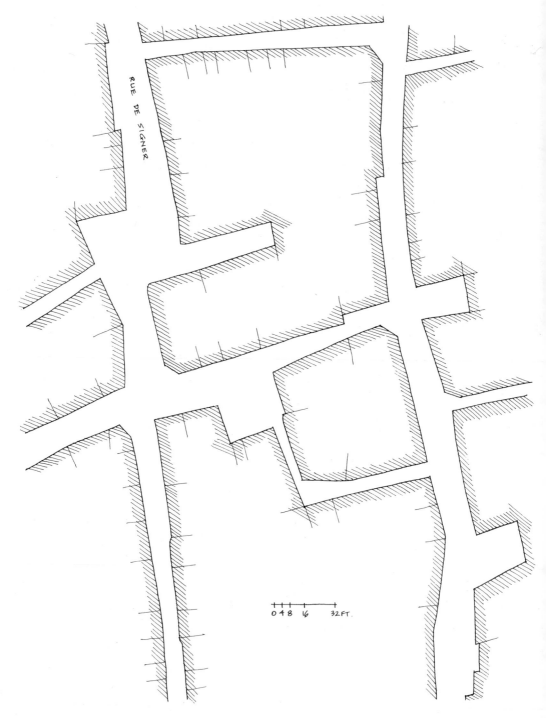

Figure 59. Laon (France) (medieval—still in existence). Based on Auzelle and Jancovic (1954), Vol. 2, Fig. 408–409.

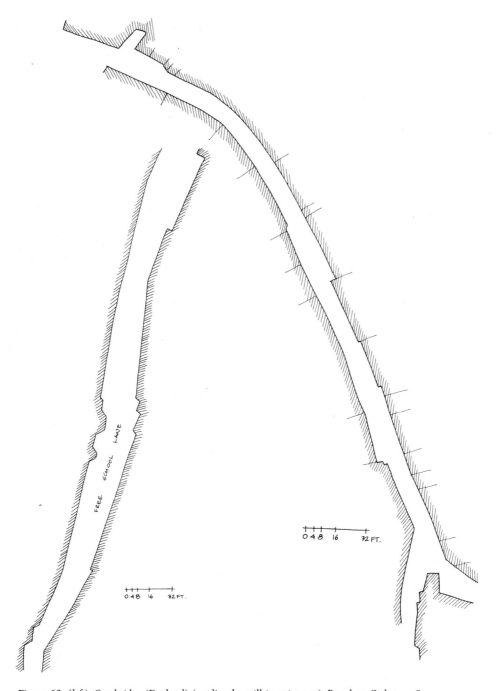

Figure 60. (*left*) Cambridge (England) (medieval—still in existence). Based on Ordnance Survey map.

Figure 61. (*right*) Cordes (France) (medieval—still in existence). Based on information provided by the School of Architecture, Royal Academy of Art, Copenhagen.

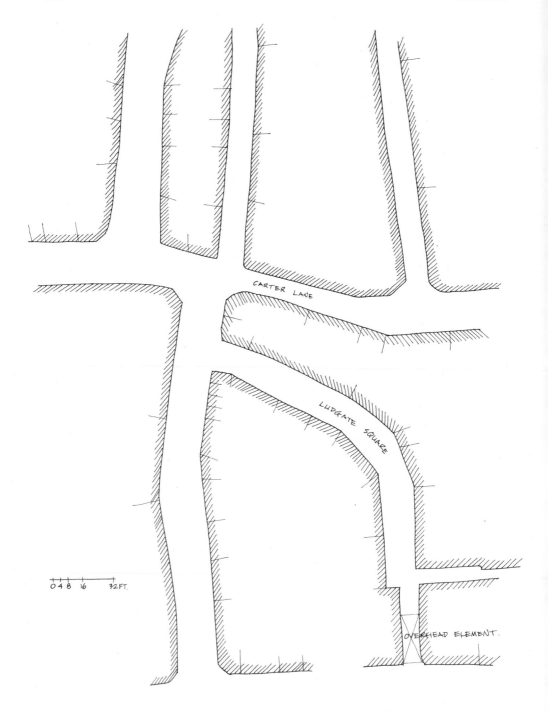

Figure 62. London (England) (City of London) (medieval—still in existence). Based on Ordnance Survey map.

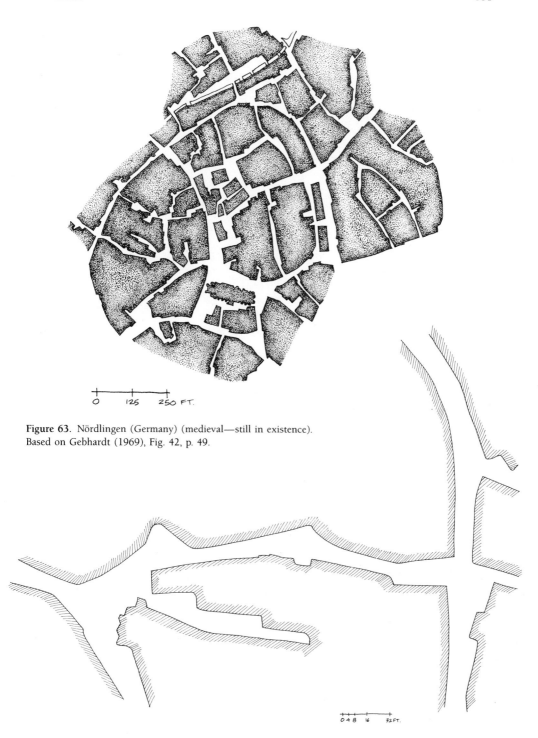

Figure 63. Nördlingen (Germany) (medieval—still in existence).
Based on Gebhardt (1969), Fig. 42, p. 49.

Figure 64. Nördlingen (Germany) (medieval—still in existence). Based on Gebhardt (1969), Fig. 42, p. 49.

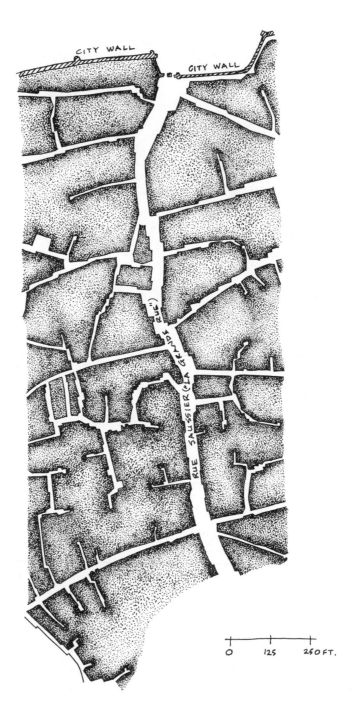

Figure 65. Kairouan (Tunisia) (still in existence). Based on Auzelle and Jancovic (1954), Vol. 2, Fig. 407. Note different character of "La Grande Rue," probably a later [colonial?] addition.)

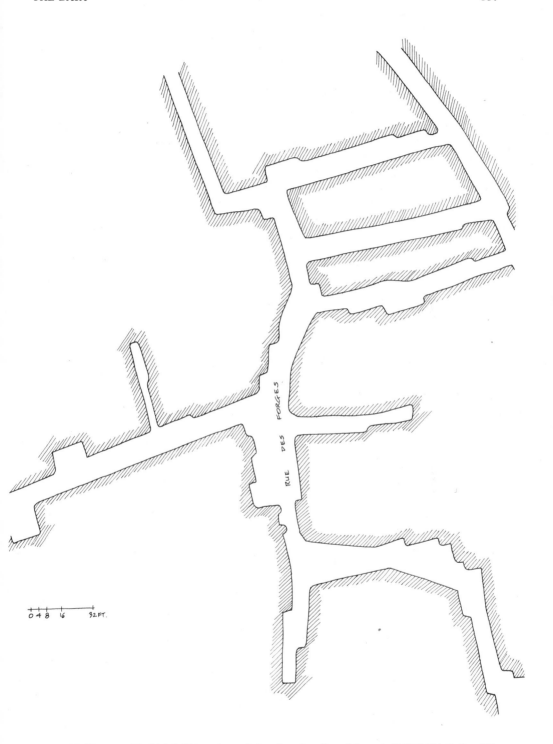

Figure 66. Kairouan (Tunisia) (still in existence). Based on Auzelle and Jancovic (1954), Vol. 2, Fig. 407.

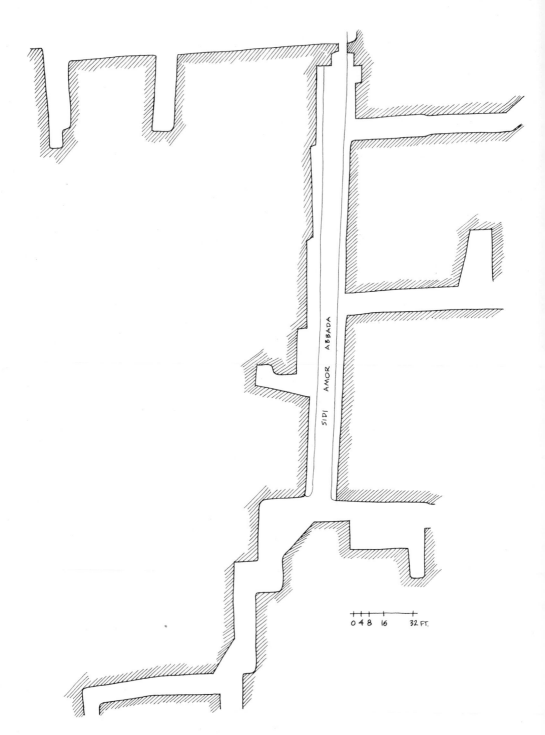

Figure 67. Kairouan (Tunisia) (still in existence). Based on Auzelle and Jancovic (1954), Vol. 2, Fig. 407.

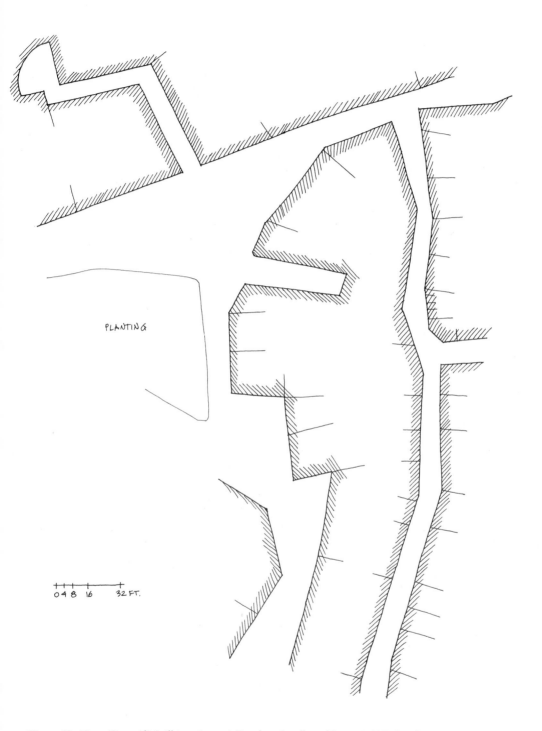

PLANTING

0 4 8 16   32 FT.

Figure 68. Viseu (Portugal) (still in existence). Based on Auzelle and Jancovic (1954), Vol. 3, Fig. 401–402.

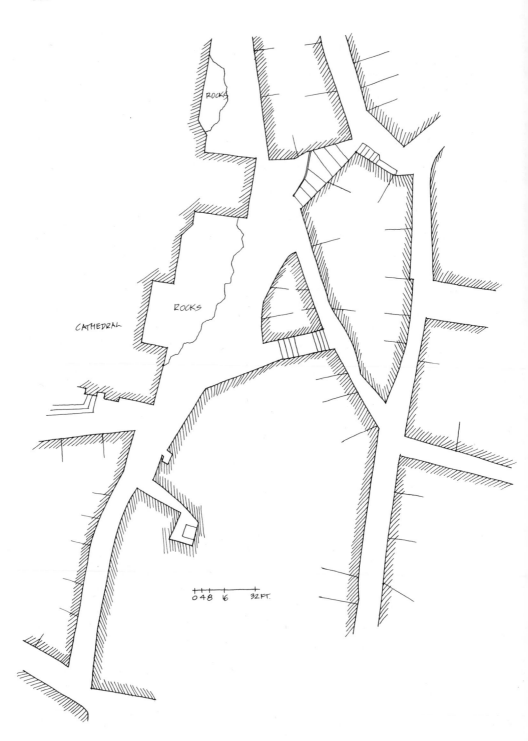

Figure 69. Viseu (Portugal) (still in existence). Based on Auzelle and Jancovic (1954), Vol. 3, Fig. 401–402.

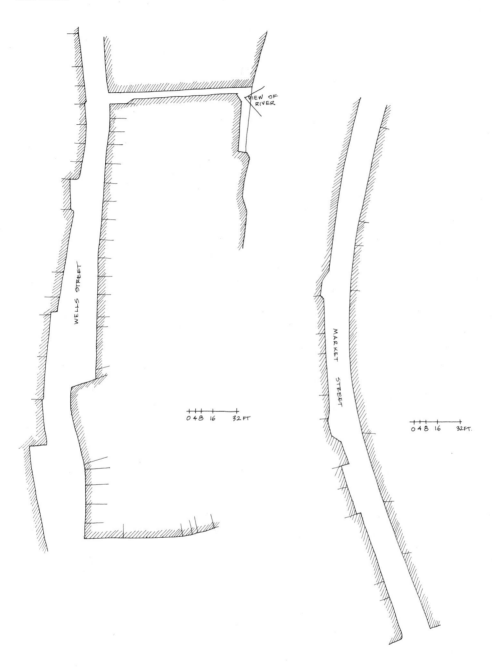

Figure 70. (*left*) Winchester (England) (still in existence). Based on Ordnance Survey map.
Figure 71. (*right*) Wells (England) (still in existence). Based on Ordnance Survey map.

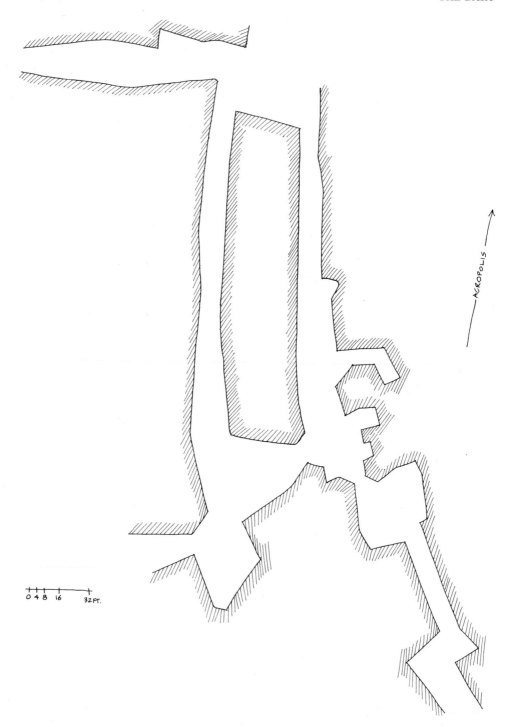

Figure 72. Athens (Greece) (Plaka area) (still in existence). Based on information provided by J. Antoniou (London) and on Antoniou (1973).

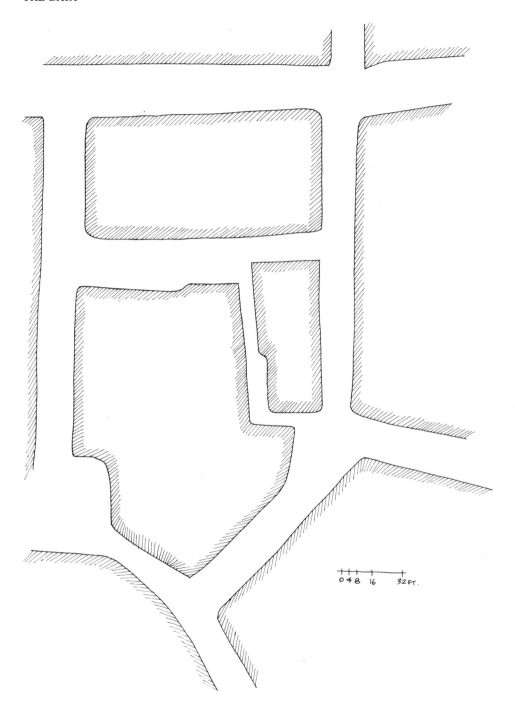

Figure 73. Athens (Greece) (Plaka area) (still in existence). Based on information provided by J. Antoniou (London) and on Antoniou (1973).

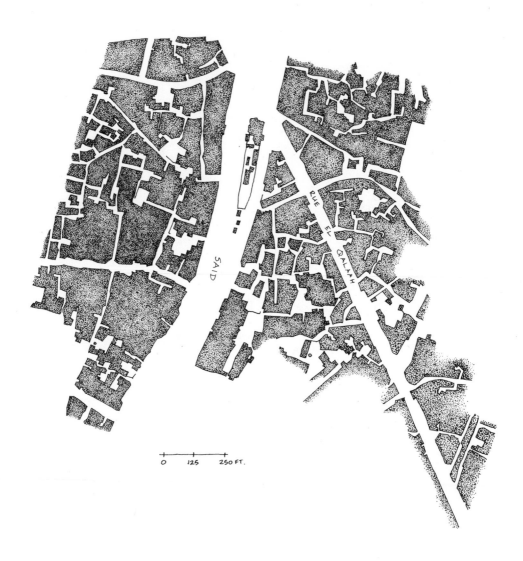

Figure 74. Cairo (Egypt) (still in existence—map drawn in 1978). Based on information provided by
J. Antoniou (London). (Note contrast with new streets and avenues).

Figure 75. Cairo (Egypt) (still in existence—map drawn in 1978). Based on information provided by J. Antoniou (London).

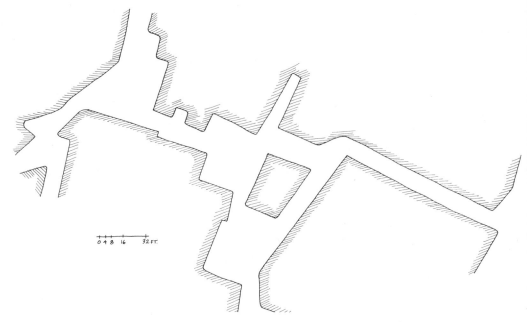

Figure 76. Cairo (Egypt) (still in existence—map drawn in 1978). Based on information provided by J. Antoniou (London).

Figure 77. Cairo (Egypt) (still in existence). Based on Lane (1977), p. 16. (Drawing by A. Gupte).

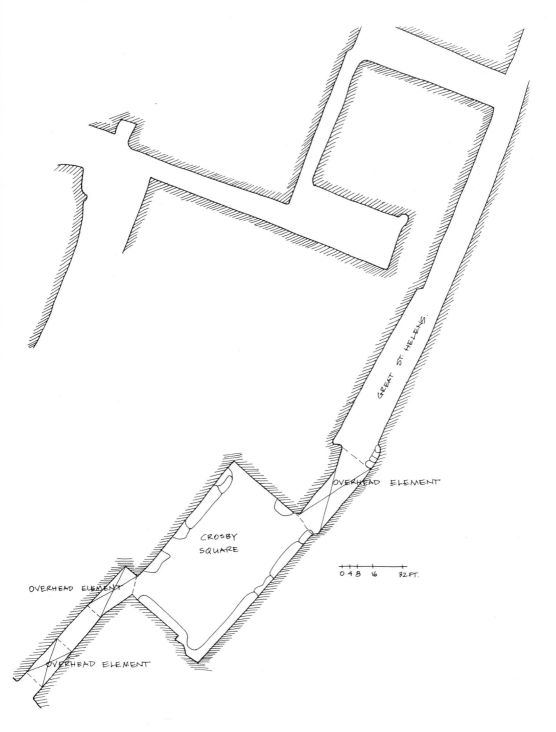

Figure 78. London (England) (Aldgate Ward) (still in existence). Based on Ordnance Survey map.

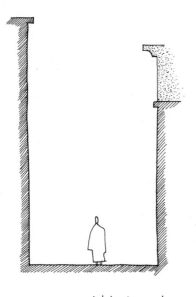

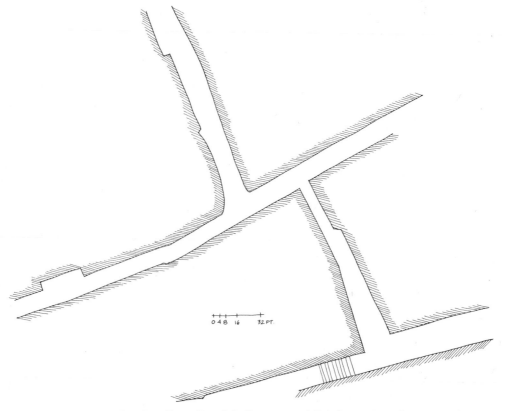

Figure 79. Kuwait (Persian Gulf) (still in existence). Based on Shiber in Wheatley (1976), p. 355.

Figure 80. Palma de Mallorca (Spain) (still in existence). Based on municipal map.

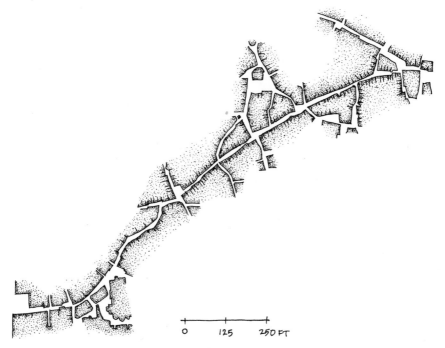

Figure 81. Stone Town (Zanzibar, Tanzania) (small segment) (still in existence). Based on Nilsson *et al.* (n.d.); (cf. Rapoport (1977), Fig. 4.23, p. 233).

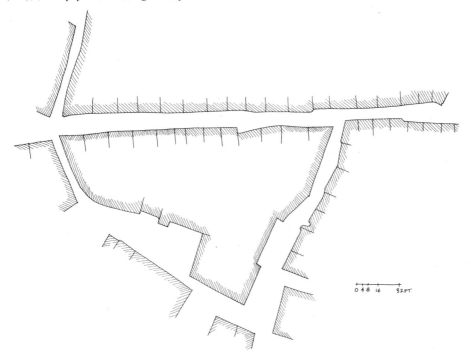

Figure 82. Stone Town (Zanzibar, Tanzania) (still in existence). Based on Nilsson *et al.* (n.d.)

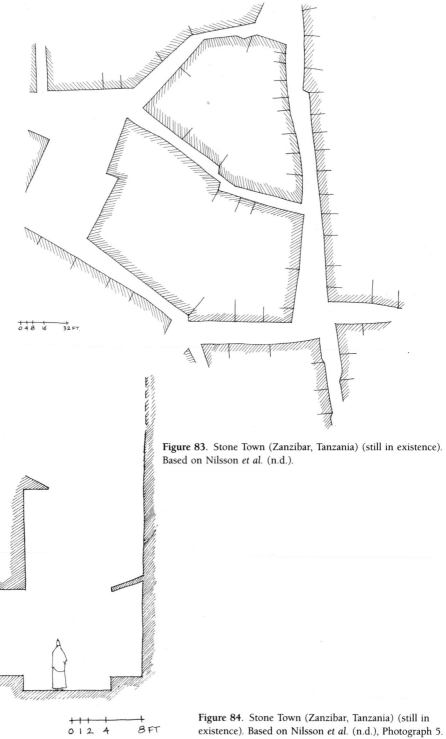

0 4 8 16   32 FT.

Figure 83. Stone Town (Zanzibar, Tanzania) (still in existence). Based on Nilsson *et al.* (n.d.).

0 1 2 4   8 FT

Figure 84. Stone Town (Zanzibar, Tanzania) (still in existence). Based on Nilsson *et al.* (n.d.), Photograph 5.

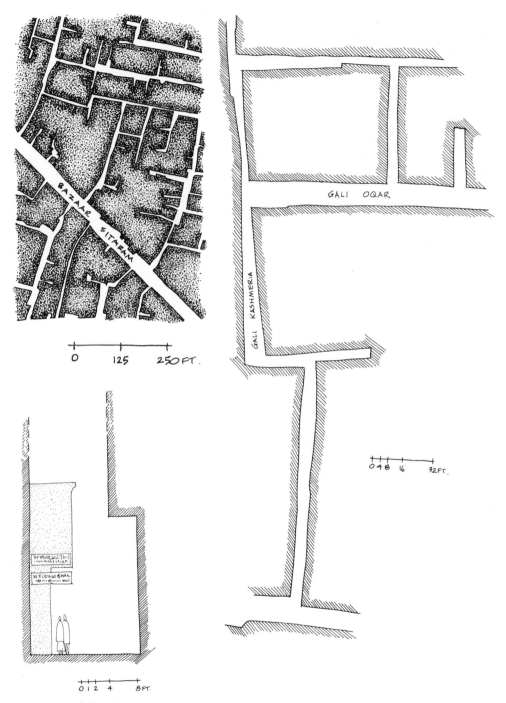

Figure 85.(*top left*) Delhi (India) (Old City) (still in existence). Based on Fonseca (1969b; 1969a, p. 21).
Figure 86. (*bottom left*) Delhi (India) (Old City) (still in existence). Based on Fonseca (1969b), p. 110.
Figure 87.(*right*) Delhi (India) (Old City) (still in existence). Based on Fonseca (1969b; 1969a, p. 21).

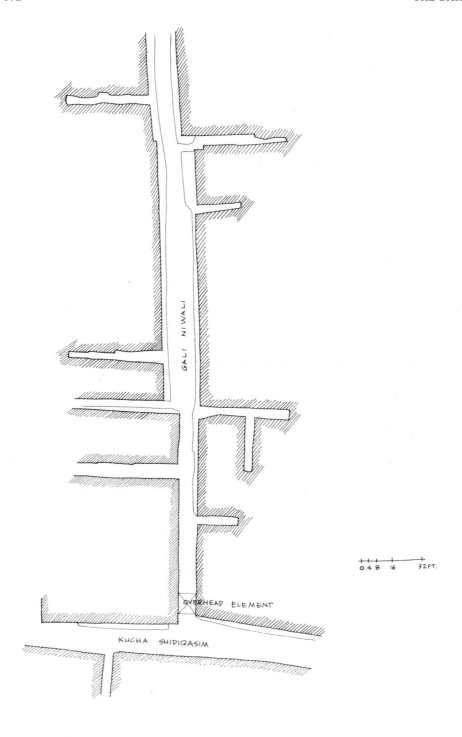

Figure 88. Delhi (India) (Old City) (still in existence). Based on Fonseca (1969b; 1969a, p. 21).

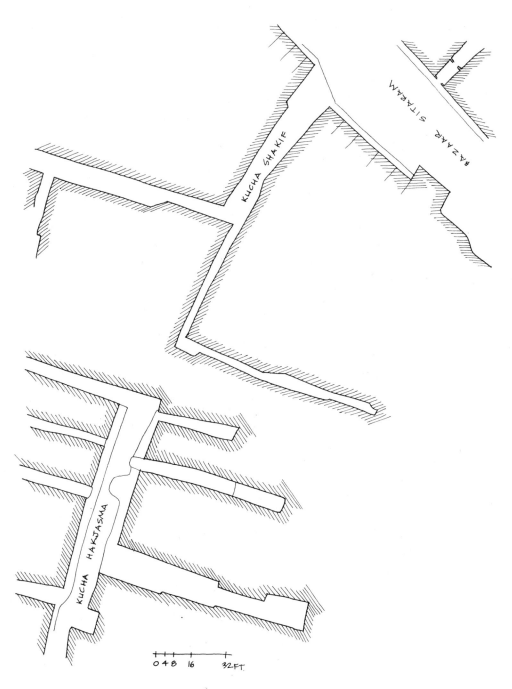

Figure 89. Delhi (India) (Old City) (still in existence). Based on Fonseca (1969b; 1969a, p. 21).

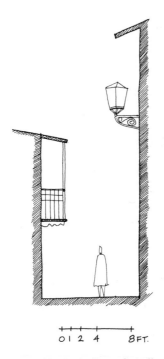

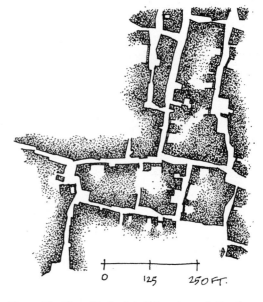

Figure 91. Afyon (Turkey) (still in existence). Based on
Aktüre (1978), Fig. 73, p. 210. Provided by
N. Inceoglu (Istanbul).

Figure 90. Gorée (Senegal) (still in existence).
Based on July (1980), Fig. 46.

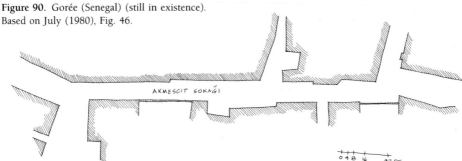

AKMESCIT SOKAĞI

Figure 92. Afyon (Turkey) (still in existence). Based on Aktüre (1978), Fig. 73, p. 210. Provided by
N. Inceoglu (Istanbul).

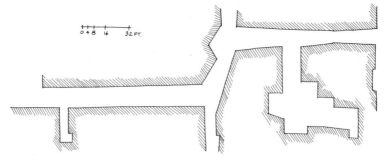

Figure 93. Afyon (Turkey) (still in existence). Based on Aktüre (1978), Fig. 69. Provided by N. Inceoglu
(Istanbul).

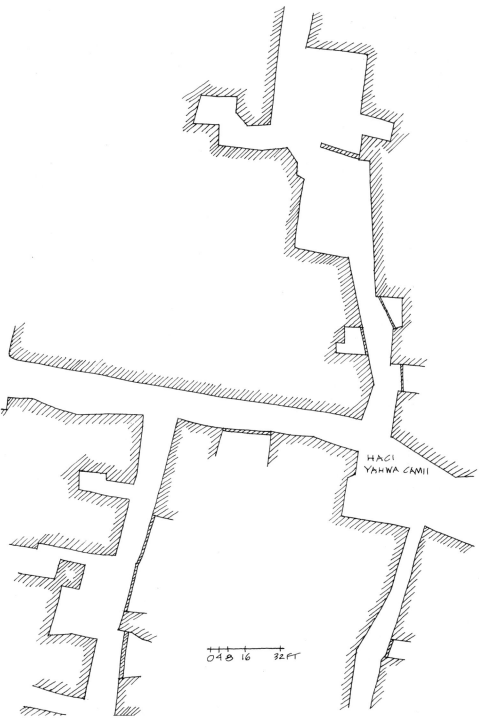

Figure 94. Afyon (Turkey) (still in existence). Based on Aktüre (1978), Fig. 73, p. 210. Provided by N. Inceoglu (Istanbul).

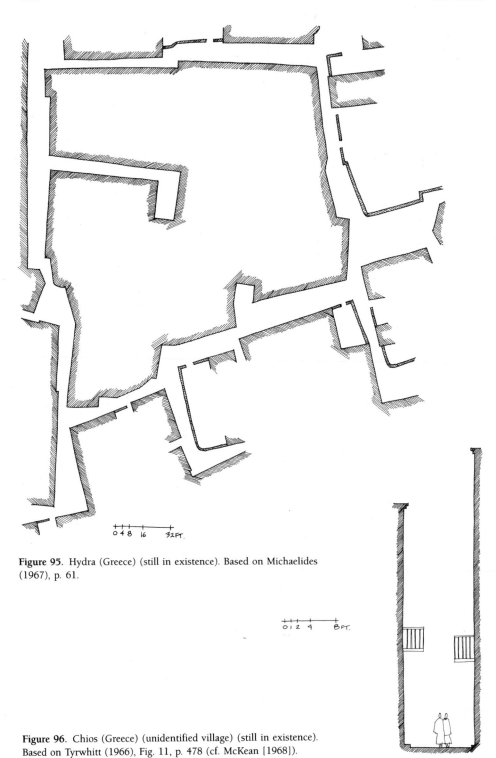

Figure 95. Hydra (Greece) (still in existence). Based on Michaelides (1967), p. 61.

Figure 96. Chios (Greece) (unidentified village) (still in existence). Based on Tyrwhitt (1966), Fig. 11, p. 478 (cf. McKean [1968]).

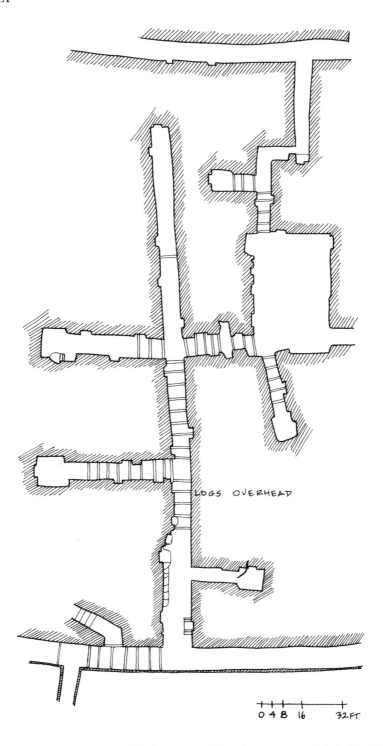

LOGS OVERHEAD

0 4 8 16   32FT

Figure 97. Tinerhir (Morocco) (still in existence). Based on Andersen *et al.* (n.d.), p. 16.

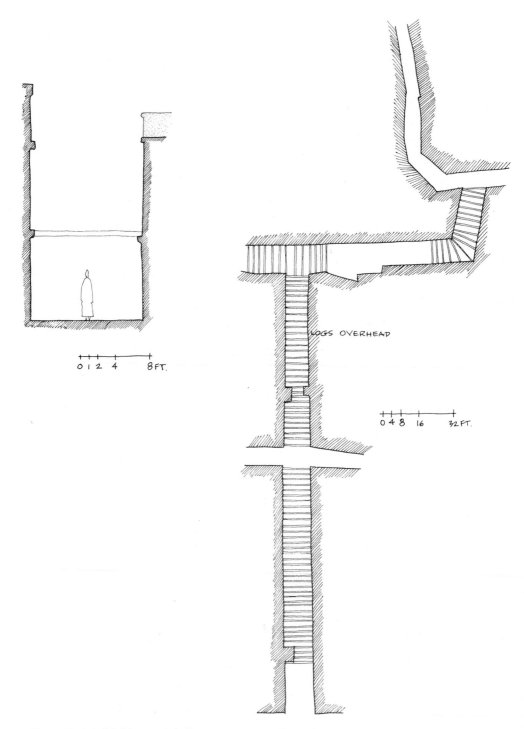

LOGS OVERHEAD

Figure 98. Kabul (Afghanistan) (still in existence). Based on information provided by S. I. Hallet (Washington D.C.) and on Hallet and Samzay (1980), p. 177.

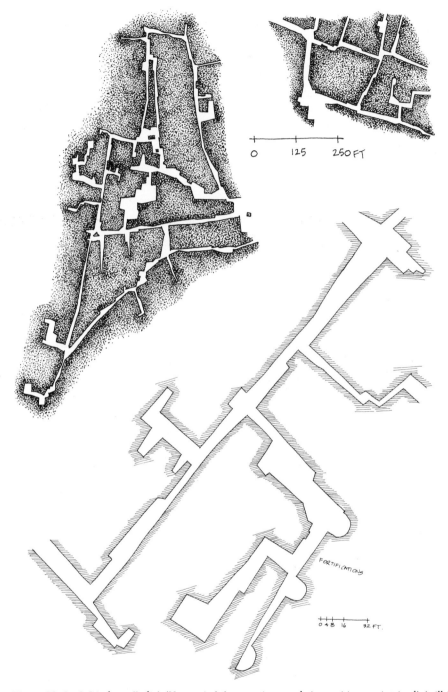

Figure 99. (*top*) Jaisalmer (India) (Upper city) (two sections—relative positions maintained) (still in existence). Based on Jain (1980a), p. 22.

Figure 100. (*bottom*) Jaisalı̄   (India) (still in existence). Based on information provided by K. Jain (Ahmedabad).

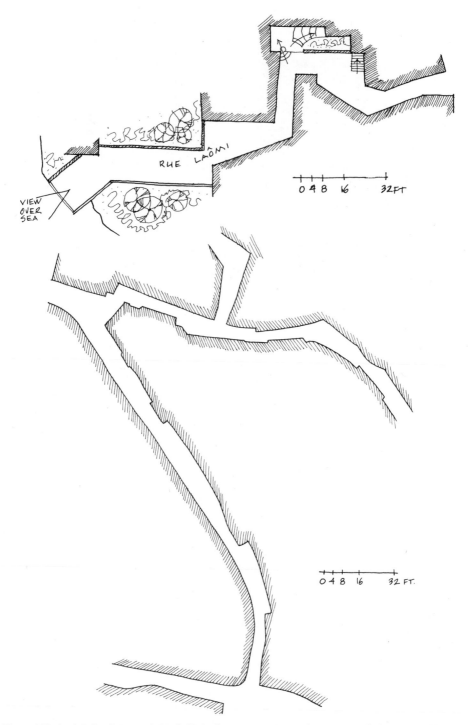

Figure 101. (*top*) Rabat (Morocco) (Kasbah) (still in existence). Based on Nijst *et al.* (1973), p. 281.

Figure 102. (*bottom*) Kula (Turkey) (still in existence). Based on Fersan (1980). Provided by N. Inceoglu (Istanbul).

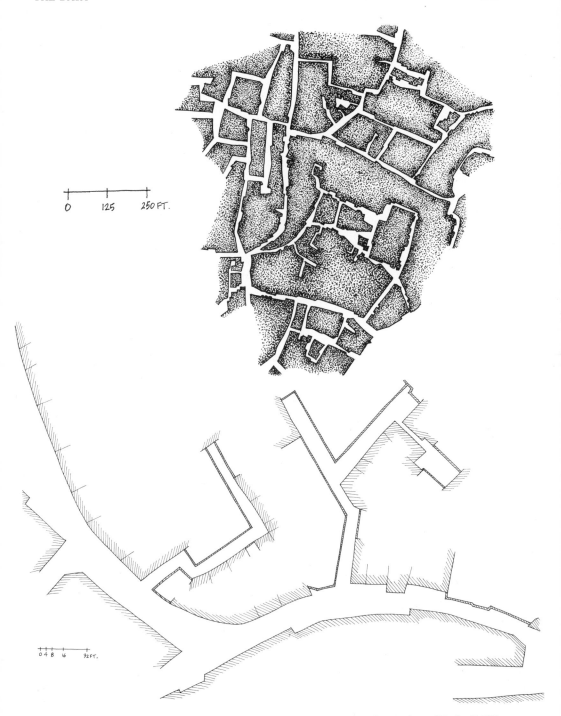

Figure 103. (*top*) Istanbul (Turkey) (Zeyrek quarter) (still in existence). Based on Butler and Butler (1976), p. 19.

Figure 104. (*bottom*) Istanbul (Turkey) (Zeyrek quarter) (still in existence). Based on Butler and Butler (1976), p. 19.

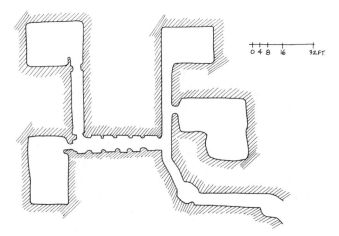

Figure 105. New Herat (Afghanistan) (still in existence). Based on information provided by S. I. Hallet (Washington, DC).

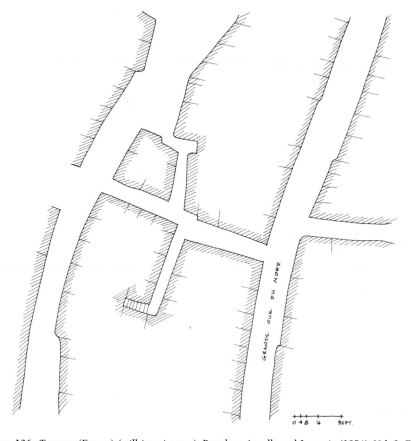

Figure 106. Tournus (France) (still in existence). Based on Auzelle and Jancovic (1954), Vol. 2, Fig. 410.

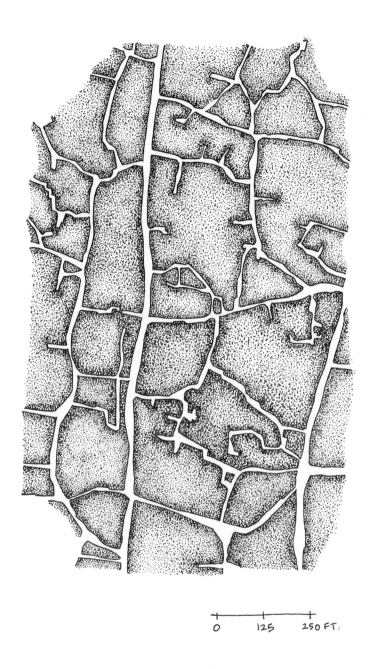

Figure 107. Midnapur (India) (still in existence). Based on municipal maps.

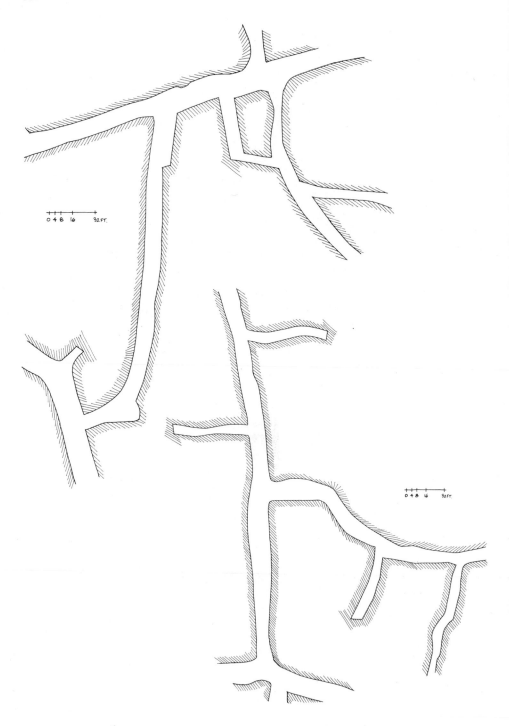

Figure 108. (*top left*) Midnapur (India) (still in existence). Based on municipal maps.
Figure 109. (*bottom right*) Midnapur (India) (still in existence). Based on municipal maps.

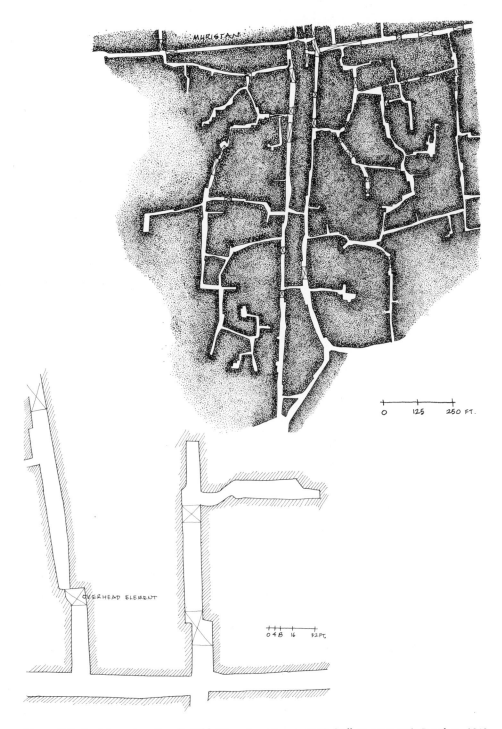

Figure 110. (*top*) Jerusalem (Israel) (Old City—Armenian quarter) (still in existence). Based on 1948 Ordnance Survey map (reprinted 1966 by Israel Survey). Provided by Y. Ben-Arieh (Jerusalem).

Figure 111. (*bottom*) Jerusalem (Israel) (Old City—Christian quarter) (still in existence). Based on 1948 Ordnance Survey map (reprinted 1966 by Israel Survey). Provided by Y. Ben-Arieh (Jerusalem).

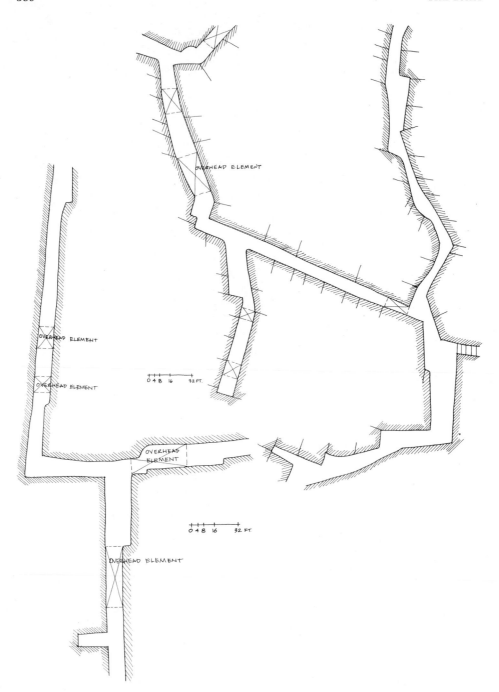

Figure 112. (*left*) Jerusalem (Israel) (Old City—Christian quarter) (still in existence). Based on 1948 Ordnance Survey map (reprinted 1966 by Israel Survey). Provided by Y. Ben-Arieh (Jerusalem).

Figure 113. (*right*) Jerusalem (Israel) (Old City—Armenian quarter) (still in existence). Based on 1948 Ordnance Survey map (reprinted 1966 Israel Survey). Provided by Y. Ben-Arieh (Jerusalem).

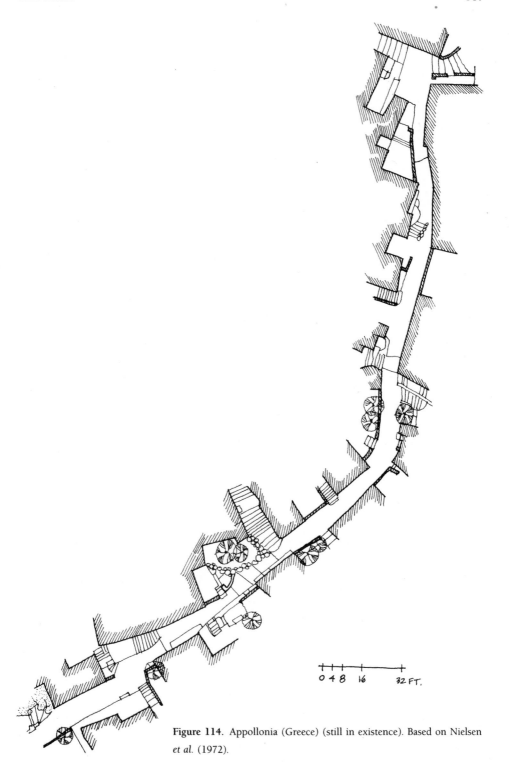

Figure 114. Appollonia (Greece) (still in existence). Based on Nielsen *et al.* (1972).

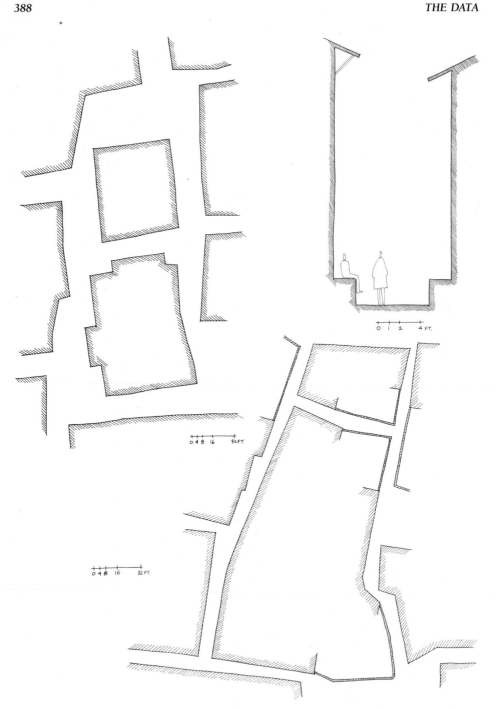

Figure 115. (*top left*) Wad Lemeid (Sudan) (still in existence). Based on Tewfik (1976), Fig. 45, p. 196.

Figure 116. (*bottom*) Yeşil Ada (Turkey) (still in existence). Based on information provided by M. Turan (Pittsburgh).

Figure 117. (*right*) Unidentified village (Nepal) (still in existence). Based on Hosken (1974), Fig. 204, p. 139; Fig. 273, p. 180, etc.

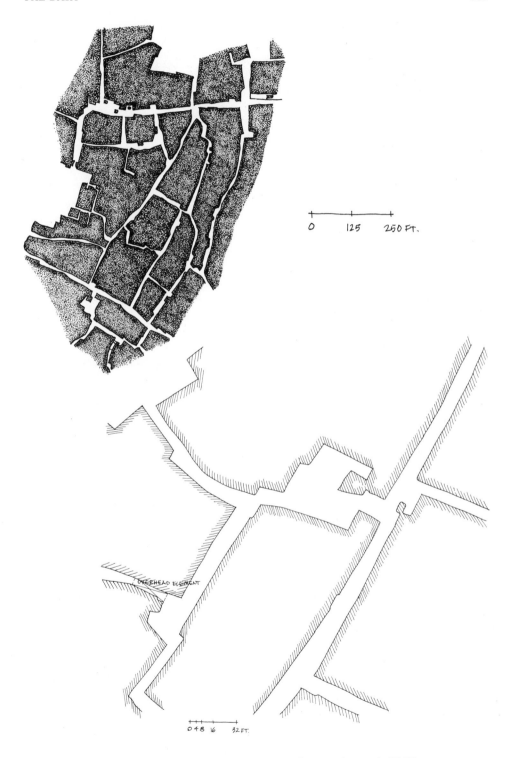

Figure 118. (*top*) Bungmati (Nepal) (still in existence). Based on Knudsen *et al.* (1969).
Figure 119. (*bottom*) Bungmati (Nepal) (still in existence). Based on Knudsen *et al.* (1969).

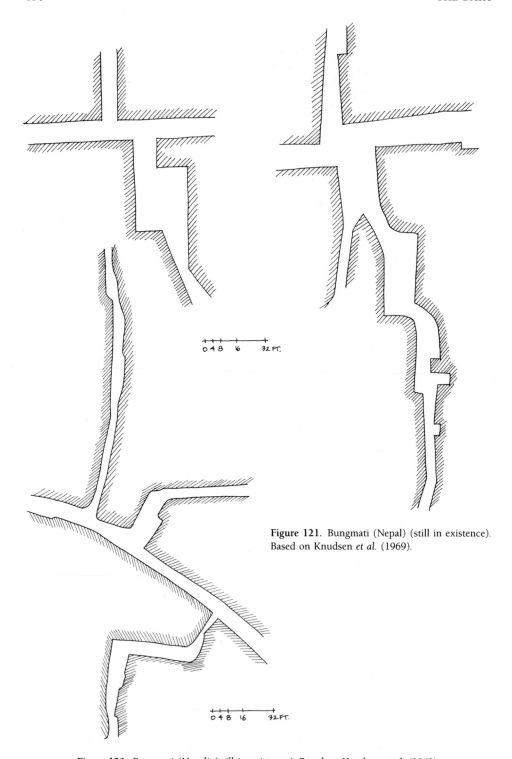

Figure 121. Bungmati (Nepal) (still in existence). Based on Knudsen *et al.* (1969).

Figure 120. Bungmati (Nepal) (still in existence). Based on Knudsen *et al.* (1969).

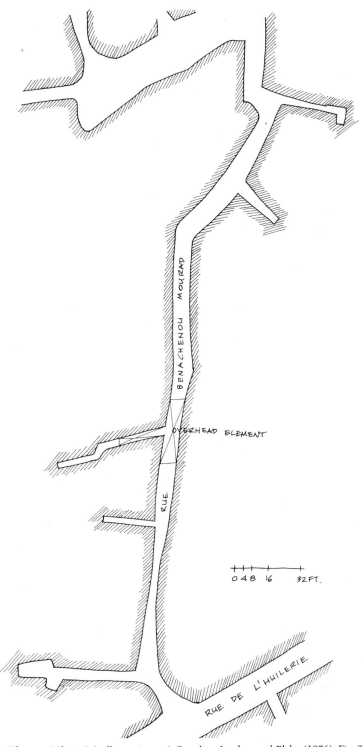

Figure 122. Tlemcen (Algeria) (still in existence). Based on Lawless and Blake (1976), Fig. 9.3, p. 102.

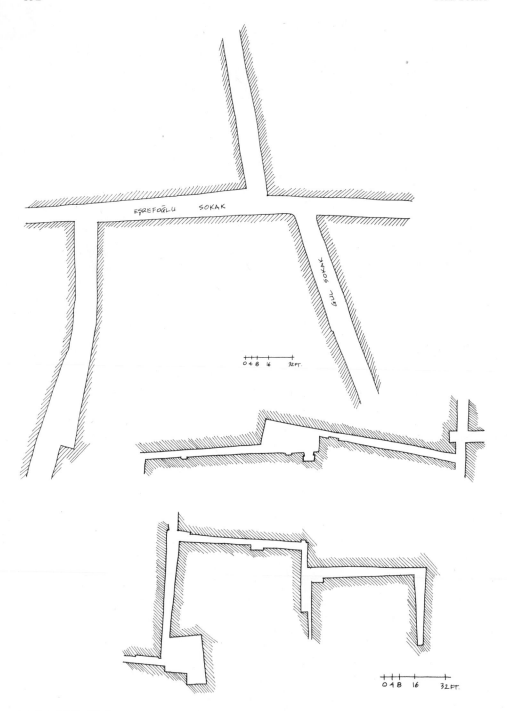

Figure 123. (*top*) Tanimlama (Turkey) (still in existence). Based on information provided by M. Turan (Pittsburgh).

Figure 124. (*bottom*) Gedi (East Africa) (still in existence). Based on Lewcock (1971), p. 85.

Figure 125. Sankhu (Nepal) (still in existence). Based on Hosken (1974), p. 308.

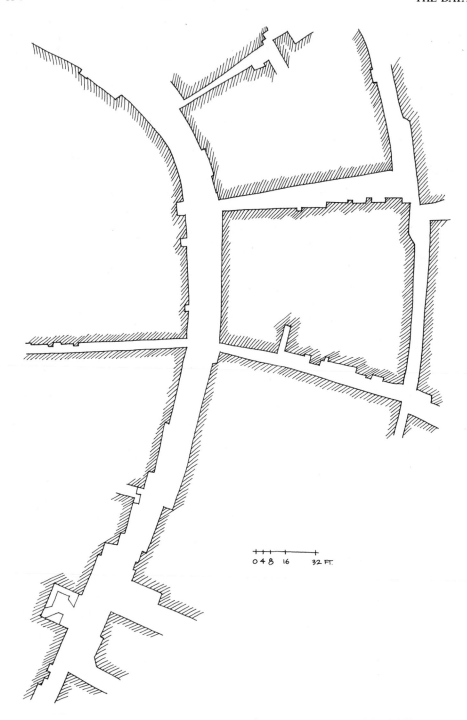

Figure 126. Takayama (Japan) (still in existence). Based on *Takayama* (n.d.), p. 15. Provided by R. Ohno (Kobe).

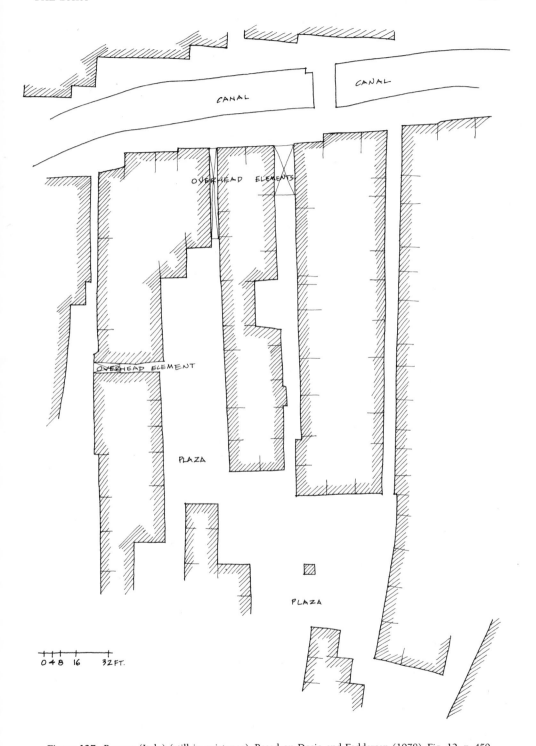

Figure 127. Burano (Italy) (still in existence). Based on Dosio and Feddersen (1978), Fig. 12, p. 459.

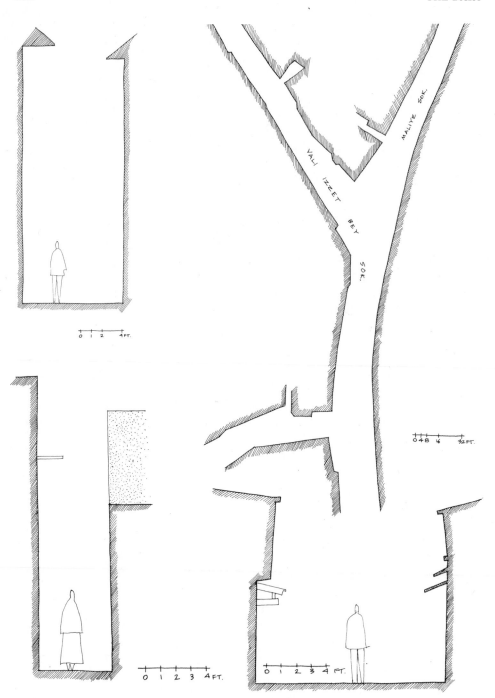

Figure 128. (*top left*) Antakya (Antioch) (Turkey) (still in existence). Based on Newman (1968), Fig. 16.

Figure 129. (*bottom left*) Konya (Turkey) (still in existence). Based on information provided by M. Turan (Pittsburgh).

Figure 130. (*top right*) Nowa (Japan) (fish market) (still in existence). Based on information provided by R. Ohno (Kobe).

Figure 131. (*bottom right*) Cairo (Egypt) (lane in Walled Babylon) (still in existence). Based on Abu-Lughod (1971), Fig. 2, p. 4.

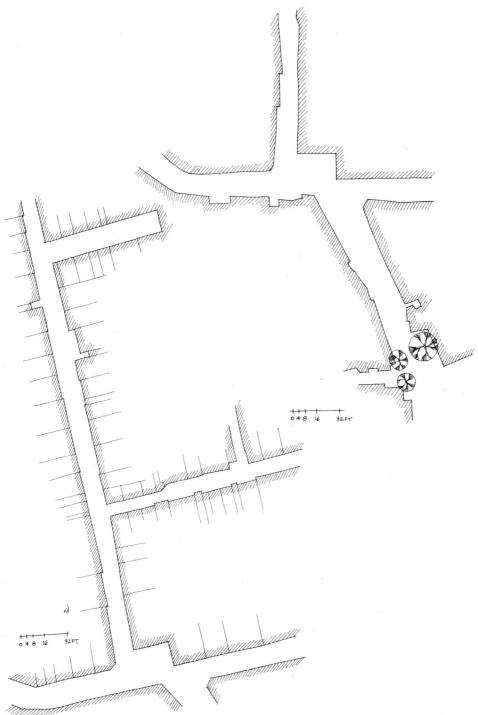

Figure 132. (*top*) Saiwacho Neighborhood center (Japan) (still in existence). Based on information provided by R. Ohno (Kobe).

Figure 133. (*bottom*) Saiwacho Neighborhood center (Japan) (still in existence). Based on information provided by R. Ohno (Kobe).

Figure 134. Cisterino (Italy) (still in existence). Based on Allen (1969a), Fig. 10, p. 147.

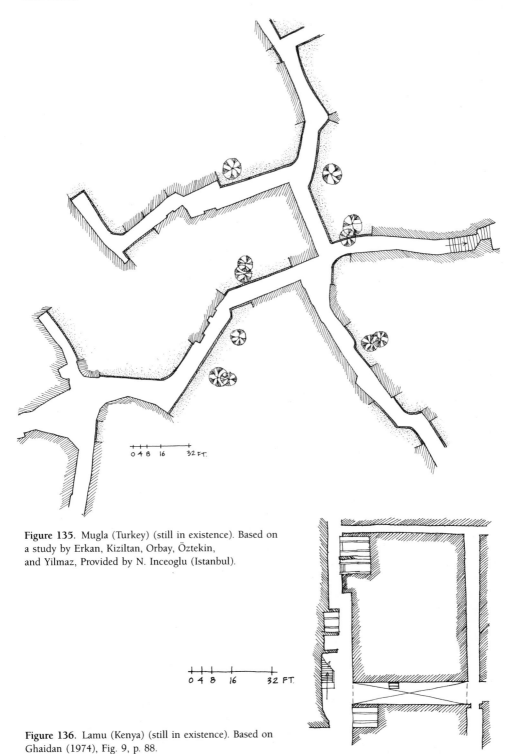

Figure 135. Mugla (Turkey) (still in existence). Based on
a study by Erkan, Kiziltan, Orbay, Öztekin,
and Yilmaz, Provided by N. Inceoglu (Istanbul).

Figure 136. Lamu (Kenya) (still in existence). Based on
Ghaidan (1974), Fig. 9, p. 88.

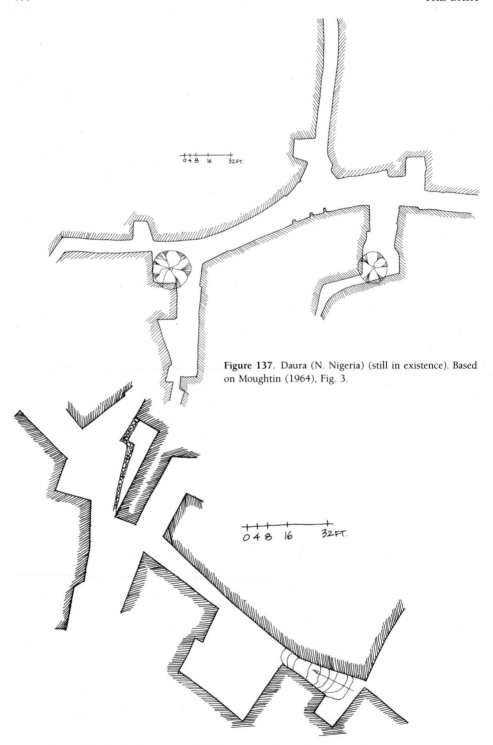

Figure 137. Daura (N. Nigeria) (still in existence). Based on Moughtin (1964), Fig. 3.

Figure 138. Aïn Leuh (Morocco) (still in existence). Based on Nijst *et al.* (1973), p. 92.

Figure 139. (*top left*) Shustar (Iran) (still in existence). Based on *Process Architecture* No. 15 (1980), p. 10 (Drawing by A. Gupte).

Figure 140. (*bottom left*) Kano (N. Nigeria) (still in existence). Based on Moughtin (1964), Fig. 10.

Figure 141. (*bottom right*) Baghdad (Iraq) (still in existence). Based on Al-Azzawi, in Oliver (1969), p. 95.

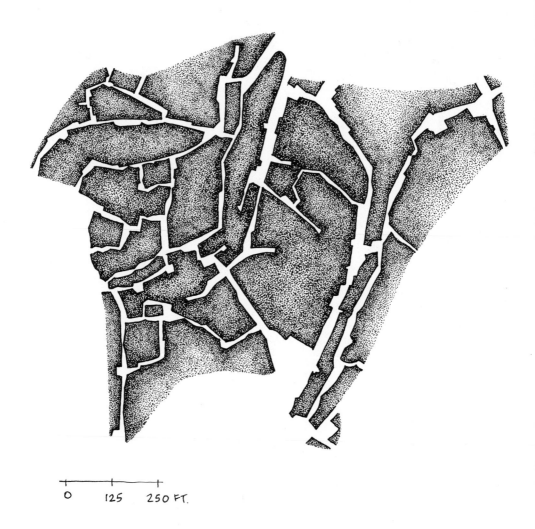

Figure 142. Antalya (Turkey) (still in existence). Based on 1977 map in a study by Tankut, Esmer, Gopur, and Balamir. Provided by M. Turan (Pittsburgh).

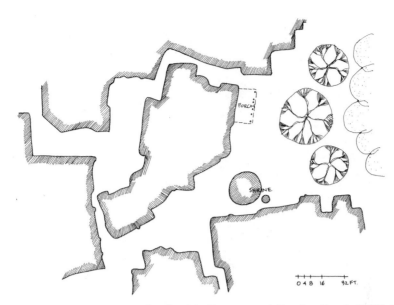

Figure 143. Sekai (N. Ghana) (an Isala village) (still in existence). Based on Prussin (1967), Fig. 4–2.

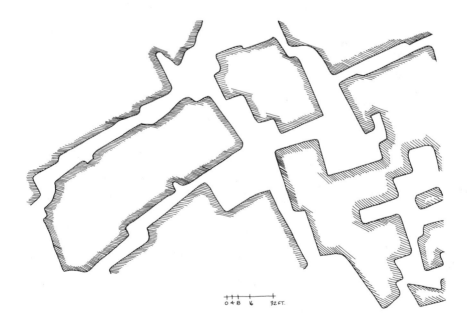

Figure 144. Sekai (N. Ghana) (an Isala village) (still in existence). Based on Prussin (1967), Fig. 4–2.

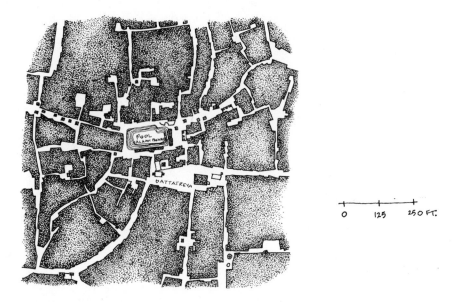

Figure 145. Bhaktarpur (Nepal) (still in existence). Based on Gutschow (1974), p. 44.

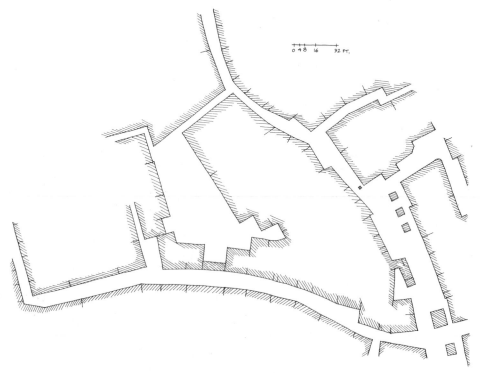

Figure 146. Bhaktarpur (Nepal) (still in existence). Based on Gutschow (1974), p. 44.

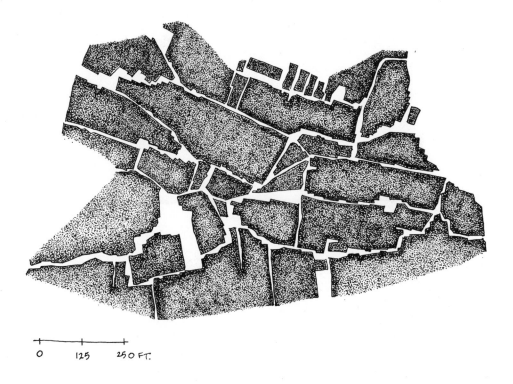

0   125   250 FT.

Figure 147. Taichung (Taiwan) (still in existence before partial change by new roads). Based on information provided by P. T. Han (Taichung).

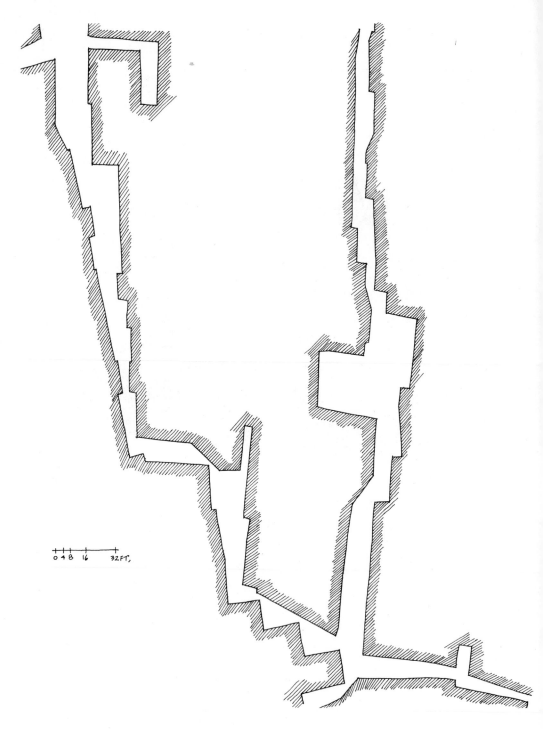

Figure 148. Taichung (Taiwan) (still in existence). Based on information provided by P. T. Han (Taichung).

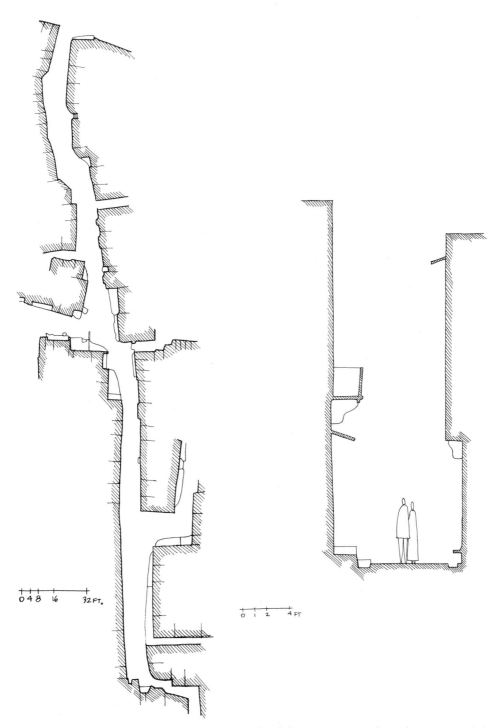

Figure 149. (*left*) Udaipur (India) (still in existence—founded in 1559 C.E.). Based on information provided by K. Jain (Ahmedabad).

Figure 150. (*right*) Udaipur (India) (still in existence—founded in 1559 C.E.). Based on *Udaipur* (1977), p. 37. Provided by K. Jain (Ahmedabad).

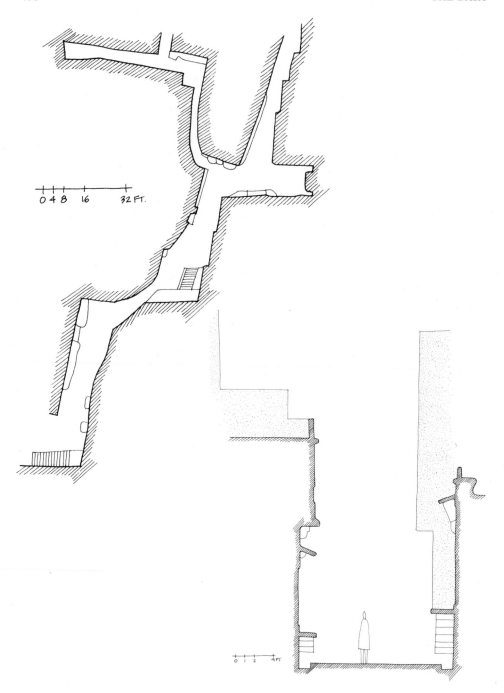

Figure 151. (*top*) Udaipur (India) (Dashora Marg) (still in existence—founded in 1559 C.E.). Based on *Udaipur* (1977), p. 44. Provided by K. Jain (Ahmedabad).

Figure 152. (*bottom*) Udaipur (India) (still in existence—founded in 1559 C.E.). Based on *Udaipur* (1977), p. 35. Provided by K. Jain (Ahmedabad).

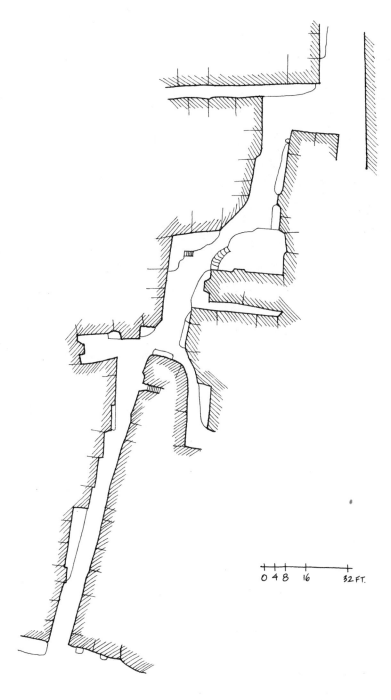

Figure 153. Udaipur (India) (still in existence—founded in 1559 C.E.). Based on information provided by K. Jain (Ahmedabad).

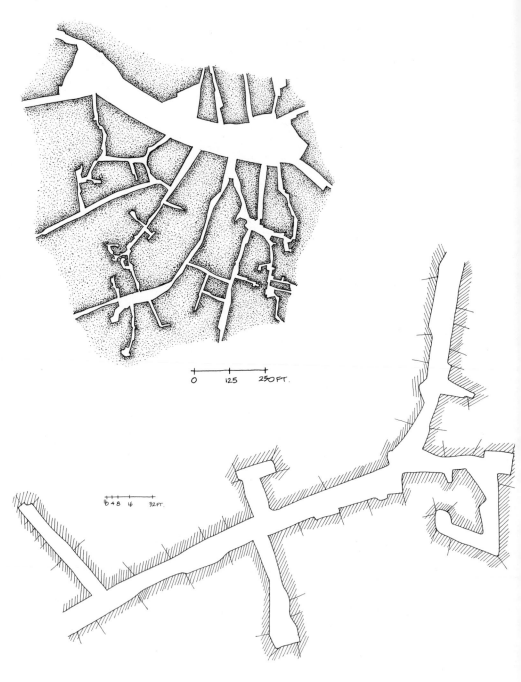

Figure 154. (*top*) Arberobello (Italy) (Rione Monti area) (still in existence). Based on Allen (1969a), Fig. 29, p. 113.

Figure 155. (*bottom*) Arberobello (Italy) (Rione Monti area) (still in existence). Based on Allen (1969a), Fig. 29, p. 113.

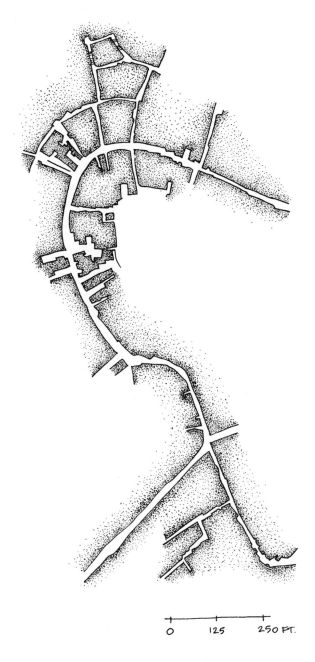

Figure 156. Murotsu (Japan) (Hyogo Prefecture) (still in existence). Based on information provided by R. Ohno (Kobe).

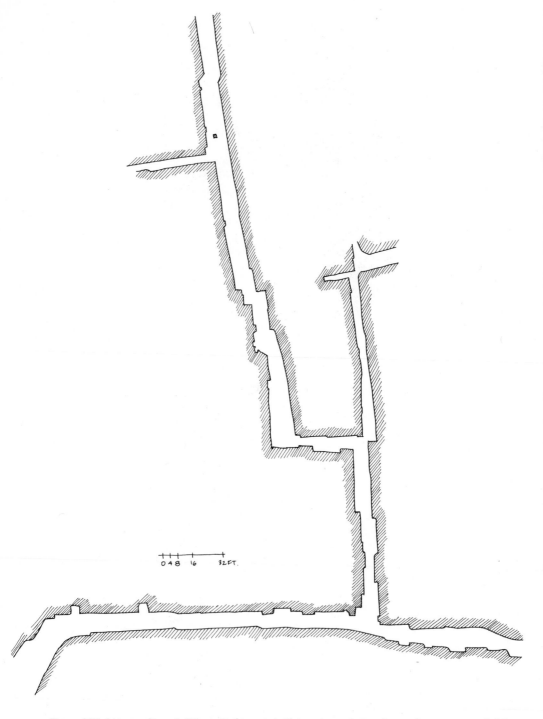

Figure 157. Murotsu (Japan) (Hyogo Prefecture) (still in existence). Based on information provided by R. Ohno (Kobe).

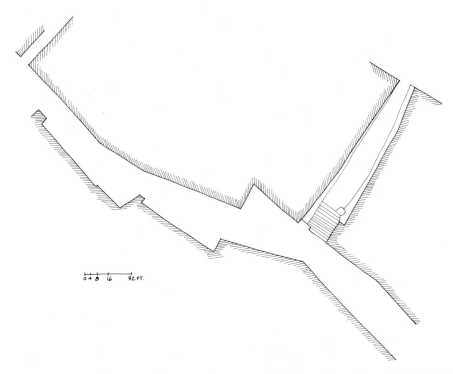

Figure 158. Zacatecas (Mexico) (16th cen.—still in existence). Based on information provided by H. Hartung (Guadalajara).

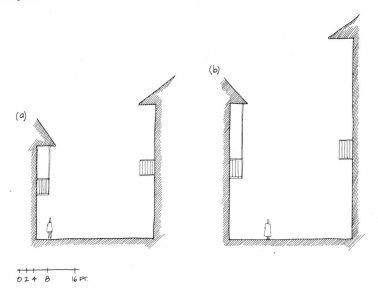

Figure 159. (a) Florián-Bogotá (Colombia) (Calle IA) (16th cen. c.e.—still in existence). Based on information provided by J. Salcedo Salcedo (Bogota). (Note that this is a "main" street, according to the Laws of the Indies). (b) Bogotá (Colombia) (Calle II) (16th cen. c.e.—still in existence). Based on information provided by J. Salcedo Salcedo (Bogotá) (Note that this is a "main" street according to the Laws of the Indies).

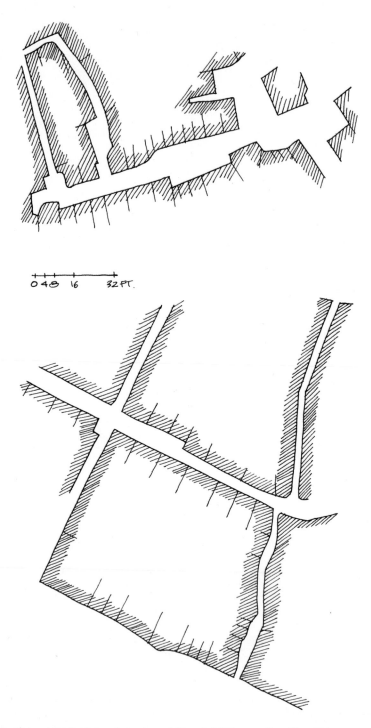

Figure 160. Bialystok (Poland) (unplanned parts) (founded 17th cen., plan of late 18th cen. C.E. city). Based on Gutkind (various), Vol. 7, Fig. 68, p. 83.

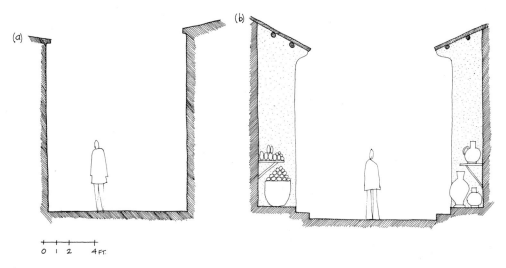

Figure 161. (a) Lukang (Taiwan) (17th–18th cen. C.E.; still in existence); (Yao Lin street). Based on Han (1981), p. 26; (b) Lukang (Taiwan) (17th–18th cen. C.E.; still in existence); (one type of shopping street). Based on Han (1981), p. 43.

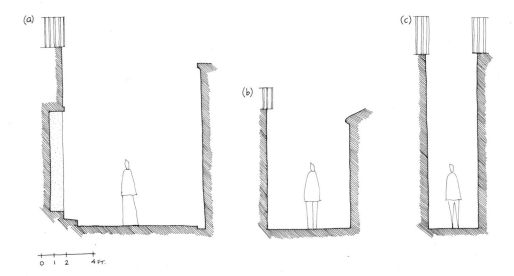

Figure 162. (a) Lukang (Taiwan) (17th–18th cen. C.E.; still in existence) (residential street). Based on Han (1981), p. 43; (b) Lukang (Taiwan) (17th–18th cen. C.E.; still in existence); (minor residential street). Based on Han (1981), p. 43; (c) Lukang (Taiwan) (17th–18th cen. C.E.; still in existence); (residential lane). Based on Han (1981), p. 43.

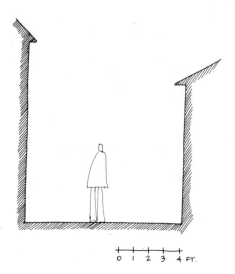

Figure 163. Tsinan (China) (Shantung province) (a "narrow, crowded commercial street") (date unknown—photographed ca. 1905 C.E.). Based on Buck (1978), p. 20.

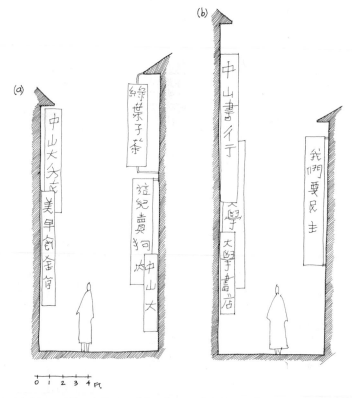

Figure 164. (a) Canton (China) (Street of the pharmacists) (Date unknown—photographed ca. 1870 C.E.). Based on Thomson (1973); (b) Canton (China) (Stationery street) (Date unknown—photographed ca. 1870 C.E.). Based on *Imperial China* (1978).

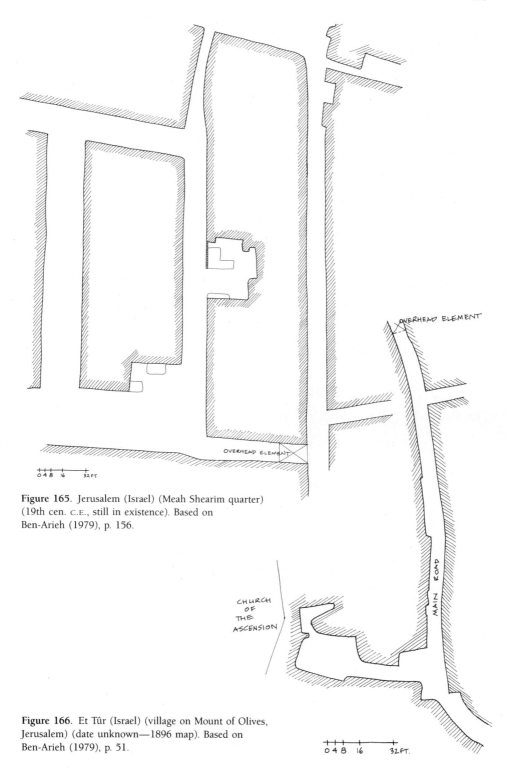

Figure 165. Jerusalem (Israel) (Meah Shearim quarter) (19th cen. C.E., still in existence). Based on Ben-Arieh (1979), p. 156.

Figure 166. Et Tûr (Israel) (village on Mount of Olives, Jerusalem) (date unknown—1896 map). Based on Ben-Arieh (1979), p. 51.

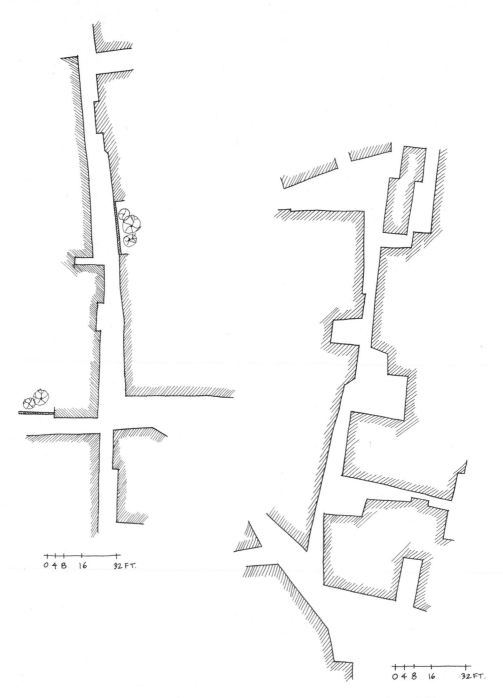

Figure 167. (*left*) Ankara (Turkey) (recent, still in existence). Based on information provided by M. Turan (Pittsburgh).

Figure 168. (*right*) Ankara (Turkey) (squatter settlement) (Early 20th cen. C.E.; still in existence). Based on information provided by M. Turan (Pittsburgh).

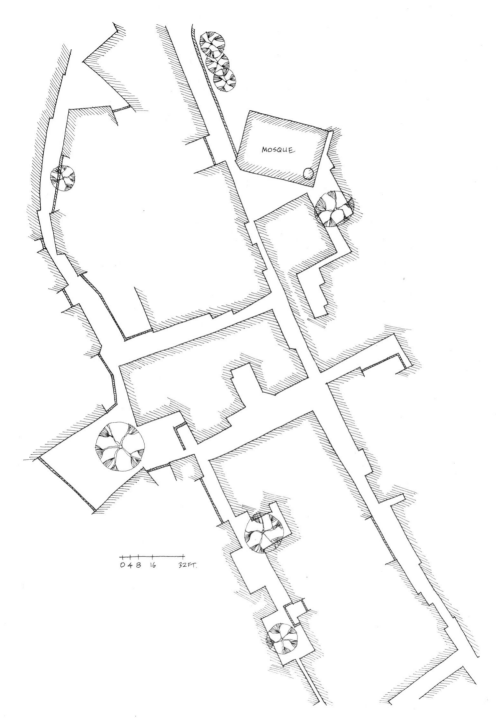

Figure 169. Ankara (Turkey) (squatter settlement in "Citadel") (Early 20th cen. C.E.; still in existence). Based on information provided by M. Turan (Pittsburgh).

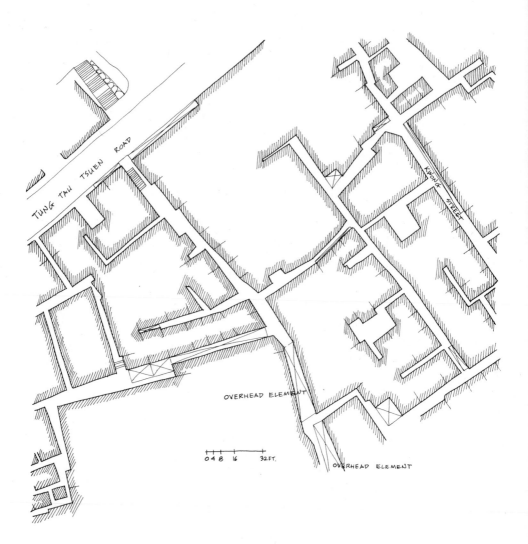

Figure 170. Hong Kong (China) (Kowloon area) (19th–20th cen. C.E.; still in existence). Based on Leeming (1977), Fig. 23.

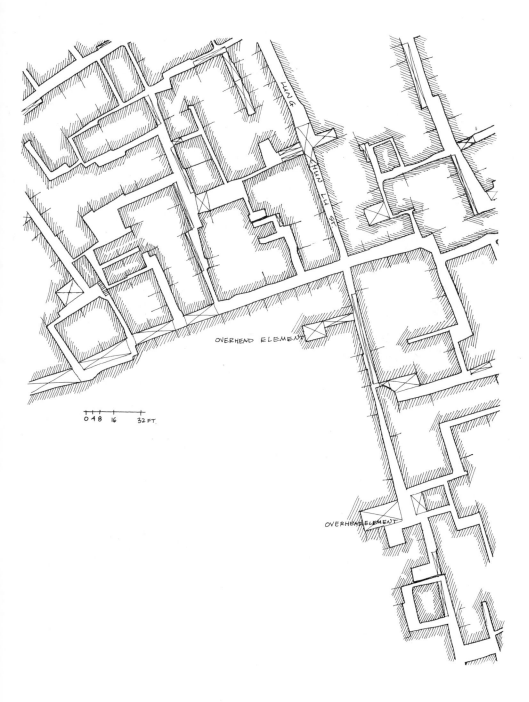

Figure 171. Hong Kong (China) (Kowloon area) (19th–20th cen. c.e.; still in existence). Based on Leeming (1977), Fig. 23.

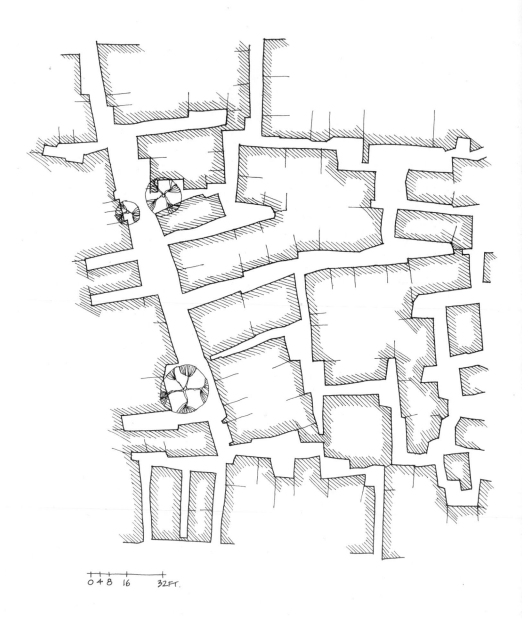

Figure 172. Rio de Janeiro (Brazil) (Favela Maré—squatter settlement) (contemporary). Based on information provided by V. del Rio (Rio de Janeiro).

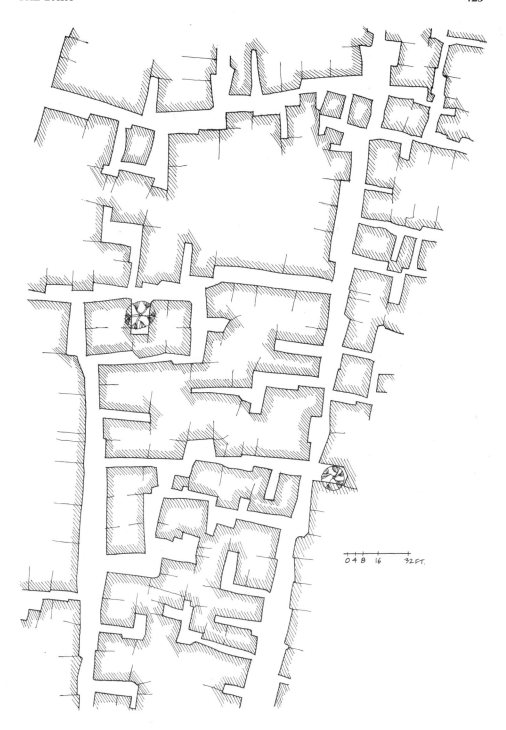

Figure 173. Rio de Janeiro (Brazil) (Favela Maré—squatter settlement) (contemporary). Based on information provided by V. del Rio (Rio de Janeiro).

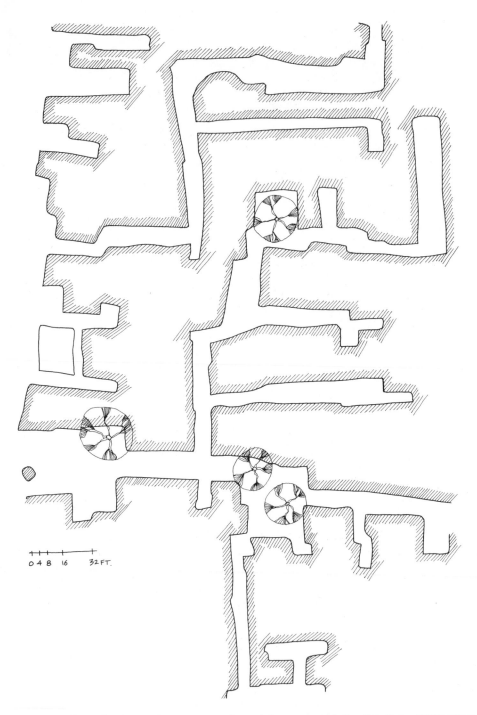

Figure 174. Delhi (India) (Rouse Ave. squatter settlement) (contemporary). Based on Payne (1979), Fig. 3.12, p. 100; Fig. 3.30, p. 112; Fig. 3.32, p. 114; Fig. 3.34, p. 115, and additional information in *Architectural Design*, Vol. 43, No. 8 (1973), p. 406.

Figure 175. Delhi (India) (Rouse Ave. squatter settlement) (contemporary). Based on Payne (1979), Fig. 3.13, p. 100; Fig. 3.30, p. 112; Fig. 3.32, p. 114; Fig. 3.34, p. 115, and additional information in *Architectural Design*, Vol. 43, No. 8 (1973), p. 406.

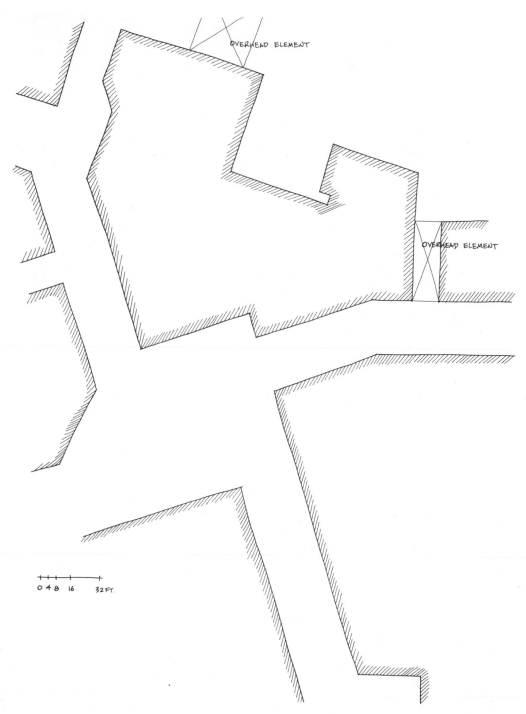

Figure 176. Louvain la Neuve (Belgium) (ca. 1970 C.E.; *professionally designed*). Based on information provided by P. Laconte (Brussels).

Figure 177. (a) Mugla (Turkey). Photograph by G. Versürer, provided by N. Inceoglu; (b) Stockholm (Sweden) (Old Town); (c) Singapore (1961 photograph); (d) Antigua (Guatemala). Photograph by H. van Oudenallen.

Figure 178. (a) Diest (Belgium) (Beguinage); (b) Solo (Indonesia); (c) Elsinore (Denmark); (d) Istanbul (Turkey) (Bazaar).

Figure 179. (a) Luoyang (China). Photograph by D. Buck; (b) Kota Gade (Indonesia); (c) Copenhagen (Denmark) (Pedestrian street); (d) Cachoeira (Bahia, Brazil).

Figure 180. (a) Durham (England); (b) Tokyo (Japan); (c) Herat (Afghanistan). Photograph by P. English; (d) Granada (Spain).

Figure 181. (a) Palma de Mallorca (Spain) (Janquet area); (b) Singapore (1961 photograph); (c) Istanbul (Turkey) (Fener neighborhood); (d) Paratí (Brazil).

Figure 182. (a) Bursa (Turkey); (b) Yogyakarta (Indonesia); (c) Rosario (Argentina) (Calle Cordobá pedestrian mall); (d) Leuven (Belgium).

Figure 183. (a) Midnapur (India); (b) Amsterdam (The Netherlands); (c) Büyükada Island (Turkey); (d) Trondheim (Norway).

Figure 184. (a) Madura (Indonesia) (Fishing village, north coast); (b) Lausanne (Switzerland); (c) Palma de Mallorca (Spain); (d) Rio de Janeiro (Brazil) (Saara area pedestrian precinct).

Figure 185. (a) Seville (Spain) (shopping street); (b) São Paulo (Brazil) (downtown pedestrian precinct); (c) Ankara (Turkey) ("Citadel" area); (d) Cordes (France).

Figure 186. (a) Gunung Kawi (Indonesia); (b) Vence (France); (c) Barcelona (Spain) (street in "Spanish village", built for World Fair in 1930s as a "typical" street); (d) Acco (Israel).

Figure 187. (a) Kam Tin (Hong Kong, New Territories); (b) Safed (Israel) (Old Town); (c) Basalú (Spain); (d) Ksar Hallal (Tunisia) (Old City). Photograph by S. Auerbach.

Figure 188. (a) Guanujuato (Mexico); (b) Chester (England) ("The Rows"); (c) San Juan (Puerto Rico) (Old City; Callejon de la Capilla); (d) Bergen (Norway).

Figure 189. (a) Basel (Switzerland) (Old City); (b) Nazareth (Israel) (Bazaar); (c) Nice (France) (Old City); (d) Ouropreto (Brazil).

Figure 190. (a) Yaffo (Israel) ("artists' quarter"); (b) Safranbolu (Turkey); (c) Villefranche sur Mer (France); (d) Sumenep (Indonesia).

Figure 191. (a) La Roquette (France); (b) Rio de Janeiro (Brazil) (Favela Maré—squatter settlement); (c) Mykonos (Greece); (d) Village (Punjab, India). Photograph by B. S. Saini.

Figure 192. (a) Marrakesh (Morocco). Photograph by T. Porter. (b) Barcelona (Spain) ("Gothic quarter"); (c) Kharagpur (India) (Bhagwanpur area); (d) Roquebrune (France).

Figure 193. (a) Geneva (Switzerland) (Old Town); (b) Jerusalem (Israel) (Old City, Christian quarter); (c) Santorini (Thera) (Greece). Photograph by G. Dix; (d) Kairouan (Tunisia). Photograph by S. Auerbach.

Figure 194. (a) Lindos (Greece; Island of Rhodes); (b) Zurich (Switzerland) (Old City); (c) Cuzco (Peru); (d) Arberobello (Italy).

Figure 195. (a) Noci (Italy); (b) Louvain la Neuve (Belgium).

Figure 196. (a) Shanghai (China) (Old City). Photograph by M. Utzinger; (b) Rabat (Morocco) (Old City [Medina]). Center for Middle Eastern Studies, University of Texas, Austin. Photograph by C. Saenz; (c) Athens (Greece) (Plaka area).

Figure 197. (a) Delft (The Netherlands). Photograph by H. van Oudenallen; (b) Isphahan (Iran); (c) Bangkok (Thailand) (shopping street).

Figure 198. (a) Quito (Ecuador) (Calle de la Ronda). Photograph by C. Bastidas; (b) Katmandu (Nepal) (shopping street); (c) Paratí (Brazil).

Figure 199. (a) Taxco (Mexico). Photograph by H. van Oudenallen; (b) Hong Kong (China) (shops, 1960s refugee housing); (c) Puno (Peru).

Figure 200. (a) Xian (China). Photograph by M. Utzinger; (b) San Francisco el Alto (Guatemala). Photograph by H. van Oudenallen; (c) Mindanao (The Phillipines) (shopping street, Moslem village). Photograph by J. Taylor.

Figure 201. (a) Unidentified Town, Kutch (India). Photograph by K. Jain; (b) Unidentified Town, Rajasthan (India). Photograph by K. Jain.

Figure 202. Florián-Bogotá (Colombia) (Calles IA and IB). From *Papel Periódico Illustrado* between 1881–1887; provided by J. Salcedo Salcedo.

Conclusion

The argument in this book has been rather lengthy and has covered much ground. This argument also seems complete and should be clear. If that is the case, conclusions are not really necessary. Moreover, any conclusions beyond the argument itself are really up to the reader. However, because of the diversity of material covered and the rather novel nature of the proposal, a conclusion, albeit short, may be in order. If, however, the argument has had any validity and been reasonably presented, there should be no need for a summary, particularly because I periodically restated the major premises.

The conclusion will, therefore, not be so much a summary but rather will emphasize the implications of the argument. In summary, it may be sufficient to point out that this book has made several major points.

1. The history of the built environment needs to be approached quite differently than current history of architecture. The subject matter of the domain should be quite different, the problems and questions need to be different, and it follows that the approach must be different. In essence, it implies a shift from a metaphor based on art to one based on science.
2. Although this proposal is quite novel in this connection, parallel developments and precedents can be found in a variety of disciplines. These include art history itself, history, and the social sciences; moreover, it can be shown that a variety of natural sciences are essentially historical and therefore bear on the argument. Recent developments in archaeology offer particularly useful parallels.
3. Like all study of the past, indeed like all science, this approach involves inference. Inference typically is analogical and inevitably implies reliance on contemporary data. Many of these data need to be based on environment–behavior studies.
4. Conversely, EBS requires historical data to supplement contemporary research, in order validly to generalize, just as it must use cross-cultural data, the whole environment, and all types of environments.
5. Only through a mutual interaction of EBS and history, in terms of concepts and theory, can the past provide valid precedents for design—precedents made necessary by the shift from selectionist to instructionist design and the need to design new kinds of environments.

453

In discussing and supporting these points, many specifics were discussed. In arguing the case and providing supporting arguments, much ground and many disciplines have been surveyed, albeit briefly, and shown to be relevant. A very preliminary case study was then tackled to illustrate the approach and to show the kinds of data needed. Some of the implications of this type of history, which can also point to the future and to further work, will now be sketched very briefly. I will do this "backwards," beginning with the case study and then, more briefly, turning to the whole book.

IMPLICATIONS OF THE CASE STUDY

Regarding the case study, three things need to be discussed:

1. What do the data show?
2. What do the data mean?
3. What are the implications—for design, for EBS, for history, and for theory.

What Do the Data Show?

One of the first things the data show is the inadequacy of published material. It is surprising how few of the hypotheses could be tested, although that is partly a function of the lack of methods for analyzing the existing material. Thus looking at pictorial examples of such streets and, even better, actually experiencing them as a pedestrian makes it quite clear that the hypotheses apply. Because personal, subjective feelings and opinions are not an adequate basis for design, however, it follows that work is needed, in the field (i.e., in surviving settlements), in archaeological settings, and on archival and library materials.

In effect, the case study turned out to be only a very preliminary test that suggests that the hypotheses are plausible and worth pursuing. Only 1 or 2 of the 36 specific hypotheses could be tested due to the lack of data. But more was involved. Not only would one need to go out into the field and go through even more archival data and measure the variables involved, there is also an urgent need to develop new methods of operationalizing these types of variables so that they can be measured and new ways of analyzing them.* Due to the lack of such methods and data I have looked at the material rather unsystematically and largely qualitatively; a more quantitative and systematic study remains to be done.

But even at this preliminary, qualitative level, before one can argue that the hypotheses are supported (quite apart from the fact that I tried to "prove" them rather than refute them), other, alternative hypotheses need to be considered. A number can easily be proposed.

1. With no wheeled traffic there is no need for wide streets. There is thus a tendency to minimize circulation distance as well as use of space and cram as

---

*For one analogous argument in archaeology, which is taken much further and has suggestive ideas for this topic, see Clarke (1978).

much building as possible into a given space. Because resources tend to be scarce, the smallest area of the city is then needed.

2. This is compounded, in the case of certain settlements, by the need for defensive walls. Minimizing the area that needs to be enclosed by such walls is an important desideratum. Because this is a researchable question, it suggests an interesting study: comparing streets in walled and unwalled settlements.

3. Again, still in terms of defense, the labyrinthine quality of street networks confuses potential enemies, whereas the specific characteristics of streets makes defense easier and more effective.

4. As an alternative social reinforcer of 1, the generally low status of the inhabitants of vernacular areas of settlements is reflected in the minimal spaces they use as compared to elite areas. The spatial character is a way of establishing and maintaining social distance. (This also suggests comparative studies to test this.)

5. Another social interpretation is that such urban fabrics and streets discourage penetration by outsiders into neighborhoods, define semiprivate spaces, and so on. This is certainly a factor in the case of the Middle East and North Africa, India, parts of Southeast Asia, and possibly elsewhere.

No doubt other hypotheses could be proposed, and readers will doubtlessly think of them. One has recently been proposed, at least implicitly (Hillier and Hanson 1984). This interprets the street pattern, particularly at what I call the "fabric" level (specific hypotheses F_{i-iv}) in terms of intelligibility, easy learning, and *attracting* people, and hence encouraging social interaction. Although the intelligibility criterion seems strange, the outcome is the same—higher levels of occupancy, that is, more pedestrians. However, the reasons for this, *the mechanisms* linking people and these settings, are rather different: Whereas I emphasize perceptual ones, theirs are based on social interaction. Moreover, with the exception of visual segmentation (e.g., specific hypotheses C_{ii-iii}, D_{ii}), they do not deal with the other characteristics that I examine. On the other hand, they provide a well-developed formal and quantitative approach that could serve as a model.

It is also important to note that these two major hypotheses are potentially complementary rather than conflicting. If Hillier and Hanson (1984) are correct, then the selectionist pressures toward the patterns found become even greater, encouraged by both perceptual and social mechanisms. Moreover, their approach provides another set of possible precedents—for settings supportive of social interaction; some of these possibilities I discuss, from a different perspective, in Chapter 7.

Social interaction mechanisms tend to be less invariant than perceptual ones, for example, in terms of desired levels of interaction and, even more, in the appropriateness of streets as settings for such interaction. This is why I tend to classify them among cultural aspects of street use. As a result, the precedents may be less directly applicable. Any precedents would depend, in the first instance, on whether social interaction is desired and whether streets are seen as appropriate settings for this (see Rapoport 1969a, 1977, 1986a). There would also be limitations due to the extent of the effect of environment on social behaviors. Finally, their hypothesis needs to be tested similarly to mine against a large sample, cross-culturally, and historically. It also needs to be tested empirically against contemporary EBS research.

What Do the Data Mean?

The first significant point is that, because work on environmental complexity began in the late 1950s and early 1960s and, so far, has been mainly limited to the United States, Canada, and other Western countries, the case study greatly broadens the body of evidence and the applicability. It suggests that the principles hold over thousands of years and all over the globe. This, of course, gives one much more confidence that an important principle and mechanism are involved. This clearly has the intended consequences for valid generalization and theory development in EBS.

The material presented and analyzed seems to support the hypotheses strongly. The material not used does also. Just in the files that I have, there are many more plans and very many photographs. It would be easy to present thousands of plans and photographs. Also, once "sensitized," I suspect that readers will come across much supporting evidence. This can be done during almost any trip to the types of environments I discuss. I recently did that informally in Jogjakarta, Jakarta, Singapore, and Hong Kong, and also during trips to Italy, the Netherlands, and Britain.

It also seems that the more purely pedestrian the setting, the more the predicted attributes are present and the more strongly they are displayed—for example Islamic cities, Nepal, Venice, and so on. Even the anomalies discussed, for example, Spanish Latin America, the Bastides, the United States, support the hypotheses and, in fact, strengthen the argument. Unlike these anomalies, however, the case of sub-Saharan Africa constitutes a more difficult counterexample.

Before discussing it briefly, however, I should point out that it proved almost impossible to obtain data, in spite of a major effort, including correspondence with and visits to libraries specializing in African material. Many of the books found on cities (not common to begin with) had nothing on streets. In one sense, like many if not most tribal societies, those of Africa engaged in entirely pedestrian movement without developing streets—but then many of them had no major settlements. In general it seems highly probable that many of the social-interactional and other functions of traditional streets occurred in compound and dwelling courtyards and other settings (cf. Rapoport 1973a). The real problem arises in those cases where settlements did exist.

In such settlements, and many of them were quite large, much pedestrian movement occurs even today but, as in the case of Tswana towns, for example, the spaces between compounds and clusters of compounds show no characteristics of streets, and none of the perceptual qualities to be expected on the basis of my argument. The situation in Yoruba towns is less clear (partly because I have not visited them). It appears that the paths among compounds do show some of the characteristics expected. This is also the case in all African cities, including those of sub-Saharan Africa, which were influenced by Islam, for example Hausa settlements, Timbuctoo, Zanzibar, Lamu, and so on. Those do show all the expected features in a very strong form as a result of diffusion.

Among the few examples that I found of non-Moslem African cities, some of the characteristics expected can be found. For example in a book by H. Ling Roth, *Great Benin,* published in Halifax, England, by King, in 1903, Figure 163, a bad photograph from 1897 appears to be a 14-ft-wide street with buildings of the same height, that is a width-to-height ratio of 1 : 1. Figures 172 and 173 also appear to show quite narrow streets. There seems to be a discrepancy between them and the verbal descriptions that refer to streets as very wide and mention 120 ft width.

T. J. Bowen, in his *Adventures and Missionary Labour in Several Countries in the Interior of Africa from 1846–1859,* on page 295 verbally describes streets as generally very narrow, crooked, and intricate. Walls are described as so close on each side that the leaves of the low, thatched roofs almost brush people as they pass. He comments that they were designed for humans and pedestrians and, because goods were carried by porters, this width was ample.

Thus the evidence is inadequate and partly equivocal. In general, however, the situation in many traditional settlements of sub-Saharan Africa is a puzzling anomaly and needs investigation.

There are many data that I collected but did not use in the case study; they also support the hypotheses. As pointed out, a rather small portion of the available examples in plans, photographs, or slides were used. However, I wish to refer to two bodies of evidence that are extremely relevant to this section of the conclusion.

The first is an eight-volume series by A. E. Gutkind, *The International History of City Development* (Glencoe, The Free Press, various dates). This covers *all* of Europe, from the Atlantic to the Urals and from Scandinavia to the Mediterranean, as well as some examples from India, Turkey, Central Asia, and so on. The time span covered is from ancient times to still existing settlements.

Many of the plans do not have scales; many were too small to measure adequately, or traditional scales were used. I counted 455 such plans! I did measure as many plans as possible that had scales and were large enough (over 100). I also examined the many hundreds, if not over a thousand, photographs. Generally, I would say, that there is hardly an example in this series, with its more than 600 plans and over 1,000 photographs, which does not almost completely support all the hypotheses. It also seems that the methods that I tried out informally offer the beginning of a more rigorous, quantitative approach. On plans I measured widths, lengths of blocks, number of units per block, length of view, and so on. In photographs, I examined the height–width ratio, number of units visible, blocked views, textures, level changes, number of overhead elements, percentage of sky visible, skyline complexity.

More specifically I conclude that:

1. The overall street pattern is as predicted as judged from all the plans, aerial photographs, and ground-level views.
2. The enclosing elements themselves are very complex, with different textures, balconies, projecting windows, and other elements, complex skylines, overhead elements, and so on.
3. Moreover, there are many such elements. By averaging the number of houses per block, and the block lengths, the width of units tended to be significantly below 9 m (27 ft), although there were exceptions. This is also clear from the hundreds of photographs that confirm the *very fine grain* (as well as high texture) of the enclosing elements. In a more detailed count of over 50 photographs, the number of units (generally houses) visible per side generally varies between 7 and 12 but sometimes goes up to 20 before a major break—

projecting tower, high-style building, or whatever. Views also appear very short.

4. The lengths of views are fairly short, partly because the blocks are short and the streets turn frequently. Although blocks range from 30 to 100 meters, most are between 30 and 70 m. Views were short; although a few went up to 300 m, most were below 100 m, many well below. This was due to angled turns, projecting elements, curves, and double curves.

5. Widths for average streets range between 8 and 12 m, whereas narrow streets are 3 to 6 m, and principal or main streets are often 20 m. These figures are encountered over and over again, although there are cases where the main street is 10 m and average side streets between 3 to 7 m.

6. Width to height ratios typically ranged between $1:1–1:5$ to $1:2–2:5$ but went as high as $1:3$.

7. In many photographs no sky was visible, and usually the sky visible was under 15% of the area of the photograph and was rarely as high as 25%. Of course, due to cropping of photographs and so on, little significance can be given this figure without more fieldwork. In many cases, the sky visible was interrupted by overhead elements—arches and the like.

8. Widths were rarely constant and tended to vary continually. This seems a particularly difficult characteristic to measure and to communicate quantitatively.

9. Strong light and shade contrasts are to be seen quite typically.

10. Underfoot level and texture changes are common and prominent.

11. It is striking how similar the characteristics tend to be across the large spatial and temporal range included. Sometimes examples in different places could be substituted for one another. It seems, however, that generally in the Mediterranean area, the streets are narrower and more enclosed, blocks are shorter, and views are shorter. This could be due to Middle Eastern/Islamic influences, to climate, or other variables.

The second body of evidence consists of 126 photographs of streets from around the world that I examined in *National Geographic* from 1888 through 1980. Once again, all of those showed almost all the characteristics predicted. Moreover, even less formal measurements and counts closely agreed with those described. Many show even less sky (many—none), higher textures underfoot, and enclosing elements; a very large number had overhead elements—many more than the Gutkind examples; more interrupting elements, and shorter views. This may be due to the fact that most of these are non-European and have been selected for their "exotic" appeal and "picturesqueness." The widest typical street in this series was about 36 ft (12 m), and most were less, at about 20 to 24 ft and even less; the narrowest was 8 ft (less than 3 m).*

---

*There were two major exceptions, both nonvernacular. One was a major avenue in Imperial Beijing which was 72 ft wide, the other a colonial Italian street in Tripoli, Libya, also 72 ft wide (compared to the average traditional street of about 10 ft). For other Chinese examples, of a more vernacular character, see early photographs in Beers (1978), Thomson (1973), and *Imperial China* (1978). Many more such photographs, from many places, can be found in books and also in magazines, archives, and collections all over the world.

Note that I have not analyzed any drawings or paintings so far, although they are as accessible and suitable as photographs.

There is another method that could be used to supplement that used and further confirm the hypotheses. This would make a good study in the future. This involves an informal (or, if desired, more formal) content analysis of written sources—travel accounts, novels, and the like (cf. Rapoport 1970a, 1977, 1982a). This has obvious methodological implications, not the least in the use of multiple methods that provide much more confidence in any results. It also greatly broadens the evidence that can be used because many travel accounts describe many more things than they illustrate and because novels, newspapers, magazines, and so on become available.

Verbal material can provide two types of data. It can often give dimensions that are not shown graphically, and it can provide descriptions of character and quality. I will give a few examples of each from archaeological and travel sources; novels and the like will not be adduced here.

Regarding dimensions, one finds Bibby (1972) referring on page 187 to early Dilmun city (2250 B.C.E.) as having a 13-ft-wide street. On pages 202–203, he mentions that Ur had narrow, winding streets averaging only about 1.6 m. On page 365, he refers to the widest avenues at Dilmun, which divided the city into precincts as being 40 to 45 ft wide. He adds that streets that wide had only been found at Harappa and Mohenjo-Daro. This not only reveals that most streets were narrow but that wide streets were rare. Similarly, in Masson (1972) we find, on pages 263–277, many examples that tell us that Neolithic settlements in Central Asia have streets that show all the characteristics that I have been discussing, with widths ranging between 0.5 m and 2.5 m. Fruin (1971, p. 6) points out that many medieval streets had a width–height ratio of 1 : 2, but that Leonardo da Vinci argued that 1 : 1 was better. Finally, one of my students (Han Chien) described to me an arcaded shopping street in the main town of the Pescadore Islands, near Taiwan, which was 5 m wide and the enclosing elements of which were 6 m tall. Other streets, lanes, and alleys were even narrower.

In a more descriptive vein, the Gutkind series often refers verbally to the "intricate web of streets" and "intricate web of streets, alleys, blind passages and squares." Doumas (1972, p. 228) refers to "corridor-like streets." Redman (1978. pp. 15–16) after mentioning that streets are narrow, 2.5 m or less, and generally at right angles, adds that they "made many short turns, so that the street plan was nowhere completely linear . . ." ". . . narrow, twisting alleyways." Russell and French (1981, p. 220) refer to "a system of narrow, intersecting passageways, often no more than 3 m wide and small irregularly shaped public spaces" and (p. 221) to a "labyrinthine system of narrow interconnecting passageways." In this case illustrations confirm this. Thus Figure 13 shows a street 5 m wide with buildings 7 to 10 m high, that is, a width–height ratio of 1 : 1.5–1 : 2. Finally, here is a series of descriptions accompanying some of the 126 photographs in *National Geographic* mentioned before.

> Ways that are narrow and far from straight . . . in which a wheeled vehicle of any size cannot pass
> Narrow torturous windings . . . of its thoroughfares . . . noisy and picturesque confusion

Narrow winding streets with steep ascent . . . picturesque . . .
Quaint, narrow streets of the residence section . . .
Twisted, narrow lanes . . .
Typical street . . . narrow, tortuous and in places vaulted over . . . under foot . . . cobblestones
Rows of tiny arched stalls, or booths . . . facing narrow streets
Vaulted thoroughfares . . . shade . . . dark alcoves [with shops]
Winding narrow streets . . . balconies . . . shuttered windows
Narrow streets . . . little elbow room . . . tall houses . . . along whose fronts the weekly wash hangs to dry [adding to texture and complexity]
Arched streets . . . splotched on sunny days with bright patches and dark shadows
Cobbled steep streets . . .
Shadowy lanes . . . too narrow and tortuous for automobiles . . . overhanging second floor . . .
Arch shades a street . . . [and breaks the view]
Only pedestrians in . . . narrow shopping lane . . .
Maze of narrow chasms . . . so constricted . . . men could touch buildings on either side . . . so crooked . . . visitors lost all sense of direction
Narrow street . . . shady . . .
Street alive with color from skyline . . . to ground itself . . . itinerant peddlers . . . [display bright things]
Narrow street . . . babble of voices, aroma of spices and displays . . . assail the senses

Enough has been cited to show that such verbal descriptions can be used to test the hypotheses. Together, such verbal and pictorial materials (plans, sections, photographs, drawings, and paintings) can lead to a grander synthesis and more confidence—particularly if one also cites the quite common verbal material on those streets and urban spaces that are *not* interesting or suitable for pedestrians. Such data could also be related to culture by studying norms of personal distance and space; acceptable contact, density, and crowding; social interaction and the like.

There seems to me to be incontrovertible evidence for *the highly culture-specific nature of design,* about which I have written much. Thus, if one finds, as one does, very strong cross-cultural and temporal regularities, these become even more significant. If *anything* remains invariant over long time periods, in many places and cultures, this makes a strong prima facie case that it is likely to be highly significant and important. That this pattern that was sought was found in the case of built environments that are expected to be highly culture specific, and hence variable, makes it even more important.

I have already discussed certain anomalies that provide counterexamples. These were areas, such as Spanish Latin America and sub-Saharan Africa, or streets such as the Champs Elysées, Dizengoff, The Ramblas and Paseo Colon, and the like. These latter I interpreted in terms of social aspects, such as people-watching. I have also interpreted them in terms of liking versus interest, that is, I have argued that such spaces, as well as plazas and other urban open spaces, may well be *liked* more. These two points apply also to some of the counterexamples that come from contemporary EBS research—because these typically deal with preference, that is liking.

For example, Funahashi (1986, p. 93), in passing, mentions work in Japan suggesting that wide, busy streets are preferred. Hill (1984) gives two apparent counterexamples. On page 37 he cites work by Blivice, which shows that people seem to prefer exposed places rather than enclosed areas such as arcades and narrow streets; to like plazas and green islands. Note that this deals with *preference* and *liking*. The same study points out that those pathways are preferred that are *thronged with people* and *lined with activities* and that pedestrian malls are chosen *in preference to quieter parks and tree-lined streets*. The evidence thus seems contradictory but seems to hint at the distinctions I made between liking versus interest and rest spaces versus movement spaces. The second study cited (on p. 41) suggests that pedestrians prefer the *simplest* path, but Hill comments that this study by Marchand is difficult to evaluate. It also does not deal with the perceptual quality of the path chosen being concerned with cognition (see the discussion of this study in Rapoport 1977).

Two further possibilities arise, specially about some of the other possible counterexamples found in Hill (1984). The first is cultural; many of the studies refer to the United States and there greenery, nature, and so on are preferred, and urban areas are generally seen negatively (e.g., Rapoport 1977; Tuan 1974). There is nothing more urban than the kinds of streets I have been discussing. The second additional factor (and hence fourth overall) is one of safety, particularly regarding crime that has become a major factor in urban space use (Rapoport 1986a). It has become so important that it has even led to major clearance of greenery in parks, along jogging and bicycle paths and trails and so on. Be all that as it may, however, these and any other counterexamples need to be considered and any contradictions worked out.

As already mentioned, the occasional counterexamples (and they are few in number) are never unequivocal. Thus regarding Japan, contrary views can be found. For example, in a lecture at EDRA 18 in Ottawa, in late May–early June 1987, Kubota (1987) argued that in Japan streets used to be mainly for pedestrians. As such, I would add, they clearly show the expected characteristics as some of the material in the case study, and my visit to Japan, would show (see also Alden 1986 later). With the spread of car ownership after the 1950s, according to Kubota, drastic changes occurred because the traditional pattern was unsuitable for car traffic. Currently, there has been a reaction, with people seeking a return to the traditional qualities. As a result, in the 1980s there is great interest in the experiential quality of traditional pedestrian streets, with much research on this beginning. As one example, I have recently received a proposed study meant explicitly to test my general hypothesis (as proposed in Rapoport 1977, 1987a) using photographic techniques (Alden 1986). It is argued that Tokyo streetscapes are among the most complex in the world and should therefore be particularly attractive, interesting, and pleasurable to pedestrians. This study then hopes to replicate the research cross-culturally and is thus complementary to this study. In any case, the author intuitively and subjectively supports Kubota's and my positions.

With this research in Japan, which clearly is trying to derive precedents for design, I am really getting into the first aspect of my third point–the implications of the case study for design.

What Are the Implications of the Case Study?

First, regarding design, one can suggest that if Kubota is correct, it is an example of something one would expect to find—intuitive responses to, and seeking of, the characteristics discussed. I would argue that many newly designed or restored settings such as the Faneuil Hall market in Boston, the Baltimore Harbor development, The Grand Avenue (Milwaukee) mall and other malls, newly popular markets and so on, all display these qualities. One would also expect to find that when areas of cities are pedestrianized, these will tend to be areas with characteristics as close as possible to the ones discussed.

This certainly seems to be the case in Germany, Denmark, Britain, Belgium and the rest of Europe, Brazil, Argentina, and so on—it is always an area with narrow streets, high enclosure, complex fabric, and so on. Even in the United States, one finds that pedestrian malls tend to be as much like that as possible, given what is available; mainly they tend to be narrow.* For example, Brambilla, Longo, and Dzurinko (1977) provide data on 76 malls in the United States and Canada. Widths vary between 24 and 160 ft, although only 8 are above 100 ft, and the widest (160 ft) is a square. 52 of the 76 are below 80 ft, and the average for all 76 is 65 ft. Moreover, many try to reduce the width by using planters, trees, sculptures, and the like. Results generally are not highly satisfactory, but it seems from my own experience, that the more successful pedestrian malls in North America tend to be the narrowest ones. Again, this could and should be studied.

It is also of interest that in the Belgian new town of Louvain-la-Neuve, which is largely pedestrian, a medieval street pattern (supposedly based on Leuven) was used. In that case the precedent seems to have been direct and also intuitive, rather than through any analysis of principles. Although the idea of the "medieval model" may be acceptable and appropriate in the context, particularly due to the meaning of the reference to Leuven, there may be other ways of achieving these goals, based on principles. Also, that would provide more certainty and confidence.

Even more important, once a precedent is in terms of *principles,* it can be applied in other contexts where a medieval form may not be appropriate, that is, one can go beyond copying a medieval plan in a pedestrian-oriented new town in Belgium. Also, one might then find that not everything needs to be copied, that only certain important components need to be incorporated, each of which may have different expressions.

Further, once *principles* are derived, they can be applied to settings other than streets, for example, to other urban spaces (Rapoport 1986a), to highways, freeways, and other transport systems (Rapoport 1977, 1987a). In terms of complexity generally, the precedents become relevant and applicable to other totally different settings, such as custodial institutions (health, mental health, geriatric, penal, etc.), schools, parks and nature trails, and many others. *It is a precedent potentially applicable to all design through EBR theory.*

There is another implication. The characteristics of pedestrian streets described by the specific hypotheses came about through *selectionist processes* whereby these settings gradually became congruent with pedestrian movement; this probably happened repeatedly and independently. Now, *instructions* are needed for design, hence principles, leading to criteria, derived from large bodies of evidence—most of it in the past; these become the precedents. It follows that such precedents need to be derived *explicitly* through research, as argued in the text. Thus the apparently "sudden" need for prece-

---

*One could, of course, also argue that such streets, and areas, are most unsuitable for car traffic, that is, that we are dealing with a push rather than a pull phenomenon.

dent is not, of course, sudden at all. Implicit precedents always existed, particularly in vernacular design that, as we have seen, has always been most of the built environment. These precedents were the traditional models used in any given situation, the unquestioned models (Rapoport 1969a, 1988b, 1989b). These changed slowly, largely through selectionist processes and adaptations, and even apparent innovations were based on them, as is the case in biological evolution (e.g., Gould 1980; Jacob 1982).

The product characteristics of vernacular design thus result from selectionism. Now instructions are needed. The potential relevance of the case study for design raises the question whether any analysis of the products of selectionism can be of help in the instructionist mode. Given that such products show congruence with human needs and wants and given that these latter are also derived independently on the basis of contemporary empirical work, the answer seems clearly to be "yes." This is particularly the case if the purpose is to provide answers to explicitly posed specific questions related to the process of instructionist design through EBR theory. In my own work, such analyses have proved useful in dealing with meaning (e.g., Rapoport 1981a, 1982a), sacred settings (e.g., Rapoport 1986b, 1982e), urban design (e.g., Rapoport 1977, 1986a), energy use (e.g., Rapoport 1986b), among others. That should also be the case more generally, as this book argues.

Finally, the implications of the case study for design also involve the question of whether designing for pedestrians is still a valid problem. It seems that it is. In any urban context, there is still a great amount of pedestrian movement. Thus Garbrecht (1978) points out that in West Germany, 27% of travel overall is still pedestrian (versus 14% public transport, 13% bicycle, and 46% cars and motorcycles). For shopping and work, there is still significant pedestrian movement; in recreation, for example en route to parks and so on, this rises to 80%. All these figures exclude strolling, promenading, jogging, and the like. He argues that the percentage could be raised—my case study suggests one major way of doing so. It is clear that walking is still significant in many other places and is still the principal means of travel in many others. Moreover, it is likely that in Western countries this will increase, given future developments in energy availability and the growing interest in physical fitness and recreational walking. Pedestrian movement, finally, still forms an essential part of any trip: There is a great deal of complex pedestrian movement in towns at either end of any other travel modes. Although these other modes have been studied (although insufficiently from an EBS perspective), pedestrian movement has been rather neglected. The case study suggests how history can play a role in such study.

The implications of the case study for EBS, for history, and for theory have already been discussed implicitly. For EBS it is an essential requirement for valid generalizations and more certain conclusions. It also forces one to consider alternative hypotheses and mechanisms and suggests how these could be studied, that is, it raises new questions and problems as we have seen repeatedly. Not only does the case study seem to confirm my argument about the value of broad and varied bodies of evidence and long time spans, but it also suggests that historical data and EBS can be mutually relevant, although that needs to be checked and demonstrated, not assumed. Moreover, *how* it becomes relevant is a matter requiring much further study.

This also relates to the relevance of the case study for the study of history and for theory. Because it was meant to illustrate the argument, its implications for those are best discussed in terms of the book as a whole. But as implied immediately before, in both these areas the case study enables one to evaluate historical and "theoretical" studies. These, whether about streets (e.g., Anderson 1978; Rudofsky 1969; Vernez-Moudon 1987), about other urban spaces (e.g., Crowhurst-Lennard and Lennard 1984; Korosec-Serfaty 1982)* or urban design generally (e.g., Cullen 1961; Krier 1979; Sitté 1965). These and many, many others could be taken as hypotheses or suggestions and tested against the type of evidence with which this book deals.

Consider the example of Korosec-Serfaty. She argues that history shows the need for sociability in urban public spaces—a common theme in the literature (e.g., Crowhurst-Lennard and Lennard 1984; Sennett 1977). I would predict, however, that using evidence of the type used in this case study and testing the argument cross-culturally and over long time spans would tend to show that concepts such as privacy, boundaries between public and private, sociability, relations between inside and outside, activities and behaviors appropriate in different settings, and so on are highly variable. In other words, the outcome would be very different from her case study and examples—and good reasons could be given for the changes that she regrets. In other words, I predict that the argument would be shown to be ethnocentric, based on personal and ideological preferences and hence too normative. There is, of course, always the possibility that my predictions would prove to be wrong; all the more reason, therefore, to test the alternatives in the ways advocated in this book.

With this, however, I have effectively turned to a discussion of the larger theme addressed by the book as a whole.

IMPLICATIONS OF THE BOOK AS A WHOLE

When one turns to the book as a whole, the implications as well as the argument seem to be quite clear. Less discussion, therefore, seems necessary than for the case study. It may be useful, however, to highlight and discuss some of the implications in terms of the same categories as used in the third point of the discussion of the case study: the implications for design, for EBS, for history, and for theory.

The first implication regarding design is that one shifted to a consideration of *environmental design* rather than, say, architecture. This is, of course, also the result of taking an approach based on EBS. The result is not only that the subject matter of the domain changes, that the concern is with systems of settings and their environmental qualities rather than just buildings, just urban spaces, and so on. It also leads from design as an analog of what has been called "private science" to design as an analog of "public science" (cf. Rapoport 1983b). The discipline of environmental design, its empirical and conceptual bases and their confirmation, refutation, and development on a cumulative base become the key, not the preferences, intuitions, taste, or psychology ("creativity") of the individual designer. Of course, as in research and science, these continue to play an important, in fact indispensable, role. By moving to a science metaphor for the study of past environments, as for the study of present-day EBR, one

*See also her presentation at EDRA 18, May–June 1987 (J. Harvey and D. Henning (Eds.), *Public Environments*, Washington, DC: EDRA 1987, p. 228).

not only emphasizes the characteristics that I have discussed in this book (explicitness, analytical and conceptual rigor, empirical support, cumulativeness, and so on). One also moves design toward becoming a *science-based profession*.

The primary goal of design then becomes the setting of explicit objectives, based on the best available knowledge and evidence (*what* to do and *why*). In other words, the objectives must be demonstrably valid—or at least reasonable. Only then does one move to the question of *how* to achieve these goals. Any design then becomes a set of hypotheses to be properly evaluated in terms of whether the objectives have been met—whereas research continues to investigate the validity of the objectives. The cumulativeness in both will then become greater, and we will also become better at achieving them and finding out whether we have achieved them. As we learn, we also learn how to learn (Shapere 1984).

Regarding EBS, the implications are that to make the body of evidence sufficiently broad and varied, the use of data from the past is as essential as the use of the whole environment, all environments, and cross-culturally representative evidence. A link will need to be established between EBS and the study of the history of the built environment: Neither can progress by itself. The conceptual and methodological implications are major and require development and critical discussion. It also means that in addition to the disciplines already involved in EBS to a greater or lesser extent, a new range of disciplines becomes relevant. In addition to history itself, those particularly relevant are the historical sciences and archaeology.* This makes it even more important to keep in touch with changes and developments in a variety of fields—as I tried to show for the case of the study of art in Chapter 3.

I have argued that in terms of history and theory, the subject matter of the domain must be changed. It is no longer the individual building as artifact, and certainly not the individual "masterpiece" building, but the cultural landscape, the relevant system of settings. This means that inevitably the study of, and the important problems in, that subject matter become relational and contextual; there is no need for much of the turgid and largely meaningless prose on these matters. At the same time, the approach advocated means that, contrary to some current trends, one moves away from the "holistic," "transactional" approaches (not to speak of hermeneutic, phenomenological and others of that ilk) to a rational approach based on explicitly defining the relevant system, decomposing it into its constituents, studying their mutual interactions and the mechanisms, and reassembling the whole at a higher level of conceptual abstraction and understanding. One immediate result is to lead one from vast, vague, and woolly pretensions to theory, to theory in the true sense of the word—although a discussion of what this means will have to be postponed to my book on that subject.

But en route to theory, and the concepts and constructs that enter into theories, generalization is essential. So is the identification of the mechanisms that operate, which can link the concepts and constructs of theory to empirical evidence. One implication of my argument is that many specific findings, studies, and insights can be generalized more easily, as well as more validly. Many studies and data can then become mutually relevant, although each may appear to be unique and *sui generis*. Because data are lacking to make many (most?) hypotheses testable, there is need for such data. This

---

*The latter also needs to realize the relevance of EBS.

not only makes it imperative that all data available be used. It is equally important to be able to combine many small, apparently unrelated and unreplicable, and hence trivial, findings and ideas. If they can be linked, they become mutually enlightening and can be used to develop higher level concepts and to identify mechanisms. Those, in turn, can be used both in the controlled analogies needed for the study of the past and also provide, via theory and EBS, the precedents relevant for design.

To give merely one small example of the many possible. Consider the Iranian submission to the UN Conference on Human Settlements held in Vancouver in June 1976 ("Habitat Bill of Rights"). One portion of it ("Illustrated Habitat Code for Iran" *Ekistics,* Vol. 43, No. 258, May 1977, pp. 260–261) discusses specific Iranian approaches to "defining place," climate, orientation, paths, and many others. Each seems so specific as to preclude generalizations and hence unusable in theory and irrelevant elsewhere. One of these (No. 4) is "gateway" that "should be used to give a sense of arrival and territoriality and divide the hidden contained spaces from exterior spaces" (p. 262). As it stands, it is only relevant to Iran.* But generalization is possible. The first step would be to look at related cultural contexts over time, where gateways play a major role: Certain more general principles may emerge. Even more important is that a more basic expansion or generalization becomes possible—in many cultures gateways may not be used, but *thresholds* seem to be important very widely and are often marked in various ways. Threshold behavior and rules can then be compared with gateways. This process quickly suggests other generalizations and higher level concepts, for example, by considering the case of nomads (such as the Bedouin) and tribal societies (such as the Australian Aborigines). One then arrives at the concept of the *passage between domains* as being the important event or act. The domains may vary; so does the repertoire of ways of marking this passage from which one selects—for one thing they may be through fixed-feature, semifixed feature, or nonfixed feature elements (i.e., behavior). All these become culture-specific expressions of a general (universal?) process, different ways of *marking the transition between domains.* In this way, the use of a broad and varied body of evidence over time can lead to a very high-level general hypothesis, which further data can make testable.

This immediately begins to suggest not only linkages among disparate data, leading to pattern recognition. It also leads one to think of mechanisms: the ways of marking transitions become *mechanisms* for doing so. The Iranian gateways to "hidden contained spaces" are equivalent to the free-standing Japanese tori, gateways without fences in Arab villages in Israel, Chinese urban axes (Rapoport 1977), the Jewish mezuzzah on doorposts, the use of other markings, such as bigger beams, varied textures (Rapoport 1979a), and Aboriginal and others' handprints at rock shelters and many others. All are specific mechanisms of marking transitions between domains. One then also realizes that these can be temporal as well as spatial domains (e.g., Rapoport 1982e).

I have already pointed out that such generalizability is also of great importance in design. In the case of the preceding example, for instance, one would not necessarily build "gateways" but think of ways of marking significant transitions. In the case of the

---

*Assuming it is correct there: It is an assertion. Moreover, many of the terms in this statement are so vague as to be of questionable utility.

case study, we have already seen that the principles derived become much more generally relevant—to other spaces, to freeways and roads, to other transport modes, to institutions, natural landscapes, and other settings. Similar conclusions, as pointed out, apply to meaning, energy use, and most if not all other questions that can be raised.

This has three major implications. First, it may provide a link with more traditional work on history and theory in the design fields. I have already referred to the way this approach could be related to studies of streets, urban spaces, and urban forms. One could also take major works that, in a way, tried to do what I suggest, such as Lewis Mumford's *The City in History,* as a set of hypotheses and test them in ways described in this book. In effect, one would evaluate them and "rewrite" them without losing the valuable work done. In this way, one form of cumulativeness would result. For theory, this could be done, for example, with Kevin Lynch's *A Theory of Good City Form.* In fact, I can hardly think of a significant work in history, theory, or EBS that could not benefit from the approach that this book recommends and test them as hypotheses.

Second, this approach links design and theory fairly directly through these generalizable findings, concepts, and mechanisms to which I have referred; theory-based design becomes inevitable. Third, as a result, the past, and hence all built environments, becomes a resource, a depository of human knowledge and experience of environment—behavior relations and environmental design. This resource then allows the past itself to be studied more effectively and makes it usable for generalizations about EBR, to build theory, and hence for valid design.

The synergistic effects of the interaction between EBS and history go beyond just precedents for design, although these are the only historical precedents useful for design. They also have major implications for giving the design fields the validity they now lack; it can turn them into disciplines that now they are not. This will be achieved by providing them with a scientific, scholarly base that provides knowledge and understanding of how things really are and work—and hence of how they can be made better.

References

Abercrombie, M.L.J. (1969). *The Anatomy of Judgement*. Harmondsworth: Penguin (originally published in 1960).

Abu-Lughod, J.L. (1971). *Cairo—1001 Years of the City Victorious*. Princeton: Princeton University Press.

Ackerman, J.A., and R. Carpenter (1963). *Art and Archaeology (Princeton Studies in Humanistic Studies in America)*. Englewood-Cliffs, NJ: Prentice-Hall.

Acking, C.A., and R. Küller (1973). "Presentation and judgment of planned environments and the hypothesis of arousal." In W. Preiser (Ed.) *Environmental Design Research (EDRA 4)*. Stroudsburg, PA: Dowden, Hutchinson and Ross, Vol. 1, pp. 72–83.

Adburgham, A. (1979). *Shopping in Style (London from the Restoration to Edwardian Elegance)*. London: Thames and Hudson.

Agha Khan Award for Architecture, various volumes (starting 1980).

Aguilera, J. Rojas, and L.J. Moreno Rexach (1973). *Urbanismo Español en America*. Madrid: Editora Nacional.

Aktüre, S. (1978). "19 Yüzyil Sonunda Anadolu 'Kenti Uckansal Yapi Cözümlemsi' " (in Turkish). Ankara: Faculty of Architecture, Middle East Technical University.

Alden, R.S. (1986). "Tokyo streets: Complexity, interest and pleasure—a photographic approach to understanding the enjoyment of walking." (English version of an article published in Japanese in the Nov. 1986 *Bulletin of the Science University of Tokyo*, mimeo.).

Alexander, C. (1964). *Notes on the Synthesis of Form*. Cambridge, MA: Harvard University Press.

Allaby, M., and J. Lovelock (1983). *The Great Extinction*. Garden City, NY: Doubleday.

Allen, E.B. (1969a). *Stone Shelters*. Cambridge, MA: MIT Press.

Allen, E.B. (1969b). "The Passagiata," *Landscape*, Vol. 18, No. 1 (Winter) pp. 29–32.

Allison, J.R. et al. (1970). "A method of analysis of the pedestrian system of a town centre (Nottingham City)," *Journal of the TPI*, Vol. 56, No. 8 (Sept/Oct).

Alpass, J. (1965). *By Center Mannecke*. Lyngby, Denmark: Institut for Center Planlaeging.

Alsayyad, N., K. Bristol, and S.F. Lin (1987). "Interpreting the form of urban space: Cross-cultural notes on open space and urban activity in a historical context." In J. Harvey and D. Henning (eds) *Public Environments (EDRA 18)*, Washington DC: EDRA, pp. 141–151.

Alstrup, I. et al. (n.d.). *Landsby i Nepal*. Copenhagen: School of Architecture, Royal Academy of Art.

Altman, I. (1975). *The Environment and Social Behavior: Privacy, Personal Space, Territory and Crowding*. Monterrey, CA: Brooks/Cole.

American Behavioral Scientist (1982). Vol. 25 No. 6 (July/August). Special issue, "The Archaeology of the Household—Building a Prehistory of Domestic Life."

Ames, K.L. (1978). "Meaning in artifacts: Hall furnishings in Victorian America." *Journal of Interdisciplinary History*, Vol. 9, No. 1 (Summer) pp. 19–46.

Andersen, H. et al. (n.d.). *Tinerhir* (2 studierejser til Syd-Maroko i 1969 og 1970) Copenhagen: School of Architecture, Royal Academy of Art.

Anderson, S. (Ed.). (1978). *On Streets*. Cambridge, MA: MIT Press.

Andreae, B., and H. Kyrieleis (Ed.). (1975). *Neue Forschungen in Pompeji*, Recklinghausen: v. Aurel Bongers.

469

Andrews, E.W. IV, et al. (1975). Archaeological Investigations on the Yucatan Peninsula, Publication 31, Middle American Research Institute. New Orleans: Tulane University.

Antoniou, J. (1968). "Pedestrians in the city." Official Architecture and Planning (August) pp. 1035–1040.

Antoniou, J. (1971a). Environmental Management. London: McGraw-Hill.

Antoniou, J. (1971b). "Planning for pedestrians." Traffic Quarterly (January) pp. 55–71.

Antoniou, J. (1973). Plaka. Athens: Lycabettus Press.

Appleton, J. (1975). The Experience of Landscape. London: Wiley.

Appleyard, D. (1970). "Notes on urban perception and knowledge." In J. Archea and C. Eastman (Eds.), EDRA 2, pp. 97–101.

Appleyard, D. (1976). "Livable urban streets: Managing auto traffic in neighborhoods." Washington, DC: Department of Transportation, Federal Highway Administration.

Appleyard, D. (Ed.). (1979). The Conservation of European Cities. Cambridge, MA: MIT Press.

Appleyard, D. (1981). Livable Streets. Berkeley: University of California Press.

Appleyard, D., K. Lynch, and J. Meyer. (1964). The View from the Road. Cambridge, MA: MIT Press.

Appleyard, D., and M. Lintell. (1972). "The environmental quality of city streets: The resident's viewpoint." AIP Journal, Vol. 38, No. 2 (March) pp. 84–101.

Aravatinos, A.J. (1963). Grossstadtische Einkkaufszentren. Essen: Vulkan-Verlag, Dr. W. Classen.

Architecture Australia (1976). "Traffic-free streets boost business," Vol. 65, No. 5 (October/November) p. 12.

Ardalan, N., and L. Bakhtiar (1973). Sense of Unity. Chicago: University of Chicago Press.

Argyris, C., R. Putnam, and D.M. Smith (1985). Action Science. San Francisco: Jossey-Bass.

Asatekin, G., and Z. Eren (1979). "Yeni-faca da anket calismasi ve sonliglari." Journal of METU, Vol. 5, No. 1 pp. 15–36.

Athens Center of Ekistics (1980). HUCO: The human community in Athens. Ekistics, Vol. 47, pp. 232–263.

Auzelle, R., and I. Jankovic (1954). Encyclopedie d'Urbanisme. Paris: Vincent, Freal.

Ayala, F.J., and T. Dobzhanski (Ed.) (1974). Studies in the Philosophy of Biology (Reduction and Related Problems). Berkeley: University of California Press.

Aydelotte, W.O. (1981). "The Search for Ideas in Historical Investigation." Social Science History, Vol. 5, No. 4 (Fall) pp. 371–392.

Ayer, A.J. (1984). Philosophy in the Twentieth Century. London: Unwin Paperback.

Bacon, E.N. (1967). Design of Cities. New York: Viking Press.

Baker, A.R.H. (Ed.). (1972). Progress in Historical Geography. Newton Abbott, David and Charles.

Baker, A.R.H. (1976). "The Limits of Inference in Historical Geography." In B.S. Osborne (Ed.) The Settlement of Canada: Origins and Transfer (Proceedings of the 1975 British-Canadian symposium on Historical Geography) Kingston, Ontario: Queen's University (February) pp. 169–182.

Baker, A.R.H., and M. Billinge (Eds.) (1982). Period and Place (Research Methods in Historical Geography. Cambridge: Cambridge University Press.

Baker, T. (1970). Medieval London. New York: Praeger.

Bakker, R.T. (1986). The Dinosaur Heresies. New York, William Morrow & Co.

Banham, R. (1971). Los Angeles (The Architecture of the Four Ecologies). Harmondsworth: Penguin.

Barash, D. (1979). Sociobiology: The Whispering Within. New York: Harper and Row.

Barfold, H.S. et al. (1972). Paros. Copenhagen: Royal Academy School of Architecture.

Barker, R.G., and L.S. Barker (1961). "Behavior Units for the comparative study of cultures." In B. Kaplan (Ed.) Studying Personality Cross-Culturally. New York: Harper and Row, pp. 457–476.

Barker, R.G., and P. Schoggen (1973). Qualities of Community Life. San Francisco: Jossey-Bass.

Barker, R.G., and H.F. Wright (1955). Midwest and Its Children. Evanston, IL: Row, Peterson and Co.

Barnard, B. (1984). "Cultural facades: Ethnic architecture in Malaysia." UFSI Reports, No. 5 (Asia) BEB2–'84.

Bartlett, Sir F. (1967). Remembering. Cambridge: Cambridge University Press (paperback). (Originally published, 1932.)

Baum, A., G.E. Davis, and J.R. Aiello (1978). "Crowding and neighborhood mediation of urban density." Journal of Population, Vol. 1(3) (Fall) pp. 266–279.

Beattie, N. (1984). "Australian house form—An empirical study." In D. Saunders (Ed.), Architectural History Papers (Australia and New Zealand). (Conference of Architectural Historians of Australia, August). Adelaide: University of Adelaide, pp. 179–184.

Becker, F.D. (1973). "A class-conscious evaluation (going back to Sacramento's mall)." Landscape Architecture, Vol. 64, No. 1 (October) pp. 448–457.

Beers, B.F. (1978). *China in Old Photographs 1860–1910.* New York: Charles Scribner's Sons.

Behrensmeyer, A.K., and A. Hill (Ed.). (1980). *Fossils in the Making (Vertebrate Tophonomy and Paleoecology).* Chicago: University of Chicago Press.

Ben-Arieh, Y. (1979). *The Beginnings of New Jerusalem* (in Hebrew). Jerusalem: Itzhak Ben Zvi Foundation.

Bendall, D.S. (Ed.). (1983). *Evolution from Molecules to Men.* Cambridge: Cambridge University Press.

Benepe, B. (1965). "The pedestrian in the city." *Traffic Quarterly,* Vol. 19, No. 1 (January).

Berelson, B., and G.A. Steiner (1964). *Human Behavior: An Inventory of Scientific Findings.* New York: Harcourt, Brace and World.

Berger, A. (Ed.). (1981). *Climatic Variations and Variability: Facts and Theories.* Dordrecht: Reidel.

Berkhofer, R.F. Jr. (1971). *A Behavioral Approach to Historical Analysis.* New York: Free Press (1st printing, 1969).

Berkowitz, W.R. (1969). *A Cross-National Study of Some Social Patterns of Urban Pedestrians.* Department of Psychology, Lafayette College (September) (mimeo).

Berlyne, D.E. (1960). *Conflict, Arousal and Curiosity,* New York: McGraw-Hill.

Berlyne, D.E. (Ed.). (1974). *Studies in the New Experimental Aesthetics: Steps toward an Objective Psychology of Aesthetic Appreciation.* New York: Halstead Press.

Best, J.B. (1963) "Protopsychology." *Scientific American,* Vol. 32, No. 208 (February) pp. 54–62.

Betzig, L.L. (1986). *Despotism and Differential Reproduction (A Darwinian View of History).* New York: Aldine.

Beurton, P. (1981). "Organismic Evolution and Subject-Object Dialectics." In U.J. Jensen and R. Harré (Eds.), *The Philosophy of Evolution.* Brighton: The Harvester Press, pp. 45–60.

Bhardwaj, S.M. (1973). *Hindu Places of Pilgrimage in India (A Study in Cultural Geography).* Berkeley: University of California Press.

Bhaskar, R. (1981) "The Consequences of Socio-Evolutionary Concepts for Naturalism in Sociology: Commentaries on Harré and Toulmin." In U.J. Jensen and R. Harré (Eds.), *The Philosophy of Evolution.* Brighton: The Harvester Press, pp. 196–209.

Bibby, G. (1970). *Looking for Dilmun.* Harmondsworth: Penguin Books.

Biblical Archaeology Review (1987). "Books in brief," Vol. 13, No. 3 (May/June) pp. 6–9.

Bickerton, D. (1983). "Creole Languages." *Scientific American,* Vol. 249, No. 1 (July) pp. 116–122.

Billinge, M. (1977). "Phenomenology and historical geography." *Journal of Historical Geography,* Vol. 3, No. 1, pp. 55–67.

Bilsky, L.J. (Ed.) (1980). *Historical Ecology (Essays on Environment and Social Change).* Port Washington, NY: Kenikat Press.

Binford, L.R. (1962). "Archaeology as anthropology." *American Antiquity,* Vol. 28, No. 2, pp. 217–226.

Binford, L.R. (1981). *Bones: Ancient Men and Modern Myths.* New York: Academic Press.

Binford, L.R. (1983a). *Working at Archaeology.* New York: Academic Press.

Binford, L.R. (1983b). *In Pursuit of the Past.* London: Thames and Hudson.

Binford, S.R., and L.R. Binford (Eds.). (1968). *New Perspectives in Archaeology.* Chicago: Aldine.

Blake, G.H. (1968). "The Form and function of Misratah's commercial centre." *Bulletin, Faculty of Arts, University of Benghazi,* No. 2, pp. 9–40.

Blanton, R.E. (1978). *Monte Albán: Settlement Patterns at the Ancient Zapotec Capital.* New York: Academic Press.

Blumenfeld, H. (1953). "Scale in civic design." *Town Planning Review,* Vol. 24, pp. 35–46.

Bock, K. (1980). *Human Nature and History (A Response to Sociobiology).* New York: Columbia University Press.

Bonine, M.E. (1979). "The morphogenesis of Iranian cities." *Annals, Association of American Geographers,* Vol. 69, No. 2, pp. 208–224.

Bonine, M.E. (1980). "Aridity and structure: Adaptations of indigenous housing in Iran." In K.N. Clark and P. Paylore (Eds.), *Desert Housing.* Tucson: University of Arizona, pp. 193–219.

Bonner, J.T. (1980). *The Evolution of Culture in Animals.* Princeton, NJ: Princeton University Press.

Bonner, J.T. (1988). *The Evolution of Complexity by Means of Natural Selection.* Princeton: Princeton University Press.

Bonta, J.P. (1975). *An Anatomy of Architectural Interpretation.* Barcelona: Gustavo Gili.

Boserup, E. (1965). *Conditions of Agricultural Growth (The Economics of Agrarian Change under Population Pressure).* Chicago: Aldine.

Boserup, E. (1981). *Population and Technological Change: A Study of Long Term Trends.* Chicago: University of Chicago Press.

Boulding, K.E. (1980). "Science: Our common heritage." *Science,* Vol. 207, No. 4433, (22 February) pp. 831–836.

Boulding, K.E. (1984). Review of F.A. Beer, *Peace against War: The Ecology of International Violence.* San Francisco: W.H. Freeman. In *Human Ecology,* Vol. 12, No. 2 (June) pp. 209–213.

Bourdieu, J. (1968). *Le Systeme des Objets.* Paris: Denoel/Gauthier.

Bower, T.G.R. (1971). "The object in the world of the infant." *Scientific American,* Vol. 225, No. 4 (October) pp. 30–38.

Boyd, R.N. (1984). "The Current Status of Scientific Realism." In J. Leplin (Ed.) *Scientific Realism.* Berkeley: University of California Press, pp. 41–82.

Boyd, R., and P. Richerson (1985). *Culture and the Evolutionary Process.* Chicago: University of Chicago Press.

Boyd, T.D. (1981). "Halieis: A fourth planned city in classical Greece." *TP Review,* Vol. 52, No. 2 (April) pp. 143–156.

Boyden, S. (1974). *Conceptual Basis of Proposed International Ecological Studies in Large Metropolitan Areas* (Mimeo).

Boyden, S. (1979). *An Integrative Ecological Approach to the Study of Human Settlements.* UNESCO (MAB Technical notes).

Boyden, S.V. (1987). *Western Civilization in Biological Perspective.* Oxford: Clarendon Press.

Boyden, S., and S. Millar (1978). "Human ecology and the quality of life." *Urban Ecology,* Vol. 3, No. 3, pp. 263–287.

Boyden, S., S. Millar, K. Newcombe, and B. O'Neill (1981). *The Ecology of a City and Its People (The Case of Hong Kong).* Canberra: ANU Press.

Bradfield, R.M. (1973). *A Natural History of Associations (A Study in the Meaning of Community).* London: Duckworth.

Brambilla, R. (Ed.). (1973). *More Streets for People.* New York: Italian Art and Landscape Foundation.

Brambilla, R., and G. Longo. (1977a). *For Pedestrians Only (Planning, Design and Management of Traffic-Free Zones).* New York: Whitney Library of Design.

Brambilla, R., and G. Longo (1977b). *Banning the Car Downtown* (Selected American cities). Washington, DC: U.S. Government Printing Office.

Brambilla, R., and G. Longo (1977c). *A Handbook for Pedestrian Action.* Washington, DC: U.S. Government Printing Office.

Brambilla, R., G. Longo, and V. Dzurinko (1977). *American Urban Malls—A compendium.* Washington, DC: U.S. Government Printing Office.

Braunfels, W. (1976). *Abendländische Stadtbaukunst.* Köln: DuMont Schauberg.

Breines, S., and W.J. Dean. (1974). *The Pedestrian Revolution (Streets without Cars).* New York: Vintage Books.

Briganti, G. (1970). *The View Painters of Europe.* London: Phaidon.

Brine, J. (1984). "Charting shifting views of Port Arthur and associated methodological problems." In D. Saunders (Ed.), *Architectural History Papers (Australia and New Zealand)* (Conference of Architectural Historians of Australia, August 1984). Adelaide: University of Adelaide, pp. 60–80.

Brine, J. (1986). "Charles Jencks's early critical stances." *Architecture Australia,* Vol. 75, No. 4 (June) pp. 21–27.

Brislin, R.W., W.J. Lonner, and R.M. Thondike (1973). *Cross-Cultural Research Methods.* New York: John Wiley.

Broda, J., D. Carrasco, and E. Matos (1987). *The Great Temple of Tenochtitlán (Center and Periphery in the Aztec World).* Berkeley: University of California Press.

Brolin, B.C., and J. Zeisel. (1968). "Mass housing: Social research and design." *Architectural Forum,* Vol. 129, No. 1 (July/August).

Bromley, J., and J. Bartagelata (1945). *Evolucion urbana de la ciudad de Lima.* Lima: Consejo Provincial de Lima.

Brooke, J. (1984). "A city's blight: Riots, crime, golf in the streets." *New York Times* (August 23).

Brown, L.C. (Ed.). (1973). *From Madina to Metropolis.* Princeton: Darwin Press.

Brown, P.J. (1986). "Cultural and genetic adaptations to malaria: Problems of comparison." *Human Ecology,* Vol. 14, No. 3 (September) pp. 311–332.

Brown, R.H., and S.M. Lyman (Eds.). (1978). *Structure, Consciousness and History.* Cambridge: Cambridge University Press.

Browne, M.W. (1987). "New findings reveal ancient abuse of lands." *New York Times* (January 13).

Bruner, J. (1951). "Personality dynamics and the process of perceiving." In R.R. Blake and G.V. Ramsey (Eds.), *Perception: An Approach to Personality*. New York: Ronald Press.

Buck, D.D. (1978). *Urban Change in China (Politics and Development in Tsinan, Shantung, 1890–1949)*. Madison: University of Wisconsin Press.

Bulliet, R. (1975). *The Camel and the Wheel*. Cambridge, MA: Harvard University Press.

Bullock, A. (1977). *Is History Becoming a Social Science? (The Case of Contemporary History)*. Cambridge: Cambridge University Press.

Bullock, T.H. (1984). "Comparative neuroscience holds promise for quiet revolutions." *Science*, Vol. 225, No. 4661 (3 August) pp. 473–477.

Bunge, M. (1983). "Speculation: Wild and sound." *New Ideas in Psychology*, Vol. 1, No. 1, pp. 3–6.

Bunkśe, E.V. (1978). "Commoner attitudes toward landscape and nature." *Annals, Association of American Geographers*, Vol. 68, No. 4, (December) pp. 551–566.

Burke, P. (1980). *Sociology and History*. London: Allen and Unwin.

Butler, M.H., and N.T. Butler (1976). *Urban Dwelling Environment: Istanbul, Turkey*. Cambridge, MA: MIT School of Architecture and Planning (June).

Butzer, K.W. (1982). *Archaeology as Human Ecology (Method and Theory for a Contextual Approach)*. Cambridge: Cambridge University Press.

Cairns-Smith, A.G. (1982). *Genetic Takeover and the Mineral Origins of Life*. Cambridge: Cambridge University Press.

Caminos, H., and R. Goethert (1978). *Urbanization Primer*. Cambridge, MA: MIT Press.

Campbell, D.T. (1969). "Reforms as experiments." *American Psychologist*, Vol. 24, pp. 409–429.

Canby, J., E. Porada, B. Ridgway, and T. Stech (Eds.). (1986). *Ancient Anatolia (Aspects of Change and Cultural Development)*. Madison: University of Wisconsin Press.

Carlstein, T., D. Parkes, and N. Thrift (Eds.). (1978a). *Making Sense of Time*. New York: John Wiley and Sons.

Carlstein, T., D. Parkes, and N. Thrift (Eds.). (1978b). *Human Activity and Time Geography*. New York: John Wiley and Sons.

Carlstein, T., D. Parkes, and N. Thrift (Eds.). (1978c). *Time and Regional Dynamics*. New York: John Wiley and Sons.

Carr, S. (1973). *City Signs and Lights: A Policy Study*. Cambridge: MIT Press.

Carr, S. and D. Schissler (1969). "The city as trip: Perceptual selection and memory in the view from the road." *Environment and Behavior*, Vol. 1, No. 1 (June) pp. 7–36.

Carter, J. (1969). Letter to A/J "Economist building revisited." *Architects Journal* (October) p. 787.

Chadwick, J. (1983). "Life in Mycenean Greece." In B.M. Fagan (Ed.), *Prehistoric Times (Readings from Scientific American)*. San Francisco, W.H. Freeman, pp. 204–212.

Chaney, R.P., and R.R. Revilla (1969). "Sampling methods and interpretation of correlation: A comparative analysis of seven cross-cultural samples." *American Anthropologist*, Vol. 71, No. 4 (August) pp. 597–633.

Chang, A. (1956). *The Existence of Intangible Content in Architectonic Form*. Princeton: Princeton University Press (and University Microfilms, Ann Arbor, MI).

Chapman, R. (1981). "Archaeological Theory and Communal Burial in Prehistoric Europe." In I. Hodder, G. Isaac, and N. Hammond (Eds.). *Pattern of the Past (Essays in Honour of David Clarke)*. Cambridge: Cambridge University Press, pp. 387–412.

Chisholm, J.C. (1983). *Navajo Infancy (An Ethological Study of Child Development)*. New York: Aldine.

Churchman, A. (1977). "Childrens' street play: Can it be accommodated?" In A.S. Hakkert (Ed.), *Proceedings of the International Conference on Pedestrian Safety, Haifa, December, 1976*. Haifa: Michlol Publishing House.

Churchman, A., and Y. Tzamir (1977). "Pedestrian behavior patterns and experiences in a shopping center combining two design alternatives for traffic segregation." In A.S. Hakkert (Ed.), *Proceedings of the International Conference on Pedestrian Safety, Haifa, December, 1976*. Haifa: Michlol Publishing House.

Ciolek, T.M. (1977). *Configuration and Context (A Study of Spatial Patterns in Social Encounters)*. Canberra: Australian National University, Ph.D. Dissertation (unpublished).

Ciolek, T.M. (1978a). "Spatial behavior in pedestrian areas." *Ekistics*, Vol. 45, No. 268 (March/April) pp. 120–122.

Ciolek, T.M. (1978b). "Pedestrian movement and stationary behavior in a public setting." Canberra: Department of Anthropology, Australian National University (mimeo.).

Ciolek, T.M. (1978c). "Some spatial features of the phenomenon of co-presence." *Sociolinguistic Newsletter,* Vol. 9, No. 2, pp. 23–24.

Clark, D.H. (1985). *The Quest for SS433.* New York: Viking.

Clarke, D.L. (1978). *Analytical Archaeology.* London: Methuen and Co. (2nd edition; 1st edition, 1968).

Clay, G. (1978). *Alleys: A Hidden Resource.* Louisville, KY: Grady Clay Co.

Clubb, J.M., and E.K. Scheuch (Ed.). (1980). *Historical Social Research.* Stuttgart: Klett-Cotta.

Cockburn, A., and E. Cockburn (Ed.). (1980). *Mummies, Disease and Ancient Cultures.* Cambridge: Cambridge University Press.

Coe, M.D. (1977, 2nd ed.). *Mexico.* New York: Praeger.

Cohen, E.A. (1985/1986). "Why we should stop studying the Cuban missile crisis." *The National Interest,* No. 2, pp. 3–13.

Cohen, H., S. Moss, and E. Zube (1979). "Pedestrians and wind in the urban environment." In A.D. Seidel and S. Danford (Ed.), *Environmental Design: Research, Theory and Application (EDRA 10).* Washington, DC: EDRA, pp. 71–82.

Cohen, J. (1967). *Psychological Time in Health and Disease.* Springfield, IL: CC Thomas.

Colbert, E.H. (1980). *Evolution of the Vertebrates (A History of the Backboned Animals through Time).* New York: John Wiley and Sons (3rd edition).

Colchester, M. (1984). "Rethinking Stone Age Economics: Some Speculations Concerning the Pre-Columbian Yanoama Economy." *Human Ecology,* Vol. 12, No. 3, pp. 291–314.

Coleman, A. (1985). *Utopia on Trial (Vision and Reality in Planned Housing).* London: Hilary Shipman.

Coles, J. (1973). *Archaeology by Experiment.* London: Hutchinson University Library.

Collett, P. (1984). "History and the study of expressive action." In K.J. Gergen and M.M. Gergen (Eds.), *Historical Social Psychology.* Hillsdale, NJ: Lawrence Erlbaum, pp. 371–396.

Collingwood, R.G. (1939). *An Autobiography.* Oxford, Clarendon Press.

Collingwood, R.G. (1961). *The Idea of History.* Oxford: Oxford University Press (original edition, 1946).

Collins, P. (1971). *Architectural Judgment.* London: Faber.

Conan, M. (1987). "Communication sur la creation de l'espace public des logements sociaux du 'Village Lobau.' " In J. Harvey and D. Henning (Eds.), *Public Environments (EDRA 18).* Washington, DC: EDRA, p. 236 (Abstract only).

Connah, G. (1981). *Three Thousand Years in Africa (Man and His Environment in the Lake Chad Region of Nigeria).* Cambridge: Cambridge University Press.

Cook, T.D., and D.T. Campbell (1979). *Quasi-Experimentation (Design and Analysis Issues for Field Settings).* Boston: Houghton-Mifflin.

Cooper, C. (1965). "Some social implications of house and site plan design at Easter Hill Village: A case study." Berkeley: University of California, Center for Planning Development Research (September).

Cooper, C. (1970). "The adventure playground: Creative play in an urban setting as a potential focus for community involvement." Berkeley: Institute for Urban and Regional Development, Working Paper No. 118 (May).

Coppa, M. (1968). *Storia dell Urbanistica (dalle Origini all Ellenismo).* Torino: Einaudi (2 volumes).

Cornell, J., and A.P. Lightman (Eds.). (1982). *Revealing the Universe (Prediction and Proof in Astronomy).* Cambridge, MA: MIT Press.

Coss, R. (1973). "The cut-off hypothesis: Its relevance to the design of public places." *Man-Environment Systems* (November), pp. 417–440.

Country Life (1955). *The First Country Life Picture Book of London.* Country Life (revised edition).

Cracraft, J. (1983). "Cladistic analysis and variance in biogeography." *American Scientist,* Vol. 71 (May/June) p. 273 ff. (and references).

Crease, R. Jr., and C. Mann (1986). *The Second Revolution.* New York: Macmillan.

Cresswell, R. (Ed.). (1979). *Quality in Urban Planning and Design.* London: Newnes-Butterworths.

Cross, J.R. (1983). "Twigs, branches, trees and forests: Problems of scale in lithic analysis." In J.A. Moore and A.S. Keene (Eds.), *Archaeological Hammers and Theories.* New York: Academic Press, pp. 88–106 (Chapter Four).

Crowhurst-Lennard, S.H., and H.L. Lennard (1984). *Public Life in Urban Spaces.* Southampton, NY: Gondolier Press.

Crozier, W.R., and A.J. Chapman (Ed.). (1984). *Cognitive Processes in the Perception of Art* (Advances in Psychology 19). Amsterdam: North Holland.

Crumrine, N.R. (1964). *The House Cross of the Mayo Indians of Sonora, Mexico*. Tucson: University of Arizona, Anthropology Papers, No. 8.

Crumrine, N.R. (1977). *The Mayo Indians of Sonora*. Tucson: University of Arizona Press.

Csikszentmihalyi, M., and S. Bennett (1971). "An exploratory model of play." *American Anthropologist*, Vol. 73, No. 1 (February) pp. 42–58.

Csikszentmihalyi, M., and E. Rochberg-Halton (1981). *The Meaning of Things: Domestic Symbols and the Self*. New York: Cambridge University Press.

Cuff, D. (1985). Symposium on "Architects' people." Summary in S. Klein, R. Wener, and S. Lehman (Eds.), *Environmental Change/Social Change* (EDRA 16). Washington, DC: EDRA, pp. 345–346.

Cullen, C. (1961). *Townscape*. London: The Architectural Press.

Cullen, C. (1964). *A Town Called Alcan*. London: Alcan Industries.

Curl, J.S. (1969). Letter to A/J "Economist building revisited." *Architects Journal* (September 17) p. 669.

Dalton, G. (1981). "Anthropological Models in Archaeological Perspective." In I. Hodder, G. Isaac, and N. Hammond (Eds.), *Pattern of the Past (Essays in Honour of David Clarke)*. Cambridge: Cambridge University Press, pp. 17–48.

Daniel, T.C., L. Wheeler, R.S. Boster, and P.R. Best (n.d.). "Quantitative evaluation of landscapes: An application of signal detection analysis to forest management alternatives" (Mimeo).

Dawkins, R. (1976). *The Selfish Gene*. London: Oxford University Press.

Dawkins, R. (1986). *The Blind Watchmaker*. New York: W.W. Norton & Co.

Deagan, K. (1982). "Avenues of inquiry in historical archaeology." In M.B. Schiffer (Ed.), *Advances in Archaeological Method and Theory*. New York: Academic Press, pp. 151–177.

de Arce, R.P. (1978). "Urban transformations and the architecture of additions." *Architectural Design*, Vol. 48, No. 4, pp. 237–266.

DeBoer, W.R. (1983). "The archaeological record as preserved death assemblage." In J.A. Moore and A.S. Keene (Eds.), *Archaeological Hammers and Theories*. New York: Academic Press, pp. 19–36 (Chapter Two).

DeBoer, W., and D. Lathrop (1979). "The making and breaking of Shipibo-Conibo ceramics." In C. Kramer (Ed.), *Ethnoarchaeology (Implications of Ethnography for Archaeology)*. New York: Columbia University Press, pp. 102–138.

Deetz, J. (1977). *In Small Things Forgotten (The Archaeology of Early American Life)*. New York: Doubleday (Anchor Books).

de Jonge, D. (1967–1968). "Applied hodology." *Landscape*, Vol. 17 No. 2 (Winter) pp. 10–11.

DeLong, A.J. (1981). "Phenomenological space-time: Toward an experiential relativity." *Science*, Vol. 213, pp. 681–683.

DeLong, A.J. (1983a). "Spatial scale, temporal experience and information processing: An empirical examination of experiential relativity." *Man-Environment Systems*, Vol. 13, pp. 77–86.

DeLong, A.J. (1983b). "Environment as code: Spatial scale and time-frames in behavior and conceptualization." In J.B. Calhoun (Ed.), *Environment and Population*. New York: Praeger, pp. 192–194.

DeLong, A.J. (1985). "Experiential space-time relativity." *Man-Environment Systems*, Vol. 15, No. 1 (January) pp. 9–14.

DeLong, A.J., N. Greenberg, and C. Keaney (1986). "Temporal responses to environmental scale in the lizard *Anolis Carolinensis* (Reptilia, Lacertilia, Ignamidae)." *Behavioral Processes*, Vol. 13, pp. 339–352.

Delson, R.M. (1979). *New Towns for Colonial Brazil*. Ann Arbor: University Microfilms.

Department of the Environment (D.O.E.). (1972). *The Estate Outside the Dwelling*. London: HMSO.

De Selm, D., and A. Ricci (1970). "Sfax: A Tunisian Medina." *Architectural Review*, Vol. CXLVIII, No. 886, pp. 364–365.

Design Council (and the Royal Planning Institute). (1979). *Streets Ahead*. New York: Whitney Library of Design.

Despres, C. (1987). "Symbolic representations of the suburban house: The case of the Neo-Quebecois house." In J. Harvey and D. Henning (Ed.), *Public Environments (EDRA 18)*. Washington, DC: EDRA, pp. 152–159.

De Wolfe, I. (1966). *The Italian Townscape*. New York: Braziller.

De Wolfe, I. (1971). *Civilia—The End of Sub-Urban Man*. London: Architectural Press.

Dissanayake, E. (1980). "Art as human behavior—ethological definition of art." *Journal of Aesthetic and Art Criticism*, Vol. 38 (No. 4) pp. 397–406.

DiVette, N. (1977). *Pedestrian Relief Areas (PRA): History, Literature Survey, Case Study, Analysis and Design Criteria*. Master of Science Thesis, Chicago: Illinois Institute of Technology (December) (unpublished).

Dixon, N.F. (1971). *Subliminal Perception: The Nature of a Controversy*. New York: McGraw-Hill.

Dobroszycki, L., and B. Kirchenblatt-Gimblett (1977). *Image before My Eyes*. New York: Schocken.

Donnadieu, C. & P.H. and J-M Didillon (1977). *Habiter le Desert*. Brussels: Mardaga.

Dosio, M.J., and P. Feddersen (1978). "Case study: The island of Burano, near Venice." *Ekistics*, Vol. 45, No. 273, pp. 453–462.

Douglas, M. (1972). "Symbolic order in the use of domestic space." In P.J. Ucko, R. Tringham, and G.W. Dimbleby (Eds.), *Man, Settlement and Urbanism*. London: Duckworth, pp. 513–521.

Douglas, M. (Ed.). (1982). *Essays in the Sociology of Perception*. London: Routledge and Kegan Paul.

Douglas, M. and Baron Isherwood (1979). *The World of Goods*. London: Routledge and Kegan Paul.

Doumanis, O.B., and M.O. Doumanis (1975). "Fit and form in the Greek village." *Design and Environment*, Vol. 6, No. 4 (Winter) pp. 42–47.

Doumanis, O.B., and P. Oliver (Eds.). (1974). *Shelter in Greece*. Athens: "Architecture in Greece" Press.

Doumas, C. (1972). "Early Bronze Age settlement patterns in the Cyclades." In P.J. Ucko, R. Tringham, and G.W. Dimbleby (Eds.), *Man, Settlement and Urbanism*. London: Duckworth, pp. 227–230.

Doumas, G.C. (1983). *Thera—Pompeii of the Ancient Aegean*. London: Thames and Hudson.

Downing, F., and U. Flemming (1981). "The bungalows of Buffalo." *Environment and Planning, B*, Vol. 8.

Doxiadis, C.A. (1972). *Architectural Space in Ancient Greece* (Ed. and Trans. J. Tyrwhitt). Cambridge, MA: MIT Press.

Doxiadis, C.A. (1975). *Building Entopia*. Athens.

Doxiadis, C.A. *et al.* (1974). *Anthropopolis—City for Human Development*. Athens.

Doxiadis, S. (Ed.). (1979). *The Child in the World of Tomorrow*. Oxford: Pergamon Press.

Drake, M. (Ed.). (1973). *Applied Historical Studies (An Introductory Reader)*. London: Methuen & Open University.

Drake, M. (1974). *Historical Data and the Social Sciences (The Quantitative Analysis of Historical Data)*. Milton Keynes: The Open University Press.

Draper, P. (1973). "Crowding among hunter-gatherers: The !Kung Bushmen." *Science*, Vol. 182, No. 4019 (October) pp. 301–303.

Dubos, R. (1966). *Man Adapting*. New Haven: Yale University Press.

Dubos, R. (1972). "Is man overadapting?" *Sydney University Union Recorder*, Vol. 52 No. 6 (April 13).

Duncan, J.S. (1973). "Landscape taste as a symbol of group identity." *Geographical Review*, Vol. 63 (July) pp. 334–355.

Duncan, J.S., S. Lindsey, and R. Buchan (1985). "Decoding a residence: Artifacts, codes and the construction of the self." *Espaces et Sociétés*, No. 47, pp. 29–43.

Dunnell, R.C. (1971). *Systematics in Prehistory*. New York: Free Press.

Durham, W.H. (1986). *Coevolution: Genes, Culture and Human Diversity*. Stanford: Stanford University Press.

Dymond, D.P. (1974). *Archaeology and History—A Plea for Reconciliation*. London: Thames and Hudson.

Ebert, L. (1979). "The ethnoarchaeological approach to reassessing the meaning of variability in stone tool assemblages." In C. Kramer (Ed.), *Ethnoarchaeology (Implications of Ethnography for Archaeology)*. New York: Columbia University Press, pp. 59–74.

Eckholm, E. (1986). "Species are lost before they're found." *New York Times* (September 16).

Edwards, J.P. (1979). "The evolution of vernacular architecture in the Western Caribbean" (mimeo).

Edwards, S.W. (1978). "Non-utilitarian activities in the lower Palaeolithic." *Current Anthropology*, Vol. 19, pp. 135–137.

Efron, E. (1984). *The Apocalyptics: Cancer and the Big Lie*. New York: Simon and Schuster.

Ehrenzweig, A. (1970). *The Hidden Order of Art*. London: Paladin Books.

Eibl-Eibesfeldt, I. (1972). *Love and Hate (The Natural History of Behavior Patterns)*. New York: Holt, Rinehart and Winston.

Eibl-Eibesfeldt, I. (1979). *The Biology of Peace and War (Men, Animals and Aggression)*. London: Thames and Hudson.

Eidt, R.C. (1971). *Pioneer Settlement in Northeast Argentina*. Madison: University of Wisconsin Press.

Ekistics (1977). Special Issue, "Cities as Cultural Artifacts," Vol. 44, No. 265 (Dec.).

Ekistics (1978). Special issue, "People-oriented urban streets." Vol. 45, No. 273 (November/December).

Eldredge, N. (1985). *Time Frames: The Rethinking of Darwinian Evolution and the Theory of Punctuated Equilibria*. New York: Simon and Schuster.

Eldredge, N. and S.M. Stanley (Eds.). (1984). *Living Fossils*. New York: Springer Verlag.

Eldredge, N., and I. Tattersall (1982). *The Myths of Human Evolution*. New York: Columbia University Press.

Elisséeff, N. (1970). "Damas a la lumière des theories de Jean Sauvaget." In A.H. Hourani and S.M. Stern (Ed.), *The Islamic City*. Oxford: Cassirer, pp. 157–177.

Ellis, M. (1972). "Play: Theory and Research." In W. Mitchell (Ed.), *EDRA 3*. Los Angeles: University of California, Vol. 1, pp. 541–545.

Ellis, W.R. (1986–1987). "Architects' people: The case of Frank Lloyd Wright." *Architecture and Behavior*, Vol. 3, No. 1, pp. 25–36.

Erikson, K.T. (1973). "Sociology and the historical perspective." In M. Drake (Ed.), *Applied Historical Studies (An Introductory Reader)*. London: Methuen & Open University, pp. 13–30.

Evans, J.G. (1975). *Environment of Early Man in the British Isles*. London: Elek.

Fagan, B.M. (1983). Introduction to *Prehistoric Times (Readings from* Scientific American*)* San Francisco, W.H. Freeman.

Fairservis, W.A., Jr. (1975). *The Threshold of Civilization (An Experiment in Prehistory)*. New York: Charles Scribner's sons.

Feinberg, G. (1985). *Solid Clues (Quantum Physics, Molecular Biology and the Future of Science)*. New York: Simon and Schuster.

Fersan, N. (1980). *Küçük Anadolu Kentlerinde Tarihsel Dokunuñ Korunmasi ile Ilgili Bir Yöntem Araştirmasi (Une approche méthodologique pour la sauvegarde dè l'ensemble historique des petites villes Anatoliennes)* (Turkish & French). Istanbul.

Festinger, L. (1983). *The Human Legacy*. New York: Columbia University Press.

Finley, M.I. (1975). *The Use and Abuse of History*. London: Chatto and Windus.

Fischer, D.H. (1970). *Historian's Fallacies: Towards a Logic of Historical Thought*. New York: Harper and Row.

Flannery, K.V. (Ed.). (1976). *The Early Mesoamerican Village*. New York: Academic Press.

Flannery, K.V. (Ed.). (1982). *Maya Subsistence (Studies in Memory of Dennis E. Puleston)*. New York: Academic Press.

Flannery, K.V., and J. Marcus (Eds.). (1983). *The Cloud People (Divergent Evolution of the Zapotec and Mixtec Civilizations)*. New York: Academic Press.

Fletcher, H. (1953). *Speech and Hearing in Communication*. New York: Van Nostrand.

Fletcher, R. (1977). "Alternatives and differences." In M. Spriggs (Ed.), *Archaeology and Anthropology: Areas of Mutual Interest*. Oxford: BAR (Supplementary Series, 19) pp. 49–68.

Fogel, R.W., and G.R. Elton (1983). *Which Road to the Past?* New Haven: Yale University Press.

Folan, W.J., E.R. Kintz, and L.A. Fletcher (1983). *Cobá—A Classic Maya Metropolis*. New York: Academic Press.

Fonseca, R. (1969a). "The Walled City of Old Delhi." *Landscape*, Vol. 18, No. 3 (Fall) pp. 12–25.

Fonseca, R. (1969b). "The walled city of Old Delhi." In P. Oliver (Ed.), *Shelter and Society*. London; Barrie and Rockliff: The Cresset Press pp. 103–115.

Forge, A. (Ed.). (1973a). *Primitive Art and Society*. London: Oxford University Press.

Forge, A. (1973b). "Style and meaning in Sepik art." In A. Forge (Ed.), *Primitive Art and Society*. London: Oxford University Press, p. 173ff.

Fox, R. (1970). "The cultural animal." *Encounter*, Vol. 35, No. 1 (July) pp. 31–42.

Fox, R. (1980). *The Red Lamp of Incest*. New York: Dutton.

Fratkin, E. (1986). "Stability and resilience in East African pastoralism: The Rendille and the Ariaal of Northern Kenya." *Human Ecology*, Vol. 14, No. 3 (Sept.) pp. 269–286.

Frederickson, G.M. (1980). "Comparative history." In M. Kammen (Ed.), *The Past Before Us*. Ithaca, NY: Cornell University Press, pp. 457–473.

Freedman, D.G. (1979). *Human Sociobiology: A Holistic Approach*. New York: The Free Press.

Freese, L. (Ed.). (1980). *Theoretical Methods in Sociology*. Pittsburgh: University of Pittsburgh Press.

Freidel, D.A., and J. Sabloff (1984). *Cozumel (Late Maya Settlement Patterns)*. New York: Academic Press.

Friedberg, M.P. (1970). *Play and Interplay*. New York: Macmillan.

Fritz, J.M., and F.T. Plog (1970). "The nature of archaeological explanation." *American Antiquity*, Vol. 35, pp. 405–412.

Fruin, J.J. (1971). *Pedestrian Planning and Design.* New York: Metropolitan Association of Urban Designers and Environmental Planners.

Funahashi, K. (1986). "A study of pedestrian path choice." In W.H. Ittelson, M. Asai, and M. Ker (Eds.), *Cross Cultural Research in Environment and Behavior,* University of Arizona, pp. 85–100.

Gablik, S. (1976). *Progress in Art.* London: Thames and Hudson.

Garbrecht, D. (1971a). "Pedestrian paths through a uniform environment." *Town Planning Review,* Vol. 42 No. 1 (January) pp. 71–84.

Garbrecht, D. (1971b). "Das Verhalten von fussgangern." *Werk 3,* pp. 161–197, 197–204.

Garbrecht, D. (1973). "Describing pedestrian and car trips by transition matrices." *Traffic Quarterly* (January) pp. 89–109.

Garbrecht, D. (1977). "Pedestrian factors and considerations in the design or rebuilding of town centers and suburbs." In A.S. Hakkert (Ed.), *Proceedings of the International Conference on Pedestrian Safety, Haifa, December, 1976.* Haifa: Michlol Publishing House.

Garbrecht, D. (1978). "Walking—Facts, assertions, propositions." Paper presented at 3rd European Colloquium on Economic Psychology, Augsburg: Federal Republic of Germany (July) (Mimeo) (summarized in *Ekistics,* November/December 1978).

Gardi, R. (1960). *Tambaran.* London: Constable.

Gardin, J-C. (1980). *Archaeological Constructs (An Aspect of Theoretical Archaeology).* Cambridge: Cambridge University Press.

Gardiner, P. (1952). *The Nature of Historical Explanation.* London: Oxford University Press.

Gebhardt, H. (1969). *System, Element und Struktur in Kernbereichen Alter Städte.* Stuttgart: Karl Kramer Verlag.

Geertz, C. (1975). *The Interpretation of Cultures.* London: Hutchinson.

Geertz, C. (1983). *Local Knowledge.* New York: Basic Books.

Geertz, C., H. Geertz, and L. Rosen (1979). *Meaning and Order in Moroccan Society.* Cambridge: Cambridge University Press.

Geist, V. (1978). *Life Strategies, Human Evolution, Environmental Design (Towards a Biological Theory of Health).* New York: Springer Verlag.

Gergen, K.J., and M.M. Gergen (Eds.). (1984). *Historical Social Psychology.* Hillsdale, NJ: Lawrence Erlbaum.

Ghaidan, U. (1974). "Lamu: A case study of the Swahili Town." *Town Planning Review,* Vol. 45, No. 1 (Jan.) pp. 84–90.

Gibbon, G. (1984). *Anthropological Archaeology.* New York: Columbia University Press.

Gibbons, G.W., S.W. Hawking, and S.T.C. Siklos (Eds.). (1983). *The Very Early Universe.* Cambridge: Cambridge University Press.

Gibson, J.J. (1950). *The Perception of the Visual World.* Boston: Houghton-Mifflin.

Gibson, J.J. (1968). *The Senses Considered as Perceptual Systems.* London: Allen and Unwin.

Gibson, J.J. (1979). *The Ecological Approach to Visual Perception.* Boston: Houghton-Mifflin.

Gilfoil, D.M., and H.M. Bowen (1980). "Behavioral effects of environmental design characteristics in public plazas." In R.R. Stough and A. Wandersman (Eds.), *Optimizing Environments (Research, Practice and Policy) (EDRA 11).* Washington, DC: EDRA p. 243 (abstract only).

Gilinski, A.S. (1951). "Perceived size and distance in visual space." *Psychological Review,* Vol. 58, pp. 460–482.

Ginsberg, Y. (1975). *Jews in a Changing Neighborhood.* New York: Macmillan.

Giteau, M. (1976). *The Civilization of Ankgor.* New York: Rizzoli.

Gittins, J.S. (1969). "Forming impressions of an unfamiliar city: A comparative study of aesthetic and scientific knowing." MA thesis, Clark University (unpublished).

Glacken, C.J. (1967). *Traces on the Rhodian Shore (Nature and Culture in Western Thought from Ancient Times to the End of the Eighteenth Century).* Berkeley: University of California Press.

Glassie, H. (1968). *The Pattern in the Material Folk Culture of the Eastern United States.* Philadelphia: University of Pennsylvania Press.

Glassie, H.H. (1975). *Folk Housing in Middle Virginia: A Structural Analysis of Historical Artifacts.* Knoxville: University of Tennessee Press.

Glassner, M.F. (1984). *The Dawn of Animal Life (A Biohistorical Study).* Cambridge: Cambridge University Press.

Gleick, J. (1986). "Solar cycles found in bands of rock." *New York Times* (Nov. 18).

Gleick, J. (1987). "Hints of more cloudiness spur global study." *New York Times* (June 30).

Glymour, C. (1980). *Theory and Evidence*. Princeton: Princeton University Press.

Godden, E., and J. Malnic (1982). *Rock Paintings of Aboriginal Australia*. French's Forest, NSW: Reed.

Goffman, E. (1963). *Behavior in Public Places*. New York: Free Press.

Goffman, E. (1971). *Relations in Public: Microstudies of the Public Order*. Harmondsworth: Penguin Books.

Gombrich, E.H. (1961). *Art and Illusion*. New York: Pantheon Books.

Gombrich, E.H. (1979). *Ideals and Idols (Essays on Values in History and in Art)*. Oxford: Phaidon.

Gonen, R. (1981). "Keeping Jerusalem's past alive." *Biblical Archaeology Review*, Vol. 7, No. 4 (July/August) pp. 16–23.

Goodsell, C.T. (1984). "The city council chamber: From distance to intimacy." *The Public Interest*, No. 74 (Winter) pp. 116–131.

Goodsell, C.T. (1988). *The Social Meaning of Civic Space (Studying Political Authority through Architecture)*. Lawrence: University Press of Kansas.

Goody, J. (1977). *The Domestication of the Savage Mind*. Cambridge: Cambridge University Press.

Gordon, M.M. (1978). *Human Nature, Class and Ethnicity*. New York: Oxford University Press.

Gottschalk, L. (Ed.). (1963). *Generalization in the Writing of History (Report of the Committee on Historical Analysis of the Social Science Research Council)*. Chicago: University of Chicago Press.

Gould, R.A. (Ed.). (1978). *Explorations in Ethnoarchaeology*. Albuquerque: University of New Mexico Press.

Gould, R.A. (1980). *Living Archaeology*. Cambridge: Cambridge University Press.

Gould, R.A., and M.B. Schiffer (Eds.). (1981). *Modern Material Culture (The Archaeology of Us)*. New York: Academic Press.

Gould, S.J. (1980). "The evolutionary biology of constraints." *Daedalus*, Vol. 109, No. 2 (Spring) pp. 39–52.

Gould, S.J. (1985). *The Flamingo's Smile (Reflections in Natural History)*. New York: W.W. Norton and Co.

Gray, J.G. (1965). *Pedestrianized Shopping Streets in Europe (A comparative study)*. Edinburgh: Pedestrians' Association for Road Safety.

Green, D., C. Haselgrove, and M. Spriggs (Eds.). (1978). *Social Organization and Settlement: Contributions from Anthropology, Archaeology and Geography*. Oxford, BAR International Series (Supplementary) 47 (i) and (ii).

Greenhalgh, M., and V. Megaw (Eds.). (1978). *Art in Society (Studies in Style, Culture and Aesthetics)*. London: Duckworth.

Gregory, D. (1978). "The discourse of the past: Phenomenology, structuralism and historical geography." *Journal of Historical Geography*, Vol. 4 (No. 2) pp. 161–173.

Grey, A.L. *et al.* (1970). *People and Downtown*. Seattle: College of Architecture and Urban Planning, University of Washington.

Gribbin, J., and J. Cherfas (1982). *The Monkey Puzzle (A Family Tree)*. London: The Bodley Head.

Grmek, M.D., R.S. Cohen, and G. Cimino (Eds.). (1981). *On Scientific Discovery*. Dordrecht: D. Reidel (Vol. 34 of Boston Studies in the Philosophy of Science).

Gruen, V. (1964). *The Heart of Our Cities*. New York: Simon and Schuster.

Guarda, G. (1968). *La Ciudad Chilena del Siglo XVIII*. Buenos Aires: Centro Editor de America Latina.

Guelke, L. (1982). *Historical Understanding in Geography (An Idealist Approach)*. Cambridge: Cambridge University Press.

Guerdan, R. (1973). *Pompei—Mort d'une Ville*. Paris: Laffont.

Guido, M. (1972). *Southern Italy: An Archaeological Guide*. London: Faber and Faber.

Guidoni, E. (1970). *Arte e Urbanistica in Toscana 1000–1315*. Rome: Mario Bulzoni.

Gulick, J. (1963). "Images of an Arab city." *AIP Journal*, Vol. 29, No. 3 (August) pp. 179–198.

Gump, P.V., and E.V. James (n.d.). "Child development and the man-made environment: A literature review and commentary" (mimeo).

Gustke, L.D., and R.W. Hodgson (1980). "Rate of travel along an interpretive trail (the effect of an environmental discontinuity)." *Environment and Behavior*, Vol. 12, No. 1 (March) pp. 53–63.

Gutkind, A.E. (various). *The International History of City Development*, Vols. 1–8.

Gutschow, A. (1974). *Bhakhtapur*, Darmstadt.

Gutschow, N. (1976). *Die Japanische Burgstadt*. Paderborn: Schöningh.

Habraken, N.J. (1983). *Transformations of the Site*. Cambridge, MA: Atwater Press.

Hacking, I. (1983). *Representing and Intervening*. Cambridge: Cambridge University Press.

Hadingham, E. (1975). *Circles and Standing Stones*. London: Heinemann.

Hadingham, E. (1987). *Lines to the Mountain Gods.* New York: Random House.

Hakim, B. (1986). *Arabic-Islamic Cities.* London: KPI.

Hall, E.T. (1966). *The Hidden Dimension.* Garden City, NY: Doubleday.

Hallet, S.I. and R. Samzay (1980). *Traditional Architecture of Afghanistan.* New York: Garland STPM Press.

Halprin, L. (1963). *Cities.* New York: Reinhold.

Hamburg, D.A. (1975). "Ancient man in the twentieth century." In V. Goodall (Ed.), *The Quest for Man.* New York: Praeger.

Hamilton, A.C. (1982). *Environmental History of East Africa: A Study of the Quaternary.* New York: Academic Press.

Hammond, N. (1972). "The planning of a Maya ceremonial center." *Scientific American,* Vol. 226 (May) pp. 82–91.

Han, P-T. (1981). *Lukang, Taiwan (Its Background, Architecture and Handicrafts).* Taiwan.

Handlin, O. (1979). *Truth in History.* Cambridge, MA, Belknap Press of Harvard University.

Hanke, L. (Ed.). (1973). *History of Latin American Civilization.* Boston: Little, Brown (2nd ed).

Hanson, F.A. (1983). "When the map is the territory: Art in Maori culture." In D.K. Washburn (Ed.), *Structure and Cognition in Art.* Cambridge: Cambridge University Press, pp. 74–89.

Hanson, R.N. (1971). *Observation and Explanation (A Guide to Philosophy of Science).* New York: Harper and Row.

Hardin, M.A. (1979). "The cognitive basis of productivity in a decorative art style: Implications of an ethnographic study for archaeologists' taxonomies." In C. Kramer (Ed.), *Ethnoarchaeology (Implications of Ethnography for Archaeology).* New York: Columbia University Press, pp. 75–101.

Hardin, M.A. (1983). "The structure of Tarascan pottery painting." In D.K. Washburn (Ed.), *Structure and Cognition in Art.* Cambridge: Cambridge University Press, pp. 8–24.

Hardoy, J.G., and R.P. Schaedel (Ed.). (n.d.). *The Urbanization Process in America from its Origins to the Present Day.* Buenos Aires: Editorial del Instituto.

Harms, R. (1979). "Oral tradition and ethnicity." *Journal of Interdisciplinary History,* Vol. 10, No. 1 (Summer) pp. 61–85.

Harré, R. (1983). *An Introduction to the Logic of the Sciences* (2nd edition). London: Macmillan.

Harré, R. (1986). *Varieties of Realism (A Rationale for the Social Sciences).* Oxford: Basil Blackwell.

Hart, R. (1979). *Children's Experience of Place.* New York: Irvington Publishers.

Hartung, H. (1968). "Ciudades Mineras de Mexico: Taxco, Guanajuato, Zacatecas." *Verhandlungen des XXXVIII Internationalen Amerikanisten Kongresses,* Vol. 4, Stuttgart-München (August 12–18), pp. 183–187.

Hassan, F.A. (1981). *Demographic Archaeology.* New York: Academic Press.

Hawkes, C. (1954). "Archaeological theory and method." *American Antiquity,* Vol. 56(1) pp. 155–168.

Hawkes, J. (1974). *Atlas of Ancient Archaeology,* New York: McGraw-Hill.

Healan, D.M. (1977). "Architectural implications of daily life in ancient Tollàn, Hidalgo, Mexico." *World Archaeology,* Vol. 9, No. 2 (October) pp. 140–156.

Heath, T. (1971). "The aesthetics of tall buildings." *Architectural Science Review,* Vol. 14, No. 4 (December) pp. 93–94.

Heath, T. (1984). *Method in Architecture.* Chichester: John Wiley and Sons.

Heckscher, A. (1977). *Open Spaces (The Life of American Cities).* New York: Harper and Row.

Hesse, M.B. (1966). *Models and Analogies in Science.* Notre Dame: Notre Dame University Press.

Hesse, M.B. (1974). *The Structure of Scientific Inference.* London: Macmillan.

Hexter, J.H. (1971). *The History Primer.* New York: Basic Books.

Hill, J.N. and J. Gunn (Eds.). (1977). *The Individual in Prehistory.* New York: Academic Press.

Hill, M.R. (1974). "The flow of pedestrians through a rectangular grid: An approach by Dietrich Garbrecht" (mimeo).

Hill, M.R. (1978). "Pedestrians, priorities and spatial games." In S. Weidemann and J.R. Anderson (Eds.), *Priorities for Environmental Design Research (EDRA 8).* Washington, DC: EDRA pp. 126–136.

Hill, M.R. (1984). *Walking, Crossing Streets and Choosing Pedestrian Routes (A Survey of Recent Insights from the Social/Behavioral Sciences).* Lincoln: University of Nebraska Studies, (New Series No. 66).

Hillier, B., and J. Hanson (1984). *The Social Logic of Space.* Cambridge: Cambridge University Press.

Himsworth, H. (1986). *Scientific Knowledge and Philosophic Thought.* Baltimore: Johns Hopkins University Press.

Hinde, R.A. (1974). *Biological Bases of Human Social Behavior.* New York: McGraw-Hill.

Hinde, R.A. (Ed.). (1984). *Primate Social Relationships (An Integrated Approach)*. Sunderland, MA: Sinauer.

Hitzer, H. (1971). *Die Strasse*. München: Callevey.

Hockings, J. (1984). *Built Form and Culture (A Case Study of Gilbertese Architecture)*. Unpublished Ph.D. dissertation, Department of Architecture, University of Queensland (Australia).

Hodder, I. (Ed.). (1978a). *The Spatial Organization of Culture*. London: Duckworth.

Hodder, I. (Ed.). (1978b). *Simulation Studies in Archaeology*. Cambridge: Cambridge University Press.

Hodder, I. (1981). "Society, economy and culture: An ethnographic study amongst the Lozi." In I. Hodder, G. Isaac, and M. Hammond (Eds.), *Pattern of the Past (Essays in Honour of David Clarke)*. Cambridge: Cambridge University Press, pp. 71–95.

Hodder, I. (Ed.). (1982a). *Symbolic and Structural Archaeology*. Cambridge: Cambridge University Press.

Hodder, I. (1982b). *The Present Past (An Introduction to Anthropology for Archaeologists)*. London: Batsford.

Hodder, I. (1982c). *Symbols in Action*. Cambridge: Cambridge University Press.

Hodder, I. (1986). *Reading the Past (Current Approaches to Interpretation in Archaeology)*. Cambridge: Cambridge University Press.

Hodder, I., G. Isaac, and N. Hammond (Eds.). (1981). *Pattern of the Past (Essays in Honour of David Clarke)*. Cambridge: Cambridge University Press.

Hodgen, M.T. (1974). *Anthropology, History and Cultural Change* (Viking Fund Publications in Anthropology, No. 52). Tucson: University of Arizona.

Hoel, L.A. (1968). "Pedestrian traffic rates in the CBD." *Traffic Engineering* (January) pp. 10–13.

Hole, F. (1979). "Rediscovering the past in the present: Ethnoarchaeology in Luristan, Iran." In C. Kramer (Ed.), *Ethnoarchaeology (Implications of Ethnography for Archaeology)*. New York: Columbia University Press, pp. 192–218.

Hole, W.V. (1965). "Housing standards and social trends." *Urban Studies*, Vol. 2, No. 2 (November).

Hole, W.V., and J.J. Attenburrow (1966). *Houses and People*. London: HMSO.

Holton, G. (Ed.). (1972). *The Twentieth-Century Sciences (Studies in the Biography of Ideas)*. New York: Norton.

Holton, G. (1973). *Thematic Origins of Scientific Thought*. Cambridge, MA: Cambridge University Press.

Holton, G. (1978). *The Scientific Imagination: Case Studies*. Cambridge: Cambridge University Press.

Holzner, L. (1970). "The role of history and tradition in the urban geography of West Germany." *Annals, Association of American Geographers*, Vol. 60, No. 2 (June) pp. 315–339.

Horvath, R.J. (1974). "Machine Space." *Geographical Review*, Vol. 66, No. 2 (April) pp. 167–188.

Hosken, F. (1974). *The Kathmandu Valley Towns*. New York: Weatherhill.

Hoskin, F.P. (1968). *The Language of Cities*. New York: Macmillan.

Hubbard, W. (1980). *Complexity and Conviction: Steps Toward an Architecture of Convention*. Cambridge, MA: MIT Press.

Hughes, B.P. (1974). *Firepower: Weapons' Effectiveness on the Battlefield, 1630–1850*. London: Arms and Armour Press.

Hunt, D.R.G., and M.R. Steel (1980). "Domestic Temperature Trends." *Heating and Ventilating Engineer*, Vol. 54, pp. 5–15.

Hyogo No Machinami (Townscape of Hyogo). (1975). Kobe: Hyogo Prefecture Education Committee—Editor and Publisher.

Iannone, C. (1987). "The fiction we deserve." *Commentary*, Vol. 83, No. 6 (June) pp. 60–62.

Im, S-B. (1984). "Visual preferences in enclosed urban spaces." *Environment and Behavior*, Vol. 16, No. 2 (March) pp. 235–262.

Imperial China (Photographs 1850–1912) (1978). New York: Pennwick/Crown (Catalogue of a Travelling Exhibition).

Ingersoll, D., J.E. Yellen, and W. MacDonald (Eds.). (1977). *Experimental Archaeology*. New York: Columbia University Press.

Ingham, J.M. (1971). "Time and space in ancient Mexico: The symbolic dimension of clanship." *Man*, Vol. 6, No. 4, pp. 615–629.

Isaac, G.L. (1972). "Comparative Studies in Pleistocene site locations in East Africa." In P.J. Ucko, R. Tringham, and G.W. Dimbleby (Eds.), *Man, Settlement and Urbanism*. London, Duckworth, pp. 165–176.

Isaac, G.L. (1983). "Human Evolution." In D.S. Bendall (Ed.), *Evolution from Molecules to Men*. Cambridge: Cambridge University Press, pp. 509–539.

Isbell, W.H. (1978). "Cosmological order expressed in prehistoric ceremonial centers," *Actes du 42 Congres International des Americanistes,* Vol. 4, Paris, pp. 267–297.

Ittelson, W.H. (Ed.). (1973). *Environment and Cognition.* New York: Seminar Press.

Jackson, J.B. (1972). *American Space.* New York: Norton.

Jackson, J.B. (1984). *Discovering the Vernacular Landscape.* New Haven: Yale University Press.

Jackson, K.T. (1985). *Crabgrass Frontier (The Suburbanization of the United States).* New York: Oxford University Press.

Jacob, F. (1982). *The Possible and the Actual.* New York: Pantheon Books.

Jäger, W.-H. (1979). "Munich—A case study in urban townscape." In R. Cresswell (Ed.), *Quality in Urban Planning and Design.* London: Newnes-Butterworths.

Jain, K. (1977). "Jaisalmer, India: Morphology of a desert settlement." *A+U* (Japan) (Feb.), pp. 13–24.

Jain, K. (1978). "Morphostructure of a planned city: Jaipur, India." *A+U* (Japan) (Aug.), pp. 107–120.

Jain, K. (1980a). "Form and function of public spaces in Jaisalmer." In J. Pieper (Ed.), *Ritual Space in India: Studies in Architectural Anthropology.* London: AARP (Vol. 17), pp. 20–24.

Jain, K. (1980b). "Satrunjaya Hill," In J. Pieper (Ed.), *Ritual Space in India: Studies in Architectural Anthropology.* London: AARP (Vol. 17) pp. 47–52.

Jain, R.K. (Ed.). (1977). *Text and Context (The Social Anthropology of Tradition).* Philadelphia: Institute for the Study of Human Issues.

Jakle, J.A. (1978). "The American gasoline station, 1920 to 1970." *Journal of American Culture,* Vol. 1, pp. 520–542.

Jakle, J.A., and R.L. Mattson (1981). "The evolution of a commercial strip." *Journal of Cultural Geography,* Vol. 1, No. 2, pp. 12–25.

James, J. (1979). *The Contractors of Chartres.* Dooralong (Australia): Mandorla Publications (Vol. 1).

James, J. (1981). *The Contractors of Chartres.* Wyong (Australia): Mandorla Publications (Vol. 2).

James, J. (1982). *Chartres, the Masons Who Built a Legend.* London: Routledge and Kegan Paul.

James, J. (1988). *The Template Makers of the Paris Basin.* Laura (Australia): West Grinstead Publishing.

Jansen, M. (1980). "Public Spaces in the Urban Settlements of the Harappan Culture." In J. Pieper (Ed.) *Ritual Space in India—Studies in Architectural Anthropology.* London: AARP (Vol. 17) pp. 11–19.

Jensen, V.J. and R. Marre (Ed.). (1981). *Philosophy of Evolution.* Brighton, England: Harverster Press.

Joardar, S.D. and J.W. Neill (1978). "The subtle differences in configuration of small public spaces." *Landscape Architecture,* Vol. (Nov.) pp. 487–491.

Jochim, M.A. (1976). *Hunter Gatherer Subsistence and Settlement (A Predictive Model).* New York: Academic Press.

Jochim, M.A. (1983). "Optimization models in context." In J.A. Moore and A.S. Keene (Eds.), *Archaeological Hammers and Theories.* New York: Academic Press, pp. 157–172 (Chapter 7).

Johanson, D.C. and M.A. Edey (1981). *Lucy—The Beginnings of Humankind.* New York: Simon and Schuster.

Johnson, G.A. (1980). "Rank-size convexity and system integration: A view from archaeology." *Economic Geography,* Vol. 56, No. 3, pp. 234–247.

Johnson, P. (1965). "Whence and whither: The processional element in architecture." *Perspecta* (The Yale Architectural Journal) 9/10, pp. 167–178.

Johnson-Laird, P.N., and P.C. Wason (Eds.). (1977). *Thinking (Readings in Cognitive Science).* Cambridge: Cambridge University Press.

Johnston, R.J. (1983). *Philosophy and Human Geography.* London: Arnold.

Jones, A. (1966). "Information deprivation in humans." In B.A. Maher (Ed.), *Progress in Experimental Personality Research* (Vol. 3). New York: Academic Press.

Judson, H.F. (1980). *The Search for Solutions.* New York: Holt, Rinehart & Winston.

July, R.W. (1980). *A History of the African People.* New York: Charles Scribner's sons (3rd ed).

Jung, C. (1964). *Man and His Symbols.* Garden City, NY: Doubleday.

Kamau, L.J. (1978–1979). "Semipublic, private and hidden rooms: Symbolic aspects of domestic space in urban Kenya." *African Urban Studies,* Vol. 3 (Winter) pp. 105–115.

Kammen, M. (Ed.). (1980). *The Past Before Us (Contemporary Historical Writing in the United States).* Ithaca, NY: Cornell University Press.

Kaplan, A. (1964). *The Conduct of Inquiry (Methodology for Behavioral Science).* New York: Harper and Row.

Kaplan, R. (1973). "Predictors of environmental preference: Designers and clients." In W.F.E. Preiser (Ed.), *Environmental Design Research* (EDRA 4), Vol. 1. Stroudsburg, PA: Dowden, Hutchinson and Ross, pp. 265–274.

Kaplan, R. (1975). "Some methods and strategies in the prediction of preference." In E.H. Zube, G.G. Fabos, and R.O. Brush (Eds.), *Landscape Assessment: Values, Perceptions and Resources*. New York: Halstead Press.

Kaplan, S. (1975). "An informal model for the prediction of preference." In E.H. Zube, G.G. Fabos, and R.O. Brush (Eds.), *Landscape Assessment: Values, Perceptions and Resources*. New York: Halstead Press.

Kaplan, S. (1982). "Where cognition and affect meet: A theoretical analysis of preference." In P. Bart, A. Chen and G. Francescato (Eds.), *Knowledge for Design* (EDRA 13). Washington, DC: EDRA, pp. 183–188.

Kaplan, S. (1987). "Aesthetics, affect and cognition: Environmental preference from an evolutionary perspective." *Environment and Behavior,* Vol. 19, No. 1 (Jan.) pp. 3–32.

Kaplan, S. and J.S. Wendt (1972). "Preference and the visual environment: Complexity and some alternatives." In W. Mitchell (Ed.), *EDRA 3*. Los Angeles: University of California, Vol. 1, pp. 681–685.

Kashiwazaki, H. (1983). "Agricultural Practices and household organization in a Japanese pioneer community in lowland Bolivia." *Human Ecology,* Vol. 11, No. 3 (Sept.) pp. 283–319.

Kearney, M. (1972). *The Winds of Ixtepeji*. New York: Rinehart and Winston.

Keene, A.S. (1983). "Biology, behavior and borrowing: A critical examination of optimal foraging theory in archaeology." In J.A. Moore and A.S. Keene (Eds.), *Archaeological Hammers and Theories*. New York: Academic Press, pp. 137–155 (Chapter 6).

Kelley, J.H. and M.P. Hanen (1988). *Archaeology and the Methodology of Science*. Albuquerque: University of New Mexico Press.

Kelsey, V., and L.D. Osborne (1943). *Four Keys to Guatemala*. New York: Funk and Wagnall's.

Kemp, B.J. (1972a). "Fortified towns in Nubia." In P.J. Ucko, R. Tringham, and G.W. Dimbleby (Eds.), *Man, Settlement and Urbanism*. London: Duckworth, pp. 651–656.

Kemp, B.J. (1972b). "Temple and town in ancient Egypt." In P.J. Ucko, R. Tringham, and G.W. Dimbleby (Eds.), *Man, Settlement and Urbanism*. London: Duckworth, pp. 657–680.

Kemp, B.J. (1977). "The city of el-Amarna as a source for the study of urban society in ancient Egypt." *World Archaeology,* Vol. 9, No. 2 (October) pp. 123–139.

Kent, S. (1984). *Analyzing Activity Areas (An Ethnoarchaeological Study of the Use of Space)*. Albuquerque: University of New Mexico Press.

Kent, S. (1987). "Understanding the use of space: An ethnoarchaeological perspective." In S. Kent (Ed.), *Method and Theory for Activity Area Research: An Ethnoarchaeological Approach*. New York: Columbia University Press, pp. 1–60.

Kepes, G. (1961). "Notes on expression and communication in the city-scape." *Daedalus,* Vol. 90, No. 1 (Winter) pp. 147–165.

Khisty, C.J. (n.d.). "Planing of pedestrian facilities using behavior circuits" (mimeo).

Khisty, C.J. (1985). "Pedestrian cross flow characteristics and performance." *Environment and Behavior,* Vol. 17, No. 6 (Nov.) pp. 679–695.

King, A.D. (1980). "Introduction." In A.D. King (Ed.), *Buildings and Society (Essays on the Social Development of the Built Environment)*. London: Routledge and Kegan Paul, pp. 1–33.

King, L.S. (1982). *Medical Thinking (A Historical Preface)*. Princeton: Princeton University Press.

Kinzer, N.S. (Ed.). (1978). *Urbanization in the Americas from Its Beginnings to the Present*. The Hague: Mouton.

Kirby, C.P. (1968). *East Africa*. London: Ernest Benn.

Knowles, R.L. (1974). *Energy and Form: An Ecological Approach to Urban Growth*. Cambridge, MA: MIT Press.

Knudsen, K. *et al.* (1969). *Landsby i Nepal* [Village in Nepal], Copenhagen: School of Architecture, Royal Academy of Art.

Köben, A.J.F. (1969). "Why exception? The logic of cross-cultural analysis." *Current Anthropology,* Vol. 8, No. 1-2 (April) pp. 3–34.

Kochetkova, V.I. (1978). *Paleoneurology*. New York: John Wiley and Sons.

Koestler, A. (1964). *The Act of Creation,* New York: Macmillan.

Kolata, G. (1986). "Shakespeare's new poem: An ode to statistics." *Science,* Vol. 231 (Jan. 24) pp. 335–336.

Konner, M. (1982). *The Tangled Wing (Biological Constraints on the Human Spirit)*. London: Heinemann.

Korn, A. (1953). *History Builds the Town*. London: Lund Humphries.

Korosec-Serfaty, P. (1982). *The Main Square* (Functions and Daily Uses of Stortorget, Malmö). Lund: Sweden: Nova.

Kosslyn, S.M. (1980). *Image and Mind*. Cambridge, MA: Harvard University Press.

Kosslyn, S.M. (1983). *Ghosts in the Mind's Machine (Creating and Using Images in the Brain)*. New York: Norton.

Kramer, C. (Ed.). (1979). *Ethnoarchaeology (Implications of Ethnography for Archaeology)*. New York: Columbia University Press.

Kramer, C. (1982). *Village Ethnoarchaeology (Rural Iran in Archaeological Perspective*. New York: Academic Press.

Krampen, M. (1979). *Meaning in the Urban Environment*. London: Pion.

Krawetz, N.M. (1977). *The Dissolution of Boundaries: A Study of Environmental Concommitants in Two Cultures*, Ph.D. dissertation in Psychology, CUNY.

Krier, R. (1979). *Urban Space*. London: Academy Editions.

Kubota, Y. (1987). "Intrinsic factors in designing street environments for pedestrians in Japan." In J. Harvey and D. Henning (Eds.), *Public Environments (EDRA 18)*. Washington, DC: EDRA, pp. 230–232.

Kurjack, E.B. (1974). *Prehistoric Lowland Maya Community Social Organization—A Case Study in Dzibilchaltun, Yucatan, Mexico*, Publication 38, Middle American Research Institute. New Orleans: Tulane University.

Lambrick, G. (1988). *The Rollright Stones: Megaliths, Monuments and Settlement in the Prehistoric Landscape* (English Heritage Archaeological Report No. 6). London: English Heritage.

Lampl, P. (1968). *Cities and Planning in the Ancient Near East*. New York: George Braziller.

Lane, E.W. (1977). *Manners and Customs of Modern Egyptians*. Cairo: East/West Publications.

Langer, S. (1966). "The social influence of design." In. L.B. Holland (Ed.), *Who Designs America?*. Garden City, NY: Anchor Books, pp. 35–50.

Lannoy, R. (1971). *The Speaking Tree: A Study of Indian Culture and Society*. London: Oxford University Press.

Laporte, L.F. (Ed.). (1978). *Evolution and the Fossil Record (Readings from* Scientific American). San Francisco: W.H. Freeman.

Lasig, K. *et al.* (1967). *Strassen und Platze (Beispiele zur Gestaltung Stadtebaulicher Raume)*. München: Callwey.

Laslett, P. (1977). *Family Life and Illicit Love in Earlier Generations (Essays in Historical Sociology)*. Cambridge: Cambridge University Press.

Laslett, P. and R. Wall (Eds.). (1972). *Household and Family in Past Time*. Cambridge: Cambridge University Press.

Lavedan, P., and J. Hugueney (1974). *L'Urbanisme au Moyen Age*. Geneva: Droz.

Laverick, C. (n.d.). "Qualitative emotional experience in towns and cities: Some possible mental mechanisms" (mimeo).

Lawless, R.I., and G.H. Blake (1976). *Tlemcen: Continuity and Change in an Algerian Islamic Town*. London: Bowker Publishing Co.

Layton, R. (1981). *The Anthropology of Art*. London: Paul Elek (Granada).

Leach, E. (1966). *Rethinking Anthropology*. London: Athlone Press (1st edition, 1961).

Leach, E. (1976). *Culture and Communication: The Logic by Which Symbols are Connected*. Cambridge: Cambridge University Press.

Leatherdale, W.H. (1974). *The Role of Analogy, Model and Metaphor in Science*. Amsterdam: North Holland.

Lee, R.B. (1979). *The !Kung San: Men, Women and Work in a Foraging Society*. Cambridge: Cambridge University Press.

Lee. R.B., and I. DeVore (1976). *Kalahari Hunter-Gatherers: Studies of the !Kung San and Their Neighbors*. Cambridge, MA: Harvard University Press.

Leeming, F. (1977). *Street Studies in Hong Kong (Localities in a Chinese City)*. Hong Kong: Oxford University Press.

LeGoff, J., and P. Nora (Ed.). (1985). *Constructing the Past (Essays in Historical Methodology)*. Cambridge: Cambridge University Press.

Leplin, J. (Ed.). (1984). *Scientific Realism*. Berkeley: University of California Press.

Leroi-Gourhan, A. (1982). *The Dawn of European Art (An Introduction to Palaeolithic Cave Painting)*. Cambridge: Cambridge University Press.

Lerup, L., R. Brambilla, and G. Longo. (1974). "Street Smarts." *Architectural Design*, Vol. 44, No. 11 (November) pp. 723–725.

LeShan, L., and H. Margenau. (1982). *Einstein's Space and Van Gogh's Sky (Physical Reality and Beyond)*. New York: Macmillan.

Levin, S. (1983). "Food production and population size in the Lesser Antilles." *Human Ecology*, Vol. 11 (No. 3) pp. 321–338.

Lévi-Strauss, C. (1955). *Tristes Tropiques*. Paris: Plon.

Lewcock, R. (1971). "Zanj, the East African Coast." In P. Oliver (Ed.), *Shelter in Africa*. London: Barrie and Jenkins, pp. 80–95.

Lewis, I.M. (Ed.). (1968). *History and Anthropology*. London: Tavistock.

Lewis, J.R. (1978). "Round the Mulberry Bush." In D. Green, C. Haselgrove, and M. Spriggs (Eds.), *Social Organization and Settlement*. Oxford: BAR International Series (Supplementary) 47, (i) and (ii) pp. 513–519.

Lewis-Williams, J.D. (1981). *Believing and Seeing (Symbolic Meanings in Southern San Rock Paintings)*. London: Academic Press.

Lewis-Williams, J.D. (1983). *The Rock Art of Southern Africa*. Cambridge: Cambridge University Press.

Ling, A. (1967). "Urban form." In D. Sharp (Ed.), *Planning and Architecture (Essays for Arthur Korn)*. London: Barrie and Rockliff, pp. 60–70.

Llewellyn-Smith, M.J. (1972). *The City Pedestrian Environment* (Unpublished MTCP Thesis), Faculty of Architecture, University of Sydney.

Lofland, L.H. (1973). *A World of Strangers (Order and Action in Urban Public Space)*. New York: Basic Books.

Longacre, W. (1981). "Kaluga pottery: An ethnoarchaeological study." In I. Hodder, G. Isaac, and N. Hammond (Eds.), *Pattern of the Past (Essays in Honour of David Clarke)*. Cambridge: Cambridge University Press, pp. 49–66.

Longworth, I., and J. Cherry (Eds.). (1986). *Archaeology In Britain since 1945*. London: British Museum Publications.

Lopreato, J. (1984). *Human Nature and Biolcultural Evolution*. Boston: Allen and Unwin.

Lowenthal, D. (1972). *Environmental Assessment* (Series of 8 reports, some with M. Riel). New York: American Geographical Society.

Lozano, E.E. (1974). "Visual needs in the urban environment." *Town Planning Review*, Vol. 45, No. 4 (October) pp. 351–374.

Lumsden, C.J. and E.O. Wilson (1981). *Genes, Mind and Culture: The Coevolutionary Process*. Cambridge, MA: Harvard University Press.

Lumsden, C.J., and E.O. Wilson (1983). *Promethean Fire: Reflections on the Origins of Mind*. Cambridge, MA: Harvard University Press.

Lynch, K. (1960). *The Image of the City*. Cambridge, MA: MIT Press.

Lynch, K. (1971). *Site Planning*. Cambridge, MA: MIT Press (2nd edition).

Lynch, K., and M. Rivkin (1970). "A walk around the block." In H.M. Proshansky *et al.* (Eds.), *Environmental Psychology*. New York: Holt, Rinehart & Winston, pp. 631–642.

Lyons, E. (1983). "Demographic correlates of landscape preference." *Environment and Behavior*, Vol. 15, No. 2, pp. 485–511.

Macaulay, D. (1979). *Motel of the Mysteries*. Boston: Houghton-Mifflin.

MacDonald, W.L. (1976). *The Pantheon*. Cambridge, MA: Harvard University Press.

Machotka, P. (1979). *The Nude: Perception and Personality*. New York: Irvington Publishers.

Mackie, E.W. (1977). *Science and Society in Prehistoric Britain*. London: Paul Elek.

Mackworth, N.H. (1968). "Visual noise causes tunnel vision." In R.N. Haber (Ed.), *Contemporary Theory and Research in Visual Perception*. New York: Holt, Rinehart & Winston.

Maertens, H. (1884). *Der Optische Masstab in den Bilden Kuensten*. Berlin: Wasmuth (2nd ed.).

Mandelbaum, M. (1977). *The Anatomy of Historical Knowledge*. Baltimore: Johns Hopkins University Press.

Maquet, J. (1979). *Introduction to Aesthetic Anthropology*. Malibu: Undenna Publications (2nd revised edition).

Maquet, J. (1986). *The Aesthetic Experience*. New Haven: Yale University Press.

Marchand, B. (1974). "Pedestrian traffic planning and the perception of the urban environment: A French example." *Environment and Planning A*, Vol. 6, pp. 491–507.

Marcus, J. (1973). "Territorial organization of the Lowland Maya." *Science*, Vol. 180 (June) pp. 911–916.

Marcus, J. (1976). *Emblem and State in the Classic Maya Lowlands*. Washington, DC: Dumbarton Oaks.

Marinatos, S. (1968–1976). *Excavations at Thera*, I-VII (7 volumes). Athens.

Marinatos, S. (1973). *Ausgrabungen auf Thera und Ihre Problemen*. Vienna: Verlag der Osterreicheschen Akademie der Wissenschaft, Vol. 287, Part 1.

Markman, R. (1970). "Sensation seeking and environmental preference." In J. Archea and C. Eastman (Eds.), *EDRA 2*, pp. 311–315.

Marshack, A. (1971). *The Roots of Civilization: The Cognitive Beginnings of Man's First Art, Symbols and Notation*. New York: McGraw-Hill.

Martienssen, R.D. (1964). *The Idea of Space in Greek Architecture*. Johannesburg: Witwatersrand University Press.

Martin, A. (1969). "Bibiography: Pedestrians in the city." *Ekistics*, Vol. 28, No. 166 (September) p. 212.

Martin, R. (1974). *L'Urbanisme dans la Grèce Antique*. Paris: Picard (2nd ed.).

Martin, R. (1977). *Historical Explanation (Re-enactment and Practical Inference)*. Ithaca, NY: Cornell University Press.

Masood, M. (1978). "The traditional organization of a Yoruba town: A study of Ijebu-Ode." *Ekistics*, Vol. 45, No. 271 (July/August) pp. 307–312.

Masson, V.M. (1972). "Prehistoric settlement patterns in Soviet Central Asia." In P.J. Ucko, R. Tringham, and G.W. Dimbleby (Eds.), *Man, Settlement and Urbanism*. London: Duckworth, pp. 263–277.

Maurer, R., and J.C. Baxter (1972). "Images of the neighborhood among black, anglo and Mexican-Americans." *Environment and Behavior*, Vol. 4, No. 4 (December) pp. 351–388.

Maxwell, M. (1984). *Human Evolution (A Philosophical Anthropology)*. London: Croom Helm.

Maxwell, R.J. (1982). *Contexts of Behavior*. Chicago: Nelson-Hall.

Mayr, E. (1982). *The Growth of Biological Thought*. Cambridge, MA: Belknap Press of Harvard University Press.

Mazumdar, S., and S. Mazumdar (1984). "Zoroastrian vernacular arthictecture and issues of status, power and conflict." In D. Duerk and D. Campbell (Eds.), *The Challenge of Diversity* (EDRA 15). Washington, DC: EDRA, pp. 261–262 (Abstract only).

McBride, G. (1972). "The nature of space control systems." Unpublished paper, Annual Conference of Australian Association of Social Anthropologists, Symposium on Space and Territory, Melbourne: Monash University.

McClelland, P.D. (1975). *Causal Explanation and Model Building in History: Economics and the New Economic History*. Ithaca, NY: Cornell University Press.

McCoy, K. (1975). "Landscape planning for a new Australian town." *Urban Ecology*, Vol. 1, No. 2/3 (December) pp. 129–270.

McCullagh, C.B. (1984). *Justifying Historical Descriptions*. Cambridge: Cambridge University Press.

McCully, R.S. (1971). *Rorschach Theory and Symbolism*. Baltimore: Williams & Wilkins.

McIntosh, S.K. and R.J. McIntosh (1980). *Prehistoric Investigations in the Region of Jenne (Mali) (A Study in the Development of Urbanism in the Sahel)*. Cambridge: Monographs in African Archaeology 2/BAR International Series 89.

McKean, J. (1968). "The social labyrinth: The town of Mesta on the Aegean island of Chios." *RIBA Journal*, Vol. 75 No. 8 (August) pp. 344–348.

McMullin, E. (1984). "A case for scientific realism." In J. Leplin (Ed.), *Scientific Realism*. Berkeley: University of California Press, pp. 8–40.

McNett, C.W., and R.E. Kirk. (1968). "Drawing random samples in cross-cultural studies: A suggested method." *American Anthropologist*, Vol. 70, No. 11, pp. 50–55.

Mead, S.M. (Ed.). (1979). *Exploring the Visual Arts of Oceania*. Honolulu: University Press of Hawaii.

Medawar, P.B. (1967). *The Art of the Soluble*. London: Methuen.

Medawar, P.B. (1969). *Induction and Intuition in Scientific Thought*. Philadelphia: American Philosophical Society.

Mehrabian, A., and J.A. Russell (1973). "A measure of arousal seeking tendency." *Environment and Behavior*, Vol. 5, No. 3 (September) pp. 315–333.

Mehrabian, A., and J.A. Russell (1974). *An Approach to Environmental Psychology*. Cambridge: MIT Press.

Meillassoux, C. (1963). "Histoire et institutions du Kato de Bamako d'apres la tradition des Niaré." *Cahiers d'Etudes Africaines*, Vol. 4, No. 14, pp. 186–227.

Mellaart, J. (1967). *Çatal Hüyük: A Neolithic Town in Anatolia*. London: Thames and Hudson.

Merlyn Jones, G. (1976). *Man, Environment and Disease in Britain (A Medical Geography of Britain through the Ages)*. Harmondsworth: Penguin (2nd edition) (1st edition, 1972).

Michaelides, C.E. (1967). *Hydra: A Greek Island Town (Its Growth and Form)*. Chicago: Chicago University Press.

Miggs, R. (1983). *Trees and Timber in the Ancient Mediterranean World*. Oxford: Oxford University Press.

Miller, A.I. (1984). *Imagery in Scientific Thought (Creating 20th Century Physics)*. Boston: Birkhauser.

Miller, D., and C. Tilley (Eds.). (1984). *Ideology, Power and Prehistory.* Cambridge: Cambridge University Press.

Millon, R. (Ed.). (1973). *Urbanization at Teotihuacàn, Mexico* (Vol. 1, Parts 1 and 2). Austin: University of Texas Press.

Mills, D.C. (1978). *Plazas for People.* Project for Public Spaces.

Milwaukee Journal (1980). "Fear is changing lifestyles, study says." (September 17).

Miner, H. (1956). "Body ritual among the Nacirema." *American Anthropologist,* Vol. LVIII, pp. 503–507.

Minsky, M. (1977). "Frame-system theory." In P.N. Johnson-Laird and P.C. Wason (Eds.), *Thinking (Readings in Cognitive Science).* Cambridge: Cambridge University Press, pp. 355–376.

Moffett, N. (1977). "Nigeria today." *RIBA Journal,* Vol. 84, No. 6 (June) pp. 244–255.

Moodie, D.W., and J.C. Lehr (1976). "Fact and theory in historical geography." *Professional Geographer,* Vol. 28, pp. 132–135.

Moore, D.D. (1981). *At Home in America (Second Generation New York Jews)* New York: Columbia University Press.

Moore, G.T., D.P. Tuttle, and S.C. Howell (1985). *Environmental Design Research Directions.* New York: Praeger.

Moore, H.L. (1982). "The interpretation of spatial patterning of settlement residues." In I. Hodder (Ed.), *Symbolic and Structural Archaeology.* Cambridge: Cambridge University Press, pp. 74–79 (Chapter 7).

Moore, J.A. (1983). "The trouble with know-it-alls: Information as a social and ecological resource." In J.A. Moore and A.S. Keene (Eds.), *Archaeological Hammers and Theories.* New York: Academic Press, pp. 173–191 (Chapter 8).

Moore, J.A. and A.S. Keene (Eds.). (1983). *Archaeological Hammers and Theories.* New York: Academic Press.

Moore, R. (1966). "An experiment in playground design." MCRP Thesis, MIT (unpublished).

Moore, T., and C. Carling. (1982). *Understanding Language (Towards a Post-Chomskyan Linguistics).* London: Macmillan.

Morgan, C.G. (1973). "Archaeology and explanation." *World Archaeology,* Vol. 4, pp. 259–276.

Morrell, V. (1987). "The birth of a heresy." *Discover,* Vol. 8, No. 3 (March) pp. 26–50.

Morris, A.E.J. (1979). *History of Urban Form.* London: George Godwin (2nd ed).

Morris, C., and D.E. Thompson (1985). *Huanuco Pampa (An Inca City and Its Hinterland).* London: Thames and Hudson.

Morris, D. (1969). *The Human Zoo.* London: Jonathan Cape.

Morris, R.L., and S.B. Zisman (1962). "The pedestrian downtown and the planners." *AIP Journal,* Vol. 28 (August).

Morris, R. (1983). *Dismantling the Universe (The Nature of Scientific Discovery).* New York: Simon and Schuster.

Moseley, M.E., and C.J. Mackey (1974). *Twenty-four Architectural Plans of Chan Chan, Peru (Structure and Form at the Capital of Chimor).* Cambridge, MA: Peabody Museum Press.

Moughtin, J.C. (1964). "The traditional settlements of the Hausa people." *Town Planning Review,* Vol. 35, No. 1 (April) pp. 21–34.

Munroe, R.L., R. Munroe and B.B. Whiting (1979). *Handbook of Cross-Cultural Human Development.* New York: Garland/STPM Press.

Munsterberg, H. (1985). *Symbolism in Ancient Chinese Art.* New York: Hacker Art Books.

Murch, G.M. (1973). *Visual and Auditory Perception.* Indianapolis: Bobbs Merrill.

Murdock, G.P., and D.R. White (1969). "Standard cross-cultural sample." *Ethnology,* Vol. 8, No. 4 (October) pp. 329–369.

Murphey, M. (1973). *Our Knowledge of the Historical Past.* Indianapolis: Bobbs-Merrill.

Nahemow, L. (1971). "Research in a novel environment." *Environment and Behavior,* Vol. 3, No. 1, p. 81ff.

Nahemow, L., and M.P. Lawton (1973). "Toward an ecological theory of adaptation and aging." In W. Preiser (Ed.), *Environmental Design Research (EDRA 4).* Stroudsburg, PA: Dowden, Hutchinson and Ross, Vol. 1, pp. 24–32.

Nairn, I. (1955). *Outrage.* London: Architectural Press.

Nairn, I. (1956). *Counterattack.* London: Architectural Press.

Nairn, I. (1965). *The American Landscape (A Critical View).* New York: Random House.

Naroll, R. (1970). "Cross-cultural sampling." In R. Naroll and R. Cohen (Eds.), *Handbook of Method in Cultural Anthropology.* Garden City, NY: Natural History Press, pp. 889–926.

Naroll, R., and G. Michik (1980). "Hologeistic Studies." In H. Triandis and J. Berry (Eds.), *Handbook of Cross-Cultural Psychology* (Vol. 2) Boston: Allyn and Bacon.

Nash, L.K. (1963). *The Nature of the Natural Sciences*. Boston: Little, Brown.

National Geographic (1888–1980).

Nayir, Z. (1978). "Arnavutkoy." *Journal of METU*, Vol. 4, No. 2, pp. 159–178.

Negev, A. (1988). *The Architecture of Mampsis (Final Report): Vol I: The Middle and Late Nabatean Period*. Jerusalem: The Hebrew University, Monographs of the Institute of Archaeology (Qedem 26).

Neill, W.T. (1978). *Archaeology and a Science of Man*. New York: Columbia University Press.

Nelson, H.J. (1963). "Townscapes of Mexico: An example of the regional variation of townscapes." *Economic Geography*, Vol. 39, No. 1, pp. 74–83.

Nersessian, N.J. (1984). *Faraday to Einstein: Constructing Meaning in Scientific Theories*. The Hague: Nijhoff.

Netting, R.M., R.R. Wilk, and E.J. Arnould (Eds.). (1984). *Households: Comparative and Historical Studies of the Domestic Group*. Berkeley: University of California Press.

New York Times (1984a). "Fossils in Kenya may link humans and apes" (August 24).

New York Times (1984b). "Theory on man's origins challenged" (September 4).

Newman, B. (1968). *Turkey and the Turks*. London: Herbert Jenkins.

Nicholas, L., G. Feinman, S.A. Kowalewski, R.E. Blanton, and L. Finsten (1986). "Prehispanic colonization of the Valley of Oaxaca, Mexico." *Human Ecology*, Vol. 14, No. 2 (June) pp. 131–162.

Nicholson, M. (1983). *The Scientific Analysis of Social Behavior (A Defense of Empiricism in Social Science)*. London: Frances Pinter.

Nielsen, E.S. *et al.* (1972). *Graeske Skitser*. Copenhagen: School of Architecture, Royal Academy of Art.

Nijst, A.L.M.T., *et al.* (1973). *Living on the Edge of the Sahara (A Study of Traditional Forms of Habitation and Types of Settlement in Morocco)*. The Hague: Government Publishing Office.

Nilsson, S.A. *et al.* (n.d.). *Tanzania, Zanzibar—Present Conditions and Future Plans* (Housing, Building and Planning in Developing Countries). London: Department of Architecture, University of London.

Noe, S. (1984). "Shahjahanabad: Geometrical bases for the plan of Mughal Delhi." *Urbanism Past and Present*, Vol. 9, Issue 2 (No. 18) Summer/Fall, pp. 15–25.

Northrop, F.S.C. (1947). *The Logic of the Sciences and the Humanities*. New York, Macmillan.

OECD (1974). *Streets for People*. Paris: OECD.

OECD (1978). *For Better Urban Living*. Paris: OECD.

Ogden, R.H. (1970). "Pedestrian Precinct in Bolton." *Journal of the TPI*, Vol. 56, No. 4 (April) p. 142ff.

Oliver, P. (Ed.). (1969). *Shelter and Society*. London: Barrie and Rockliff.

Oliver, P. (Ed.). (1971). *Shelter in Africa*. London: Barrie and Jenkins.

Oliver, P., I. Davis, and I. Bentley (1981). *Dunroamin: The Suburban Semi and its Enemies*. London: Barrie and Jenkins.

Orme, B. (1981). *Anthropology for Archaeologists: An Introduction*. London: Duckworth.

Ortony, A. (Ed.). (1979). *Metaphor and Thought*. Cambridge: Cambridge University Press.

Otten, C.M. (Ed.). (1971). *Anthropology and Art (Readings in Cross-Cultural Aesthetics)*. Garden City, NY: Natural History Press.

Paden, J. and E. Soja (Eds.). (1970). *The African Experience*, Vol. III, Bibliography. Evanston: Northwestern University Press.

Pagels, H.R. (1985). *Perfect Symmetry (The Search for the Beginning of Time)*. New York: Simon and Schuster.

Papineau, D. (1978). *For Science in the Social Sciences*. London: Macmillan.

Papineau, D. (1979). *Theory and Meaning*. Oxford: Clarendon Press.

Parkes, D., and N. Thrift (1980). *Times, Spaces and Places: A Chronogeographic Perspective*. Chichester: John Wiley and Sons.

Parr, A.E. (1967). "Urbanity and the urban scene." *Landscape*, Vol. 16, No. 3 (Spring) pp. 3–5.

Parr, A.E. (1969a). "Speed and community." *Journal of High Speed Transportation*, Vol. 3, No. 1 (January) (no page nos.).

Parr, A.E. (1969b). "Problems of reason, feeling and habitat." *Architectural Association Quarterly*, Vol. 1, No. 3 (July) pp. 5–10.

Parsons, J.R. *et al.* (1982). *Prehistoric Settlement Patterns in the Southern Valley of Mexico* (Memoirs of the University of Michigan Museum of Anthropology, No. 14). Ann Arbor: University of Michigan, Museum of Anthropology.

Passmore, J. (1978). *Science and its Critics*. New Brunswick, NJ, Rutgers University Press.

Paul, C. (1980). *The Natural History of Fossils*. London: Weidenfeld and Nicolson.

Pavlides, E. (1985a). *Vernacular Architecture in Its Social Context: A Case Study of Eressos, Greece*. Ph.D. dissertation, University of Pennsylvania.

Pavlides, E. (1985b). "Architectural change in a vernacular environment, A case study of Eressos, Greece." In S. Klein and R. Wener (Eds.), *Environmental Change/Social Change* (EDRA 16). Washington, DC: EDRA, pp. 57–74.

Pawley, M. (1971). *Architecture vs. Housing*. New York: Praeger.

Payne, G.K. (1979). *Urban Housing in the Third World*. Boston: Routledge and Kegan Paul.

Perocco, G., and A. Salvadori (1975). *Civiltà di Venezia*. Venice: La Stamperia de Venezia Editrice, 3 volumes (3rd ed).

Pettit, P. (1977). *The Concept of Structuralism: A Critical Review*. Berkeley: University of California Press.

Phelan, J.G. (1970). "Relationship of judged complexity to change in mode of presentation of object shapes." *Journal of Psychology*, Vol. 74 (1st half) (January) pp. 21–27.

Phillips, D.C. (1976). *Holistic Thought in Social Science*. Stanford, CA, Stanford University Press.

Phillips, P. (1980). *The Prehistory of Europe*. London: Allen Lane.

Pieper, J. (Ed.). (1980a) *Ritual Space in India: Studies in Architectural Anthropology*. London: AARP (Vol. 17).

Pieper, J. (1980b). "The spatial structure of Suchindram." In J. Pieper (Ed.), *Ritual Space in India: Studies in Architectural Anthropology*. London: AARP (Vol. 17) pp. 65–80.

Piggott, S. (1961). *Prehistoric India*. Harmondsworth: Penguin Books.

Piggott, S. (1978). *Antiquity Depicted (Aspects of Archaeological Illustration)*. London: Thames and Hudson.

Pilbeam, D. (1984). "The descent of hominoids and hominids." *Scientific American*, Vol. 252, No. 9 (March) p. 84ff.

Pipes, R. (1986). "Team B: The reality behind the myth." *Commentary*, Vol. 82, No. 4 (Oct.) pp. 25–40.

Pitt, D.C. (1972). *Using Historical Sources in Anthropology and Sociology*. New York: Holt, Rinehart & Winston.

Platt, C. (1978). *Medieval England (A Social History and Archaeology from the Conquest to A.D. 1600)*. London: Routledge and Kegan Paul.

Platt, J.R. (1964). "Strong inference." *Science*, Vol. 146, pp. 347–353.

Plog, F. (1982). "Is a little philosophy a dangerous thing?" In C. Renfrew, M.I. Rowlands, and B.A. Segreaves (Eds.), *Theory and Explanation in Archaeology (The Shouthampton Conference)*. New York: Academic Press.

Pollock, L.S. (1972). "Relating urban design to the motorist: An empirical viewpoint." In W. Mitchell (Ed.), *EDRA 3*, Los Angeles: University of California, Vol. 1, p. 11-1-1–11-1-10.

Popper, K.E. (1964). *The Poverty of Historicism*. New York: Harper Torchbooks.

Porter, D.H. (1981). *The Emergence of the Past (A Theory of Historical Explanation)*. Chicago: University of Chicago Press.

Porter, T.J. (1964). *A Study of Path Choosing Behavior*. Berkeley, University of California Dept. of Architecture; B. Arch. thesis (unpublished).

Portugali, J. (1984). "Location theory in geography and archaeology." *Geographic Research Forum*, Vol. 7, pp. 43–60.

Prak, N.L. (1977). *The Visual Perception of the Built Environment*. Delft: Delft University Press.

Prak, N.L. (1984). *Architects, the Noted and the Ignored*. Chichester: John Wiley and Sons.

Price, D. de Solla (1981). "The analytical (quantitative) theory of science and its implications for the nature of scientific discovery." In M.D. Grmek, R.S. Cohen, and G. Cimino (Eds.), *On Scientific Discovery*. Dordrecht: Reidel (Vol. 34 of Boston Studies in the Philosophy of Science) pp. 179–189.

Price, V.A. (1982). *Type A Behavior Pattern (A Model for Research and Practice)*. New York: Academic Press.

Prigogine, I., and I. Stengers (1984). *Order out of Chaos: Man's New Dialogue with Nature*. New York: Bantam Books.

Prince, H.C. (1971). "Real, imagined and abstract worlds of the past." In C. Board *et al.* (Ed.), *Progress in Geography* (Vol. 3). London: Edward Arnold, pp. 4–86.

Process Architecture No. 15 (1980). Special Issue, "Indigeneous Settlements in S.W. Asia" (May).

Prussin, D.H. (1967). *The Architecture of Northern Ghana*. Berkeley: University of California Press.

Pugh, G.E. (1977). *The Biological Origins of Human Values*. New York: Basic Books.

Pulliam, H.R., and C. Dunford (1980). *Programmed to Learn (An Essay on the Evolution of Culture)*. New York: Columbia University Press.

Purcell, A.T. (1984). "Aesthetics, measurement and control." *Architecture Australia,* Vol. 73, No. 4, pp. 29–38.
Purcell, A.T. (1986). "Environmental perception and affect (a schema discrepancy model). *Environment and Behavior,* Vol. 18, No. 1 (January) pp. 3–30.
Purcell, A.T. and R. Thorne (1976). *Spaces for Pedestrian Use in the City of Sydney.* University of Sydney, Department of Architecture, Architectural Psychology Research Unit.
Pushkarev, B.S., and J.M. Zupan (1975). *Urban Space for Pedestrians.* Cambridge: MIT Press.
Pyron, B. (1971). "Form and space diversity in human habitats (perceptual responses)." *Environment and Behavior,* Vol. 3, No. 2 (December) pp. 382–411.
Pyron, B. (1972). "Form and diversity in human habitats (judgmental and attitude responses)." *Environment and Behavior,* Vol. 4, No. 1 (March) pp. 87–120.
Quale, G.R. (1988). *A History of Marriage Systems.* New York: Greenwood Press.
Radford, J.P. (1979). "Testing the model of the pre-industrial city: the case of ante-bellum Charleston, South Carolina" *Transactions, Institute of British Geographers* (New Series) Vol. 4, No. 3, pp. 392–410.
Rapoport, A. (1957). "An approach to urban design." M. Arch. Thesis, Rice University (May) (unpublished).
Rapoport, A. (1964–1965). "The architecture of Isphahan." *Landscape,* Vol. 14, No. 2 (Winter) pp. 4–11.
Rapoport, A. (1967). "*Whose* meaning in architecture?" *Interbuild/Arena* (October) pp. 44–46.
Rapoport, A. (1968a). "The personal element in housing: An argument for open-ended design." *RIBA Journal* (July) pp. 300–307.
Rapoport, A. (1968b). "Sacred space in primitive and vernacular architecture." *Liturgical Arts,* Vol. 36, No. 2 (February) pp. 36–40.
Rapoport, A. (1969a). *House Form and Culture.* Englewood Cliffs, NJ, Prentice-Hall.
Rapoport, A. (1969b). "The notion of urban relationships." *Area* (Journal of Institute of British Geographers), Vol. 1, No. 3, pp. 17–26.
Rapoport, A. (1969c). "Housing and housing densities in France." *Town Planning Review,* Vol. 39, No. 4 (January) pp. 341–354.
Rapoport, A. (1969d). "The Pueblo and the Hogan: A cross-cultural comparison of two responses to an environment." In P. Oliver (Ed.), *Shelter and Society.* London: Barrie and Rockliff, pp. 66–79.
Rapoport, A. (1969e). "Facts and Models." In G. Broadbent and A. Ward (Eds.), *Design Methods in Architecture* (Architectural Association Paper No. 4). London: Lund Humphries, pp. 136–146.
Rapoport, A. (1970a). "An approach to the study of environmental quality." In H. Sanoff and S. Cohen (Eds.), *EDRA 1.* Raleigh, NC: pp. 1–13.
Rapoport, A. (1970b). "The study of spatial quality." *Journal of Aesthetic Education,* Vol. 4, No. 4 (October) pp. 81–96 (Reprinted in *Architectural Quarterly,* Vol. 19, No. 1, 1978, pp. 45–50).
Rapoport, A. (1971a). "Designing for complexity." *Architectural Association Quarterly,* Vol. 3, No. 1 (winter) pp. 29–33.
Rapoport, A. (1971b). "Programming the housing environment." *Tomorrow's Housing,* Department of Adult Education, University of Adelaide, Publication No. 22.
Rapoport, A. (1972). "Environment and people." In A. Rapoport (Ed.), *Australia as Human Setting (Approaches to the Designed Environment).* Sydney: Angus and Robertson, pp. 3–21.
Rapoport, A. (1973a). "Some thoughts on the methodology of man-environment studies." *International Journal of Environmental Studies,* Vol. 4, pp. 135–140.
Rapoport, A. (1973b). "The city of tomorrow, the problems of today and the lessons of the past." *DMG-DRS Journal,* Vol. 7, No. 3, pp. 256–259.
Rapoport, A. (1974). "Urban design for the elderly." In T.O. Byerts (Ed.), *Environmental Research and Aging.* Washington, DC: Gerontological Society, pp. 97–110.
Rapoport, A. (1975a). "Australian Aborigines and the definition of place." In P. Oliver (Ed.), *Shelter, Sign and Symbol.* London: Barrie and Jenkins, pp. 38–51.
Rapoport, A. (1975b). "An anthropological approach to environmental design research." In B. Honikman (Ed.), *Responding to Social Change* (EDRA 6). Stroudsburg, PA: Dowden, Hutchinson and Ross, pp. 145–151.
Rapoport, A. (1975c). "Towards a redefinition of density." *Environment and Behavior,* Vol. 7, No. 2 (June) pp. 133–158.
Rapoport, A. (1976). "Sociocultural aspects of man-environment studies." In A. Rapoport (Ed.) *The Mutual Interaction of People and Their Built Environment: A Cross-cultural Perspective.* The Hague: Mouton.
Rapoport, A. (1977). *Human Aspects of Urban Form.* Oxford: Pergamon Press.

Rapoport, A. (1978a). "Culture and the subjective effects of stress." *Urban Ecology,* Vol. 3, No. 3 (Nov.) pp. 341–361.

Rapoport, A. (1978b). "Nomadism as a man-environment system." *Environment and Behavior,* Vol. 10, No. 2 (June) pp. 215–246.

Rapoport, A. (1978c). "Culture and Environment." *The Ecologist Quarterly,* No. 3 (Winter) pp. 269–279.

Rapoport, A. (1979a). "On the cultural origins of architecture." In J.C. Snyder and A.J. Catanese (Eds.), *Introduction to Architecture.* New York: McGraw-Hill, pp. 2–20.

Rapoport, A. (1979b). "On the cultural origins of settlements." In A.J. Catanese and J.C. Snyder (Eds.), *Introduction to Urban Planning.* New York: McGraw-Hill, pp. 31–61.

Rapoport, A. (1979c). "An approach to designing Third World environments." *Third World Planning Review,* Vol. 1, No. 1 (Spring) pp. 23–40.

Rapoport, A. (1979d). "On the environment and the definition of the situation." *International Architect,* Vol. 1, No. 1, pp. 26–28.

Rapoport, A. (1980a). "Cross-cultural aspects of environmental design." In I. Altman, A. Rapoport, and J.F. Wohlwill (Eds.), *Environment and Culture (Vol. 4 of Human Behavior and Environment).* New York: Plenum Press, pp. 7–46.

Rapoport, A. (1980b). "Environmental preference, habitat selection and urban housing." *Journal of Social Issues,* Vol. 36, No. 3 pp. 118–134.

Rapoport, A. (1980c). "Towards a cross-culturally valid definition of housing." In R.R. Stough and A. Wandersman (Eds.), *Optimizing Environments—Research, Practice and Policy* (EDRA 11). Washington, DC: EDRA, pp. 310–316.

Rapoport, A. (1980d). "Vernacular architecture and the cultural origins of form." In A.D. King (Ed.), *Buildings and Society.* London: Routledge and Kegan Paul, pp. 283–305.

Rapoport, A. (1980/1981). "Neighborhood heterogeneity or homogeneity." *Architecture and Behavior,* Vol. 1, No. 1, pp. 65–77.

Rapoport, A. (1981a). "Identity and Environment: a cross-cultural perspective" in J.S. Duncan (Ed.), *Housing and Identity (Cross-Cultural Perspectives)* London, Croom Helm, pp. 6–35.

Rapoport, A. (1981b). "On the perceptual separation of pedestrians and motorists." In H.C. Foot, A.J. Chapman, and F.M. Wade (Eds.), *Road Safety (Research and Practice).* New York: Praeger, pp. 161–167.

Rapoport, A. (1981c). "Some thought on units of settlement." *Ekistics,* Vol. 48, No. 291 (November/December) pp. 447–453.

Rapoport, A. (1982a). *The Meaning of the Built Environment (A Nonverbal Communication Approach).* Beverly Hills, CA: Sage Publications. (Updated edition, 1990, University of Arizona Press.)

Rapoport, A. (1982b). "An approach to vernacular design." In J.M. Fitch (Ed.), *Shelter: Models of Native Ingenuity.* Katonah, NY: Katonah Gallery, pp. 43–48.

Rapoport, A. (1982c). "Design, development and man-environment studies." *Environments,* Vol. 14, No. 2, pp. 1–8.

Rapoport, A. (1982d). "Urban design and human systems—on ways of relating buildings to urban fabric." In P. Laconte, J. Gibson, and A. Rapoport (Eds.), *Human and Energy Factors in Urban Planning: A Systems Approach)* (NATO Advanced Study Institute Series D: Behavioral and Social Sciences, No. 12). The Hague: Martinus Nijhoff, pp. 161–184.

Rapoport, A. (1982e). "Sacred places, sacred occasions and sacred environments." *Architectural Design* 52(9/10) pp. 75–82.

Rapoport, A. (1983a). "Development, culture change and supportive design." *Habitat International,* Vol. 7, No. 5/6, pp. 249–268.

Rapoport, A. (1983b). "Debating architectural alternatives." *Transactions 3* (Royal Institute of British Architects). London, pp. 105–109.

Rapoport, A. (1983c). "The effects of environment on behavior." In J.B. Calhoun (Ed.), *Environment and Population (Problems of Adaptation).* New York: Praeger, pp. 200–201.

Rapoport, A. (1983d). "Environmental quality, metropolitan areas and traditional settlements." *Habitat International,* Vol. 7, No. 3/4, pp. 37–63.

Rapoport, A. (1983e). "Studious questions." *The Architects' Journal,* Vol. 178, No. 43 (23 October) pp. 55–57.

Rapoport, A. (1984a). "Review of A.W. Spirn, *The Granite Garden: Urban Nature and Human Design,* New York, Basic Books, 1984." In *Urban Ecology,* Vol. 8, No. 4, pp. 364–366.

Rapoport, A. (1984b). "Culture and the urban order." In J.A. Agnew, J. Mercer, and D.E. Sopher (Eds.), *The City in Cultural Context*. Boston: Allen and Unwin, pp. 50–75.

Rapoport, A. (1984c). "Review of Lorna Price: *The Plan of St. Gall in Brief (An Overview of the Three Volume Work by Walter Horn and Walter Born).*" *Journal of Architectural and Planning Research*, Vol. 1, No. 3 (October) pp. 222–225.

Rapoport, A. (1985a). "Review of *Environmental Perception and Behavior*, T.F. Saarinen, D. Seamon, and J.L. Sell (Eds.), Research Paper 209, Dept. of Geography, University of Chicago, *Urban Ecology*, Vol. 9(2) pp. 224–226.

Rapoport, A. (1985b). "Thinking about home environments: A conceptual framework." In I. Altman and C.M. Werner (Ed.), *Home Environments (Vol. 8 of Human Behavior and Environment)*. New York: Plenum Press, pp. 255–286.

Rapoport, A. (1985c). "On diversity." In B. Judd, J. Dean, and D. Brown (Eds.), *Housing Issues 1: Design for Diversification*. Canberra: Royal Australian Institute of Architects, pp. 5–8.

Rapoport, A. (1985d). "Designing for diversity." In B. Judd, J. Dean, and D. Brown (Eds.), *Housing Issues 1: Design for Diversification*. Canberra: Royal Australian Institute of Architects, pp. 30–36.

Rapoport, A. (1986a). "The use and design of open spaces in urban neighborhoods." In D. Frick (Ed.), *The Quality of Urban Life (Social, Psychological and Physical Conditions)*. Berlin: de Gruyter, pp. 159–175.

Rapoport, A. (1986b). "Settlements and energy: Historical precedents." In W.H. Ittelson, M. Asai, and M. Ker (Eds.), *Cross-Cultural Research in Environment and Behavior*. Tucson: University of Arizona, pp. 219–237.

Rapoport, A. (1986c). "Culture and built form—A reconsideration." In D.G. Saile (Ed.), *Architecture in Cultural Change (Essays in Built Form and Culture Research)*. Lawrence: University of Kansas, pp. 157–175.

Rapoport, A. (1986d). "Human ecology and environmental design." In R.J. Borden (Ed.), *Human Ecology: A Gathering of Perspectives*. College Park, MD: Society for Human Ecology, pp. 94–106.

Rapoport, A. (1986e). "Statement for the ACSA 75th Anniversary (Jubilee) issue of JAE." *Journal of Architectural Education*, Vol. 40, No. 2, pp. 65–66.

Rapoport, A. (1987a). "Pedestrian street use—culture and perception." In A.V. Moudon (Ed.), *Public Streets for Public Use*. New York: Nostrand-Reinhold, pp. 80–92.

Rapoport, A. (1987b). "On the cultural responsiveness of architecture" *Journal of Architectural Education*, Vol. 41, No. 1 (Fall) pp. 10–15.

Rapoport, A. (1988a). "Levels of meaning in the built environment." In F. Poyatos (Ed.), *Cross-cultural Perspectives in Nonverbal Communication*. Toronto: C.J. Hogrefe, pp. 317–336.

Rapoport, A. (1988b). "Spontaneous settlements as vernacular design." In C.V. Patton (Ed.), *Spontaneous Shelter*. Philadelphia: Temple University Press, pp. 51–77.

Rapoport, A. (1989a). "On the attributes of 'tradition.'" In N. Al Sayyad and J-P Bourdier (Eds.), *Dwellings, Settlements and Tradition* Lanham, MD: University Press of America, pp. 77–105.

Rapoport, A. (1989b). "On Regions and Regionalism." In N.C. Markovich, W.F.E. Preiser, and F.G. Sturm (Eds.), *Pueblo Style and Regional Architecture*. New York: Van Nostrand Reinhold, pp. 272–288.

Rapoport, A. (1990). "Systems of activities and systems of settings." In S. Kent (Ed.), *Domestic Architecture and Use of Space (An Interdisciplinary Perspective)*. Cambridge: Cambridge University Press.

Rapoport, A. (in press). "Defining vernacular design." In M. Turan (Ed.), *Vernacular Architecture: Paradigms of Environmental Response*. Aldershot: Gower.

Rapoport, A., and R.E. Kantor (1967). "Complexity and ambiguity in environmental design." *AIP Journal*, Vol. 33, No. 4 (July) pp. 210–221.

Rapoport, A., and R. Hawkes (1970). "The perception of urban complexity." *AIP Journal*, Vol. 36, No. 2 (March) pp. 106–111.

Rapoport, A., and N. Watson (1972). "Cultural variability in physical standards." In R. Gutman (Ed.), *People and Buildings*. New York: Basic Books, pp. 33–53. (Originally published in *Transactions of the Bartlett Society*, No. 6, 1967–1968).

Rappaport, R.A. (1979). *Ecology, Meaning and Religion*. Richmond, CA: North Atlantic Books.

Rasmussen, S.E. (1951). *Towns and Buildings*. Cambridge, MA: MIT Press.

Reason, P. and J. Rowan (Eds.). (1981). *Human Inquiry (A Source-book of New Paradigm Research)*. Chichester: John Wiley.

Redman, C.L. (1978a). "Multivariate artifact analysis: A basis for multidimensional interpretations." In C.L. Redman *et al.* (Eds.), *Social Archaeology (Beyond Subsistence and Dating)*. New York: Academic Press.

Redman, C.L. (1978b). "Qsar Es-Seghir: An Islamic port and Portuguese fortress." *Archaeology,* Vol. 31, No. 5 (September/October) pp. 12–23.

Redman, C.L. (Ed.). (1978c). *Social Archaeology (Beyond Subsistence and Dating).* New York: Academic Press.

Regional Plan Association. (1969). *Urban Design Manhattan.* London: Studio Vista.

Renfrew, C. (1982). *Towards an Archaeology of Mind* (Inaugural lecture, Nov. 30). Cambridge: Cambridge University Press.

Renfrew, C. (1983). "The social archaeology of megalithic monuments." *Scientific American,* Vol. 249, No. 5 (Nov.), pp. 152ff.

Renfrew, C. (1984). *Approaches to Social Archaeology.* Edinburgh: The University Press.

Renfrew, C., M.J. Rowlands, and B.A. Segreaves (Eds.). (1982). *Theory and Explanation in Archaeology (The Southampton Conference).* New York: Academic Press.

Rhodes, E.A. (1973). "The human squares: Athens, Greece." *Ekistics,* Vol. 35, No. 208 (March) pp. 124–132.

Richardson, J., and A.L. Kroeber (1940). "Three centuries of women's dress fashions: A quantitative analysis." *Anthropological Records,* Vol. 5, No. 2.

Ritter, P. (1964). *Designing for Man and Motor.* New York: Macmillan.

Robertson, J. (1973). "Rediscovering the street." *Architectural Forum* (November).

Rosch, E. (1978). "Principles of categorization." In E. Rosch and B.B. Lloyd (Eds.), *Cognition and Categorization.* Hillsdale, NJ: Erlbaum, pp. 27–48.

Rosch, E., and B.B. Lloyd (Eds.). (1978). *Cognition and Categorization.* Hillsdale, NJ: Erlbaum.

Rossi, A.S. (1977). "A biosocial perspective on parenting." *Daedalus,* Vol. 106, No. 2 (Spring) pp. 1–32.

Rowe, C., and F. Koetter (n.d.). *Collage City.* Cambridge, MA: MIT Press.

Rowntree, L.B., and M.W. Conkey (1980). "Symbolism and the cultural landscape." *Annals, Association of American Geographers,* Vol. 70, No. 4 (December) pp. 459–474.

Royse, D.C. (1969). *Social Inferences via Environmental Cues.* Cambridge, MA: MIT, PhD in city planning (unpublished).

Rubbo, A. (1978). "Culture, class and shelter: Peasant and state housing in the Cauca Valley, Colombia." Paper given at *EDRA 9* (April) (mimeo).

Rubenstein, H.M. (1979). *Central City Malls.* London: Wiley.

Rubin, B. (1979). "Aesthetic ideology and urban design." *Annals, Association of American Geographers,* Vol. 69, No. 3 (September) pp. 339–361.

Rudofsky, B. (1969). *Streets for People (A Primer for Americans).* Garden City, NY: Doubleday.

Rudwick, M.J.S. (1972). *The Meaning of Fossils (Episodes in the History of Palaeontology).* London: Macdonald.

Ruse, M. (1979). *Sociobiology: Sense or Nonsense.* Dordrecht: Reidel.

Ruse, M. (1986). *Taking Darwin Seriously (A Naturalistic Approach to Philosophy).* Oxford: Basil Blackwell.

Russell, J., and J. French (1981). "Jeddah renewal." *Architectural Review,* Vol. 169, No. 1010 (April) pp. 217–222.

Rybczynski, W. (1986). *Home (A Short History of an Idea).* New York: Viking.

Rykwert, J. (1976). *The Idea of a Town.* Princeton, NJ: Princeton University Press.

Saad, H.T. (1981). *Between Myth and Reality: The Aesthetics of Traditional Architecture in Hausaland,* Unpublished D.Arch. dissertation, University of Michigan.

Saarinen, T.F., D. Seamon, and J.L. Sell (Eds.). (1984). *Environmental Perception and Behavior: An Inventory and Prospect,* Research Paper 209, Dept. of Geography, University of Chicago.

Sabloff, J.A. (Ed.). (1981). *Simulations in Archaeology* (A School of American Research Book). Albuquerque: University of New Mexico Press.

Sadalla, E.K., and S.G. Magel (1980). "The perception of traversed distance." *Environment and Behavior,* Vol. 12, No. 1 (March) pp. 65–80.

Sadalla, E.K., and L.J. Staplin (1980a). "The perception of traversed distance: Intersections." *Environment and Behavior,* Vol. 12, No. 2 (June) pp. 167–182.

Sadalla, E.K., and L.J. Staplin (1980b). "An information storage model for distance cognition." *Environment and Behavior,* Vol. 12, No. 2 (June) pp. 183–194.

Salk, L. (1973). "The role of the heartbeat in the relations between mother and infant." *Scientific American,* Vol. 228, No. 5 (May) pp. 24–29; Refs. p. 129.

Salmon, M.H. (1976). "Deductive versus inductive archaeology." *American Antiquity,* Vol. 41, pp. 376–380.

Salmon, M.H. (1982). *Philosophy and Archaeology.* New York: Academic Press.

Sancar, F.H. (1985). "Tourism and the vernacular environment." In S. Klein and R. Wener (Eds.), *Environmental Change/Social Change* (EDRA 16). Washington, DC: EDRA, pp. 47–56.

Sanders, D.H. (1985). "Ancient behavior and the built environment: Applying environmental psychology methods and theories to archaeological contexts." In S. Klein and R. Wener, (Eds.), *Environmental Change/Social Change* (EDRA 16). Washington, DC: EDRA, pp. 296–305.

Sanders, J. (1975). "Between man and city: An evidential inquiry into the existence of neighborhoods for specific ancient settlement contexts." Term paper in Arch. 790, Department of Architecture, University of Wisconsin–Milwaukee (unpublished).

Sanders, R.A. (1984). "Some determinants of urban forest structure." *Urban Ecology,* Vol. 8, No. 1/2 (Sept.) pp. 13–27.

Sandström, C.I. (1972). "What do we perceive in perceiving." *Ekistics,* Vol. 34, No. 204 (November) pp. 370–371.

Sarna, J.D. (1985). "The German period" (Review of N.W. Cohen, *Encounter with Emancipation: The German Jews in the United States, 1830–1914).* Philadelphia: Jewish Publication Society (1984) *Commentary,* Vol. 80, No. 3 (September) pp. 72–76.

Scardino, A. (1986). "Designing tomorrow's city (Cooper sets modest scale)." *New York Times* (April 9).

Scarre, C. (Ed.). (1983). *Ancient France (Neolithic Societies and Their Landscapes (6000–2000 B.C.)* Edinburgh: The University Press.

Schak, D.C. (1972). "Determinants of childrens' play patterns in a Chinese city: An interplay of space and values." *Urban Anthropology,* Vol. 1, No. 2 (Fall) pp. 195–204.

Schank, R.C. and R.P. Abelson (1977). "Scripts, plans and knowledge." In N. Johnson-Laird and P.C. Wason (Eds.), *Thinking (Readings in Cognitive Science).* Cambridge: Cambridge University Press, pp. 421–432.

Schavelzon, D. (1981). *Arquelogia y Arquitectura del Ecuador Prehispánico.* Mexico: Universidad Nacional Autónoma de México.

Scheflen, A.E. (1976). "Some territorial layouts in the United States." In A. Rapoport (Ed.), *The Mutual Interaction of People and Their Built Environment.* The Hague: Mouton, pp. 177–221.

Schiffer, M.B. (Ed.). (1982). *Advances in Archaeological Method and Theory.* New York: Academic Press.

Schmandt-Besserat, D. (1983). "The earliest precursor of writing." In B.M. Fagan (Ed.), *Prehistoric Times* (Readings from *Scientific American).* San Francisco: W.H. Freeman, pp. 194–203.

Schnapper, D. (1971). *L'Italie Rouge et Noire (les Modeles Culturels de la vie Quotidienne a Bologne).* Paris: Gallimard.

Schoder, R.V. (1963). "Ancient Cumae." *Scientific American* (December) (pp. 109–121).

Schön, D.A. (1969). *Invention and the Evolution of Ideas.* London, Tavistock Publications (originally published 1963).

Schön, D.A. (1983). *The Reflective Practitioner.* New York: Basic Books.

Schopf, T.J.M. (Ed.). (1972). *Models in Paleobiology.* San Francisco: Freman, Cooper and Co.

Schwartz, J.H. (1986). *The Red Ape (Orang-Utans and Human Origins).* Boston: Houghton Mifflin.

Scully, V. (1962). *The Earth, the Temple and the Gods.* New Haven: Yale University Press.

Seddon, G. (1970). *Swan River Landscape.* Nedlands: University of Western Australia Press.

Sekler, E.F. (n.d.). "Seminar on historic urban spaces." Cambridge, MA: Harvard University Graduate School of Design.

Seligmann, C. (1975). "An aedicular system in an early twentieth century American popular house." *Architectural Association Quarterly,* Vol. 7, No. 2 (April/June) pp. 7–11.

Sennett, R. (1977). *The Fall of Public Man.* New York: Knopf.

Seymour, Jr., W.N. (1969). *Small Urban Spaces.* New York: New York University Press.

Shafer, E.L. Jr., *et al.* (1969a). "Natural landscape preferences: A predictive model." *Journal of Leisure Research,* Vol. 1, No. 1 (Winter) pp. 1–19.

Shafer, E.L., Jr. (1969b). "Perception of natural environments." *Environment and Behavior,* Vol. 1, No. 1 (June) pp. 71–82.

Shanks, H. (1988). "Two early Israelite cult sites now questioned." *Biblical Archaeology Review,* Vol. XIV, No. 1 (Jan./Feb.), p. 48.

Shapere, D. (1977). "Scientific theories and their domains." In F. Suppe (Ed.), *The Structure of Scientific Theories.* Urbana: University of Illinois Press (2nd ed.), pp. 518–573.

Shapere, D. (1984). *Reason and the Search for Knowledge.* Dordrecht: Reidel.

Share, L.B. (1978). "A.P. Giannini Plaza and Transamerica Park: Effects of their physical characteristics on users' perceptions and experiences." In W.E. Rogers and W.H. Ittelson (Eds.), *New Directions in Environmental Design Research* (EDRA 9). Washington, DC: EDRA, pp. 127–139.

Sharp, T. (1968). *Town and Townscape,* London: John Murray.

Shaw, J.W., P.P. Betancourt, and L.V. Watrous (1978). "Excavations at Kommos (Crete) during 1977." *Hesperia,* Vol. 47, No. 2, (April/June) pp. 111–170.

Shaw, T. (1970). *Igbo-Ukwu.* London: Faber and Faber.

Siegel, B. (1970). "Defensive structuring and environmental stress." *American Journal of Sociology,* Vol. 76, pp. 11–32.

Sieveking, G. de G., *et al.* (Eds.). (1976). *Problems in Economic and Social Archaeology.* London: Duckworth.

Sieverts, T. (1967). "Perceptual images of the city of Berlin." In *Urban Core and Inner City.* Leiden: Brill.

Sieverts, T. (1969). "Spontaneous architecture." *Architectural Association Quarterly,* Vol. 1, No. 3 (July) pp. 36–43.

Siksna, A. (1981). "Understanding Australian urban space: The lesson of country towns." *Proceedings, 1981 Conference of the Australian and New Zealand Architectural Science Association.* Canberra (October), pp. 121–127.

Simmons, I., and M. Toole (Eds.). (1981). *The Environment in British Prehistory.* London: Duckworth.

Sims, B., and N. Jammal (1979). "What makes downtown pleasant?" In A.D. Seidel and S. Danford (Eds.), *Environmental Design: Research, Theory and Application* (EDRA 10). Washington, DC: EDRA, p. 113 (abstract only).

Sitté, C. (1965 [originally 1889]) (Trans. G.R. and C.C. Collins). *City Planning According to Artistic Principles.* New York: Random House.

Sivengard, S. (1968). *Lokalisering Av Overgangstalle I Gangtrafiksystem,* Lund: Lund Institute of Technology, Department of Traffic Planning and Engineering.

Smets, G. (1971). "Pleasingness vs. Interestingness of visual stimuli with controlled complexity: Their relationship to looking at time as a function of exposure time." *Perceptual and Motor Skills,* Vol. 40, No. 1 (February) pp. 3–10.

Smets, G. (1973). *Aesthetic Judgment and Arousal.* Leuven: Leuven University Press.

Smith, B.D. (1977). "Archaeological inference and inductive confirmation." *American Anthropologist,* Vol. 79, No. 3, pp. 598–617.

Smith, B.D. (1978). *Prehistoric Patterns of Human Behavior (A Case Study in the Mississippi Valley).* New York: Academic Press.

Smith, B.D. (1982). "Explanation in archaeology." In C. Renfrew, M.J. Rowlands, and B.A. Segreaves (Eds.), *Theory and Explanation in Archaeology (The Southampton Conference).* New York: Academic Press, pp. 73–82.

Smith, P.F. (1972). "The pros and cons of subliminal perception." *Ekistics,* Vol. 34, No. 204 (November) pp. 367–369.

Smith, R.C. (1973). "Colonial towns of Spanish and Portugese America." In L. Hanke (Ed.), *History of Latin American Civilization.* Boston: Little, Brown (2nd ed.) pp. 269–277.

Smith, R.L. *et al.* (1975). "Reactivity to systematic observation with film: A field experiment." *Sociometry,* Vol. 38, No. 4, pp. 536–550.

Sokal, R.R. (1977). "Classification: Purposes, principles, progress, prospects." In P.N. Johnson-Laird and P.C. Wason (Eds.), *Thinking (Readings in Cognitive Science).* Cambridge: Cambridge University Press, pp. 185–198.

Sokal, R.R., and P.H.A. Sneath (1963). *Principles of Numerical Taxonomy.* San Francisco: Freeman.

Solomon, D. *et al.* (n.d.). *Change without Loss (Residential Development and Preservation for San Francisco Neighborhoods).* Prepared for the San Francisco Department of City Planning. Berkeley: Department of Architecture, College of Environmental Design, University of California.

Sopher, D. (1969). "Pilgrim circulation in Gujarat." *Ekistics,* Vol. 27, No. 161 (April) pp. 251–260.

Sordinas, A. (1976). "Traditional building materials in rural Corfu, Greece (a technological and socio-economic analysis)." Paper presented at the 4th International Congress of Agricultural Museums, Reading (England) April 5–10 (mimeo).

Sorrell, A. (1981). *Reconstructing the Past.* London: Batsford.

Southworth, M. (1969). "The sonic environment of cities." *Environment and Behavior,* Vol. 1, No. 1 (June) pp. 49–70.

Sözen, M. and O.N. Dülgerler (1978). "Mimar muzafferin konya ogretmen lisesi." *Journal of METU,* Vol. 4, No. 1, pp. 117–134 (and Figures 16–25, no page nos.).

Specter, D.K. (1974). *Urban Spaces*. New York: New York Graphic Society.

Spier, R.F.G. (1967). "Work habits, postures and fixtures." In C.L. Riley and W.W. Taylor (Eds.), *American Historical Anthropology (Essays in Honor of Leslie Spier)*. Carbondale, IL: Southern Illinois University, pp. 197–220.

Spreiregen, P. (1965). *Urban Design*. New York: McGraw-Hill.

Spriggs, M. (Ed.). (1977). *Archaeology and Anthropology: Areas of Mutual Interest*. Oxford: BAR (Supplementary Series) 19.

Staeger, P.W. (1979). "Where does art begin in Puluwat?" In S.N. Mead (Ed.), *Exploring the Visual Arts of Oceania*. Honolulu: University Press of Hawaii, pp. 342–353.

Stanislawski, D. (1961). "The origin and spread of the grid-pattern town." In G.A. Theodorson (Ed.), *Studies in Human Ecology*. Evanston, IL: Row, Patterson, pp. 294–303.

Stanley, S.M. (1981). *The New Evolutionary Timetable: Fossils, Genes and the Origin of Species*. New York, Basic Books.

Staski, E. (1982). "Advances in urban archaeology." In M.B. Schiffer (Ed.), *Advances in Archaeological Method and Theory*. New York: Academic Press, pp. 97–149.

Stea, D., and D. Wood (1971). *A Cognitive Atlas: Explorations into the Psychological Geography of Four Mexican Cities*. Chicago: Environment Research Group, Place Perception Research Report No. 10.

Steadman, P. (1979). *The Evolution of Designs*. Cambridge: Cambridge University Press.

Stebbins, G.L., and F. Ayala (1985). "The evolution of Darwinism." *Scientific American*, Vol. 253, No. 1 (July) pp. 72–82.

Steensberg, A. (1979). *New Guinea Gardens (A Study of Husbandry with Parallels in Prehistoric Europe)*. New York: Academic Press.

Steinberg, S. (1969). Cartoon in the *New Yorker* (April 12), p. 43.

Steinitz, C. (1968). "Meaning and congruence in urban form and activity." *AIP Journal*, Vol. 34, No. 4 (July) pp. 233–248.

Stilitz, I.B. (1969). *Behavior in Circulation Areas*. University College Environmental Research Group, University College, London: Environmental Design Research Unit (June).

Stoks, F.G. (1983). "Assessing urban space environments for danger of violent crime—especially rape." In D. Joiner *et al.* (Eds.), *Conference on People and Physical Environmental Research*. Wellington, NZ: Ministry of Works and Development pp. 331–343.

Stone, T.R. (1971). *Beyond the Automobile (Reshaping the Transportation Environment)*. Englewood Cliffs, NJ: Prentice-Hall.

Stretton, H. (1985). "LeCorbusier and the faeces count" (Review of A. Coleman, *Utopia on Trial*, London, Hilary Shipman, 1985). *The Adelaide Review*. Adelaide: South Australia, No. 14 (June) pp. 4–5.

Stuart, D.G. (1968). "Planning for pedestrians." *AIP Journal*, Vol. 34, No. 1 (January) pp. 37–41.

Sudman, S. (1976). *Applied Sampling*. New York: Academic Press.

Sumner, W.M. (1979). "Estimating population by analogy: An example." In C. Kramer (Ed.), *Ethnoarchaeology (Implications of Ethnography for Archaeology)*. New York: Columbia University Press, pp. 164–174.

Sun, The (1971). (Sydney, Australia). "Remember when you were a little boy" (December 29).

Sunday Times, The (1968). "Bringing good cheer to pubs" (October 27).

Swanson, C.P. (1983). *Ever Expanding Horizons: The Dual Informational Sources of Human Evolution*. Amherst: University of Massachusetts Press.

Swedner, H. (1960). *Ecological Differentiation of Habits and Attitudes* (Lund Studies in Sociology). Lund: CWK Glerup.

Szalay, F.S., and E. Delson (1979). *Evolutionary History of the Primates*. New York: Academic Press.

Szamosi, G. (1986). *The Twin Dimensions (Inventing Time and Space)*. New York: McGraw-Hill.

Takayama (A Study of Town and Townscape) (n.d.). Nara (Japan), Nara National Properties Research Institute, No. 24.

Tanghe, J., S. Vlaeminck, and J. Berghoef (1984). *Living Cities (A Case for Urbanism and Guidelines for Reurbanization)*. Oxford: Pergamon Press.

Tanner, N.M. (1981). *On Becoming Human*. Cambridge: Cambridge University Press.

Tavassoli, M. (1977). *Architecture in the Hot Arid Zone* (in Farsi).

Taylor, C. (1983). *Village and Farmstead (A History of Rural Settlement in England)*. London: George Philip.

Taylor, R.B. (1988). *Human Territorial Functioning*. Cambridge: Cambridge University Press.

Tewfik, M. (1976). "Aspects of regional planning and rural development affects by factors of physical environment (dust storms and desert creep)" (Eastern Sudan). Department of Architecture II A; University of Lund, Sweden (Ph.D. dissertation).

Thakudersai, S.G. (1972). "Sense of place in Greek anonymous architecture." *Ekistics,* Vol. 34, No. 204 (November) pp. 334–340.

Thiel, P. (1961). "A sequence-experience notation for architectural and urban spaces." *Town Planning Review,* Vol. 32, No. 2 (April) pp. 33–52.

Thiel, P. (1970). "Notes on the description, scaling, notation and scoring of some perceptual and cognitive attributes of the physical environment." In H. Proshansky *et al.* (Eds.), *Environmental Psychology,* New York: Holt, Rinehart & Winston, pp. 593–618.

Thompson, B., and P. Hart (1968). "Pedestrian movement: Results of a small town survey," *Journal of the TPI,* Vol. 54, No. 7 (July/August) pp. 338–342.

Thompson, D.W. (1961). *On Growth and Form.* Cambridge, Cambridge University Press (Abridged edition, Ed. J.T. Bonner).

Thompson, P. (1978). *The Voice of the Past—Oral History.* Oxford: Oxford University Press.

Thomson, J. (1973). *Illustrations of China and Its People.* London.

Thorne, R. (1987). "The environmental psychology of theatres and movie-palaces 1902 to 1930." In D.V. Canter, M. Krampen, and D. Stea (Eds.), *Ethnoscapes.* Aldershot: Gower.

Tiger, L. (1969). *Men in Groups.* New York: Random House.

Tiger, L., and R. Fox (1971). *The Imperial Animal.* New York: Delta Books.

Tiger, L., and J. Shepher (1975). *Women in The Kibbutz.* New York: Harcourt Brace Jovanovich.

The Times (London). (1982a). "Maxey Bronze Age Cleanliness" (Nov. 8).

The Times (London). (1982b). (Dec. 2).

The Times (London). (1983). "Early link in chain of whales found" (28 April).

Tilley, C. (1982). "Social formation, social structure and social change." In I. Hodder (Ed.), *Symbolic and Structural Archaeology.* Cambridge: Cambridge University Press, pp. 26–38 (Chapter 3).

Todd, I.A. (1976). *Çatal Hüyük in Perspective.* Menlo Park, CA: Cummings Publishing Co.

Todd, W. (1972). *History as Applied Science (A Philosophical Study).* Detroit: Wayne State University Press.

Tolman, E.C. (1948). "Cognitive maps in rats and men." *Psychological Review,* Vol. 55, pp. 189–208.

Traffic in Towns (1963). London: HMSO.

Tricaut, J. (n.d.). *Cours de Geographie Urbaine.* Paris: Centre de Documentation Universitaire.

Tuan, Y.F. (1968). "A preface to Chinese cities." In R.P. Beckinsale and I.M. Houston (Eds.), *Urbanization and Its Problems.* Oxford: Blackwell, pp. 218–253.

Tuan, Y.F. (1974). *Topophilia (The Study of Environmental Perception, Attitudes and Values).* Englewood Cliffs, NJ: Prentice-Hall.

Tunnard, C., and B. Pushkarev (1963). *Man-made America: Chaos or Control?* New York: Yale University Press.

Tversky, A., and D. Kahneman (1974). "Judgement under uncertainty: heuristics and biases." *Science,* Vol. 185, pp. 1124–1131.

Tyler, S.A. (1978). *The Said and the Unsaid (Mind, Meaning and Culture).* New York: Academic Press.

Tyrwhitt, J. (1966). "Chios." *Architectural Review* (June).

Tyrwhitt, J. (1968). "The pedestrian in megalopolis: Tokyo." *Ekistics,* Vol. 25, No. 147, pp. 73–79.

Ucko, P.J. (Ed.). (1977). *Form in Indigenous Art (Schematization in the Art of Aboriginal Australia and Prehistoric Europe).* (Canberra, Australia: Institute of Aboriginal Studies, Prehistory and Material Culture Series No. 13.) London: Duckworth.

Udaipur (The City and Its Elements). (1977). School of Architecture, Ahmedabad, Monograph No. 1.

Udy, S.H., Jr. (1964). "Cross-cultural analysis: A case study." In P.E. Hammond (Ed.), *Sociologists at Work.* New York: Basic Books.

Uesugi, T. *et al.* (n.d.). *"Streetscape"—A Summary of Field Notes on Japanese Towns.* Pomona: California State Polytechnic University, Department of Landscape Architecture.

Ulmer, M.J. (1984). "Economics in decline." *Commentary,* Vol. 78, No. 5 (November) pp. 42–46.

Uphill, E. (1972). "The concept of the Egyptian palace as a 'ruling machine.'" In P. Ucko, R. Tringham, and G.W. Dimbleby (Eds.), *Man, Settlement and Urbanism.* London: Duckworth, pp. 721–734.

Urban Design Newsletter (1979). Vol. I, No. 2 (July).

Vallet, G. *et al.* (1970). "Megara Hyblaea." *Annales,* Vol. 25, No. 4.

Van den Berge, P.L. (1975). *Man in Society—A Biosocial View.* New York: Elsevier.

Van Zantwijk, R. (1985). *The Aztec Arrangement (The Social History of Pre-Spanish Mexico)*. Norman: University of Oklahoma Press.

Vansina, J. (1965). *Oral Tradition (A Study in Historical Methodology)*. London: Routledge and Kegan Paul.

Vansina, J. (1985). *Oral Tradition as History*. Madison: University of Wisconsin Press.

Vayda, A.P. (1983). "Progressive contextualization." *Human Ecology,* Vol. 11, No. 3 (September) pp. 265–282.

Venturi, R. (1966). *Complexity and Contradiction in Architecture*. New York: Museum of Modern Art.

Venturi, R. *et al.* (1972). *Learning from Las Vegas*. Cambridge, MA: MIT Press.

Vernez-Moudon, A. (1980). Presentation at session, "The development of San Francisco Zoning Legislation for residential environs" (at EDRA 11, Charleston, SC, April).

Vernez-Moudon, A. (1986). *Built for Change (Neighborhood Architecture in San Francisco)*. Cambridge, MA: MIT Press.

Vernez-Moudon, A. (Ed.). (1987). *Public Streets for Public Use*. New York: Van Nostrand-Reinhold.

Verschure, H. (1977). "A comparative evaluation of housing responses: Case studies Bandung (Indonesia) and Kigali (Rwanda)." *Proceedings, International Conference on Low Income Housing—Technology and Policy,* Bangkok, Thailand (June) pp. 119–139.

Vigier, F. (1965). "An experimental approach to urban design." *AIP Journal,* Vol. 31, No. 1 (February) pp. 21–31.

Vinnicombe, P. (1976). *People of the Eland (Rock Paintings of the Drakensberg Bushmen as a Reflection of Their Life and Thought)* Pietermaritzburg: University of Natal Press.

Vogt, E.Z. (1968). "Some aspects of Zanacantan settlement patterns and ceremonial organization." In K.C. Chang (Ed.), *Settlement Archaeology*. Palo Alto, CA: National Press, pp. 154–173.

Vogt, E.Z. (1976). *Tortillas for the Gods (A Symbolic Analysis of Zinacanteco Rituals)*. Cambridge, MA: Harvard University Press.

von Frisch, K. (1974). *Animal Architecture*. New York: Harcourt Brace Jovanovich.

von Schilcher, F., and N. Tennant (1984). *Philosophy, Evolution and Human Nature*. London: Routledge and Kegan Paul.

von Simon, O. (1953). *The Gothic Cathedral*. New York: Pantheon.

Wachter, K.W., A.E. Hammel, and P. Laslett (1978). *Statistical Studies of Historical Social Structure*. New York: Academic Press.

Wagner, J. (1979). "Avoiding Error." In J. Wagner (Ed.), *Images of Information (Still Photography in the Social Sciences)*. Beverly Hills: Sage, pp. 147–159.

Walker, E.L. (1970). "Complexity and preference in animals and men." *Annals of the New York Academy of Sciences,* Vol. 169, Article 3 (June 23) pp. 619–653.

Walker, E.L. (1972). "Psychological complexity and preference: A hedgehog theory of behavior." Paper given at NATO symposium, Kersor, Denmark (June) (mimeo).

Wallace, W.L. (1983). *Principles of Scientific Sociology*. New York: Aldine.

Ward, A. (1977). *Adventures in Archaeology*. London: Hamlyn.

Warren, K.B. (1978). *The Symbolism of Subordination (Indian Identity in a Guatemalan Town)*. Austin: University of Texas Press.

Washburn, D.K. (Ed.). (1983). *Structure and Cognition in Art*. Cambridge: Cambridge University Press.

Watson, P.J. (1979). "The idea of ethnoarchaeology: Notes and comments." In C. Kramer (Ed.), *Ethnoarchaeology (Implications of Ethnography for Archaeology)*. New York: Columbia University Press, pp. 277–287.

Watson, P.J., S. LeBlanc, and C. Redman (1971). *Explanation in Archaeology: An Explicitly Scientific Approach*. New York: Columbia University Press.

Watson, P.J., S.A. LeBlanc, and C.L. Redman (1984). *Archaeological Explanation (The Scientific Method in Archaeology)*. New York: Columbia University Press.

Webster, D., and G. Webster (1984). "Optimal hunting and Pleistocene extinction." *Human Ecology,* Vol. 12(No. 3) pp. 275–289.

Weichardt, C. (1898). *Pompeii* (Trans. H. Brett). Leipzig: Koehler.

Weinreich, M. (1980). *History of the Yiddish Language*. Chicago: University of Chicago Press.

Weiss, R.S., and S. Boutourline (1962). *Fairs, Exhibits, Pavilions and Their Audiences* (mimeo report).

Wells, B.W.P. (1965). "The psycho-social influence of building environment: Sociometric findings in large and small office spaces." *Building Science,* Vol. 1, pp. 153–165.

Werner, C.M., I. Altman, and D. Oxley (1985). "Temporal aspects of homes: A transactional perspective." In I.

Altman and C.M. Werner (Eds.), *Home Environments (Vol. 8 of Human Behavior and Environment)*. New York: Plenum Press, pp. 1–32.

Westerman, A. (1976). *On Aesthetic Judgment of Building Exteriors* [in Swedish] Doctoral dissertation, Department of Architecture, Royal Institute of Technology, Stockholm.

Wheatley, P. (1971). *The Pivot of the Four Quarters,* Chicago: Aldine.

Wheatley, P. (1976). "Levels of space awareness in the traditional Islamic city" *Ekistics,* Vol. 42, No. 253 (December) p. 354–366.

Wheeler, L. (1972). "Student reactions to campus planning: A regional comparison." In W. Mitchell (Ed.), *EDRA 3,* Los Angeles: University of California, Vol. 1, pp. 12-8-1–12-8-9.

Wheeler, M. (1967). *History was Buried (A Sourcebook in Archaeology).* New York: Hart Publishing Co.

Whitehouse, D. (1970). "Sirāf: A Medieval Port on the Persian Gulf." *World Archaeology,* Vol. 2, No. 2.

Whyte, W.H. (1968). *The Last Landscape.* Garden City, NY: Doubleday.

Whyte, W.H. (1980). *The Social Life of Small Urban Spaces.* Washington, DC: The Conservation Foundation.

Widgren, M. (1978). "A simulation model of farming systems and land use in Sweden during the early Iron Age, c. 500 B.C.—550 A.D." *Journal of Historical Geography,* Vol. 4 (No. 2) pp. 161–173.

Wiedenhoeft, R. (1975). "Downtown pedestrian zones: Experiences in Germany" *Urban Land,* Vol. 34, No. 4 (April) pp. 3–11.

Wigley, T.M.L., M.J. Ingram, and G. Farmer (Eds.). (1981). *Climate and History (Studies in Past Climates and Their Impact on Man)* Cambridge: Cambridge University Press.

Wilford, J.N. (1985). *The Riddle of the Dinosaurs.* New York: Alfred A. Knopf.

Wilford, J.N. (1987). "New fossil find alters view of man's evolution." *New York Times* (May 21).

Wilk, R.R. (1984). "Households in process: Agricultural change and domestic transformation among the Kekchi Maya of Belize." In R.M. Netting, R.R. Wilk, and E.J. Arnould (Eds.), *Households: Comparative and Historical Studies of the Domestic Group.* Berkeley: University of California Press, pp. 217–244.

Wilk, R.R. (1989). "House, home and consumer decision-making in two cultures." *Advances in Consumer Research.*

Wilk, R.W., and W. Ashmore (Eds.). (1988). *Household and Community in the Mesoamerican Past.* Albuquerque: University of New Mexico Press.

Wilkins, B.T. (1978). *Has History Any Meaning? (A Critique of Popper's Philosophy of History).* Hassocks, Sussex: The Harvester Press.

Willems, E.P., and H.L. Raush (Eds.). (1969). *Naturalistic Viewpoints in Psychological Research.* New York: Holt, Rinehart & Winston.

Willer, D., and J. Willer (1973). *Systematic Empiricism: Critique of a Pseudoscience.* Englewood Cliffs, NJ: Prentice-Hall.

Williams, R. (Ed.). (1981). *Contact: Human Communication and Its History.* London: Thames and Hudson.

Wilmott, P. (1963). *The Evolution of a Community.* London: Routledge and Kegan Paul.

Wilson, A. (1985). "The molecular basis of evolution." *Scientific American,* Vol. 253, No. 4 (Oct.) pp. 164–173.

Wilson, E.O. (1975). *Sociobiology: The New Synthesis.* Cambridge, MA: Harvard University Press.

Wilson, E.O. (1978). *On Human Nature.* Cambridge, MA: Harvard University Press.

Wittkower, R. (1962). *Architectural Principles in the Age of Humanism.* London: Tiranti.

Wobst, H.M. (1977). *Stylistic Behavior and Information Exchange.* In C.E. Cleland (Ed.), *For the Director: Research Essays in Honor of James B. Griffin,* University of Michigan, Museum of Anthropology, Anthropological Papers 61, pp. 317–342.

Wohlwill, J.F. (1968). "Amount of stimulus exploration and preference as differential functions of stimulus complexity." *Perception and Psychophysics,* Vol. 4, No. 5, pp. 307–312.

Wohlwill, J.F. (1971). "Behavioral response and adaptation to environmental stimulation." In A. Damon (Ed.), *Physiological Anthropology.* Cambridge: Harvard University Press.

Wohlwill, J.F. (1976). "Environmental aesthetics: The environment as a source of affect." In I. Altman and J.F. Wohlwill (Eds.) *Human Behavior and Environment,* Vol. 1. New York: Plenum Press, pp. 37–86.

Wohlwill, J.F. (1983). "The concept of nature: A psychologist's view." In I. Altman and J.F. Wohlwill (Eds.), *Behavior and the Natural Environment* (Vol. 6 of Human Behavior and Environment). New York: Plenum Press, pp. 5–37.

Wolfe, T. (1981). *From Bauhaus to Our House.* New York: Farrar, Straus, Giroux.

Wolff, J. (1982). *Aesthetics and the Sociology of Art.* London: George Allen and Unwin.

Wolff, M. (1973). "Notes on the behavior of pedestrians." In A. Birenbaum and E. Sagarin (Eds.), *People in Places*. London: Nelson, pp. 35–48.

Wolpert, L. (1978). "Cells." *Scientific American*, Vol. 238, No. 4 (October) pp. 154–164.

Wood, D. (1969). "The image of San Cristobal." *Monadnock*, Vol. 43, pp. 29–45.

Wood, E.W., Jr., S.N. Brower, and M.W. Latimer (1966). "Planners' people." *AIP Journal*, Vol. XXXII, No. 4 (July) pp. 228–234.

Wooley, L. (1955). *Excavations at Ur*. London: Ernest Benn.

Worskett, R. (1969). *The Character of Towns*. London: Architectural Press.

Worswick, C. (1979). *Japan—Photographs 1854–1905*. New York: Pennwick/Knopf.

Worth, S., and J. Adair (1972). *Through Navajo Eyes (An Exploration in Film Communication and Anthropology)*. Bloomington: Indiana University Press.

Wycherley, R.E. (1962). *How the Greeks Built Cities*. London: Macmillan.

Wylie, M.A. (1982). "Epistemological issues raised by a structuralist archaeology." In I. Hodder (Ed.) *Symbolic and Structural Archaeology*. Cambridge: Cambridge University Press, pp. 39–46 (Chapter 4).

Yellen, J.E. (1977). *Archaeological Approaches to the Present (Models for Reconstructing the Past)*. New York: Academic Press.

Young, J.Z., E.M. Jope, and K.P. Oakley (Eds.). (1981). *The Emergence of Man* (Joint symposium, Royal Society and British Academy, held March 1980). London: Royal Society and British Academy.

Zaltman, G., K. Lemasters, and M. Heffring (1982). *Theory Construction in Marketing (Some Thought on Thinking)*. New York: John Wiley.

Zeisel, J. (1981). *Inquiry by Design: Tools for Environment-Behavior Research*. Monterrey, CA: Brooks-Cole.

Ziman, J. (1978). *Reliable Knowledge*. New York: Cambridge University Press.

Ziman, J. (1984). *An Introduction to Science Studies*. Cambridge: Cambridge University Press.

Index